RADIO DAZE

RADIO DAZE

STORIES FROM THE FRONT IN CLEVELAND'S FM AIR WARS

MIKE OLSZEWSKI

THE KENT STATE UNIVERSITY PRESS   KENT & LONDON

© 2003 by The Kent State University Press, Kent, Ohio 44242
All rights reserved
Library of Congress Catalog Card Number 2003006833
ISBN 0-87338-773-2
Manufactured in the United States of America

Unless otherwise specified, accompanying photos and illustrations are courtesy of the author from his personal collection.

07 06 05 04 03   5 4 3 2 1

Library of Congress Cataloging-in-Publication Data
Olszewski, Mike, 1953–
  Radio daze : stories from the front in Cleveland's FM air wars / Mike Olszewski.
    p. cm.
Includes bibliographical references and index.
  ISBN 0-87338-773-2 (alk. paper)
  1. Radio broadcasting—Ohio—Cleveland Region—History.
  2. Rock music—Ohio—Cleveland Region—History and criticism.
  I. Title.
PN1991.3.U6O43 2003
384.54'09771'32—21

                                                          2003006833

British Library Cataloging-in-Publication data are available.

This book is dedicated to my wife, Janice,

        who has always been, and always will be,

                my greatest source of inspiration.

Contents

Foreword by Tom Batiuk **ix**

Preface and Acknowledgments **xi**

Prologue **xix**

Part 1: A Generation Finds a New Voice: 1968–1975
The Birth of Progressive Rock Radio 1 The Electric Renaissance **25** The Times, They Are a-Changin' **37** Bass Makes His Move **53** The New Beginning **58** Rock-and-Roll Television **67** Coffee's On **70** Expanding the Horizon **71** Ziggy Invades Cleveland **79** Sold! **83** Shake, Rattle—And Roll On **89** Where Do We Turn? **90** The Long Battle Ahead **101** Winds of Change, Winds of War **108** Feeling Lucky, Punk? **117**

Part 2: FM Rock Comes of Age: 1975–1983
The Pot Begins to Boil **123** Transitions **128** Bowie Goes Pop **134** Buzzards, Penguins, and All Kinds of Animals **135** The Start of a Dynasty **139** Happy Birthday—From the Boss! **150** Television, Togas, Playboy, and the Buzzard's Enemies List **153** Gateway to the Eighties **159** Dominance and Decay **176** Baseball Buzzards **180** The Long March Continues **182** Another One Bites the Dust **187**

Part 3: Benchmarks: 1983–1988
Drugs **211** Payola **216** News **217** Exclusives **219** Get Down! **225** Mr. Leonard **229** The Buzzard **233** How Did That Happen? **239** The Battle Rages **242** Beat It **244** The Line in the Sand **245** Stages **251** Live Wire **253** The Legend Grows **256** The New Face of the Enemy **266** The Fight for the Rock and Roll Hall of Fame **269** All Aboard the North Coast Express **278** Bar Blasters, Browns, and the Revenge of the Buzzard **293** The Buzzard Droppings Hit the Fan **313**

Part 4: The Changing of the Guard: 1988–1999
Baby, I Was Born to Run **335** The Numbers Game **339** Taylor Hall, Beatle Paul, and Paint Falls **350** The Buzzard Flies West **369** The Heart of Rock and Roll **377** The New Kid in Town **384** Transition **393** Enter the Inner Sanctum **401** Days of Uncertainty **405** The Long Goodbye **406** Howard Stern in Cleveland **421** Aftermath **429** The New Zoo **432** Buzzard Paloozas **434** Thirty Days in the Hole **435** Change Again **438**

Epilogue 446

Bibliography 450

Index 457

Foreword

The life of a cartoonist is similar to that of a hermit. Jules Feiffer once described how he spent his youth alone in a room, drawing pictures while dreaming of becoming a world-famous cartoonist and how, once he became a world-famous cartoonist, he still spent all of his time alone in a room, drawing pictures.

Having spent a large part of my youth and most of my adult life in a similar fashion, I found that I was saved by two things: First, by the fact that I actually happen to enjoy sitting alone in a room all day long and, second, by my radio. The radio I listened to as a teenager, as a student at Kent State, and later as an adult not only entertained me, it empowered me, informed me, and became a part of the fabric of my life and my work.

Initially it was the wonderfully eclectic AM radio of the early sixties, where you could hear everything from the Beatles to Bob Dylan to Brubeck on WIXY, WHK, and WKYC. The Beatles opened my mind to the endless possibilities before me, Dylan filled it with images, and Brubeck filled it with feeling. But it wasn't until I became a student at Kent State and FM radio seemingly arrived from nowhere that radio became a political and artistic force to be reckoned with.

A new generation with a new attitude was emerging, and my radio brought it to me. It shaped my attitudes, ambitions, and tastes. Billy Bass, Denny Sanders, and Kid Leo introduced me to new friends like David Bowie, Roxy Music, Mink DeVille, and a guy named Bruce. I wasn't at the tenth-anniversary concert for WMMS at the Agora; I never made it to a *Coffee Break Concert,* and I'm the only member of my generation willing to admit that he wasn't at Woodstock. I was too busy sitting alone in my room, working on becoming a world-famous cartoonist, but thanks to my radio and the personalities and music that it brought into that room, I was a part of it all. For a brief time radio transcended itself and went from being a purveyor of pop tunes to a true artistic form in and of itself.

The bad news is that, like all good things, it didn't last. The good news is that the book you're holding brings it all back again with an almost aching nostalgia. Not only that, but it takes you behind the scenes of that radio revolution, revealing the personalities and machinations that helped bring it all about. Along the way it supplies engaging anecdotes and reveals numerous secrets that are only now seeing the light of day, beginning on page one.

Tom Batiuk

Preface and Acknowledgments

Perhaps one of the hardest challenges for a journalist is to tell a story that involves them, yet still keep it strictly at an objective level. That's the problem I'm facing here with this story, the history of progressive rock radio in Cleveland.

Oh sure, it would be very easy to simply write this as if they were my memoirs, but unfortunately that wouldn't work. By the time I came into the fold at WMMS, the station already had a rich history involving people, conflicts, and changing times. The only way I would be able to write all of that on a first-hand basis is if I was about ten or fifteen years older, and I'm not ready to try to accomplish that just yet.

Fortunately though, media is a great subject for history. Even if you're not involved with it right away, you can still see or hear it progress and evolve.

As I said, I'm telling this story partially from the point of view of a WMMS staff member, but believe me I'm trying to be as unbiased as possible. It's not easy, especially since many of these people are my friends or heroes, but I'll try really hard to be fair in this book. This is their history, warts and all.

I joined the WMMS staff in early March 1988, in the middle of one of the station's biggest controversies, but I was trying to get my foot in the door almost two decades before.

It was early spring of 1971. One of those bright, sunny Saturdays that you hardly ever saw in Cleveland. A nice day for a walk, and man, did I walk. All the way from Public Square to East 50th and Euclid. I was a man on a mission.

Well, "man" may not be the best word. I was still a senior in high school, and I had been thinking long and hard about the future—especially how I was going to make a living. I needed something that would satisfy my creative urge and somehow make use of a talent that I was pretty sure I had. (I just wasn't sure what it was yet.)

In 1967 CBS aired a comedy series called *Good Morning World*. It starred Ronnie Schell and Joby Baker playing a couple of disc jockeys in Los Angeles who called themselves Lewis and Clarke. Radio looked like a job that offered lots of creativity, was full of energy, and, best of all, I didn't see how anyone could call what they did "work!" Here in Cleveland, the jockeys at WHK, WIXY, and KYW always sounded like they were having a great time, and that sounded pretty good to me. Radio seemed like a very distinct career opportunity, though I never really pictured myself as a Top-40 deejay. However, some other options were becoming evident very quickly.

Starting in 1968 new kinds of stations were popping up on the FM dial, where serious deejays played progressive "music for the head" and weren't afraid to comment about the stuff that was on everyone's mind: Vietnam, Nixon, the draft, drugs, the social order, but most of all the music. That option started to look a lot more attractive as both the format and I matured.

Now and then the disc jockeys would ask listeners to give them a call to let them know that they were being heard, and I would happily oblige. They told me that the best way to get a job in radio was through an internship, and that's what brought me to that building on East 50th and Euclid: the doors of WMMS.

It's now the site of the Agora rock club, but back then it was home to the powerhouse WHK-AM and its ugly FM stepchild, 100.7. The elevator rattled every inch of the way to the third floor and slowly opened to a beat-up hallway. I walked up to the door, took a deep breath, and gave a loud knock.

No one answered.

I gave another couple of sharp knocks, without success. After a half-hour of beating on the door, I decided to regroup and call one of the disc jockeys from home. He told me that they had a skeleton staff outside of regular business hours. It never occurred to me that a radio station wouldn't have a receptionist on Saturdays.

It would be quite a while before I would return to WMMS.

In the meantime, I did what a lot of kids were doing: getting a job, going to school, and just hanging out. In 1977 I landed a job at all-news station WERE-AM, where I learned the trade from a fine staff of serious journalists. But the management seemed more concerned with promoting its sister station, WGCL-FM, and the purse strings started getting tighter on the AM side. A lot of good people such as Carl Monday, John Webster, Eleanor Hayes, Ken Prewitt, and Charlie Steiner left for better opportunities, but I stayed on until I felt a little more confident about my own options.

The years went on, equipment fell into disrepair, and the engineers were hard pressed to keep the station "alive and kicking." I remember a conversation I had in the early 1980s with one of the engineers, as he was lying flat on his back under a workstation. His name was Ric Bennett, and after failing to resurrect the dead with some very creative cursing, he slid out in the open, looked at me, and said, "There's gotta be a better place than this!" He left the station soon after.

By 1986 I was a news anchor and host of the afternoon drive show, but I was ready for a change. It was great working with my partner Cliff Baechle, who is still a close friend (and who would later share some memorable times with me in the newsroom at WHK and WMMS), but it was obvious to me that management had no intention of doing outside promotion for anything but the morning drive slot, weren't in touch with the audience or the times, and seemed intent on making the station as boring as it could possibly be. It was time for me to look elsewhere.

In May of that year I was part of the Cleveland media pool that flew with the state's political bigwigs to New York City for the official announcement about where the Rock and Roll Hall of Fame would be located. Cliff and I broke the story on our show the previous Friday that Cleveland would get that designation after we had cornered board member Clive Davis of Arista Records live on WERE's *Michael Jackson Show*. WMMS's Ed "Flash" Ferenc was in New York as well, so I decided to introduce myself.

I walked up to Flash and asked in true New York fashion, "Do you have the time, or should I go screw myself?" At times Flash had a laugh that could be likened to that of a hyena, and as we stood in front of Radio City Music Hall he screamed out a guffaw that turned heads and stopped traffic dead, even in Manhattan. We struck up an acquaintance and promised to stay in touch. At the same time, wheels were turning to sell both WERE and WGCL, and that almost always means staff changes. Sure enough, despite assurances that no changes were in store, just after the stations switched ownership I was called into a meeting in November 1986. I left WERE by mutual agreement. They said, "Get out!" and I said, "Don't try to stop me!"

After a few weeks of vacation, I went to work at WRMR/WLTF for a year. Just a couple of days before signing a contract that would have kept me there a long time, fate decided to lend a hand. I got a call at home from Flash. He said, "We've got an opening. Wanna talk?" Flash knew that I was itching to go to WMMS ever since I left WERE, but full-time vacancies there were few and far between. Part-time work was available, but I had a mortgage to worry about, and he knew that I would be very interested in the first full-time gig that opened up.

The next day I sat down with him and Lonnie Gronek, the station's general manager, at the New Yorker Deli on East 12th. We agreed on a start date. At first I would do news on WHK's morning show, with additional work at WMMS. I went back to WRMR and put in my notice.

Just days later the story broke about stuffing the ballots for *Rolling Stone*'s readers' poll for favorite radio station. When I visited the station to do the paperwork for my benefits, I was escorted to Gronek's office. He said, "I hope all this publicity hasn't affected your decision to sign on with us." I just smiled. This was the one station I wanted to work for, and after working for far less successful companies, the Buzzard's "killer instinct" inspired me. Nothing short of a papal decree would have stopped me from coming over, and even then I would have appealed if it came down to that.

On the appointed start day, I rode the elevator to the twelfth floor of the Statler Office Tower. I walked into the lobby, past the carpeted walls, and said hello to Verdell Warren, longtime receptionist and local industry favorite. Gronek's secretary, Jo Ellen Ferrington, came out to fetch me, and I could just feel the energy as she walked me down the halls. There were so many faces I recognized from the papers, and the voices that were a part of my life. I passed the air studio and spotted Matt the Cat laughing with Jeff Kinzbach. Ruby Cheeks chatted with Kid Leo in the hall—and then I saw another familiar face. It was Ric Bennett, now known as Rocco the Rock Dog! He smiled and said, "I told you there was a better place!" Then he grabbed me in a bear hug that I'm still recovering from.

I had finally arrived.

There are tons of special memories that I treasure from my time at WMMS. I cannot begin to tell how proud I am of some of the projects I had done for the station, such as the work Pat Artl and I did producing the *Source Report* for Westwood One about the twentieth anniversary of the Kent State shootings. It was especially moving to hear the stories from those involved as I manned the remote site at the school during the live broadcast that day.

Pat and I would work again for the one-hour Christmas Eve news special, *We Built This City on Rock and Roll,* focusing on the Rock and Roll Hall of Fame's benefits to the city. I wrote it, and Flash and I narrated, with skillful production by Pat. It was the result of two weeks of careful editing, and it was completed just hours before it was scheduled to air. We all tuned in to hear the finished product, and when it was over Flash beat me to the phone and called to share congratulations on a very satisfying project. Then he asked me to hang on so someone else could say hello, and it was his father-in-law, radio legend Ronnie Barrett. He was equally kind in his comments, and it was a sincere honor to have shared those few minutes on the phone with such a gracious man.

Flash and I had already shared a rather special moment in 1988. At that time I had been manning the WMMS/WHK afternoon news desk for a few weeks, and that included chasing down stories and tapes for the following day. Part of the routine was to call Flash at home, tell him what he had lined up for the next morning's newscasts, and catch up on any station gossip I may have missed. On December 15, 1988, he dropped a bombshell. Kid Leo was leaving the station the next day.

Another fellow journalist I had some memorable experiences with was Cliff Baechle. Cliff has a great sense of humor, but when it came to the newsroom, he's all business. As a result, every effort was made to get a laugh out of him, including a recurring tradition of trying to give away his watch to guests who would come in for interviews. It wasn't easy to make him crack up, and he wasn't always very receptive.

We noticed that Cliff could carry on a conversation and type letters without even looking down at the keys. Then he would grab one of his business cards, drop it in the envelope, and mail it off to thank his guests. I snagged one of his cards and had a new batch made up. But instead of the words "News Anchor" under the WHK logo, the new box read "Assistant to Mr. Olszewski."

The new cards were interspersed randomly in the box at his desk, and quite a few went out before he glanced at one while stuffing an envelope. The glasses came off, Cliff calmly walked across the newsroom to where Carmen Tedesco and I were having a conversation, and he calmly asked, "Have you seen my new cards?"

Starting to perspire a bit, I mentioned how very touched I was that he would send out cards with that inscription, which drew a somewhat intense stare. He shook his head and said, "I don't mind that you reprinted them, but the copies look better than the originals!"

But perhaps the most thrilling experience I had at the station was when I got to interview Paul McCartney during his Flowers in the Dirt Tour. The process and lengths we went to in order to get an exclusive are detailed later in the book, but just to quickly sum up, it took a helluva lot of schmoozing and bombarding McCartney and his crew with WMMS information before he arrived in town.

During the week of the show, I had foregone my vacation in order to do live cut-ins with Rocco from the Stadium. At this point we still had no word if McCartney was going to speak to anyone in the press while in town.

Rocco and engineer Mike "Mondo" Knisely had set up a satellite studio in Malrite's stadium loge. While we were waiting for the station to make final arrangements to secure an exclusive with McCartney, Rocco and I spoke to everyone from the sound crews to the staff dietician. When we interviewed the guitar technician, he actually let us touch McCartney's famous Hofner bass. It still had the Beatles' playlist taped to the side. We treated it like a precious ancient relic, and as friendly and trusting as he was, the guitar tech kept his eyes riveted to us the entire time.

Word got back to us from a local television crew that McCartney's publicist announced that the singer would only speak to us. We quickly headed back to the loge for confirmation, as Brad Hanson was dialing furiously to find out if those stories were true. When Brad got the official okay, we all broke out in loud cheers.

But there was still a problem. We had to get the signal back to the station, but all of the steel and concrete at the Stadium would make that very difficult. Mondo and Rocco snuck around backstage for some quick site checks, and they determined that we had only one option.

Tour security would allow only two people backstage for the interview, so Mondo gave Brad a quick lesson in handling the remote unit, and we headed to the rendezvous point. The publicist met us there and asked us for two favors: McCartney was open to any questions, but we were asked to be sure to mention his involvement with the environmental group Friends of the Earth. Secondly, when he walked in, we were to break into programming so he could make maximum use of his time at the venue. No problem.

Paul McCartney casually walked into the room with a self-introduction, and we took our places. I sat with McCartney on the couch with a wireless microphone, and Brad aimed the remote unit out a nearby door toward the company loge. But that signal was very narrow, and he was forced to stand perfectly still the entire time, with a five-pound remote-relay unit about six inches under his chin. From there, the signal bounced to at least five other points before it went out to the radio audience.

McCartney was pretty amazed at the set up. Before we went on the air he looked at Brad, and then he quietly asked me, "You think he'll be all right?"

I nodded yes, adding, "I think he works out in a gym."

McCartney shrugged, and we were ready to roll. Just before we switched on the remote, I asked McCartney how long we had with him, and he said we'd just play it by ear. He was having a little problem with his voice, but he had a cough drop in his mouth and didn't seem concerned. We agreed that he would give me a signal when he wanted to wrap it up. We broke into Len Goldberg's show for an exclusive interview with the man of the hour.

McCartney was so easy to talk to. We discussed the tour, Linda's Cleveland roots ("Shaker Heights, the posh bit"), his relationship with John Lennon, the Rock Hall, and even how hot it was back at his Washington, D.C. show on the Fourth of July (which is where we cornered him in order to arrange this chat).

He was quick with a joke, at one point turning the tables and interviewing me. We could have gone on for hours. But after about twenty minutes, I noticed his hand came up to his throat, which I took as the signal to bring the interview to a close.

We wrapped it up, and he smiled at me and said, "Hey! You're good!" He then looked over at Brad again, who was standing stiff as a statue. He told Brad, "And you are amazing! How did you stand still for so long? Come over here, kid!"

Brad joined us, and I mentioned that I noticed his hand signal, but McCartney said, "Oh that! My cough drop was going down me throat! We could have kept going!" At that point, he sat back, put his feet on the coffee table, and said, "So tell me about Cleveland!"

We all talked for a while. At one point McCartney, ever the gracious host, asked, "Say, would you guys like a beer or something?" We quickly shook our heads no, realizing that if McCartney left, he probably wouldn't come back.

I took the opportunity to pull out a rare promo version of the *Yesterday and Today* album, with the infamous so-called "butcher" cover that was printed but scrapped before retail release. When Paul took a look at it, he did a double take and asked, "Where did you get this? I remember when we picked this photo! I didn't even think it got to this point!"

He described what each Beatle was doing when that shot was taken, "Ringo was bored, I was asking John something about the photographer, and George was grumbling how he had to leave to get his air conditioner fixed!" McCartney took a pen and signed it, but as he was handing it to me, he paused and took it back, saying, "Wait a minute. I'm a vegetarian! Look at all that raw meat!"

My heart sank. A truly unique Beatles collectible, and I thought he was going to keep it. Instead, he wrote "Go Veggies" under his signature and said, "There you go! Much better!"

Finally his publicist came by. After many thanks, Brad and I wished McCartney a great show. We just sat there for a few moments smiling and looking at each other, realizing we had just seen one of the true highlights of both of our careers! In what we later found out to be a ritual among his other interviewers, we gathered as many souvenirs as possible, with Brad getting his water glass, and I picked up the pen to pass on to another rabid McCartney fan.

But anyway, this isn't a story about me. It's about radio stations that sprung from humble starts to inspire a then-dying Rust Belt city and gave it something to hold onto. Something to be proud of, that helped it weather a long storm of national jokes and civic malaise. It was a time when the Indians couldn't get out of the basement, the Browns couldn't get into the post-season, and the city of Cleveland couldn't pay its bills. But we had rock and roll—and great rock and roll radio! It showed people how to dream and how to make those dreams come true. It inspired a city to new heights and became a part of every one of our lives.

Because of the whole Cleveland FM rock scene, a lot of us ended up in a better place.

This book is not meant to be a comprehensive, all-encompassing history of Cleveland radio, nor is it the final word on the era it covers. It simply tells the story about some key personalities and stations who waged an "air and ground assault" for Northeast Ohio's rock radio audience. The people and stations profiled all played critical roles in the story presented here. Some disc jockeys, even a few popular ones, are not featured a lot or are just mentioned in passing, but that's only because they didn't play an integral role in the progressive history. This is by no means an attempt to edit or to whitewash Cleveland's entire radio history, especially when the extensive contributions of jockeys in the urban format are considered, but this is just one specific story in a vast history.

This project came together through the help and encouragement of some very special people. First and foremost is the incredible group at The Kent State University Press: Joanna Hildebrand Craig, Kathy Method, Will Underwood, Christine Brooks, and Susan Cash. They shared my vision for a book about FM radio and made it a reality. This wasn't the book I expected, and I'm glad, because it turned out so much better with their guidance.

Special thanks also go to Richard Berg; Cleveland State University Library-Special Collections (Bill Barrow, Bill Becker, and Joanne Cornelius); Sarah Crump; Alexandra Diotalevi; Sabatino Diotalevi; Bob Fiorelli; Rick Funk; Eric Funk; Brad Funk; Andrew Heinzman; Eric Heinzman; Jennie Kingsolver; Jack Kluznik; Bianca Kontra; Tom Kontra; Erik and Theresa Leas; Greg Method; Tony Misciagna; Peter Motz; Lee Peters; Linn and Laura Sheldon; Jim, Matt, and Dan Swingos; Bill Tash, and every person whose life was touched by the sound of FM rock radio.

The stories in this book came alive through the photographs of the following talented individuals: Brian Chalmers, the longtime Buzzard artist who documented many of the station's most historic moments at shows, in the studio, and even overseas; Bob Ferrell, who currently works with agencies like Belkin Productions, Playhouse Square, and various sound companies and who says, "I only work with people I like"; Joyce Halasa, who documented the early days of Cleveland's counterculture and currently teaches photography at Lorain Community College; Janet Macoska, whose work has appeared in *Creem, Circus, Hit Parader, Rolling Stone, 16, New York Times, London Times, People, Entertainment Weekly, American Photo,* and other major publications and whose photographs have been exhibited in private and public galleries around the world, the National Portrait Gallery in London, the Rock and Roll Hall of Fame and Museum, the Akron Art Museum, and her classic photos in Hard Rock Cafes worldwide; and Fred Toedtman, who described himself as "a frustrated commercial photographer looking for a new outlet," offered his photographic skills to a number of alternative publications back in the 1970s, including the *Great Swamp Erie da da Boom, Star, Cuyahoga Current,* and *Exit* and whose love of music led him to a job in record promotion for Elektra-Asylum for several years.

Prologue

The year was 1968.

The Beatles were still together. Brian Jones, Jimi Hendrix, Janis Joplin, and Jim Morrison were all still alive. Lyndon Johnson was president. Vietnam was raging on, and the peace movement was growing rapidly. Assassins gunned down Robert Kennedy and Martin Luther King Jr. The top television shows were *The Mod Squad, The Smothers Brothers,* and *The Hollywood Palace.* Drew Carey and Halle Berry were just kids growing up in Cleveland. Customers could still walk into Record Rendezvous and talk to Leo Mintz, who sponsored Alan Freed and may well have been the person who really labeled the music "rock and roll."

Elsewhere, four students named Allison, Jeffrey, Sandra, and Bill were dreaming about a future at Kent State University and beyond.

Front-page headlines in the *Cleveland Plain Dealer* warned "Hippies Given It Straight: Language Is Going to Pot." Joe Eszterhas (who broke the My Lai massacre story and went on to fame as a Hollywood screenwriter), pointed to hippies as "a minority group opposed to the establishment, a protest-like group which has developed its own defensive terminology specifically designed not to be understood by the establishment which is protested against." He defined bizarre phrases like "dig some sides," "groove in my own scene," and "the whole scene is a wipe out." He labeled the new slang "semantic psychedelirium tremens." Those headlines sent a ripple of fear through the parents, while the kids turned to the comics page. But there was an underlying theme, and that was the new way young people were learning to communicate.

Radio stations WIXY, WHK, and WKYC were battling for those listeners on the AM dial. For many stations, their FM counterparts were a couple of turntables and some sales pamphlets at the far end of the building. Very few people had FM tuners in their cars, so advertisers went for listeners and money on the AM band, with some stations simply opting to simulcast their AM signals. In the mid-60s the FCC decided that practice would end, and FM stations would have to air mostly original programming.

The seeds of FM rock had already been sown at college radio stations and with enterprising disc jockeys who later bought time on ethnic outlets like WZAK and WXEN. There was even a hint of things to come on the AM dial. There was a new music, a new attitude, and a new lifestyle surrounding them. What they needed was a venue to bring it all together.

PART ONE

A GENERATION FINDS A NEW VOICE

1968–1975

The Birth of Progressive Rock Radio

The grand experiment would begin in Cleveland in August 1968. Progressive rock stations were starting to emerge in cities such as San Francisco, where disc jockey Tom Donohue was a pioneer of the format by playing a new, cerebral type of sound to a segment of society that challenged just about everything the previous generation held dear.

The Summer of Love in 1967 had shown the strength of the numbers who would listen to lengthy, musical jams at shows such as those at People's Park in Berkeley and the Haight-Ashbury district in San Francisco. The Beatles had stunned the music world with *Sergeant Pepper,* and other groups were challenging the boundaries of commercial music with extended cuts designed to be heard in the context of an entire album rather than quick hits on Top-40 radio.

The time was ripe for stations and formats to bring that music to the public. Some called it the "gathering of tribes," and Donohue's vision was an audio beacon calling out to the masses who embraced the concept of "tune in, turn on, drop out." It was a time of resistance—to a faraway war that few wanted to fight and that was cutting down young men for a cause no one really understood, to a political system that angrily squashed the ideas of any who questioned it, to the morals and ideals imposed by an older generation out of touch with the realities of modern-day youth. The late 1960s was a time that drew together and embraced those who allied themselves in the search for artistic, social, and political freedom and vehemently protested against anyone who would deny them. FM radio had pretty much been nonexistent, but there were visionaries such as Donohue and his staff who explored its possibilities and made it a forum for the emerging counterculture. Stations KMPX-FM and KSAN-FM were doing it in the San Francisco Bay Area, as were KMET-FM in Los Angeles and WNEW-FM and WBAI-FM in New York. Some in Cleveland thought Northeast Ohio was ripe for a similar experiment.

Oddly enough a big part of Cleveland's FM-rock scene was born on WIXY 1260, an AM rock station that dominated the ratings and held a tight grip on the lucrative teenage radio market. It was an era that saw the emergence of some spectacular music, such as that coming out of Motown, the Beatles, and the rest of the British Invasion—and WIXY played it all. The station, the brainchild of partners Norman Wain, Robert Weiss, and Joe Zingale, was so popular that its program directors would chew out the disc jockeys if ratings fell below 50 percent of the listening audience, a mighty big order considering there were over twenty stations serving Northeast Ohio at the time. If WIXY added a song to its playlist, it was more than likely on its way to being a hit.

Born in Brooklyn, New York, Wain was an avid fan of radio, and he yearned for a job on the air since he was very young. After graduating from Brooklyn College he started work as an announcer. He continued his career path when he was called to serve in the U.S. Army during the Korean War, where he worked for Armed Forces Radio. He eventually relocated to Cleveland in 1954 for several positions at Wyse Advertising, WEWS-TV, and WDOK-AM.

Weiss was raised in White Plains, New York, after which he came to Cleveland in the late 1950s to work as a salesman at WHK, an AM station that also had an FM sister with the same call letters. He recalls, "It was easy to get a job as a salesman there, because nobody could sell any time!" It was at WHK that Weiss teamed up with fellow salesmen Wain and Zingale, a native Clevelander who had seen phenomenal success selling radio ad time there, to buy their own radio stations.

People thought the three were crazy in November 1965 when they took over WDOK, changed the call letters, and started playing Top-40. But big promotions, fast-talking deejays, and a shrewd business sense made them the kings of Cleveland radio in a relatively short time. In fact their two biggest rivals, WHK and WKYC-AM, had been playing the same type of music but couldn't compete with the upstart station, so both changed their formats in short order. A few years after Beatlemania hit the scene, music was taking a dramatic turn toward more serious fare, heavy in folk and blues roots, although it wasn't getting much airplay.

News about the new musical styles was pretty much spread by word of mouth among Northeast Ohio's emerging counterculture, and Barry "Buttons" Weingart was one of the people who heard the call. A member of local band The Case of E. T. Hooley, Weingart had spent time with Jimi Hendrix when he played Cleveland. The story goes that Weingart met Hendrix at his hotel and saw him bewilderedly staring at a menu posted on the wall. Hendrix turned slowly to Weingart and asked in a soft voice, "What does that thing mean? What's the meaning of this?" Weingart then knew that he and Hendrix would be getting along just fine. They were together for a radio interview on WKYC to promote Hendrix's performance at Cleveland Music Hall. When the guitar master paid cash for a new Corvette at local car dealer Blaushield Chevrolet, Weingart knew Hendrix could be playing much bigger houses. So he took it upon himself to help make that happen.

Weingart didn't really want to be a disc jockey. He just loved the new music, had a great ear for it, and wanted to get it out to a larger audience. He approached the management of WZAK-FM, the ethnic "station of the nations," in hopes of producing a program. They told him that they didn't have much going on overnight, so if he could get advertising to pay for it, Weingart could have the time to play whatever kind of music he liked. He was a good friend of the owner of the now-defunct Giant Tiger department stores, who offered to back the show. After failing to find a jockey to man the microphone, Weingart was told by some friends in radio about Steve Nemeth, who had the same idea about broadcasting but didn't

Doc Nemo (1968). Nemo's flair for drama, keen ear for music, and bizarre sense of humor made for a show that could not be ignored.

have anyone to put him on the air. He rang Nemeth up to make arrangements to meet, and that phone call in late 1967 helped spark a revolution.

Nemeth was a crazy biker and drummer out of suburban Warrensville Heights who once played with the Rivingtons. The group had a hit with "Papa-Oom-Mow-Mow," which gained greater local notoriety when Ernie "Ghoulardi" Anderson started playing it on his late-night television horror show. Nemeth would later say that Anderson and disc jockey Pete "Mad Daddy" Meyers were his biggest influences, and he used a lot of both in his on-air delivery. Weingart told

Nemeth that they could play lots of great music and maybe smoke some pot and meet some chicks in the process—that was enough to win Nemeth over. Looking back, Nemeth remembers driving to WZAK's studios that first night. "I decided my real name was too drab for an underground show," he recalls, "and one of my favorite characters from the Old West was Doc Holliday. So I took the Doc from Holliday, Nemo from my last name, and there it was!" He walked into the studio, turned on the mike, and "Doc Nemo" was born.

As Doc Nemo, Nemeth was bold, brash, and very funny, as he sometimes dropped hints about vacations at "the Cooley Farm," the Warrensville workhouse, a minimum-security jail. Using his flair for acting he had a voice and attitude that demanded attention, and listeners embraced him as one of their own. It came easily to Nemeth because the Doc Nemo character, nicknamed "The Mad Hungarian," was simply an extension of his real life persona.

It didn't take long before Weingart and Nemo started to get a following. Weingart recalls that they wanted to move on to the big time in Cleveland radio, and there was an opportunity to do Sunday nights on WIXY. He says, "We figured we'd have a much bigger audience on WIXY than on the FM, because back then no one had FM in their cars. It wasn't that big of a thing, so we took the plunge." They convinced the management at WIXY that they could produce a cutting-edge rock show for the station. But Weingart points out, "WIXY, of course—being the idiots they were—wouldn't let us stay on both their station and the FM. We should have stayed on FM and the Sunday night thing."

FM radio back in the late 1960s was sort of like the Internet would be thirty years later. It was virgin territory, and almost anything imaginable could be put on. But this was not so at WIXY. Owner Wain listened to the premiere show and heard Nemo introduce the very first record, a song by Steppenwolf called "The Pusher." Singer John Kay groaned that he "smoked a lot of grass" and "popped a lot of pills," but it was the line "God damn the pusher man" that sent Wain to the phone. The red line rang, and only certain high-ranking management types had that number. Weingart took the call only to hear a voice yelling, "What the hell are you playing on my radio station?" He tried to argue the importance of playing those kinds of lyrics, but he realized within minutes that his relationship with Wain would be a lot different than the one Wain shared with Weingart's parents, who had been longtime friends at the same country club. Weingart simply didn't seem to care.

Today Wain likens Nemo to "a missionary" but also admits that he didn't really understand what the show was all about. "He [Nemo] was really weird," Wain recalls, "and it got to be a very popular show, but of course, he was very, very difficult to deal with. He was a star. He wanted things his way, and he wouldn't read certain commercials. He didn't even want us to sell the show."

The program turned into a freewheeling romp through late-1960s psychedelia. Weingart and Nemo traded jabs on the air, sometimes fooling the audience into thinking that they hated each other. Nemo recalls it as "theater of the mind—for whatever shape your mind might be in at that late-night hour." In actuality it was the relationship between Weingart and Wain that was strained at best, with Nemo recalling "a huge, near-violent confrontation between the two, where Wain threw Barry out of the office, which wasn't doing my career any good."

Weingart and Nemo had their radio voice, and the time had come to get people to come out and experience that music live. The two hatched a plan to stage a show in downtown Cleveland. They rented the Palace Theater, at that time a crumbling shadow of its former glory and a prime candidate for the wrecking ball in the near future. Weingart and Nemo had hoped to present Cream, the pioneering blues trio featuring Eric Clapton. Clapton's stunning guitar solos had many fans worshiping him like some kind of deity, as it was very common to see graffiti around town reading "Clapton Is God." But so much attention was being paid to Cream, thanks to chart-toppers such as "Sunshine of Your Love" and the groundbreaking cover of Robert Johnson's blues classic "Crossroads," that the band had already been locked up for another show somewhere else in Northeast Ohio.

Nemo and Weingart were instead offered three other acts—Blue Cheer, which had a hit with a power-metal version of "Summertime Blues"; Traffic, featuring a very young Steve Winwood of the Spencer Davis Group; and Clapton's old band the Yardbirds, who scored a string of hits such as "Over Under Sideways Down" and "Shapes of Things." The problem was that this version of the Yardbirds didn't have a big-name player like Clapton or even like the other guitarist who had helped put the band on the map, Jeff Beck—both of whom played on the band's biggest hits. This incarnation of the Yardbirds was headed in an entirely different and far more surreal musical direction, led by a studio guitarist who had played bass during Beck's final days with the group. His name was Jimmy Page, and he was honoring old Yardbirds contracts before resuming work with his new band, Led Zeppelin.

Nemo and Weingart were offered all three bands, each doing two shows, for a package price of $12,000, serious money in 1968. The two talked it over and decided that it was worth the risk. Weingart borrowed the money from another retail friend named Chuck Avner, who headed up the Man Talk stores, and from his old acquaintance at Giant Tiger, and soon Day-glo posters popped up all over town advertising DOC NEMO'S MIND BLOWING CONCERT #1. It would not be an easy ride.

Perhaps it was a breakdown in communications, but the promotional support they expected from WIXY didn't work out the way they had hoped. Weingart remembers a heated exchange with Wain, who told him that he would have to pay commercial money if they plugged the concert to any extent on their radio show. "If it was a WIXY promotion, you sure couldn't tell," Weingart later recalls.

On the night of the show the marquee of the Palace Theater was lit with the names of the groups, drawing hippies like moths to a flame. For years, a film of this marquee was featured during the introduction of the *Hoolihan and Big Chuck* late-night show on WJW-TV. Unfortunately, because of budget constraints security was a bit lax. The concertgoers were laid back enough, but Weingart points out, "The Palace was like a Swiss cheese. You would have needed fifty people to guard all the exits. People were pouring in from every door, including the back." There were too many to contain, the hippies who were looking for free music far outnumbered the ticket-holding patrons, and mayhem ensued. The only option was to let the show go on as planned.

Blue Cheer delivered an earsplitting set of high-decibel rock that was brought to an early end when the soundman cut their mikes. Nemo says the reason was simple: "They just said 'fuck' too many times." These kinds of obscenities were an unwelcome novelty to straight-laced audio engineers, many of whom got their starts doing sound in vaudeville in far simpler and more innocent times. Traffic came through with a very tight performance, highlighted by their British hit, "Paper Sun," which was later incorporated into a Coca-Cola commercial. Steve Winwood's reprisal of the Spencer Davis hit "Gimme Some Lovin'" brought the crowd to its feet. The Yardbirds then took the stage, fueled by Jimmy Page's six-string pyrotechnics as he coaxed some of the eeriest sounds ever to emanate from a Gibson Les Paul, playing at times with a violin bow. The crowd heard different takes on Yardbirds' standards such as "You're a Better Man Than I" and new songs by Page such as "I'm Confused," which was later released on Led Zeppelin's first album as "Dazed and Confused." By evening's end the crowd had witnessed a truly memorable night in Cleveland's rock and roll history. It was also memorable for Weingart and Nemo since they had to pay back the money they borrowed to book the bands. Even after the take at the door and their profits were factored in, they were still down a thousand bucks.

Nemo and Weingart blamed a lot of their financial problems on WIXY's lack of promotional muscle behind the show. They believed that it was fair that the station should have shouldered some of the financial burden. A few days later the two sat in the WIXY studios mulling over their options while the records played, but there weren't many and the situation looked understandably grim. However, fate lent a hand. With statutes of limitations most likely having expired a long time ago, thirty years had passed before Nemo and Weingart revealed how they were able to repay the loan.

During that time WIXY was sponsoring what is known in the industry as a "forced listening" contest. It was a "high-low" game, in which listeners would have to guess how much money, to an exact penny, they could win by calling the deejays at an appointed time. For example, a listener would call in and guess the amount at $10,000. The jockey would most likely say, "Sorry, that's too high." Another might call in to guess it was $500, and be told that it was too low. The key to guessing the exact amount involved the listener having to hear all of the earlier estimates. It was a very effective contest.

During their late-night show Nemo and Weingart were sifting through commercial copy when they found a black book used by the daytime deejays, in which they had written the exact amount of the prize. It was suddenly clear to Weingart what they needed to do. That Monday Weingart recruited a number of people, including his girlfriend, to stand by a radio during Larry Morrow's show. The previous on-air guess had been about $3,000, and when Morrow called again for entries, Weingart's girlfriend got through. "Hi! WIXY! What's your guess?" Morrow asked. She calmly announced, "$1,017.34," the exact amount. Morrow was stunned, but sure enough that was the total. She was told to come down to the station, where a confused promotions director asked a few questions, handed over the money, and sent her on her way—to deliver the much-needed thousand dollars to Weingart.

Nemo recalls, "I was driving down the street in my beat-up old Caddy when I heard her win, and I damn near drove off the road. I regained my composure and raced to a pay phone to call an attorney to cover my ass." They chose not to stay at WIXY much longer, and both would find places in a new radio format.

In Cleveland there were other breeding grounds for progressive-rock radio. One was the classical station WCLV-FM, broadcasting from high atop the Terminal Tower. Deejay Martin Perlich worked at Disc/Discount Records, and his knowledge of music could be exhilarating and sometimes intimidating to anyone who questioned him or his credentials.

Perlich had grown up listening to many of the greats of Cleveland radio, including Alan Freed and Bill Randle, but he never saw himself entering the radio field. He didn't really care for rock and roll for that matter. Besides classical, he preferred the music of people like Thelonius Monk, John Coltrane, and Charlie Parker. Artists like Ritchie Valens and the Four Seasons had no impact on him. It wouldn't be until the late 1960s, during the height of Vietnam War, when artists such as the Beatles, Jefferson Airplane, and particularly Bob Dylan would change his musical tastes.

He was a prep school graduate, having attended the Western Reserve Academy. He later earned his impressive credentials with degrees from Columbia University and the University of Chicago Graduate School, as well as studying at the Julliard School of Music. In 1963 Perlich applied for a job at WDGO-FM, which would soon become WCLV. He won a spot auditioning new classical releases in a Sunday-night show called *The Audition Booth*, which drew a lot of critical praise. But his concept of programming was far from the usual radio fare of the day.

Perlich approached WCLV co-founder Bob Conrad in 1965 with an idea for a late-evening show that would stretch the boundaries of audio perception, a freewheeling program where nothing was off-limits, and he could play and say whatever he wanted. Conrad gave him that slot, initially on Friday nights, and it evolved into the *Perlich Project*.

Perlich was not a garden-variety hipster. He admits that he came from a privileged background and in some ways considered himself an "Ivy League hippie, a sojourner rather than a disc jockey," though his show could be appreciated on a number of different levels.

He might mix bits of Dvorak with McKendree Spring and Phil Ochs, include liberal helpings of the Beatles, Quicksilver Messenger Service, Woody Herman, and maybe a few interview segments with Judy Collins or the Cleveland Orchestra's acclaimed music director George Szell, with some of his own comments on the political scene to round things out. The amazing thing was the way it flowed, a stream of musical consciousness and thought that took the listener on a sonic journey into their own inner worlds.

Perlich was quoted in the *Cleveland Press* as saying that he highlighted underground music on his *Perlich Project* to "fill a void created by Cleveland's Top-40 stations. The underground sound is interesting and creative, and the music is excellent throughout." WCLV became an unlikely source for music by the Velvet Underground and Nico, the Mothers of Invention, Love, and the Butterfield Blues Band. He also presented works such as *Howl* by Allen Ginsberg and supported Cleveland's "poet laureate," D. A. Levy, by putting Discount Record ads in Levy's underground paper, *The Buddhist 13th Street Third Class Junk Mail Oracle*. No one could question Perlich's commitment to the underground scene.

Perlich also interviewed artists such as the Blues Project or whoever was playing at venues such as the hip Cleveland club La Cave, and he appealed to the intelligence of his audience. It helped that WCLV listeners were generally very serious about their equipment, as when the programming was showcased on some excellent home systems it would bring out the subtlest nuances of the music. The only unfortunate part of listening to *The Perlich Project* was the way his mixing board was wired at WCLV. After a particularly moving set of titles Perlich's voice would come on lower than the music and usually just on the left stereo channel, which would cause a listener to have to continually adjust their expensive setup back and forth in order to hear both the music and his comments. But it was only a minor inconvenience for an audience hungry for this type of programming.

There was also college radio where student disc jockeys broke down the boundaries of commercial radio to forge new ideas about music and the way it was presented. What they lacked in polish they more than made up with ambition and creativity. Back then, college radio wasn't held to the same strict programming guidelines as its commercial counterpart, and deejays were more open to making bold statements without fear of reprisal. Since those stations had only a fraction of the power and coverage of professionally run stations, they had little fear that the small audience, who would only know of them by searching to find the signals or by word of mouth, would object. Besides, the audience most likely shared those very same views.

Pat McCoy (1967). His vision for WHK-FM's progressive format helped fuel a massive change in local rock radio and the way people listened to music.

The underground was about to become mainstream due to a local promoter at Mainline Record Distributors named Pat McCoy, who was deeply entrenched in the local and national music scene. Cleveland's biggest underground newspaper, the *Great Swamp Erie da da Boom*, profiled McCoy's connection to local music, pointing out that his wife was a singer in a cocktail lounge and that McCoy had worked in radio for a time as well. He saw the new music coming through Cleveland but also recognized that there was no station to help him promote it. McCoy devised a plan and knocked on the door of his old station, WHK. Metromedia was owner of WHK at the time, and McCoy pointed out the success of the underground format at other Metromedia stations. To his surprise the company said it would give his idea a chance.

On August 15, 1968, at 7:00 A.M., the plug was pulled on WHK-FM's old easy-listening format, and the starting gun sounded for a new concept in local commercial radio. Plans were made for WHK to try the untested waters of progressive rock.

The lineup included Victor Boc, from college Case Western Reserve's WRUW-FM, on morning drive; Rick D'Amico, the former Dick Michaels from Ashtabula's WREO-AM, doing middays; and McCoy as "Daddy Mack" manning afternoon drive and weekends. A record-store manager named Billy Bass, who ran the city's first progressive music store Music Grotto on East 4th, was named to do the evening shift, followed by Nemo, who left WIXY to work nights at the new station. A simulcast with WHK-AM filled out five hours in the overnight slot. Perlich had a Sunday show that was often scheduled around the Cleveland Browns' games, also simulcasted on WHK-AM. He would later quit when it was suggested that the jockeys use playlists, and he went back to WCLV. The liners and bumps, the taped IDs and station imaging, announcing the new FM format were voiced by the young WHK-AM news director Tim Taylor, who went on to a distinguished career in television news at Cleveland's WEWS and WJW.

The very first song played on the new station was a track from an astrology LP, *Sounds of the Zodiac,* an almost classical piece in nature, "Aquarius." That was followed by more current material, though the playlist would depend on the ear and imagination of each individual deejay.

Boc was an Avon Lake High School graduate and a multitalented musician, even writing a symphony at the age of thirteen that was performed by the school's orchestra. He studied electrical engineering at Case Tech, Cuyahoga Community College, and Western Reserve. His philosophy was that "music isn't music if you can do dishes by it or read a book along with it." Boc demanded involvement with his show and the material he presented. He had gotten some mainstream press earlier in the year when the *Cleveland Plain Dealer* showcased his late-night *Head Show* on WRUW as one of the early examples of underground radio. His musical tastes leaned to groups such as Pearls before Swine, The 13th Floor Elevators, and Velvet Underground. Boc would occasionally do "poetic patterns" between the records to maintain the momentum. The show was on seven days a week, and Boc told the paper that he would "seldom say the name of the song. Too distracting. But I'm likely to say where we'll be a thousand years from now." He added that his "ultimate aim is to let people hear the sounds that really are." His show started through an echo cham-

Victor Boc (1968). Going all the way back to his days in college radio, Boc's show took on a more cerebral and literate tone, challenging the listener to concentrate on the music rather than simply use it as background noise. Cleveland Press Collection/Cleveland State University

ber and always ended the same way, by reciting lyrics from *The Long Ships* by Charles Albertine: "Go to bed tonight with hope in your heart, and if you are willing, there will be a tomorrow. And no matter what tomorrow may bring, above and beyond all, love." WHK-FM was his chance to spread love on commercial radio.

Boc had known Nemo from his work at WZAK and WXEN-FM, and the two had even simulcasted their shows to try for a larger audience. "[Nemo's] musical tastes were slightly different than mine," according to Boc, "as he gravitated a little more toward Hendrix and Steppenwolf, and I leaned more toward the Doors. There was a little difference between the two of us, which seemed a little more magnified at the time than it really was. I got together with him and Billy Bass, and the three of us basically marched in to WHK-FM and discussed the idea [with station management and McCoy]."

Bass grew up on Cleveland radio, which is quite interesting since his mother was deeply religious and avoided rock music for its supposed dark influences. He would often say that if a kid didn't listen to "The Moondog," Alan Freed on WJW-AM, there was no need to go to school the next morning as that was what all the kids would talk about. Bass also had a discerning ear for the new music coming through his stores, the R & B records at his location near Public Square, and the "underground stuff" he was selling at the Music Grotto across from Cleveland State University. Bass was a fixture on the local scene and was often seen checking out new acts in clubs and concerts. He named his record store Music Grotto in tribute to the legendary Otto's Grotto, which had become a landmark venue for cutting edge music in Cleveland. When WHK's offer came his way, he decided to take a crack at the industry that so influenced his childhood.

In October 1968 the *Plain Dealer* had an advertisement for a "New Groove" on the recently christened WMMS-FM, which had changed call letters from WHK-FM the previous month. Boc says, "The New Groove was Pat McCoy's invention, and none of us liked it. He came up with the phrase, and neither Nemo, nor myself, nor Bass liked it, but we had to use it." The new call letters stood for "Metro Media Stereo" to reflect its parent company, as a sister station in Philadelphia changed formats and call letters right around the same time to WMMR-FM, which represented "Metro Media Radio." The new staff set to work creating a new sensibility about the music and lifestyle WMMS would come to reflect. Bass would enter the studio, light some incense and scented candles, and put the volume at ten, "just like home." He enjoyed the extra money, and the radio position soon became the most important job in his life.

Boc didn't stay on mornings very long, and that suited him just fine. "The kind of music I like to play was actually more suited for night," according to Boc, "and I agreed that mornings should be a little softer, at least in the early days of the station. The music at that time of day should be a little more digestible and maybe mix in a little Top-40."

The deejays read their own spots, pushing head shops, record stores, draft counseling, and involvement in the "freek" community, which was purposely misspelled by the underground press to distinguish longhairs and flower children from the sideshow acts that many in the older generation likened them to. From the start, the station started to break new music such as the Steve Miller Band's "Sailor," Earth Opera's "American Eagle Tragedy," and former James Gang front man Glenn

Billy Bass (1968). One of the original WHK-FM progressive jocks, Bass combined his love and knowledge of music with a deep commitment to the community and the counterculture. Cleveland Press Collection/Cleveland State University

Schwartz's new group Pacific Gas and Electric. Most AM stations, especially in Cleveland, wouldn't have touched those artists if they hadn't been played elsewhere first. The station aired an eclectic mix of artists ranging from blues legends Muddy Waters and Albert King, to folkies Collins and Tom Rush, and long-form rockers such as Cream, Pearls before Swine, and Iron Butterfly, the last of which were already WHK-FM favorites by granting interviews and recording liners prior to the format change. Before the call letter change, the station received early support from the La Cave club on Euclid Avenue that promoted "the sounds of La Cave on WHK-FM, 7 A.M. to 2 A.M." on its promotional postcards. The station personalities made themselves very visible, hosting shows by Jefferson Airplane at the Allen Theater, Sly and the Family Stone at Leo's Casino, Country Joe and the

Fish at Cleveland Music Hall, and the Moody Blues at the Grande Ballroom. It was the Moody Blues' show that drew the curiosity of *Plain Dealer* reporter Karl Burkhardt.

In a story titled WMMS FLIPS OUT, Burkhardt described the "audio abstract" of progressive rock and the strange habits and dress of the people who followed it. He wrote about plaster being shaken from the walls of the Grande Ballroom by the ear-shattering music of the Moody Blues and the bizarre light show projected behind the band. Burkhardt complained of an "electronic hangover . . . complete with earache, eyestrain, and a throbbing cranium." While only in his late twenties, Burkhardt was stunned by the wide range of people "grooving alone" and "grooving together," clad in "capes . . . headbands . . . and mini clothes," with some of the women looking "as if they just stepped off the cover of *Glamour* magazine, and others resembling circus performers—clowns, that is." He ended the piece with the sage advice, "Like, if that's your bag, do your thing, 100.7 on your FM dial."

The point he may have missed is that people were really listening, a lot of them in fact, and WMMS saw it in phone calls, response at shows, and in the mail. Bass once received more than 4,000 postcards in a Beatle-album giveaway, which no one at the station ever expected. Another Beatle promotion was centered on the group's much anticipated new release. Without letting on about the sleeve art, listeners were asked to send in what they believed the cover of the new LP would look like, and the winners would receive the group's complete thirteen-album library. *The White Album,* as it came to be known, was released the first week in December and featured a blank cover with only the words "the Beatles" and a serial number embossed on its front. One of the ten winners was an artist from Grafton, Ohio, named Tom Batiuk, who would later find fame with the newspaper comic strips *Funky Winkerbean* and *Crankshaft.*

Batiuk recalls that he was working on a couple of paintings in his Kent State University dorm room when he first heard about the contest, quickly writing down the station's address. Later that evening he switched over to the campus radio station, and the deejay gave a description of the cover to his listeners. He explains, "I just sent in a blank, white sheet with what I thought the embossing would look like, and it was a winner." With a smile Batiuk adds, "I still have that album, with a hole in the corner to show it was a promotional giveaway." WMMS would later get nods in both of his nationally syndicated comic strips.

Besides being fodder for Burkhardt, another concert at the Grande Ballroom was also the scene of an awakening for the station's salespeople, many of whom were still in the dark about this whole cultural upheaval that WMMS was leading. Capitol Records brought in Quicksilver Messenger Service for a free show, and the station filled every seat. The sales staff stood in amazement at the diverse segment of society that came out for the show.

Rick D'Amico (1968). A student of the greats in early rock and roll radio, D'Amico brought a higher-energy delivery to the usually more reserved FM band.

The music had a less positive reaction to those at the actual station. D'Amico remembers the first days of the new format as culture shock for old timers, especially the engineers who ran the control board from a separate booth: "They couldn't figure out what the hell was going on. They ran the board, and we ran the turntables and equipment, and one engineer named Fred was totally bored by the whole concept, so every now and then, he would just pick up and walk away. We could cue up the records, but he was the guy who put them on the air. I'd have to open the microphone, go on the air, and yell for Freddie to come back and give us a hand."

The engineers also clashed with the personalities, especially their lifestyles. D'Amico recalls, "I came to work about quarter to ten one morning, and you had to walk around the engineer to get into the studio. The guy in the booth was really upset. I asked him what the problem was, and he said, 'What the hell is going on in there?' I looked in the studio, and all the lights were off, and he pushed me in and said, 'You better go in and see if he's alive!'" D'Amico stumbled around the dark studio to find Boc lying on the floor, looking at the ceiling. "I said, 'Victor! It's me, Rick! You okay?' and Boc just casually looked over at me and softly said, 'Oh. Hi, Rick!' I just flipped on the lights, Victor packed up and left, and it was another day at the office."

Boc wasn't the only one with a more-festive spirit. Nemo's twisted sense of humor was a highlight of those early days.

D'Amico recalls when some corporate "suits" from Metromedia visited Cleveland and gathered the staff in General Manager Dick Jansen's office. (Jansen would go on to head Scripps-Howard Television.) The deejays listened patiently to what their station was supposed to sound like, with the Metromedia people suggesting that they focus on topics of interest to Clevelanders. The level of boredom was reaching new heights when one of them suggested how Lake Erie's pollution might be an issue with pretty high interest. Nemo slowly put up his hand and in his own flamboyant style told the company representatives, "I'll tell you why that lake's polluted. Because fish fuck in it!" The room fell quiet, with the exception of the muted laughter from some of the disc jockeys, and the corporate types realized they weren't making any headway with this group. They mumbled a few things among themselves and then shuffled out of Jansen's office to let the staff sort out its own programming.

Another of Nemo's pranks was to light some exotic type of incense in the studio, fan the smoke toward Jansen's office, and watch him run frantically into the studio asking, "What are you guys doing? You smoking something? What's going on in here?"

But the station's original lineup stayed in place for only a few weeks. Boc left WMMS that November, and Nemo left about four months after he first signed on. Today Nemo laughingly says his idea of programming may have been "too underground for the underground station." He had always been something of a free spirit, and he never really forged any strong bonds with the other deejays, though he does speak highly of Boc. Nemo went to California for a role in the upcoming film, *Wild in the Streets*, the story of Max Frost and the Troopers who secured voting rights for eighteen-year-olds and turned the United States into a huge hippie enclave (featuring a very young Richard Pryor in a supporting role). Nemo's role saw him riding his brand new, "candy apple blue," 450 Honda, as he was appropriately cast as a "hippie biker." "I got to whisk the heroin away in a tight situation," he recalls, and that was enough to put the acting bug into him. That movie also spawned a hit soundtrack, including the song "The Shape of Things to Come." Nemo would later continue his film career and star in *The Lobster Man from Mars* and *Robocop* (in which he proclaimed the line, "I'll buy THAT for a dollar!"), among others.

Nemo would return to Cleveland just a few weeks later, but not to WMMS. The station replaced him with twenty-five-year-old Victor Trapp, a Kent State graduate and martial arts enthusiast who had just relocated from California.

Perhaps the strangest departure was Boc's. Although the personalities were encouraged to pick their music, Boc's tastes ran to the extreme ends, and he felt that some of the station's music didn't belong in the new format. In February 1969 Bass and a few of the other deejays decided to add an Aretha Franklin song to the music library, which Bass remembers as "I Never Loved a Man (The Way I Love You)," but Boc is sure it was "Respect." Whichever it was, they wanted it on the air because, despite the fact it was pretty heavy R & B, it was also a great, great song. There was a similar disagreement over Marvin Gaye's "I Heard It through the Grapevine." Boc strongly objected, saying those songs just didn't fit with progressive rock. He was so incensed that he did his Saturday show in the persona of a stereotypical fast-talking Top-40 deejay with a wildly exaggerated voice, playing crossover tunes also heard on the AM dial. He then walked out of the station in a huff, never to return.

His loss had an immediate impact. Bass explains, "He brought Pink Floyd, he brought all that kind of consciousness, and when Vic walked, it left a vacuum. We didn't know how to pick the hits in that genre the way he did. He wouldn't go for Aretha Franklin, and it came down to that one song." Years later Bass and Boc met again and had a conversation about that incident. Boc had come to the realization that he had made a mistake by leaving in that way for that reason, but Bass saluted him for having principles and sticking to them.

As WMMS was gaining respect, the AM stations such as WIXY and WKYC continued to battle over the lion's share of the audience. This competition resulted in stations trying to affiliate themselves with bigger and bigger acts. WMMS would join with a concert promotion company called "The Grape" to sponsor groups such as Canned Heat and the Velvet Underground at Public Hall and Blood, Sweat, and Tears and the Psychedelic Stooges at the El Grande. WKYC personality Chuck Dunaway got high profile assignments such as Janis Joplin at Public Hall. In December, Nemo joined Don Allen and Dick Liberatore at WZAK, offering exclusive interviews with Eric Clapton, Ian and Sylvia, and Iron Butterfly. Liberatore, who was Alan Freed's son-in-law, also managed a college club on East 18th called Socrates Cave and, in an odd arrangement with a competing station, hosted a weekend show heard on WHK and WMMS.

It was tough for WMMS to compete. Looking back, Boc now says the format change was too much for sales and management. "They never got it," he says, "because they were looking at the bottom line and wanted to see the numbers come up, and the jocks just wanted to be on the air and show what was happening in the United States in those days. We really thought we could be the number one station in six months. How could it be any different? Well, we were a little naive, as it took years and years for WMMS to come up to significant ratings. There were little inklings of slight movement in the ratings back then, but not much, and we couldn't understand what was going on, because it seemed like everyone was tuning in to us." In reality, the growth of ratings was slow because it takes years for stations and formats to establish themselves, and this was at a time when very few cars had FM tuners to begin with. Drive times took a beating because AM was still king of the road. It was one thing to go to concerts or talk to the disc jockeys at events and record stores, but it took a real commitment for someone to sit down and fill out a ratings survey. Many others, if they did get a survey, were already too preoccupied with drugs to be able to concentrate on a project like that.

Things were looking up in late 1968 when promoter Belkin Productions joined forces with WMMS to showcase some of the better underground acts such as Judy Collins, Steppenwolf, Arlo Guthrie, and Gordon Lightfoot. Bass hosted the Steppenwolf show and showed up in a tuxedo and white Cadillac for Collins's performance at Severance Hall, which was more accustomed to black-tie patrons as the home of the world-renowned Cleveland Orchestra.

Still the revolving door continued at WMMS. Lee Andrews came from WREO to take over the midday shift, WTTO-AM's Bob Knight followed in the afternoon drive and evening shows, and D'Amico took over a time slot later in the day.

In early 1969 Nemo made another move, this time back to WXEN. There he would sometimes take his show on the road, emceeing and broadcasting live from events such as the biker shows at the Polish Women's Hall on Broadway, later the scene of a bloody gang war immortalized in the pages of *Rolling Stone*. WXEN was the latest to try the progressive rock format, not only with Nemo's show from 2:00 to 3:00 P.M. on Monday through Wednesday and going an extra hour on Thursday and Friday, but also with Captain Sly and his *Mad Pad* show for an hour after Nemo's program.

Belkin Productions was making impressive gains on the local concert scene, offering acts such as Joplin; Country Joe and the Fish, sponsored by WMMS; the Association, sponsored by WIXY; Petula Clark, sponsored by WJW-AM; and the Temptations, sponsored by WJMO-AM. Ted Nugent and the Amboy Dukes were playing Chanel High School dances, while Glenn Schwartz and Pacific Gas and Electric were booked at La Cave and did autograph sessions with the help of WXEN.

After Detroit promoter Russ Gibb had ended his stay at the Cleveland El Grande and The Grape faded into oblivion, a group called Established Management Unlimited took up residence at 5000 Euclid Avenue, with plans to call the hall "Freedom." Jeanne Preston, who was known for booking big name shows, headed this new group. They lined up all the food and music permits, but they made little impact. A group calling itself the Second Foundation Music Series brought in Hello People and the Nazz to the Cleveland Heights Skating Pavilion, which was followed by announced dates for the Moody Blues, Youngbloods, Fleetwood Mac, Procol Harum, and the Buddy Miles Express for the spring and summer of 1969. Unfortunately they wouldn't be getting the initial support they expected from WMMS.

It was clear that changes were coming to the station, as artists such as the Who, Bob Dylan, the Beatles, and the Jeff Beck Group were sharing airtime with Brenda Lee, the New Colony Six, and even Henry Mancini. It billed itself as the station where "Music Means Satisfaction," with "20 or more records per hour," and "Million Dollar Memories." The station also promised a "steady stream of hits with 32,000 watts of current" on "Northeast Ohio's constant cooker," and D'Amico told listeners of his *Heavy Jock Show* to tune in to the "highly controversial" Pat McCoy later in the day.

Although the progressive format was a critical success, it was a financial disappointment. There were only so many head shops, record stores, and kids' clothing shops that could afford radio advertising, and those that were able to survive were finding a bigger audience on the AM side. The problem took its toll on McCoy, as he found it nearly impossible to get sponsors for the station. He left town shortly after, not realizing that he had laid the groundwork for a radio revolution.

On May 27, 1969, WMMS had abandoned most of the "new groove" music for a more adult contemporary format. It had been rumored for weeks, with engineers at the station tipping off the staff as they installed automation equipment. The general manager called the deejays to his office to announce the change, and he uttered the phrase that would ring in their ears to this very day: "Rock and roll can never work on FM." Strangely enough that was the consensus of opinion among many in radio at the time.

Despite Bass still hosting a progressive show at night, fans of the format were complaining about the change, including one memorable gripe, "Rick D'Amico sounds like he's enjoying it!" Bass would later tell *Record World Magazine*, "WHK-FM got off to a really good underground start; it was never to be taken seriously enough as anything more than an underground station, in terms of market ratings. But as one, it was breaking new artists, taking a chance on new records; it succeeded culturally as a radio station. Or counter-culturally, to be more accurate."

The format change took its hits in the national press. *Billboard* magazine's "Vox Jox" column took aim at the station in the July 5 issue, calling it a "very bad programming mistake, in my opinion. It's like throwing away an FM signal. If you can't use it properly, for God's sake give it away to someone who can!"

As WMMS was losing its focus, another FM outlet announced plans for a rock format. In July WNOB-FM (107.9) cited a "void in programming" that they intended to fill, with station president Phil Kerwin saying it was a choice between "rock or commercial religion," and rock won out. Offering what it called a "mature Top-40," WNOB went on the air with Jim Kelly, formerly of WTOD-AM in Toledo, in the morning drive slot; WMMS's Andrews working afternoon drive, in addition to being named music director; and Randy Scott, fresh from WZAK, doing five hours until midnight. Station manager Tom McCormick hosted the syrupy weekend music show *Sunday Serenade*, and WNOB offered extensive local and national news along with sports programming.

WXEN was also trying to maintain a presence on the progressive scene with the addition of a midnight-to-2:00 show hosted by Chuck Avner (of the Man Talk stores) and Barry Weingart, who was continuing his career in rock radio. Other media outlets were paying attention to the new music trend as well. WCLV's Martin Perlich told the *Cleveland Press* that rock was a tool "to radicalize the young." He spoke about the revolutionary and tribal qualities of the music, as well as the energy, vitality, and spirituality it offered to the listener. Perlich called the "thumping, erotic rhythms" the spark of a populist feeling among the young, likening their attention to the new pop music to a political act—much in the same vein as smoking pot.

Bass continued to plug away WMMS both on the street as well as on the air, hosting a special appearance by folk blues artist Taj Mahal at Roxboro Junior High in Cleveland Heights, opening the Kent State Winter Festival, and generally keeping his face before the public. Still Bass wasn't happy, and he saw a different sort of radio on his horizon.

Having seen the staff decimated and the format trashed, Bass made the unlikely move to the evening shift at the top-rated AM rock station WIXY. The station had already flirted with harder-edged rock at certain hours, airing a midnight Underground Hour with songs that weren't likely to make it to the daytime playlist. Hendrix's "The Stars That Play with Laughing Sam's Dice" and similar fare got airtime in the middle of the night, and the addition of Bass was seen as another move toward greater credibility with that growing audience of intellectual listeners. Years later Bass told an audience at the Rock and Roll Hall of Fame that Cleveland might not have been ready for that type of radio at the time. It wouldn't be long before it was.

Bass had to learn how to be a Top-40 deejay at a high-profile station, and he credits another WIXY mainstay, Chuck Knapp, for helping him reach that goal. Dunaway offered his expertise as well, based on his history at WABC-AM, one of the country's legendary Top-40 outlets. He joined a WIXY staff in Cleveland that included Dick Kemp, Jack Armstrong, and Mike Reineri—some of the biggest names in popular radio at the time. Bass was under pressure to learn a whole different type of format, but he learned quickly and got noticed a lot. That gave him a different kind of credibility when he moved on to his next station because Bass would now be looked at as a real disc jockey.

As he had done at WMMS, Bass was again put in front of the public by his new station. He emceed the massive Youth Cares rally at Cleveland Municipal Stadium that July and the Appreciation Day at Geauga Lake Park later that month with Smokey Robinson, Three Dog Night, and Tommy Roe. He also hosted a four-hour progressive show on Sunday nights starting at 9:00, featuring extended mixes and similar "head music."

In some ways it was a much different era in broadcasting. Bass had been doing nights the entire time he had been at WIXY, so one day he walked into Norman Wain's office and asked, "Hey, Norm. Why can't I have a show in drive time, in the middle of the afternoon?" Wain said, "For God's sake. Don't you realize why? You're black! That's why." Bass paused for a second and said, "Oh! Nobody ever told me that!" Both parties laugh about the incident now, but Wain has gone on record expressing his shame that that was the cold reality of Top-40 radio in a predominantly white, blue collar, ethnic town like Cleveland. He was afraid they wouldn't be able to attract advertisers. At the same time Perlich was impressed that Bass would make the inroads he did at a successful station like WIXY because he was black, the very point that concerned Wain.

No matter his on-air limitations, Bass often went above and beyond the call of duty for his show. He scored a coup by airing the Beatles' *Abbey Road* album on his show before its release, and the same week he featured a bootleg recording of the *Let it Be* soundtrack a year before it hit the stands. The latter recording came from what Bass would later call "a friend of a friend of a friend," had no song titles listed, and was followed shortly after by a "cease and desist" order from Capitol Records. When the "Paul Is Dead" rumor was making its rounds, selling lots of Beatle records in the process, Bass hosted a special Sunday-night program centering on a voiceprint made in a New Jersey sound lab that was meant to serve as a type of audio fingerprint to prove the theories about McCartney's demise.

Not everything Bass did on the air was as silly as indulging dead Beatle rumors. He tried to get a message across as well. Bass played Crosby, Stills, Nash, and Young's "Ohio" in heavy rotation as soon as it was released to mobilize the audience and to show his outrage over the Kent State shootings. Many saw this as a bold political move in the often-oppressive days of Nixon's America, and it was done with the full endorsement of Wain.

Bass continued to keep his progressive image alive. He sat on a panel show on John Carroll's WABU-FM with Perlich and Mike and Jules Belkin. Bass also emceed the Country Joe concert at Case Western Reserve University that was sponsored by the school's Committee to End the War in Vietnam.

The station also published "Billy Bass and Friends," a progressive, top-twenty listing in the *Plain Dealer Action Tab* every Friday, with songs such as Rhinoceros's "Satin Chickens," Taj Mahal's "Giant Step," and Tommy Flanders's "Moonstone." None of the songs were ever played in regular rotation and were pretty much there to catch the attention of progressive rock fans. The problem was that Bass wasn't especially happy where he was.

He told the *Great Swamp Erie da da Boom* that even though he was working there, he never listened to WIXY at all—not even his own show. He told the paper, "You know, those tunes are nice and they have a place ... people dig them and they're fine, except it wasn't my kind of music. I wanted to come home and listen to a radio station that I dug." It was anyone's guess how long he could hang on at WIXY, but there was still plenty of room for movement on the FM dial.

WNOB was in for yet another change. Just three months after dumping middle-of-the-road pop for a mature Top-40, the station announced that it would switch to a computer-driven, syndicated tape service on October 1. WNOB called it progressive rock, but there was doubt that it was exactly that. The Reverend John Rydgren, as Brother John, guided the so-called Love Format. Rydgren had already been heard on WABC and was formerly the national director of radio and television films for the American Lutheran Church. WNOB was the ninth station to sign on to his new format, which had been selling well in other parts of the country. A good part of the current WNOB staff was dismissed, but Kelly, Johnny Kaye, and Mark Donley were asked to stay on to do production.

An example of the Love Format came with a radio-show "special" called *The Ultimate Rock Concert,* a lame attempt to capture the energy of Woodstock that featured an imaginary lineup of the Beatles, the Rolling Stones, the Doors, and other superstars. Rydgren and his partner Bob Lewis, as Bob-a-Loo, seemingly did stage announcements and taped backstage interviews. But in actuality the production consisted of canned applause on studio recordings, and the interview answers varied wildly in sound quality from the pair's questions, supposedly from the same mike (sort of like *Space Ghost: Coast to Coast,* but boring). It was theater of the mind gone sour.

Despite Rydgren's religious leanings WNOB lent its call letters to a decidedly antireligious promotion. In an event cosponsored by weekly entertainment newspaper *Cleveland after Dark* and Belkin Productions, the station offered what it called "a Psychic Rock Phenomenon" at Public Hall, with black magic, sorcerers, astrologers, and assorted demonology. The featured bands were probably chosen on the strength of their names alone and included Raven, Cold Blood, Coven, the Inner Sanctum, and Cleveland's Damnation of Adam Blessing, but really had nothing to do with the occult. If that wasn't enough dark entertainment for one evening, the event also promised a witch named Sandra Sennes, hypnotist Ed Baron, ESP with David Hoy, magician and modern-day escape artist Houston, and Santani Demon, the fire-eating devil. A character named Uncle Dirty emceed the festivities, but credibility was not an issue here. As a perfect capper to the bizarreness surrounding the event, the show was booked for the day after Christmas.

WMMS was in yet another transition of its own. Its change the previous May to a Top-40 format soon became big-band programming, but it quickly became evident that it wasn't going to work. The intended audience very likely wasn't even aware of the FM band, let alone have the equipment to get it. Something had to be done fast, and WMMS general manager Ken Gaines chose a syndicated music service developed by "format doctors" Bill Drake and Gene Chenault, who had helped develop CKLW-AM's programming. It was called Hit Parade '69, and although it was a little more upbeat than the big-band sound, it made light rock sound like heavy metal.

The new format premiered on the air November 7, 1969, with a forty-eight-hour *History of Rock and Roll* rockumentary, featuring interviews and music. With the sole exception to broadcast Browns games, the format went fulltime on November 12. The Drake-Chenault Hit Parade was in fifteen other markets, reportedly had great ratings success, and Metromedia was itching to try it in other cities too. Basically the service was shipped to each station on ten-inch reels of tape and recycled around-the-clock by engineers. WMMS would still simulcast its sister station WHK-AM from midnight to 6:00 A.M. A name from WMMS's past, Pat McCoy, was given the program director's title.

The Hit Parade format appealed to a much older crowd, and someone only had to look at their top-twenty listings in the newspaper for proof. As the "Billy Bass and Friends" list from WIXY offered music from Zephyr, Ten Wheel Drive, Cold Blood, and King Crimson, WMMS's similar "Big 100" column listed B. J. Thomas, Ferrante and Teicher, Englebert Humperdink, and even the Archies. Clearly there was a place for a full-time commercial FM station playing serious rock, and a major player in the radio business was thinking about doing just that.

The Electric Renaissance

Major changes were in the air for the Cleveland listening public. The *Plain Dealer*'s Karl Burkhardt predicted in 1969 that "in five years, AM radio will be dying and nearly forgotten," saying that the technically superior FM would be the giant of the industry. The Federal Communications Commission (FCC) was considering a requirement that all radio receivers have FM capabilities, similar to the ruling that all television sets carry both VHF and UHF bands. Burkhardt noted that such receivers were more expensive, but he pointed to the shift in well-known personalities on the FM band, starting with Tom Armstrong, who took a good segment of his following to WDOK-FM. The stumbling block at that time was the far greater number of AM receivers, especially in cars, but automakers were also starting to take notice. Burkhardt would soon be proven right.

With the possible future capabilities of a wider FM audience, no one was exactly sure what WGAR-FM was going to do, but plans were being made to take a step closer to that future. The station was expected to start broadcasting in stereo on April 1, 1969. Two weeks later on April 16 it joined in an experiment with WCLV-FM, the fine-arts station, to broadcast a Cleveland Orchestra concert from Boston in four-track stereo, an early attempt at quadraphonic sound. It was a one time only simulcast, but it was enough to confuse the public about the new direction of WGAR-FM.

Cleveland's WGAR-AM and FM had been considered sleeping giants at the time. For a while, the FM station only saw two hours of airtime on Sundays to keep its license, simulcasting WGAR-AM. The rest of the week the signal was dark, other than occasional tests by the engineers. It was future general manager Jack Ambrozic's assignment to "crank them up" and program a little excitement into both stations to give the sales departments some ammunition. But parent company Nationwide Communications gave him a side trip en route to Cleveland to a

Richmond, Virginia, station, where Ambrozic was named general sales manager. He saw it as a logical step to something that was much bigger, and that's exactly what it turned out to be. Thinking back Ambrozic says, "The underground sound was happening. I knew we couldn't play it there, but it was going to burst. All the record guys were telling us it was happening, so I figured FM was going to be a perfect place for this sound."

Soon Ambrozic was given marching orders to Cleveland, and he found himself on the road to the new WNCR-FM. "They were going to go the easy-listening route," he recalls, "but I convinced them with sales. I assured them that most of the 'buys' from the major sales agencies were going to eighteen–thirty-four-year-olds. Easy listening was dying. We had a meeting with the Nationwide board of directors, and we won, though I put my neck on the line."

It came down to marketability, and what kinds of formats could be sold effectively. Ambrozic explains, "In those days, they expected to do about a quarter million dollars in sales, and they would have been happy. Of course, those were different dollars than we're talking today." But Ambrozic saw the market and suggested that his salary be based on 10 percent of the station's gross profit, and he would be happy. By the end of the first year the station had reached seven figures. Soon after, Jack Thayer showed up at sister station WGAR-AM and not long after, so did Don Imus. Nationwide wanted a winner in Cleveland and spent nearly a million dollars promoting the AM station and Imus. Their gamble soon paid off with a number one rating, but they wanted to see what the FM signal could do.

Applications had been filed with the FCC to change the call letters for WGAR-FM, and internal memos from that time show a debate over the new call letters as early as October 23, 1968. The station's engineer Jim Holston suggested to broadcast-division head George W. Campbell that they "should reflect location, corporate identity or basic station format, if possible." He ruled out anything close to WCLV or WCUY-FM because of a possible misunderstanding on the air, though Holston suggested, "Corporate identity yields the possibility of WNCC, which might establish a pattern to be followed over the years in other cities. How about WNAT?" All call letters would have obviously hinted at the station's owner, Nationwide Communications.

Holston went on to say, "It's a little early to discuss format-oriented calls, but I'm playing with an all show-tune idea, which would suggest calls such as WSHO, WBIZ, WMUS, possibly even WACT." He ended the memo with a smile, hinting, "and then there's always WGWC," obviously reflecting Campbell's initials. The memorandum shows a hand scrawled "WNEO," standing for North East Ohio, above the signature, but the debate was far from over. On August 25 general manager Carl George wrote to Campbell suggesting, "WNIC, which would stand for Nationwide Insurance Companies." It was finally decided to change the calls when the license renewal came up. Eventually in a letter to FCC secretary Ben Waple dated February 12, 1970, a formal request was made to change the call letters of WGAR-FM to WNCR.

Anticipating the change in format, Holston outlined an "action summary" to George in June 1970. It was still not certain that WNCR would go to rock, but a meeting with Nationwide officials in Columbus would nail down a decision in short order. For new personnel Holston suggested, "Six deejays, one of whom could serve as FM Production Manager, to coordinate all production, pick and supervise programming of music." He also wanted to add, "One newsman to augment the present news staff, which would then serve both stations on a staggered schedule basis 24 hours per day." He concluded that, pending the decision in Columbus, live disc jockeys would be needed if a rock format were adopted. Holston stressed that a launch date for the new station could not be estimated until, "a) we have the programming decision (and the money to implement it), b) we have (or know when we will have) necessary production facilities, and c) we begin the staffing process." He noted that "music acquisition will take at least a few days."

When the decision was made to try a progressive format, station management set its budget goals, paying disc jockeys $150 a week, which was not bad for the time. It was also decided that the new format would be unique to Cleveland and not reflect Nationwide's Columbus station, WNCI-FM, which was pretty much Top-40, but that wasn't what the main headquarters downstate was led to believe. Walt Tiburski was in sales at WNCR and recalls,

> According to Nationwide corporate, the station was supposed to follow the same format as WNCI—chicken rock, or light rock, bubblegum or whatever they called it at the time. Instead, we took a little different formatic direction. There were a lot of "42 longs" walking around. Those were guys who should have been selling insurance policies who found themselves transferred to the radio division, and didn't know that much about the radio business, so we were able to take that different direction. It later became very interesting when Jack Thayer got involved and got heavy handed, because he didn't like the fact that the station was starting to make noise. He was running WGAR-AM, and that was populated by other "42 longs," and the sales people there used to refer to us unflatteringly as the hippies down the hall. "You don't want to buy time from those guys because they're drug addicts, and their audience is drug addicts." They were doing everything they could to deep six all of us who were in that new era of music. But we prevailed!

As for personnel, two names from the WMMS experiment were getting a close look. Handwritten notes from a meeting on June 8 showed Martin Perlich and Billy Bass to be prime candidates, but there was some hesitation. Most of the concerns centered on their politics. Both were cited as "controversial members of the 'Peace movement' who would want complete say regarding program content." On the lighter side both Bass and Perlich were seen as "realistic hippies" and "go go swingers."

Holston did some detective work. In a confidential memo to George, dated June 9, 1970, Holston described a call from WCLV's Bob Conrad about Perlich. It described Perlich this way:

- Very, very good, both as administrator and performer.
- Thoroughly familiar with both classics and progressive rock.
- Knows and lives in the world of the "youth movement."
- Would make an excellent key man around which a very successful progressive-rock operation can be built.
- Is emotionally oriented and can be a bit difficult to control at times. Best way to operate is to lay down ground rules at the beginning and hold him to them.
- Sometimes would get depressed . . . "Wonder if anyone is listening?" Cure was to run a request show which would put him right back on Cloud Nine.
- Definitely politically oriented . . . member of the Peace Movement. Doesn't enjoy having to maintain on-air balance by interviewing conservatives; would prefer somebody else handle this, probably.

He went on to say,

It occurs to me that operation of this kind of radio station must be very similar to a classical station. Audiences are young, hip, intelligent, quick to identify a put-on of any kind. Maybe this is what happened to WMMS—they tried old rock approach with a lot of gimmicks. Think we should have substantial newscast content; contests if used should not be just gimmicks or people will turn off. Perlich might be the answer; he should be aware of this through his WCLV association.

For whatever reason Perlich didn't come on board until a bit later.

Some new personnel did sign on right away. Ambrozic was named to the newly created post of general sales manager for WNCR, and later general manager, and Nationwide Communications announced the station would dump its "good music" format in favor of adult rock by mid summer. The broadcasting giant sent Jerry Dean from WCOL-FM in Columbus to assume the new post of program administrator to oversee programming, the airstaff, and the music. Dean said the new format would be heavy on album cuts and target the twenty-to-thirty-year-old crowd. At first it would be broadcasted in stereo only four hours a day until the FCC approved the new transmitter's "proof of performance." The kick off date was set for the Fourth of July.

Five new deejays were announced for the station, most of them from Nationwide's Columbus home base. The new lineup included Strongsville's Mitch Michaels on morning drive, New Jersey native Dave Elmore from 10:00 A.M. to 2:00 P.M., and Steve Scott from the capital city's WCOL doing afternoon drive. Another Columbus area native, nineteen-year-old Ginger Sutton (in actuality Steve Sinton) did the 6:00-to-midnight shift, and Chris Gray from that same city was pegged to do overnights. WGAR's veteran news director and anchor Charles Day did the early newscasts for what was now being billed as "The Electric Renaissance."

Sutton was already a graduate of Arlington High School when he met a guy at a beer blast at Ohio State University. The gent had a pirate radio station in Cincinnati and he was planning to start another one for the state capital. He asked Sutton to be part of the airstaff, but the skeptical teenager thought "it was the keg talking." But sure enough, when Sutton visited him the next day, he saw an entire studio in his basement, including some equipment stolen from a junior high school. Soon 1610 AM was up and running, a 16-watt station covering the north side of Columbus. It became an underground sensation, as people would drive to locations where they could pick up the signal to listen to progressive rock that wasn't being played on WCOL. The FCC warned the group to shut down the station after only five weeks. Faced without an outlet, Sutton accepted Dean's invitation to join him in Cleveland.

Sutton remembers the Statler Hotel at East 12th and Euclid as "an archaic hotel that had an archaic radio station. It was a dinosaur with one foot in the grave, but it was palatial! We were in a studio where orchestras used to play. You could walk out on the roof and see the Cleveland skyline." WNCR's staff shared a sincere love for the music and was deeply political. Sutton explains, "I thought it was weird that our parents had done everything they could to encourage the next generation to think for themselves, and then were trying to step on it when we did. Our generation happened for a number of reasons. Everybody came together at that time because of the music, because of politics, because of the Vietnam War, and we all smoked dope and got high. One of the few places we could all meet was on the radio. We were the huge, electronic backyard fence for Cleveland's counterculture."

On July 6 the penthouse of the Hotel Statler Hilton came alive with a new sound, encompassing Indian raga to rock. The daytime programming leaned toward folk and more laidback artists such as Judy Collins, Curved Air, and Phil Ochs, while the night side blasted the mind-shattering sounds of Grand Funk Railroad and Bloodrock. The station used jingles and had a fairly polished sound, and occasionally the disc jockeys would acknowledge their listeners' lifestyles with comments about the availability of certain illegal substances on the street, and how they might affect the way they listened to music.

The deejays were street people too, and change was definitely happening on the street. Cleveland Barber's Local 129 warned its members to treat "freeks" with more respect. The barbers often picked on the long hairs because of their looks, but even the older folks were waiting longer between haircuts, and business was down nearly 40 percent. They were told to treat every customer with respect, no matter how long their hair, and not to try anything smart like purposely cutting it too short when they came in for a trim. The customer would make it a point not to come back, and word would spread quickly after that.

Acceptance in visual appearance wasn't the only growing trend. Some surveys showed that one in six high school students at the time had tried pot, as did half the kids in college. Hard drug use on campuses was still fairly low, but 13 percent of students were lighting up on a regular basis. To reflect this trend, now and then a disc jockey might sound "distracted" and maybe a bit giggly. There were no complaints. Real progressive rock had returned to Cleveland airwaves. The risk paid off, because within just a few months of going on the air, WNCR would be nudging up against WGAR-AM in some key age groups among the listening audience.

Even though the FM rock stations weren't really showing up in the overall ratings, there was a growing audience for the kind of music and banter that WNCR was airing. In fact one listener was so taken by the new format and lineup that he sent in a check to say "thanks." Station management thanked him and returned the check. The artists took notice as well.

WIXY-AM sponsored the Crosby, Stills, Nash, and Young show at Public Hall that summer, but the band refused to walk on stage until the station took down a banner with its call letters it had draped across the stage. After the concert they hung out with the WNCR deejays back at the Statler and decided they would all meet for breakfast the next day.

The Kent State shootings had occurred just a few weeks before, and the group was riding the popularity of "Ohio," the anthem that Neil Young wrote after seeing the coverage in *Life* magazine. The disc jockeys and musicians piled into cars and headed to the university to see where the carnage had taken place.

When they arrived Young looked at the Commons and asked, "How could this have happened? It's so quiet and beautiful. Why?" They headed back to Cleveland, a lot more subdued than when they left that morning.

Perhaps the initial popularity of WNCR, and to a point WMMS before it, was based on a growing sense of a community built from the ashes of rejection. "We were all we had," said Sutton, "because the Establishment perceived us as a threat. They couldn't take our barbs, and they couldn't take our commentary. We were the focal point of everything that angry adults hated. We were the reason their kids had long hair, smoked dope, evaded the draft, and got pregnant. Nobody blames themselves. Turn on Jerry Springer. He's made a fortune from that! Everybody's a victim! It all got hung on us, and anything more, no matter how incorrect, was a bonding force." It was a deep fraternity, with WNCR as a jungle drum. People were broke, scared about the war, and shunned by straights, but the underground media let them know that they weren't alone.

The WGAR-AM staff was spearheaded by morning man Imus, who had just been axed by Sacramento's KXOA-AM for running an Eldridge Cleaver Look-Alike Contest, where the winner got five years in jail. WGAR was spending big money to promote Imus, with billboards and even a television show, and he more than lived up to the hype. He also enjoyed hanging with the WNCR staff, and they could often be seen on the Statler's roof watching the planes take off from Burke Lakefront Airport. At first most of the management's attention focused on pushing Imus and crushing crosstown rival WIXY, which suited the FM deejays just fine because they were left alone to do the kind of radio they did best. They were kids let loose in Cleveland without a road map, and the audience liked the direction they were going. It was pretty much based on instinct, and few of the disc jockeys thought of radio as any kind of lasting career. But they were enjoying it as long as they could.

In the next few weeks the format started to solidify even more, and the staff was very goal-oriented, but budget money was tight. Ambrozic says the station was able to do a lot with smoke and mirrors. "I had no budget," he recalls, "and they gave WNCR absolutely nothing. I had to trade out bus cards and promotional posters that we handed out." There was a lot of tinkering with the air sound. An August 11 memo from Ambrozic to Dean and Gray focused on previewing the new Joe Cocker album and Michaels playing an entire side of that LP the following morning. It read,

> Suggest that new albums of obvious importance should be played in its entirety in the evening, and then reintroduced cut by cut throughout the former for several days immediately following the introduction. In other words, one cut off the album played each two hours during the entire day format.
>
> Ginger is still playing the entire albums using commercials as the only breaks. If a person of our audience is not into that particular artist, we are risking a tune-out factor which at this point we cannot afford.

Then, he made a startling assessment:

> Playing these albums should be considered like having sex with a broad. If you use short fast strokes, you basically have the same result, but without complete satisfaction. The object is to prolong and tease as much as possible. The teasing part should be worked on starting off with a possible three cuts of the album keeping the mood and the flow of the album intact. Coming back with "We'll get into more of this album in the next half hour," and then get into something that is compatible with the mood that was created by the artist previewed. Approximately a half hour or 45 minutes later, play two more cuts of this same album, coming out of it saying "We'll be coming back with more cuts in the next half hour, so stay tuned." I think you get the idea now of teasing and prolonging.

Also would suggest that Saturday night be promoted throughout the week as WNCR's Stereo Concert Night, and each Saturday night WNCR should air a stereo concert by one of the more obvious and prominent artists, i.e., Beatles night, Stones night, Airplane, Doors, Who, (B. B. King/Sly/Chambers Brothers) (Joplin/Cocker) (Poco/Crosby Stills) (Muddy Waters/Howling Wolf) (Collins/Donovan) (Dylan/Baez) (McCartney/Winwood). I feel that this will be extremely effective and already has judging from some of the concerts Ginger has done in building a listening audience on Saturday night which is usually considered date night.

Other stations took notice. WIXY was still the ratings leader in Cleveland, but one of the deejays watching WNCR's rapid ascent was Bass, and at times he didn't like what he heard. Sutton says, "We would kid Top-40 formats, and do recorded bits about WIXY poking fun at them, but Billy took it personally. After all, he was the original guy who put 'MMS on the air! He was, and always will be, the vanguard of hip, but Billy let it be known that comments about his station or him were not appreciated." Sutton and Bass remained friends for many years after that incident. Soon there would be a lot more competition, and all was not well at the "Renaissance" either.

At first the station forged ahead with as much enthusiasm as its disc jockeys could muster, and they were certainly dedicated to the cause. The station staged an Appreciation Day show a few weeks after going on the air with a daylong show at Southgate Shopping Center in Cleveland suburb Maple Heights, featuring local bands such as Natchez Trace, Wild Butter, and the North River Street Rock Collection, among others. That same month WNCR flirted with "tri-sonic" broadcasting. It joined with its sister station, WGAR-AM, for a combined 100,000 watts to simulcast the new release from Cleveland's Damnation of Adam Blessing, *Second Damnation*, as well as the Moody Blues' *Question of Balance*, the *Woodstock II* album, and a few cuts from Orson Welles's *Begatting of the President* for good measure. The idea was to simulate a concert hall sound by placing two stereo speakers facing the listener and placing the AM receiver in back of the room. WGAR delayed its signal by 1/30 of a second to achieve the hall effect.

It was a remarkable time, not only for Northeast Ohio radio but also for the entire music scene. It was still an "age of innocence," with many artists themselves a part of the counterculture community that they played to and shared the same concerns with. If only for a few short years, the sense of brotherhood really did exist, and perhaps no better example can be made than the night of August 29, 1971. The Allman Brothers Band had played the Musicarnival tent in Warrensville Heights, with the Pure Prairie League opening for them. After the show Duane Allman stood at the front gate and shook the hands of fans leaving the venue, thanking them for attending the show. That same hand of friendship was extended to the hundreds of fans who crashed the show by running under the tent.

Another example was seen at the Led Zeppelin show at Public Hall. Bassist John Paul Jones's grandfather had died, and even though radio spread the word that the show would start a couple hours earlier than scheduled, he still had to leave to catch a plane for Scotland. The rest of the band assembled on the stage to do some acoustic numbers from their new release, *Led Zeppelin III,* and Robert Plant asked if there was a musician in the audience who could fill in on bass. Soon, after a quick set of instructions from Jimmy Page, a kid from Northeast Ohio had his moment in the sun, if only for a few minutes, as a member of one of the hottest rock bands ever to grace a stage.

Backstage security was usually lax at best, with people walking in and out of restricted areas with relative ease. When the Moody Blues played Public Hall, Sutton walked back after the show to say hello and found the bandmembers sitting around holding their instruments. He thought it was because they were that much into their music, but they said, "Not really. We're just making sure no one steals our gear!"

Meanwhile there were rumblings of another kind of change at East 50th and Euclid. Hit Parade '70 just wasn't cutting it and would soon be history on WMMS. Instead a new lineup of live disc jockeys had been announced to go on the air in early September. David Moorehead was the new vice president and general manager, and he had been planning the shift for at least a couple of months. Moorehead was from Chicago but went to John Carroll University, so he was well aware of the popularity of WIXY. He brought on two of that station's former deejays as cornerstones of the new station. Lou "King" Kirby had been out of work for about three months and spent the time traveling around Australia and New Zealand. He was tapped to do morning drive, with another familiar Cleveland name, Dick "the Wilde Childe" Kemp, holding down the 7:00-to-midnight shift. Kemp was often quoted as saying that he was the kind of guy who "could go out for a cup of coffee, and come back 180 miles later," and that's just what he did. Kemp had left Cleveland for about eighteen months to work at WIXY's sister station in McKeesport, Pennsylvania, but he was itching to return to the shores of Lake Erie. Kemp and his wife Georgette had lived on a riverboat during his time in Pennsylvania and were more than happy to settle into a more stable townhouse in Painesville.

But Kirby and Kemp didn't come without a fight. Lawyers for WIXY and Metromedia went nose-to-nose over non-compete clauses in each deejay's previous contracts that would prevent them from going to another station within a certain amount of time. Both sides worked down to the wire. Kirby and Kemp were able to get the green light just hours before the debut of the new format.

They were joined by Ted Ferguson out of Knoxville, Tennessee, and Mike Griffin from Indianapolis. Griffin had quite a varied background, having worked as a songwriter, a club owner, and a police-beat reporter in South Bend, Indiana. Rounding out the lineup was twenty-two-year-old Gary Edwards (a.k.a. Charles Eduardos), a Glenville High grad who played bass in the hard rock band Checkered Demon, a

Dick "the Wilde Childe" Kemp and Lou "King" Kirby (1971). It took some last-minute legal work to clear the way for these longtime WIXY favorites to make their FM debut on WMMS. Cleveland Press Collection/Cleveland State University

group named after S. Clay Wilson's classic underground comic character. Edwards brought some experience with him, having already been on the air in Cleveland radio for two years before donning his cap and gown, and most recently as a newsman and deejay at WJMO-AM. He was quoted as saying "ethnic radio was dead," and his addition to the staff was like "pepper to salt," adding to the spice of the city's underground rock scene.

The station promised that it would do its part for the ecology movement by targeting polluters. WMMS encouraged listeners to phone in reports to its "pollution solution" line, and Moorehead would fly his plane over the sites, take photos, and presumably turn them in to the Environmental Protection Agency. It's anyone's guess whether this project ever produced any photos, let alone results, but the good intentions were there.

Simulcasting overnight programming with WHK-AM was out, as taped programs from stations in New York, Los Angeles, and San Francisco were now filling out the time from midnight to 6:00. William "Rosko" Mercer's syndicated show, from Metromedia's WNEW-FM, was among those aired in that time slot. The station also adopted a rainbow as its symbol, though the cost of reproducing it on stationery restricted the colors to red and black.

In the early morning hours of September 11, 1970, WMMS's Hit Parade '70 package was still playing artists such as Steve Lawrence and the Classics IV. At 6:00 A.M., as Hit Parade's announcer was back with a tune by Brenda Holloway, the tape quickly faded out and a new, upbeat jingle came ringing out, singing, "The heavy music keeps coming on! Ladies and gentleman, Lou Kirby, WMMS—Stereo—Cleveland."

Kirby opened the microphone to say, "Well, good morning! Brand new show. Brand new station, and a brand new time. Dig this!" and swept right into Neil Diamond's "Cracklin' Rosie." It might have seemed like an odd choice now to kick off a new format, but at the time Diamond was still embraced by a huge cross-section of an audience respectful of his songwriting skills. Nonetheless, it was the end of "jocks in a box" piped-in music, a fresh start, and a definite move toward reestablishing credibility with the counterculture audience. That same morning Kirby's choice of music reflected a serious turn from the saccharin sounds of the taped shows, airing harder-edged tracks from Steppenwolf, Free, and Sugarloaf and even some AM crossovers such as the Assembled Multitude's version of the overture from Tommy and Clarence Carter's "Patches." Ferguson recalls, "WHK-AM was in the same studio as us. These guys were all real old pros with an MOR [middle of the road] format, and they thought we were all crazy with our long hair and weird music, while they were groovin' to Mel Torme and Patti Page!" The experiment had begun anew.

Up the street at WIXY Bass had been given a new slot, switching from late-nights to the high-profile 6:00-to-10:00 evening shift, and a whole new segment of the audience was about to discover what he was all about. The same day that was announced, all six deejays at WNCR walked off the job at 12:40 on the morning of Friday, September 18, coincidentally the same day Jimi Hendrix died.

They demanded changes in management, personal contracts, and the freedom to program their individual shows. The six—Dean, Lansing, Gray, Sutton, Michaels, and Elmore—blamed new station manager Jack Thayer for their discontent. They told the *Plain Dealer*'s Jane Scott in a full-page story that even though they had put WNCR in a strong ratings position in just a short time, Thayer was offering them no respect. Sutton was quoted as saying, "What we did was build up a radio station, and got ripped off for it." Lansing added that they feared a change when Thayer said he had the "dubious distinction" of being the new general manager and said that it appeared "the inmates were running the asylum." Michaels told Scott that the station wanted a new music format, with programming written out in advance. Departed music director Gray admitted that the staff strayed from the original format because that's what the audience wanted. He also noted that WNCR was the first station in the country to play Sugarloaf and Free, and despite the occasional on-air goof the six felt they were part of their audience and felt a kinship with that generation. Dean claimed WNCR was the most successful venture that parent company Nationwide Communications ever had within such a short time, booking over $100,000 in sales for the progressive format. He said the station was in the black within just six weeks. Dean took a trip to Columbus to air his gripes to Nationwide management but didn't get anywhere. When he returned to Cleveland he demanded a six-month contract for himself and the staff, with a little more money, which was promptly rejected. That was the final straw.

Nationwide reportedly felt that Ambrozic was too close to the WNCR staff to negotiate on the company's behalf, so they sent in Thayer. The group met at midnight in the conference room, with Imus sitting in as mediator. The disc jockeys said that they were happy the station was doing so well, with a topnotch sales department selling like mad even though there was less and less room for the music and events of the day. They said, "The radio station is starting to pig on the community that helped it take off. There are too many spots, and we have to cut back!" Of course radio management took a different view of the economics of the time and flat-out refused. Then they brought up the issues of contracts and compensation for the airstaff.

Thayer reportedly told the six that they would remain at their present $130 a week, or they could go find jobs elsewhere. They chose the latter and walked out the next day.

The six went on record to say that they should have joined AFTRA, the American Federation of Television and Radio Artists. Sutton claimed the WIXY deejays made four times as much as they did, which was understandable since that station had a long run as a ratings powerhouse. The staff also wanted general manager Thayer "off their backs." That was a big order, but as the six told Scott, "It was more than a job. It was our life itself. Our blood was flowing down those hallways."

The disc jockeys said that they wanted "a chance to prove that we can blow anybody off the air in our progressive category." Their final shows together would be a few days later. At WUJC-FM's invitation, they gathered at John Carroll University to explain their position on the walkout, to give their farewell shows, and to pepper the airwaves with the kind of music and language they felt the audience related to.

Thayer had been on the job for just ten days, and he didn't blink. He promised a new staff would be on the air within a few days, and he lived up to his promise, but in a bizarre way. The new deejays started September 28 and were only identified on air by their astrological signs. As the *Plain Dealer*'s Ray Hart put it, "What in the world is radio coming to?" In a memo to the staff dated the 28, George Campbell wrote:

> I want each of you to know how much I appreciate your outstanding efforts that kept WNCR on the air through the chaotic week-end beginning Friday, September 18.
>
> It was, of course, disappointing to know that misunderstandings could develop to the point where six disc jockeys would walk out. It seems obvious that they had been operating under some wrong assumptions. But that is past.
>
> It is encouraging and heartening, however, that we do have a nucleus of dedicated, loyal, and enthusiastic people, who could step into the breach and carry on.
>
> My personal thanks to each of you for both your attitude and the hard work you put in. I am grateful for your support and understanding.

Ambrozic now says that it was extremely difficult to see that staff leave and he wanted to hire Sutton, but Nationwide rejected that move. There was a new staff and, they hoped, a new attitude toward the company and broadcasting in general.

The first chapter of the Electric Renaissance only lasted about nine weeks.

The Times, They Are a-Changin'

WNCR's ploy to identify its disc jockeys with astrological signs ended soon after the new lineup took to the air, though the men behind the signs still identified themselves that way along with their names. Jim Allen "Aquarius" did mornings, "Sagittarius" Jerry Allen followed him, and veteran "Virgo" Lee Andrews did afternoon drive. Mark Eddinger, "the Gemini," was the night man.

Martin Perlich (1971). A pioneer in FM rock radio, the *Perlich Project* helped introduce an entirely new audience to coming trends in music, along with insight and commentary on various issues of the day. Cleveland Press Collection/Cleveland State University

The management at WNCR decided to take a hard look at the audience, and they approached researchers at Kent State University to profile the average person who was listening to their station. Jack Ambrozic says,

We asked Kent to survey its students about their lifestyles, and their musical tastes—not only about progressive rock, but other kinds of music as well. It resulted in something now called "psychographics," which is still being taught in many universities. A lot of people in that eighteen-to-twenty-four-year-old range had similar lifestyles, but different musical tastes. The results were surprising.

It showed that the number two selection behind prog rock was jazz! Light jazz. Number three was classical! Apparently those music appreciation classes did the trick. We called Martin Perlich.

The station management took another listen to Perlich's style, mixing various musical genres, with even the sounds of whales, into long, compelling pieces that made listening to his show a unique personal experience.

Perlich took over Eddinger's slot later that October, but he still did special projects and interviews for WCLV-FM. He also made it clear that he was not going to identify himself by his astrological sign.

His life had taken a few bizarre turns since he first brought the underground consciousness to Cleveland. Much like his audience Perlich's hair was longer, and he had a full, long beard, and more than one person thought he resembled a modern-day Rasputin. For years he had been providing insightful interviews with some of the giants of the classical music field, which were played during intermissions of broadcasts of the Cleveland Orchestra. In November 1970, he got a call from orchestra management saying, "Your appearance is inconsistent with the image of the Cleveland Orchestra. You're out."

But not long after, his interview with orchestra conductor George Szell did appear on a memorial album, and Perlich received a royalty check for eighteen dollars. Perlich was especially proud of the Szell interviews because the great conductor had something of a reputation for being difficult. He found Szell to be a "charming, gracious man . . . who treated me as a younger colleague, almost a grandson."

Part of Perlich's mission at WNCR was to "keep the cultural vehicle in the hands of the people." He went on to say, "Black people, for instance, have felt culturally isolated for years because of the white controlled, socio-political apparatus and cultural media. Avant-garde jazz, with its constricted, arcane atonal language is an expression of that feeling of isolated separatism. A considerable amount of black participation in pop culture has occurred recently, but is still disproportionate to the overall influence that black music, for example, has had on white music. It is sad that the Rolling Stones and the Beatles were the major vehicles by which black music was transmitted to today's pop music." Perlich gave off the impression on his show that he would not be a good person to argue with. He rode his motorcycle to work, wrote books, loved his family, and knew his music. The audience's job was to sit back, listen, think, and enjoy. Thirty years later Billy Bass fondly remembers Perlich as the "conscience of FM rock, even though he was a communist!"

The format was still very much like the old WNCR. The station aired cuts from Frank Zappa's *Chunga's Revenge,* Derek and the Dominoes, and the cast album of *Jesus Christ Superstar,* among many others. But the addition of Perlich brought a renewed sense of urgency for the music and the artists. His Saturday-night show gathered daytime requests, and he wove the tunes into an eclectic mix of hard rock, folk, jazz, and classical that flowed into a central theme. It was often like a single piece of music, an audio tapestry, that challenged the listener to relax in any way they saw fit but still have the presence of mind to think about what was happening around them, and have the courage to speak out and try to change what they knew was wrong. A typical show might include selections from Roberta Flack and John Mayall, Buddy Miles and Bob Seger, Badfinger and the Guess Who. It was often spiced with Perlich's thoughts on politics, Richard Nixon and the "establishment," and sometimes his own family. In addition WNCR opened its mikes to voices such as the American Civil Liberties Union and the gay liberation movement and similar groups, even airing a program called *Teach Your Children* that centered on the civil rights of school-aged kids. In some ways Perlich saw the station as a "community campfire" at which thoughts and ideas were exchanged. It was a distinct change from the type of programming heard on other stations. But there was a changing of the guard over at WMMS too, and new battle lines were being drawn in the process.

The media in the 1970s slowly grew as the decade progressed. Even though few cars had FM tuners at that time, eight-track tapes were still the main alternative, there weren't many in the "freek" community who could bring themselves to listen to AM's music formats. Most of the FM tuners were in the home, and in an age when cable was a distant dream, there was very little television the "heads" could watch, let alone embrace. This was at a time when tuning in, getting off, and having a conversation were a lot more relevant than *Marcus Welby* or *The Doris Day Show*. Dick Cavett was going up against Johnny Carson with shows featuring Janis Joplin, Jimi Hendrix, and Joni Mitchell, but the mainstream still defined programming, and Cavett went down after a good fight. PBS aired specials with the Jefferson Airplane or Leon Russell with Furry Lewis, but that didn't draw a huge audience. The so-called counterculture was a lot more focused on radio back then. A continuing problem with ratings remained that young people generally didn't fill out the ratings books, and numbers fueled the industry, especially the sales departments. Even so, word was spreading about the new face of radio, and so often that is the best form of advertising. It's slow, but eventually it can be effective. To a point, it was starting to be for WNCR.

A report in *Billboard* in December 1970 cited WNCR as "one of the most valuable showcases for artists and records in the city, with groups featured such as Blood, Sweat and Tears, and Crosby, Stills, Nash and Young going on the air live, as well as the Byrds, Badfinger, Moody Blues and Emmit Rhodes, Poco, Three Dog Night, Dewey Martin and the James Gang." It quoted Walt Tiburski, who at that time guided the promotions department as well as co-managed the station, as saying the "showcase image developed inadvertently." He pointed out that even though WIXY-AM sponsored the Crosby, Stills, Nash, and Young show earlier that year, WNCR staff members who made their way backstage told the band what their station was all about and invited them to come over to the studio after the show, which the group did, staying for about an hour.

That happened just as WNCR was changing to its progressive rock format, which gave the station an instant shot of credibility with that audience. Neil Young did a benefit concert in Cleveland some time later and made it a point to stop by the station and sat down from midnight to 2:00 to chat with the overnight deejay. Members of the Byrds did the same, staying from 1:00 to 4:00 A.M. and talking about everything from ecology to politics.

Perhaps the biggest coup came from the bond WNCR established with one of Cleveland's hometown groups, the James Gang. The band opened for the Who at Public Hall, and Pete Townshend was so taken with Joe Walsh's guitar playing and the overall sound of the band that he dedicated his band's set to the opener. Their reputation caught fire after that, and they received invitations to tour with supergroups such as the Moody Blues and Led Zeppelin, whom they opened for the previous year at Musicarnival, a huge outdoor performance tent, on the night of the first moon landing. Loyal to their friends, the James Gang joined the Who's European tour, even appearing in the opening slot at the show that produced the classic *Live at Leeds* album. On that tour members of the James Gang phoned back daily reports to WNCR about their adventures and even grabbed Townshend for an exclusive. Upon the James Gang's return WNCR hosted Walsh, Jim Fox, and Dale Peters for a three-hour show. It scored a major boost for the fledgling station.

WIXY still had big muscle and did a great job in promotions. The sheer number of listeners assured the AM station a good degree of attention, and in the late fall of 1970 it put it to work behind Ron "Ugly" Thompson from Milwaukee. The station gave very little information about the new guy at night except to say he was over six-feet tall with dark brown hair and blue eyes, was high on the Rolling Stones, and was against being "forced' to take any drug, including aspirin. WIXY promoted a contest asking artistically inclined listeners to send in sketches of what they believed Thompson looked like. The top prizes were Polaroid cameras, and even though Thompson was going to be in WIXY's annual Christmas parade, he would wear a bag on his head so no one had an unfair advantage. The sketches were published in the *Plain Dealer,* and no one even came close. Thompson was a better-than-average-looking guy with what appeared to be a long future at WIXY. That is until he accidentally uttered a forbidden word on the air. There were many apologies, but he didn't stay on WIXY long after that.

There was another change at WIXY that same December. Bass had announced his resignation to take over the program director's slot from Andrews at WNCR, though Andrews would stay on afternoon drive. Bass was due to start a couple of days after Christmas. On top of that, WIXY promotional director Bill Sherard tendered his resignation, and even though he veiled his departure in secrecy, he soon turned up at WNCR too.

A lot of people can lay claim to reviving the spirit of progressive radio that began in 1968, and Peter Schliewen, who owned Record Revolution on Coventry Road, is one of them. He recalled listening to the first Yes album, thinking how great it was, and lamenting that few people were buying it. He told *Record World*, "They weren't aware of it. There was no single on that album that could be played on WIXY, which at the time was the only game in town. It was easy to see what was needed, but hard to convince a radio station to take up that kind of format."

He was among those who suggested to WNCR's longtime general manager Carl George that Bass, then at WIXY, would be the person to successfully man the helm of a successful FM format.

Bass came to WNCR with a vision, one that began a few years back at the original WMMS, and the time was ripe to bring it to full blossom. He envisioned a community-based station, People's Radio, that would embrace all that was good about the Greater Cleveland area and call attention to and motivate people to change anything that wasn't. He told the *Great Swamp Erie da da Boom* that, "WNCR is a woman and a man. I'd like WNCR to be sensitive in that female sort of way that can feel Cleveland, and respond to where it's at. Also, I see WNCR as a cat with more balls than I would ever hope to have—a person that's willing to speak up and do all those things that I would like to do." He continued saying, "WNCR has to be a clown: we've got to be entertaining. Music and theater have always worked together. Look at the Jefferson Airplane. Look at the Who. Theater is a part of everything we do."

One of his first moves was to welcome aboard the recently deposed Thompson. Thompson and his wife Kaye had done commercials together in San Diego, mostly for the infamous Pussycat Theater, and they stopped by to pitch their work to Bass at WNCR. Bass hired them on the spot. Bass took the bold move of hiring women for air shifts, a brave decision for a programmer in the late 1960s and early 1970s. There was still an "old boys network" mentality that Bass struck down in short order. But if it didn't work, he would have fallen on his sword.

Kaye Olbrys was a musician who had seen a good amount of success with a group called Brothers and Sisters, which she joined fresh out of high school in Milwaukee. She even played Caesars Palace in Las Vegas, opening for Red Buttons, but eventually returned to her hometown in Wisconsin. Thompson was a friend of her brother, and she got to know him better when he emceed one of her shows. They were married a short time later and started the nomadic life of a disc jockey, traveling around the country in a 1963 Chevy Belvedere. They set up housekeeping near Shaker Square with their two cats Omaha and San Diego, named after cities where Thompson had previously worked.

They took over the midday slot on January 4, 1971, as one of the few husband-and-wife radio teams in the country. There were no playlists, so the Thompsons played an upbeat mixture of progressive music, though their tastes were distinctly different. While Ron favored high-energy rock from the likes of the Beatles, Santana, and Chicago, Kaye leaned toward Elton John, Donovan, Van Morrison, and even Barbra Streisand, albeit secretly. Both agreed that acid rock would not be played on their show. Requests for Bloodrock's "D.O.A." would be saved for the night crew.

Along with a passion for the music, they joined with Bass in his mission to air a true station for the people. In fact they introduced *People's Want Ads* on their show, aimed at selling goods and services for listeners who didn't have the money to advertise any other way. It became a staple of WNCR even after the Thompsons had moved on.

Both WMMS and WNCR had similar playlists and also went through the industry dance of courting record people and promoters. WNCR's mix included artists such as Dory Previn, McDonald and Giles, and Lovecraft, along with more mainstream favorites such as Jefferson Airplane; Emerson, Lake, and Palmer; and Mountain. WMMS included Gordon Lightfoot with Uriah Heep and Rastus, and both stations had deejays that delivered the music in a laid-back but thoughtful manner.

WNCR also took its People's Radio tag very seriously. A typical weekend might see disc jockeys talking up the station at record stores or Hot Pants clothing outlets, and they pushed a free film series at local theaters with sneak previews of movies, usually at midnight. Occasionally these premieres would ask for donations to benefit local charities and causes. Their showing of *Celebration at Big Sur* required only a one-dollar donation to help the Free Clinic raise a $100,000 operating budget and open a West Side location.

The station also reached out to a generation frustrated by the threat of Vietnam and the inability to get jobs because of long hair. Some listeners even called WNCR in desperation because of bad acid trips, and many may have had their lives saved because the deejays and engineers put them in touch with people who could help them.

The street slang at the time was far less politically correct than in later years. Deejays Dick "Wilde Childe" Kemp and Gary Edwards were pals, and Kemp jokingly referring to him as "Super Spade," and the name stuck. Once the station held a contest with front row concert tickets as the prize if you could pick up the clues and "Find the Super Spade." Edwards says it was all done in good fun, and a kid from the East Side took the prize. "He was at Cleveland State," he recalls, "we talked a while, and he told me he wanted to be a disc jockey." Edwards also remembers the kid wore a really ugly hat.

The cornerstone of Cleveland's FM revolution was creative programming, offering the listener something different and relevant to their lifestyle. WMMS offered *Revolution through Poetry,* a series of programs hosted by Michael J. Griffin comparing the works of Joni Mitchell and Bob Dylan to Robert Frost and T. S. Eliot. It centered on a different theme every day, including loneliness, poverty, war, and peace. The segment on love included Leonard Cohen's "Suzanne," "Colors" by Donovan, and Joan Baez singing "Turquoise." Griffin spoke about how the "poets" were inspiring and fueled a social revolution, and while the series didn't draw a huge audience, it was indicative of the bold attempts at innovative programming happening on the FM dial.

But there was turmoil at the station too. Lou "King" Kirby, who reopened the door to live programming at WMMS, left the station in April 1971 after just eight months on the air. Ted Ferguson left the same day. Griffin took over the morning show, Kemp was on at night, and in May of that year a familiar voice, Ginger Sutton, joined the staff to do afternoon drive.

Sutton called his show *The Journey through Middle Earth,* and it kicked off every night with the live version of the Rolling Stones' "Midnight Rambler." His delivery was very relaxed, low key, and casual, almost as if he had been enjoying a puff or two. But he still came across as being very friendly, smooth, and knowledgeable while introducing the music, no matter what style it was. His mellowness would seem like a sharp contrast if it led into records such as "Summertime Blues" from the Who's *Live at Leeds.* Not long after Sutton's return, he stepped outside the WMMS studios at 50th and Euclid and was approached by a young kid with long hair carrying a five-inch reel of tape. He said, "Ginger! Man, you gotta hear this. You gotta hear this!" It was his audition tape from Cleveland State University's radio station. They spoke for a while and Sutton gave the kid some advice and wished him luck. It wouldn't be the last time this kid with the really ugly hat would come by with his tapes.

Another important name added to the lineup was David Spero, who had come from a family with close ties to the local music scene. He helped his dad, Herman Spero, produce the *Upbeat Show,* which was syndicated out of the WEWS-TV studios. It was not unusual for David to come home and see Simon and Garfunkel or Terry Knight and the Pack sitting at the family dinner table. The stars all signed a wall in the Spero home that the house's current owners have kept intact to this day. The young Spero had spun records at his high school, practiced radio at the station he set up in his bedroom (WDAS, his initials), and done some radio time with Barry "Buttons" Weingart and Chuck Avner at WXEN-FM. Spero was to become an important player on the local radio scene, and he started by taking over Andrews's Saturday-night shift on WNCR, but that wouldn't last long. He was just nineteen years old at the time and still living with his parents. Andrews got an earlier weekend shift from six to noon, and Spero got a lucky break when WGAR-AM's Don

David Spero (1971). An entertainment industry veteran while still in his teens, Spero was rubbing elbows with recording artists long before getting into radio. Photo by Joyce Halasa

Imus, who had worked with the elder Spero on a television project, cornered the general manager and gave the "new kid" a glowing if somewhat exaggerated review. "He's great!" Imus said, "I've worked with him in Detroit. I've worked with him in California. A real pro!" Spero of course hadn't worked anywhere but Cleveland and wasn't even sure yet how to run the board. Imus thought he was doing him a favor, but to Spero's horror they made him the morning man because of all his "experience." He learned the board pretty quickly after that.

Bill Barrett at the *Cleveland Press* gave a listen to what was happening at WNCR and wrote that when he was asked what kind of station it was he replied, "Dadburned if I know!" In a series of articles for the paper Barrett said, "If you remember Glenn Miller and Gabriel Heatter, you really ought to tune in and turn on with the likes of Kyle, who sings, and Billy Bass, who introduces the news with the phrase: 'And now WNCR lays the hard stuff on you.'"

He went on to write, "If you still get haircuts, dad, or permanent waves, mom, what you hear on this station will make you gasp maybe. But it will bring into sharpest focus what your kids are thinking about, worrying about, singing about. Whether you agree or not is beside the point. What is happening at WNCR is something you ought to know about."

Barrett commented on the news, which he likened more to editorials, which the WNCR staff characterized as, "dramatization, a sort of little theater of the news." He looked at the leftward lean of most news items, read to a musical background, and he stated, "It is likely that a WNCR listener who got his news only from this liberal station would get a slanted view of the world around him. But the same might be said, too, for the listener who took his news only from conservative Paul Harvey on WGAR, which happens to be WNCR's sister station on the AM side."

The article quoted a staff member who said, "There is a leaning, a point of view —not intentional, but inevitable because we're young here." The station's Tiburski added, "On most rock stations, the big tune-out factor has always been the news. We decided to make the news a strong tune-in factor." Barrett also pointed out that the music on People's Radio or the Electric Renaissance could be described as "generally loud ... gut blues stuff with the traditional flavor of black New Orleans. Or corny camp stuff ("Don't Call Me Honey When Your Mother's Around"), or folk rock. Or—most often—loud and strident and taut with protest."

Barrett had always had a reputation among broadcasters as a fair and honest journalist, a regular and friendly guy who was a lot more hip than one would imagine for his age. But he pointed out, "musical crudities [are] just barely masked. That ultimate four letter word, for example, is a favorite with the lyricists who write anti-war ballads for the crooners of the Alternative Community. The word is snipped out on WNCR—but it's a near miss. 'We don't want your —king war,' screams the chanteuse in one such ditty. 'Bring our brothers home.'"

Jack Ambrozic added, "Sometimes our engineer reverses the word—turns it around altogether." In truth, that was the version that was expected to be played on the air, but many years later, at the Rock and Roll Hall of Fame's Don't Touch That Dial forum, Bass confronted Martin Perlich to ask if he did indeed use that version of Jefferson Airplane's "We Can Be Together." "Because if you didn't," Bass said with a laugh, "you lied to me!" Chances are if Perlich did play that version, there weren't many complaints. Barrett rightfully pointed out that it wasn't "the kind of material you'd expect to find on a radio station owned by an insurance company," though Tiburski countered by saying, "We're pioneers in a new sound. We're in the position the Top-40 stations were in ten years ago. We're developing the stars and the music that WIXY will be playing in years to come." In that respect he hit it right on the head.

Station management told Barrett they didn't play "acid" or "underground" music but rather "adult rock," aiming their programming at "the mass audience

Shauna Zurbrugg (1972). Her air shift with partner Jeff Gelb evolved over several months into a topflight morning show, complete with the free *People's Want Ads*. Collection of Joyce Halasa

of young people—the college students, the young married, the veterans just returned to civilian life, the people making their first start."

The experiment with the Thompsons didn't last long, as they were replaced four months after signing on at WNCR. Jeff Gelb and Shauna Zurbrugg, a couple of twenty-year-olds with local ties, stepped in. Gelb was a student at Kent State, and Zurbrugg graduated from Marlington High in Alliance. She also went to broadcasting school after a couple of stabs at other forms of higher education, and she walked into WNCR to see how big-time radio worked. Five months later Bass asked Zubrugg to submit a tape. He liked it, and she was offered a job, if she could find someone to co-host the show.

Gelb and Zurbrugg first met through a mutual friend. Zurbrugg asked Gelb if he was willing to put his studies at Kent State aside for a time to start their broadcasting careers because she was anxious to get moving. Starting his career in a market like Cleveland sounded pretty good to Gelb, and they debuted at WNCR in May 1971.

Zurbrugg once said that she pursued a career in broadcasting because, "In FM radio you don't have to go through the hassle of fighting off the advances of your boss. I originally wanted to be an actress, but I was too immature and thought I could get messed up." Despite having the confidence of an actress, she also recalls the terror of doing their first show on WNCR. "At that time women weren't on the air in top markets unless it was in talk radio. The bottom line is when Jeff and I went on the air, the only way a woman could really get on the air full time—rock and roll, or any other format—was to have a 'team show.' We went on green, novices on our first day, and when we opened the mikes for the first time, we were tongue tied. As far as we were concerned, it was probably one of the worst shows in history!" Gelb was new to big-time radio and in the early days would sometimes mutter things into an open mike that were better left off the air. Gelb has a tape of that show, and he and Zurbrugg still get a laugh out of it when they get together today.

Despite their self-criticism Gelb and Zurbrugg had a relaxed air style and worked well together. They continued the *People's Want Ads* that had become a hallmark at the station. They also did solo weekend jobs, with Gelb on Saturdays and Zurbrugg spinning records on Sunday. When he looks back now, Gelb recalls that they were "thrown together as two people who only knew each other cordially, through a mutual friend, and the audience benefited from the fact that they were all becoming friends, and got to know each other all at the same time. It was like watching a relationship evolve day to day." He also points out that, "Luckily, we got on really well, and there were no real points of diversion, where one person was totally against something and the other person was into it."

Another hallmark was thoughtful conversation, with Bass often joining Perlich to host *Renaissance Outlook,* the public service show that ran on Sunday nights. Guests ranged from Jane Fonda and Isaac Asimov to Leonard Bernstein and Buzzy Linhart, and talk centered on a wide range of topics. It satisfied part of the FCC's public-service requirement while helping to restate and define the mission of WNCR as the People's Radio station. Listeners embraced the spirit of the station as well. After Linhart taped his interview with Perlich, he was so fascinated by Gelb and Zurbrugg's *People's Want Ads* that he helped answer phones for nearly two hours.

Disc jockeys from WMMS and WNCR were always on the streets talking to people and promoting their stations. A typical weekend might see Bass doing his show in a live remote from Higbee's, while the WMMS deejays were at a store like Pant Talk doing giveaways. Word of mouth was still an important form of advertising, and it was not uncommon to see personalities from both stations hanging around Record Revolution, Bill Jones Leather, or the other stores on Coventry Road talking to fans and each other.

As on-air music continued along nicely, the local concert scene was going through changes. Belkin Productions partnered with WNCR to stage a fundraiser concert with Mountain at Cleveland Arena that June. Tickets were only two bucks, and all proceeds were earmarked to benefit the Free Clinic.

Belkin had emerged as Cleveland's primary concert promoter, but one of its top lieutenants had defected from its ranks. Roger Abramson, a former vice president at Belkin, had started A Friend Productions and was booking shows at the Musicarnival tent in Cleveland suburb Warrensville Heights and other locations. He also produced WNCR's first birthday bash, The Summer Cleveland Learned to Boogie, on Wednesday, July 7, 1971, a free outdoor concert at Edgewater Park that drew a crowd estimated at more than 25,000 at its peak. The show offered acts such as Jim Glover, Hessler Court, Eli Radish, Mushroom, the Pure Prairie League, and Savage Grace. It resulted in a huge traffic jam on the West Shoreway during the evening rush hour, with some motorists abandoning their cars on the rim of the freeway. More than a hundred of them were ticketed. As the *Great Swamp Erie da da Boom* described it, "Free Coke was dispensed on one side of the field, while Pepsi insisted on 25 cents a hit on the other." WNCR also provided ice water and suntan lotion.

The following day WNCR solicited a group of volunteers to clean up the grounds, and after a few hours the only evidence of the previous day's event were eighty-nine garbage bags filled to capacity.

The friendly rivalry between WNCR and WMMS began to heat up, fueled in part by underground publications such as the *Great Swamp Erie da da Boom*. As a rule, hippies just didn't like corporate America, and the paper helped create the image of the Belkins as symbols of big business. (Legendary promoter Bill Graham of Fillmore fame got a lot of the same kind of press.) Belkin Productions answered the critics charge for charge and would grow into one of the biggest concert promoters in the world, but they still took a lot of heat from the *Great Swamp Erie da da Boom*.

The chess game of big-time radio continued through the year, with a shuffling of air shifts at WMMS aimed at strengthening its position against crosstown rival WNCR. As a result Kemp was moved into the midday slot. It seemed as though People's Radio was promoting a lot more concerts such as Elton John, Black Sabbath, and the Who, while WMMS had the jump on new releases and aired a lot more special programming, covering artists such as John Lennon, Bob Dylan, and Joni Mitchell. The stations were going out of their way to get their disc jockeys more press, and WNCR was very effective in that area. Bass once traveled to McCrea, Louisiana, filing reports live from the Celebration of Life Festival. Those were the days of huge outdoor shows, and every one of them had the potential to become a new Woodstock.

At least one of the WMMS deejays wasn't faring quite as well. Sutton left the station on a warm July night to find that his car had been stolen. He rang up his friend, Marty Mooney with Columbia Records, to ask him for a lift, but Mooney's car had been stolen a half an hour before. Looking back, Sutton says, "Calling it a guarded parking lot at WMMS was an oxymoron. Like Peacekeeper Missiles! Some old drunk in a guard's booth passed out every night, and some of our people even got smacked with lead pipes and rolled! You were nobody at WMMS until your car was driven out the gate of the guarded lot, and everybody had their car stolen out of that lot! One night I went out, and my car was gone." The cops soon showed up and were actually pretty sympathetic considering the fact that as many as eighty cars were being stolen in Cleveland every day.

Whoever snagged Sutton's car knew whom they were dealing with. He got a note the next day composed of letters cut from a magazine that said, "I took your car and I'm proud of it! I hope this teaches all of your smart mouthed degenerate freaks a lesson." It did. The car was found stripped at East 72nd and Hough Avenue, and Sutton parked his car in a different spot after that. He left WMMS a short time later.

The summer of 1971 saw the start of free shows from WMMS at the WHK Auditorium, again asking those who could afford it for donations to benefit charities such as the Pakistani Relief Fund. Groups such as Damnation, Eli Radish, and Tiny Alice all donated their time, though musicians often had a hard time dealing with the midday hours. WNCR took live music to the streets, hosting its second annual Appreciation Day at Southgate Shopping Center. It was just local bands and a fashion show, but the magic word "free" drew a pretty sizable crowd.

Meanwhile the store Music Grotto spawned another FM deejay for WNCR. Lynn Doyle opened the store at 24th and Euclid with Bass some years before, and Bass brought him on in early 1971 to take over the all-night shift from Spero, who played fast and loose with the rules and regulations they were supposed to meet to stay on the air. One day Bass told Spero that the FCC wanted the station to air a religious show to meet their public service requirements. Spero had a spiritual side but didn't see himself as particularly religious at the time, but Bass found a way around that. He sent away to a mail-order religious sect; Spero became a card-carrying minister of some obscure church and went on the air with *Our Lady of the Beatles*. Basically he would just recite Beatle lyrics, and since the FCC didn't have the time to listen to every show on every station, the name alone satisfied their requirements. Spero was a longtime Beatles fan, even insisting that all the disc jockeys play a song by the band at least once during their air shifts, so it was a perfect fit. However, WNCR's crew did a lot more public service as People's Radio than just about any other station in town, maybe even the whole country, so fudging a bit on the religious show wasn't going to send them to hell.

There was a great camaraderie among the underground deejays at both stations, even though they were technically in competition. One disc jockey might call another to compliment him on a great segue, and it was not uncommon to tip one's hat on the air to a friend at a rival station. During a slow shift, deejays might spend hours talking to each other on the phone during their shows between taking requests from listeners, back when requests were taken on the spot. One afternoon Spero rang up Gelb for a chat and noticed that they had both played an identical song at the same time. The two decided to air their respective spots, but each would play the same tunes for the rest of their shifts. Phones started ringing from confused listeners asking, "Do you know that WMMS or WNCR is playing the same stuff you are?" They would laugh and say something to the effect of, "That can't be true, but who cares? It's still great music!"

Gelb says they were all that close because most of the deejays at each station got into the business for the same reason: they loved the music. He explains, "It wasn't really about being a personality. Back then, it was about being a person who loved music enough that he wanted to share it with his friends, and that's what we did when we went on the air. We all hung out together whether you were from WMMS, WNCR, or wherever. There was very little sense of competition back then, which has been lost over the years." The AM stations weren't too fond of their FM cousins because AM was formatted to playlists, and they may have been a bit jealous of the freedom enjoyed by the other side of the dial. But there was a great sense of community in general and teamwork at each station. According to Spero,

> We were just having a good time. There weren't any radio stars then. We were just there to play the music, to have a good time, to have our "Parties in the Park," to do our yearly birthday concert, and you were there all day long! It wasn't a job! I would do the morning show and get off at ten o'clock, and I probably wouldn't go home until six or seven at night! We'd do those news bits, and everyone was part of everybody else's show. The cool thing was to find the "cool" song! That was the competition. You'd find a [tightly harmonized folk duo] Batdorf and Rodney! "Oh, boy! What a great song!" Then you'd wait until your next show, and spill it on everybody.

At WMMS, and especially WNCR, one requirement of all the deejays was to learn to segue smoothly from one song to the next. The audience came to expect it, and just as importantly, so did the program director. Gelb remembers,

> The ideal segue was to make two songs match so perfectly that the ending of one and the beginning of the next song would gel to the point where you didn't know where one ended and the next one started. It was a point of pride with the deejays at both stations that they did perfect segues. You would do incredible sets of music that might go from a Joan Baez track, to something from an early Genesis record, to a piece of classical music, and then back to the Beatles.

We were trying to bring in the widest cross section of audience that might not have heard of Genesis, but loved the Beatles. Or maybe someone who liked classical music who was across the dial at that point and tuned in to stay around for the rest of the set. It was the "be all" and "end all" for those radio stations.

It was also something that Perlich had perfected years before on WCLV's *Perlich Project.*

Making a living wasn't a real consideration to a disc jockey. As Spero recalls, "I don't even remember what we got paid. I don't even remember worrying about what we got paid! I was naive—I mean, obviously they paid us. I just couldn't believe they were giving us money to play records, get free t-shirts and concert tickets, meet people—like having Neil Young sit in your studio—and they paid us for that! That was amazing! I guess we never thought about how long it would go on, or if it would go on. It just did."

Despite the good feelings between the two stations, WNCR was still held in a bit higher regard by listeners. It stemmed from a couple of years before, when WMMS started progressive commercial radio, then backed off, and tried it again when WNCR started to make waves. WNCR had even cracked the top ten in ratings. The *Great Swamp Erie da da Boom* speculated that WNCR had built "long lived Cleveland friendships and audience identification" and that listeners had been "betrayed by WMMS in the past." There may have been something to that, because at the time Metromedia had FM stations in four other markets that were doing great, but WMMS wasn't doing nearly as well. On a lesser note, late-night listeners could still ring up a disc jockey for a request at WNCR, while WMMS had a taped show overnights. WNCR gained greater fame by promoting the popular Crooked River Festival in the Flats with Starshine Medias that August, with rock bands, folksingers, and craftsmen lining the river for a three-day festival.

Even though there was an apparent edge held by WNCR, the situation was about to change very rapidly.

A major transition was brewing that would have far-reaching effects on Cleveland's FM scene for many years to come. In September 1971, Bass sat down with a reporter from the *Great Swamp Erie da da Boom* for a lengthy interview about the state of Cleveland radio. When the interview was published on September 23, Bass came in to WNCR on Thursday morning and laid a copy of the paper on the desk of every management person at the station. Then he walked out.

The front-page story: BILLY BASS QUITS WNCR.

Bass Makes His Move

The interview hit like a bombshell. Billy Bass told the paper that he had been relieved of his program director duties a couple of weeks before when general manager Jack Ambrozic took him out to breakfast and said the staff had concerns about the direction WNCR was heading. Ambrozic claimed Bass made certain members of the staff uptight, and a few found him difficult to work with. At that point, Ambrozic suggested Lee Andrews might be more suitable to lead WNCR toward the goals he had in mind, and that was targeting the entire eighteen-to-thirty-four-year-old age group.

Bass countered by saying, "We would all like to get as many straight people listening as possible. I'd like to get as many as possible for sure to turn them on and make them aware of certain things that we do on WNCR. [But] because of certain problems with Belkin, and because of the way I look, and because of the way I dress, and because of what I do, Jack Ambrozic thinks it's impossible for the straight community to look at me, or associate with me, and want to be identified with what I do. I think that's the basic fear of most straight cats." Bass suggested, "If we take my image, my goals and put them up in front of the public, I think Jack Ambrozic is a little worried that the public would reject it." He pointed out that Andrews's image was "very clean cut and mod," and Ambrozic chose him because he was "trying to be very safe."

But Bass also admitted that Ambrozic was, at times, a "brilliant man, a genius." He pointed to Ambrozic's sales techniques that brought money to WNCR before they really had any numbers to sell, basing it on what the station was about rather than the number of listeners. As examples, he mentioned a waterbed outlet and Record Revolution on Coventry Road as businesses that needed a place to advertise, aimed at that segment of the community. It worked, and ratings started to back up their promises. Bass said, "We had a frightening percentage of the 18-to-24-year-old market—frightening to other radio stations. They couldn't believe it was true! In 18–24 males, we were number one all day long—[but] as you start thinking about numbers, it just ruins you."

Revenue keeps stations on the air, and it is up to sales departments to get their share of the money, but ultimately, sincere, creative programming begins with the program director and airstaff, and sometimes the departments clash. It was no different at WNCR.

"We had two situations on the air at first that nobody liked creatively, but we had to do 'em because we needed the revenue. That's really important," Bass said. The first concerned the J. P. Snodgrass stores, which dragged on for months. He said, "The Ed Speizel Advertising Agency handles Cook Coffee, which owns J. P.

Billy Bass at WNCR (1971). His strong will and vision for a true People's Radio station would sometimes lead to confrontations with the station's management and sales staff. Photo by Joyce Halasa

Snodgrass. He came up with this idea of having J. P. Snodgrass, this old geezer from back in the old days, doing newscasts and all kinds of things from that period and he would talk in this absurd voice. Cook Coffee tried it—and the success of J. P. Snodgrass initially was unbelievable." Bass thought the chain's success was the result of WNCR being such a credible source for the "freek" community: "People just say, 'Well, if WNCR is advertising it, maybe we ought to look it over." As the *Great Swamp Erie da da Boom* pointed out, it could have also been the right product at the right place at the right time.

Whatever the reason, the battle continued to escalate, with Bass telling the sales department, "We want our products to be believable; we want your approach to be believable; we want it to be as believable as we think we are." His thought was that advertisers didn't usually do that and that sales people are paid commissions so their priorities were different. He also charged that every radio and television station, plus the newspapers in the straight community were "primarily owned and controlled by business interests that tend to fall in line with the sales department rather than their own creative department. They have to turn in a sheet at the end of every month that shows a PROFIT. They can't turn in a P&L sheet that says WNCR sounded good, but they didn't make any money." These were strong words, especially for the record, but a lot of people felt exactly the same way. Once again, distrust of government and corporate America fanned the fires of rebellion.

The interview went on to show Bass's passion for WNCR and his philosophy on sales and how a station could be run. He noted that WNCR ran eight spots an hour, but they were so repetitious and "aesthetically so inferior" to the rest of the programming that they were inappropriate. Bass wanted the jocks to read live copy and no produced spots, but advertisers didn't want to hear that. A commercial for the Paul Harris clothing-store chain finally brought the situation with the sales department to a boil. The programming staff said that the ads were the worst things they'd heard on the station and told the salesman to go back to Paul Harris and explain that the way the spots were presented stood to annoy their audience. Surprisingly the salesman agreed, but when he met with the Paul Harris representative he was reportedly told it was the most successful campaign they'd ever run, with sales up over 500 percent. With results like that, pleas to change the spots fell on deaf ears.

Another key factor in Bass's decision to leave was a spot for Saran Wrap, which was made by Dow Chemical. He told the *Great Swamp Erie da da Boom* that Dow also made napalm, though the paper was quick to point out the company dropped that contract to "get out of the public hot seat—though Saran Wrap burns you ecologically." Bass questioned whether a station could "sacrifice aesthetics [programming] for the dollar?" He said no. "I believe there is such a thing as a compromise that can be worked out. But I don't think that I should ever put a commercial on the air that will annoy anybody. If we say, 'From a point of view of aesthetics, we can't run this the way it is,' and they say, 'We have to run it: we need the bread,' they win every time." But he also nodded to Larry Robinson, the "Diamond Man," who was open to that sort of discussion.

Then Bass took aim squarely between the eyes of station management, particularly Ambrozic, saying,

> [Ambrozic] seems to think that the only people that can associate with Billy Bass are dope heads, revolutionaries, irresponsible people that hang out on Coventry. Those are not his words. I mean that figuratively, not literally. But that seems to be his attitude.
>
> I don't think Jack Ambrozic understands, nor does any general manager in this business understand yet, that people do not have to smoke dope, or burn the flag, or fly a Viet Cong flag in order to appreciate Judy Collins, or to appreciate a radio station involvement with the issues of today. Like racial prejudice. Like pollution. Like the environment. Like alternative education. People like that exist all over the city. They exist in Parma. They exist in Lakewood. They exist in downtown Cleveland. There are a lot of people sitting at home watching that idiot box, knowing that there's got to be more to life than that, and they want an answer to it.

Our big problem as creative programmers in this industry is convincing those managers that all people's goals and interests are not directed by Madison Avenue! That's what our biggest challenge is.

The interview with the *Great Swamp Erie da da Boom* covered a wide range of topics. Bass had always been known as a man who stood by his convictions, and his beliefs reflected the general attitude of a large and growing segment of society. He battled against dress codes at the station, saying, "I'm going to be the same programmer no matter what I wear." Friends around the Statler Hilton pulled him aside suggesting he not be so hardheaded, but Bass stood firm. When it was suggested the staff should try to dress in a more business-like fashion because they "have to represent Nationwide Insurance Company—the owners of the radio station," Bass countered by saying he made it crystal clear when he took the job that he did not represent them. He said, "I represent only this radio station. I have no aspirations in the Nationwide Communications industry. Like, I don't want to be anything for them! I don't want to program any other station in any other market. I'm talking about WNCR in Cleveland and that's all I represent!"

Despite a threat that Nationwide could close the doors of the station no matter how successful WNCR was unless the image improved, Bass held his ground: "Martin Perlich has never made a concession to Nationwide Insurance. What you hear on the air . . . comes straight from the hearts and minds of the people on the air." Commenting on a dispute over spots for Zapple soft drinks, Bass said,

That arose as a result of the boycott for better working conditions for migrant workers. There was a boycott against the makers of Zapple at the time, and Martin Perlich refused to run the commercial. Jack, of course, took the position that he, as general manager, would decide whether the ad was run or not. And he decided we weren't fighting anyone's causes, and that the ad should be aired. As it turned out, that boycott was called off the next day and that situation was all settled. But that could have developed into a major problem in that Martin Perlich's job was on the line. We could easily have just run those commercials and not said anything about it, and chances are, nobody would ever know. Except then you wouldn't be able to believe this WNCR person any more.

Bass then disclosed that he had been fired in August 1971 but was asked to come back. He listened to Nationwide Communications' position, decided it wasn't in line with his thinking, and turned in his resignation. He told the paper it disturbed him when he saw "the sales department and the general manager taking control of the programming. . . . Their decisions were final. Jack sent out a memo saying, 'From now on I am taking a position of dictation on this radio station. Everything I say will be final. If you don't like it, leave.' That was his memo. It was the first thing I read when I got back from vacation." There was a philosophical difference over some staff assignments as well.

Ambrozic objected to Bass's plan to eliminate Andrews from the airstaff. It was nothing personal against Andrews, but Bass saw that it was in the best interest of the station. As expected Ambrozic didn't buy it. Bass claimed, "He thinks that Lee Andrews has what every woman in the world wants to hear, and that's a sexy voice from the radio. I cannot make him understand that people who want to hear that do not fall into the category of our audience. They are not our listeners. People do not listen to WNCR to hear a sexy voice." He also suggested that he wasn't trying to fight Ambrozic, though that was the likely perception:

> I think together we could make a good team. I just think that he has to let me be, and I have to let him be. I can't say, "Why are you giving all the profits to Nationwide Insurance Company? You should turn the profits back into the community" Although I feel that way, he can't. I don't think he should come to me and say, "You should play a record to make everyone in the world love us."
>
> I'm leaving WNCR, but I'm leaving WNCR in, hopefully, good hands. I'm splitting because I can't take any more emotionally. I wanna be a human being, and exist, and laugh, and do whatever I have to do. But I'm leaving WNCR in the hands of a communal program directorship, made up of the WNCR family on the air. I think together, if these cats can organize and speak from a position of power, I think they can get control of the programming department again. Unfortunately, if Jack continues to have that dictator sort of attitude, then that's going to negate anything that they might do. However, if they are together, and strong, then it's going to be very hard for him to act as a dictator over their wishes. I think they will make a lot more concessions than I would make, but I think that's because as a group, they are a lot more reasonable than I am as an individual. I think aesthetically they will make good programming decisions for WNCR. All they have to do is be strong.
>
> I think they have to make sure that there's no one subverting them. They have to keep everything up front so they know where everybody stands. They can't be concerned about losing one job, they have to be concerned about losing all jobs.

And Bass offered his advice to the staff he was leaving behind:

> There's a lot more to talk about than dope and music. The informers that police use to arrest kids using a little marijuana [rather than to arrest smack dealers], look just like you. You've got to remember that, and work on people's heads the way you did in 1967, when you gave someone a flower and talked to them. We have to get into energy again.

I think we have the right people in Cleveland, but I don't think we have the right energy. The *da da Boom,* and WNCR, and WMMS, and the Free Clinic, and the Peace Coalition—all of these things are helping. If we can get rid of the "Cleveland eats shit" attitude, and replace it with a "Cleveland could be outasight" attitude, in two or three years we could have the kind of city we want.

In the 1990s Bass recalled that interview as being a bit heavy-handed but very sincere in expressing his frustration with station management. He also expressed no regrets about leaving WNCR at that time.

Bass tipped his hand a bit in the interview about a new concept he wanted to try called rock and roll television, which would come to fruition in a relatively short time and years before MTV. The interview in the *Great Swamp Erie da da Boom* was his parting shot at WNCR management. He left and didn't look back, and he wouldn't be the last to leave.

The New Beginning

WNCR continued to promote the local concert scene, airing a live Grateful Dead show from the Allen Theater in October 1971 and putting a big push behind Pink Floyd's Cleveland debut at Case Western Reserve University's Emerson Gym later in the month. Soon after, the station began live Sunday-night broadcasts from Agency Recording on East 24th, with folksinger Alex Bevan on the premiere broadcast. They also hired a new morning man, Digby Welch, who had a "veddy" British accent, which was his hook when he graduated from broadcasting school.

Down the street, a familiar face walked through the doors of 5000 Euclid Avenue. Although WMMS had plenty of talent, it wasn't making any significant ratings gain against WNCR, and that proved to be very frustrating to general manager David Moorehead. He contacted Billy Bass about returning to the place where he started his radio career. Bass agreed and brought David Spero and Martin Perlich along with him. As Spero recalls, "We left WNCR on a Friday, and started at WMMS the following Monday." Bass also took the job as program director, and Perlich and Spero were more than happy with that arrangement. Years later Perlich remembers Bass as being anything but the typical radio management type saying, "Billy Bass is a genius. He's got that energy, that incredible charismatic energy. He wasn't an executive. He was there in shorts, with his scrotum hanging out!" An exaggeration perhaps but still very typical of the loyalty and respect his coworkers had for Bass. The general philosophy was, "Your air shift is yours—play what you like," though on rare occasions someone might casually mention, "maybe we play too much Dave Van Ronk or Miles Davis," or something of that nature. It was a reflection of the times, and the airstaff knew how to best relate to its audience, though Perlich likened it to "a guy with a divining rod. A conduit. Let the music play through you. Find the energy, define it, and exploit it for the good of the audience." Most everyone involved remembers the early days with Bass especially as being a wonderful experience in radio, with mutual respect among the disc jockeys, and a program director that "stayed out of your ca-ca."

Another deejay had cracked the mike at WMMS for the first time that very same month, on October 1. He was a twenty-year-old kid from Boston named Denny Sanders, who had been working in radio since 1966 as a teenager at WTBS-FM in Cambridge, Massachusetts. Two years later Sanders took additional work at WNTN, one of the few AM progressive stations in the country. It was there he met music director John Gorman.

Sanders brought with him a style of radio much akin to what Perlich had been doing on the *Perlich Project*. It was a varied mix of rock, jazz, folk, spoken word, and just about anything else that fit the mood of the evening. At the time, and outside of what Perlich and Sanders were doing, most of the underground formats centered on Jimi Hendrix, Cream, Traffic, and others examples of psychedelic rock. Both Sanders and Perlich had roots in public radio, while Bass came from record stores and for a time Top-40, and Spero had links to pop with Cleveland's syndicated version of *American Bandstand, Upbeat*. All were very good at what they did, but Sanders's eclectic spin added a new perspective to the way people listened to the station.

Spero and the rest spent endless hours at both WNCR and then at WMMS. He recalls, "We had shifts, but no set hours. You came into the station early, came in late and on your days off. It wasn't a job. It was a lifestyle."

The people behind local radio impressed artists. Humble Pie appeared on the WMMS program *Electric Tongue* and gave Spero credit for breaking them locally, and Marc Bolan of T. Rex played a few acoustic numbers for Shauna Zurbrugg during his visit to Cleveland, and then stuck around to help pull records for her show. The public response was huge.

Another move that lit up the phones came out of necessity rather than creativity, but it was still a magic moment. A representative from Atlantic Records had just dropped off the new Led Zeppelin LP. The previous *Led Zeppelin III* was a departure from the first two, had a lot of acoustic material, and was something of a disappointment in sales, so the company put a big promotional push behind this fourth album. Spero had yet to hear the record, but as it so often happens during a long air shift, nature called and wasn't about to wait. He picked up the album, cued up a long cut, and raced to the bathroom. When Spero returned to the studio the song was just finishing, and the phones had lit up like a Christmas tree. He had just premiered "Stairway to Heaven," and he knew by the response that it had to be a great song, even though he really didn't get to hear it in the usual way.

The musical mix at WMMS remained a discriminating blend of jazz, rock, and blues. Music logs from June 1972 show that a typical program might include Harvey Mandel, Larry Coryell, Richie Havens, and the New York Rock and Roll Ensemble, along with a healthy dose of Beatles, Rolling Stones, and Faces thrown in to round out the overall sound. Artists such as Marc/Almond *(The City)*, Thelonious Monk *(Monk's Point)*, and Les McCann *(Generation Gap)* would end up in the same show as Leon Russell, Procol Harum, and Elton John.

Bass made a surprise move that December and gave the West Coast a taste of Cleveland's FM scene. On a vacation in Los Angeles, Bass paid a visit to sister station KMET-FM, opened the microphone, and showed the audience there what Northeast Ohio was all about. He was back in Cleveland soon after but not before letting California know they didn't have a corner on great music and equally great radio.

The departure of Bass, Perlich, and Spero from WNCR may have caused some concern about programming over there. As WMMS gained in stature among listeners, management at WNCR started to wonder if they were headed in the right direction. Jack Ambrozic and program director Bill Garcia decided to answer that question by going directly to the public, asking listeners to call in the *Renaissance Outlook* program late on a Sunday night to see if the station was meeting their listening needs. It was a rare chance to offer questions and comments to radio station management in a public forum. It was a bold move, considering the kind of press the station got when Bass split, and was most likely a public-relations move to show that the station really was concerned about its audience. The two handled questions about the departing deejays and WNCR's role in the community. The station moved to fill holes left in the lineup by adding Seth Mason and Steve Capen, who had worked at different stations in Chicago.

Both FM rock stations didn't make any concerted effort to try to be different. It was gut feeling, "seat of your pants," broadcasting, with the disc jockeys using their best judgments to pick the music. Programmers in New York and Los Angeles didn't understand what Cleveland was doing, and the deejays here didn't care if they did. They were serving Northeast Ohio the way they saw fit. Zurbrugg had the freedom of choosing the music for her show based on her mood. Looking back she says,

> As unprofessional as that may seem now, it worked very well because our audience was very loyal. Some of the other jocks did a similar thing. You couldn't take it too far, but if it was a rainy day and you were feeling a little blue, bluesy music worked very well. Throw in a couple of up things here and there, but bluesy music worked nice on a rainy day. On a sunny day, you'd hear a great Beatles song. You know, like "Here Comes the Sun." We went with the mood of the heart. There was only one trade publication back then that we even took a look at. It was called *Walrus,* and it went out of business years ago, but it was the first

"tip sheet" for AOR [album oriented rock] and progressive radio stations. It did present playlists from other radio stations across the country. However, in Cleveland, nothing was based on anything other than gut. It was listening to a record and saying, "I gotta share this with people who listen to me! This is great music!" We'd have music meetings where anyone on the staff who wanted to come was welcomed. There would be heated discussions about whether a record should go in or not, but if somebody really believed in something, you could play it.

It was that self-determination and heartfelt conviction on the part of the staff that helped define WMMS at that time. "We weren't into big, as much as we were hip," according to Bass. "We really wanted to be the coolest thing going, and if we only had ten listeners, we felt they were the ten most important people in Cleveland. We programmed the station that way." There were few rules, and what there were hung above the console on what Bass called a "boogie sheet," the show guidelines. It simply read, "Never play three electric tunes in a row. Never play three acoustic tunes in a row. Be real. Be good. Have fun. Know the music. Be tight."

That's not to say there weren't some outside influences. Bass points out that other stations around the country were doing similar programming, and they all influenced each other:

You know, we picked up a lot from KSAN [in San Francisco], and a lot from WNEW [New York]. They were sister stations, all Metromedia. WMMR, too, in Philadelphia, but we weren't as influenced by Philadelphia as they were influenced by 'MMS. We did pick up a lot of things like Stoneground, Grateful Dead, Jefferson Airplane—we picked up a lot of that from KSAN, and New York was giving us a lot of Allman Brothers and that sort of thing. We would get that sort of influence from them. Believe it or not, that influence we were getting from the East and the West Coast really helped define Cleveland. Martin, and David, and I would meet and say, "We don't want to do this. We don't want to sound like every other hippie radio station, or the next thing you know we're going to have Marshall Tucker, followed by Lynyrd Skynyrd, followed by the Allman Brothers, and we don't want to sound like that."

When out-of-town radio people visited the stations, they didn't understand how a station like WNCR or WMMS could play "Rhapsody in Blue" and follow it with the Rolling Stones. The important thing was for the listeners to understand, and by and large they did.

A key player emerged in November 1971. Metromedia Radio sold WMMS and WHK-AM to Malrite of Ohio, Inc., part of a Detroit group called Malrite Broadcasting Company. It had radio stations in Milwaukee, St. Paul, and Rochester, and the sale price was $3.5 million. Milton Maltz, a businessman from Mount Clemens,

Michigan, headed the group, and it was anyone's guess what directions the stations would take after the sale. Initial word was that Malrite wanted WHK and that WMMS was an added value. WHK was well known to Maltz, who once auditioned for a job there when he was in Cleveland in 1958 waiting out a federal hearing on a bid to buy a station in Tiffin, Ohio. Now he was coming back as an owner. Malrite's AM outlets were mostly country-western stations, while the FMs broadcasted soft pop and "beautiful music." There was a very distinct possibility that WMMS was about to make another drastic programming change.

Shortly after the sale was announced, Malrite President Milton Maltz told the *Plain Dealer*'s Raymond Hart that, "In the past, we've mostly purchased 'sick' stations, and made them well." Of the impending takeover, he said, "It will be the first time we've acquired what we consider to be a healthy and capably staffed broadcast property. Along with WHK, we're more than pleased to acquire WMMS-FM, because our company has great faith in the future of stereo broadcasting, and we are looking forward to strengthening the FM band." Maltz couldn't announce any definite plans for the FM format until the sale was approved, but Malrite planned a community survey, and Maltz promised, "We will interview people from all aspects of the community. We want to find out what they expect of broadcasting in Cleveland." The "freeks" were mobilizing to let him know what they expected of rock radio in their fair city, and as it would turn out Maltz was willing to listen. But it would take a while before the two sides could get together.

Word of the sale sent up a red flag for Cleveland's counterculture. As the *Great Swamp Erie da da Boom* pointed out:

> The powers behind the airwaves have usually been in it only for the money. Cleveland's rock stations are no exception to all this. WIXY (and WDOK-FM) are now owned by the people who own the Globetrotters Basketball Team. WNCR-FM (and WGAR) are owned by Nationwide Insurance Co., and WMMS (and WHK) have been owned by Metromedia—part of a vast conglomerate that also owns billboards and movie studios.
>
> Metromedia has never been commercially successful here with WMMS. Their WHK was last a real moneymaker with a "color radio" rock format ten years ago. Only sporadically (Browns games, etc.) has either station shown up in radio ratings in the last few years. However, Billy Bass recently took over programming duties at WMMS, and at *da Boom* we've been hoping for significant change in our town's radio shows. New ARB ratings are rumored to show WMMS with the biggest chunk of the young adult audience.
>
> It's highly probable that programming changes will come to both WHK and 'MMS when the sale becomes final.
>
> If you are not happy with this possibility, you can do something about it. If you have a million bucks, you can offer to buy WMMS yourself. Or, you can

write and complain to the FCC (especially to Commissioner Nicholas Johnson). . . . Or, you can write and bitch to your Congressman, especially if you just registered to vote.

It will take more than six months for a changeover. In this time the FCC will investigate the details of the sale, and make a cursory needs-of-the-community investigation. If you let yourself be heard in this same period, if you support the station in its next half year, it can make an awful lot of difference in what you'll hear a year from now.

The paper continued, saying,

Bass and his WMMS staff are stoically determined to persevere in the meantime. Bass says he wants to put together a station without worrying about money—to aesthetically coordinate a lot of real culture services in the community.

WMMS's Transcendental Sunday . . . was certainly one of the nicest community vibes things since Crooked River Revival and the Edgewater Beach WNCR Birthday Party (both of which events were organized by people now at WMMS). The station plans similar Sunday afternoons as long as they stay with us.

The direction of Cleveland rock radio is a big question mark right now. Community control looks doubtful. . . . Support meaningful alternatives like the current WMMS, and pray for Cleveland.

There were some who didn't think prayer would be enough, and Henry Speeth Jr. was among them. While Speeth was very much in tune with the counterculture, he was politically savvy and well connected, and started to devise a plan to save the station's format.

Another important arrival on the Cleveland radio scene was John Lanigan from KRLD-AM in Dallas. He came to Cleveland in December 1971 to take over morning drive duties at WGAR-AM from Don Imus, who was leaving for the greener pastures of New York City. A San Diego native raised in Nebraska, Lanigan started his career in radio when he was just sixteen. Following a couple of years at the University of Nebraska's campus station, Lanigan continued to find work in Scotts Bluff, Denver; Albuquerque; and Dallas. Lanigan was not a progressive rock disc jockey, but he would prove to be a major player and a thorn in the side of other stations, especially WMMS, in years to come.

Then there was the new promotions director at WMMS, Joyce Halasa, a native Clevelander who ended up having an air shift because the station wanted a voice connected to the person working community events. Halasa learned the ropes from Spero by sitting in on his show that December before finally going solo on December 31, 1971. She later said, "Bass had promised to be there to help me with the music, but he overslept. I was just thrown into it." Halasa and Bass were forging a

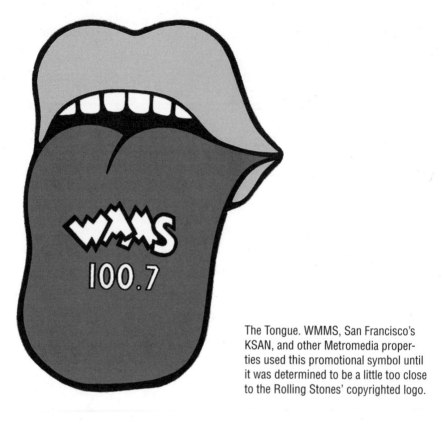

The Tongue. WMMS, San Francisco's KSAN, and other Metromedia properties used this promotional symbol until it was determined to be a little too close to the Rolling Stones' copyrighted logo.

tight working relationship that would eventually result in some spectacular events for the city and with little or no budget.

WNCR moved its studios from the Statler down the street to the Stouffer Building, and were welcomed by the Playhouse Square Association during a performance of the Sierra Leone Dance Company at the Allen Theater. WNCR was starting to reshape itself into a contender, but apparently there were those at the station's Playhouse Square offices who didn't want the FM to succeed.

The long war between general managers at WMMS and WNCR continued. While there was a great spirit of community and cooperation between the airstaffs, the top management was willing to break wiretapping laws to get a competitive edge. Ambrozic says the old rivalry between himself and Jack Thayer took a bizarre turn:

> WMMS, at that time, was under Metromedia, and the g.m. there, and a top engineer, and Jack Thayer were all old buddies from Metromedia. With or without Thayer's permission—I don't know, but I have to believe he was involved—they bugged my office at the Stouffer Building. They had a small portable transmitter and a microphone in there, so that all of our plans were known to them. We had meetings in my office with the jocks and [Walt] Tiburski, and we would lay out some great schemes, and a week before we were ready to

The Imp. Given out free at record stores and events, this sticker would later command high prices in Internet trading.

implement them, 'MMS would have them. They would jump on our plans, and it looked like we were copying them! I knew we had a leak somewhere, and I was trying to find out who inside the organization was that leak. We couldn't find the source of that problem.

More live music made its way onto WMMS as well, with a free concert on January 2 from the WHK Auditorium featuring Eli Radish, Tiny Alice, and Crabby Appleton. The ticket, billed on the air and in newspapers as a "cup that cheers," was made of white Styrofoam with a very familiar looking logo—the so called "Big Mouth," WMMS's call letters emblazoned across a big red tongue. It was the same symbol used by all of Metromedia's FM outlets. The problem was that it looked an awful lot like the tongue and lips adopted by the Rolling Stones for their new record label that premiered some time before on their *Sticky Fingers* LP. It was withdrawn by Metromedia soon afterward and replaced with a green "imp" having a smoke underneath a mushroom, which would remain the station's logo in some form for years to come. That same logo was seen many years later in the Cameron Crowe film, *Almost Famous*, which had several key scenes set in 1973 Cleveland.

On the day of the show the strains of Mountain faded down at the appointed hour, the signal switched over to the concert hall, and Bass welcomed the crowd and introduced morning man Spero. He took the microphone, told the crew to fade down the house lights, and warned the band, "You can't say dirty things, man. We're on the radio." He then announced, "Elektra Records brought to us, and brings to you, Crabby Appleton," and with that, the lead singer approached the mike to say, "Hello Cleveland! How are y'all? You all wasted a little bit? Well, we're drunk on our collective ass, and we're going to start out with a little something called 'Smokin' in the Morning.'"

The on-air shuffle at WNCR and WMMS continued with a seemingly endless stream of disc jockeys. Bass got dozens of tapes from deejay "wanna-bes," and one kid in particular at Cleveland State University kept popping up again and again. It was the same kid who ran up to Ginger Sutton a few months before. There were only so many slots to fill, but this guy just wasn't going to give up, even though his only on-microphone experience was broadcasting to the college cafeteria.

Disc jockey Carolyn Thomas was given the nod to host WNCR's Sunday night broadcasts from Agency Recording atop the Cleveland Agora, as well as the overnight duties at the station. The station got some heat in early 1972 when a caller told Thomas that he heard on the AM station that President Nixon had been shot. Commenting on that call, Thomas noted that clairvoyant Jeanne Dixon had predicted a "dark cloud over the White House." She then checked with United Press International, which assured Thomas the report was not true, and she noted that on the air. Even so, nearly a hundred calls were made to the station about the broadcast, which didn't look good for a deejay who had only been a full timer for about two weeks.

Not long after, Bass welcomed another former WNCR staff member when Zurbrugg came over to do a solo gig, first in a soft, acoustic morning show, and later taking over Perlich's evening slot when he left for Los Angeles. She remembers WMMS at that time as a station that "broke convention," saying, "We wouldn't play anything at that time that you heard on WIXY, or any other AM station around the country. We played Carole King's *Tapestry* album before anyone knew of her. Same thing for David Bowie. We played strange, obscure groups whose names I don't even remember, but it was very serious rock and roll. Acoustic, too—a lot of Buffalo Springfield, Joni Mitchell kind of things. Even though they had the numbers, we felt we weren't in competition with AM. We were going for the progressive, underground thing."

Buddy Maddock held down the weekend shift that Gary Edwards left at WMMS, and another addition at the station would be a vital cog in the machinery helping to shape the emerging new sound of WMMS, even though he would only be there a relatively short time. He was a tall, lanky nineteen-year-old with long dark hair named Tom Kelly. Calling himself "Tree" after WIXY's John Roberts supposedly made a comment about his height, Kelly was a master of production, had a great sense for music, and took over the 1:00-to-6:00-A.M. slot at the station.

Meanwhile, there were more arrivals due on the horizon, as Malrite Broadcasting Company announced that it would move its corporate offices from Detroit to Cleveland's WHK building after the FCC approved its takeover of WMMS and WHK. Malrite's Maltz also announced plans to expand the company from its new home base. There was still a lot of speculation on the street and in the press about what might happen to the stations' formats. Despite the uncertainty, the Cleveland State kid kept calling the station and sending in tapes.

Rock-and-Roll Television

Until Malrite's plans for the station were etched in stone, it would be business as usual at WMMS, or as usual as it could ever be. But Billy Bass had an idea that had been brewing for some time, and it went beyond the boundaries of radio. Rock and Roll TV.

It wasn't really a new idea. Tommy Dorsey and Milton Berle presented Elvis Presley on their shows in the 1950s, and Ed Sullivan made history by showcasing the Beatles and other up-and-coming acts every Sunday night. *The Hollywood Palace* usually had at least one rock act, although the Rolling Stones were none too pleased with host Dean Martin's comments about them during their American debut on that show, on which he begged the audience not to leave him alone with them and said that an acrobat was actually the group's father who was trying to kill himself. Programs such as *The Smothers Brothers Comedy Hour* and *The Jonathan Winters Show* also showcased pop music stars. Eric Clapton's Derek and the Dominoes were featured on *The Johnny Cash Show,* and in Cleveland David Spero's father Herman had syndicated the locally produced *Upbeat* around the country. But Bass had a different take on it. He wanted a live televised concert to be simulcast in stereo on WMMS.

Bass had the talent readily available. The band chosen for the initial broadcast would be Glass Harp, the Youngstown-based trio that stole the show at the daylong Edgewater Park festival. The band had been getting rave reviews for its live performances, even winning over the hardened New York City audiences at Carnegie Hall when they opened for the Kinks. When the band played the Fillmore East, a very tough venue for a new act, they walked on stage to tepid applause and heard one patron snidely call out, "You better be good!" By the end of the show, fans were standing on their seats and screaming for more.

The members of Glass Harp had been doing live radio broadcasts around the country since the release of their first album. One evening in San Francisco proved to be especially noteworthy and somewhat unsettling at the same time.

The legendary KSAN-FM broadcasted a weekly Sunday-night show live from the Bay Area's Pacific High Recording with acts that might be passing through town. Sometimes there would be a couple of groups showcased, or if no one was available the station put out a call to members of the Grateful Dead, Jefferson Airplane, Quicksilver, and other local musicians who might like to sit around in an informal jam or debut some work in progress. On the night they visited Pacific High, Glass Harp had the good fortune to be opening for one of bandmember Phil Keaggy's long-time guitar heroes, Mike Bloomfield. As the show aired, the band noticed the renowned guitar slinger enter the studio, and inspired by the presence of his mentor, Keaggy tore into what he would later jokingly call his "best Mike Bloomfield impression." The performance was inspired, and as Keaggy's fingers tore into rapid-fire notes backed by Dan Pecchio and John Sferra's rhythm juggernaut, even the KSAN deejays hoped aloud that Bloomfield might join the band for a few numbers. Instead Bloomfield meekly looked around the studio, put his guitar case near the amps, and made himself scarce until it was time for him to plug in. Keaggy, in particular, was crestfallen but shook off his disappointment and finished a dazzling performance that became the stuff of legend in a city that prided itself on its homegrown, cutting-edge talent.

A few years later, after leaving Glass Harp for a solo career, Keaggy ran into Bloomfield and introduced himself. Bloomfield smiled and said, "I know who you are. You're the guy who played his ass off in San Francisco that night!" It was a long-overdue compliment that helped ease any misgivings Keaggy had about one of his favorite artists.

Glass Harp had been getting that response all over the United States. Despite the heavy spiritual tone of its recorded work, the band's strong presentation was winning a growing audience that crossed most musical boundaries and won over even some of the toughest critics. Their first album was recorded at Electric Lady Studios in New York, which had been custom-built by Jimi Hendrix.

A rumor quickly circulated that Hendrix said in a late-night television interview, possibly on the *Tonight Show*, that eighteen-year-old Keaggy was "the best young guitarist in America," but today Keaggy himself is the first to downplay that story. As Keaggy points out, Hendrix died the week Glass Harp's first LP was released, and it was unlikely that he would have even heard the tapes in the mixing stage since he wasn't in the New York area during the last few weeks of his life. However, researchers now say it's very likely that comment was made about ZZ Top's Billy Gibbons, whose band Moving Sidewalks had opened for Hendrix on various concert dates. Even so, a lot of people were very willing to believe that remark about Keaggy based on the strength of Glass Harp's performances. Bass himself had praised the three musicians during a live radio performance as "probably one of the most important new bands in America," and he was sure they could admirably pull off his plans for the live broadcast.

Pecchio recalls that, "We were really busy doing a lot of things, and we had just come back from doing one of the albums. Things happened so quickly then, and a lot of it came down from record companies. The Carnegie Hall and Fillmore

East shows happened that way. We happened to be in New York at the time, so they said, 'Well, let's try to do something while they're in town.' We also played a McGovern fundraiser in Central Park that way, and a lot of people in the New York scene got to know us from those shows." The benefits of doing the WVIZ-TV broadcast seemed obvious.

Bass joined with Michael Kerwin, under the name Idyllic Productions, and armed with a grant from Decca Records planned to stage the event before a live audience at the WHK Auditorium on February 25, 1972, with WVIZ broadcasting the video portion of the show. There were a couple of concerns. One, it was to be broadcast the same night as Marc Bolan and T. Rex's Cleveland concert debut at the Yorktown Theater in Parma. WMMS had been playing a lot of the *Electric Warrior* LP, and a free televised show might have cut into the box office for the band's two performances.

Second, there were too many potential technical snags that could have arisen while broadcasting from the WHK Auditorium, and with the publicity already being generated about the show, it would not be a time to experiment. The venue was changed to the studios of WVIZ on Brookpark Road in Parma, with a small but enthusiastic group on hand to witness this groundbreaking event.

"I remember our studio engineer, Ralph Moss, was with us at the time," according to Pecchio, "and he did the mix for television. It was pretty crude. They had maybe four little Shure mixers stacked on top of each other. They were each four channels, and they tried to make a mixing board for what we were planning to do. In those days, sound wasn't a big thing in TV. The video equipment at WVIZ was okay, but it wasn't state-of-the-art stuff." Even so, like the kids in the Our Gang series putting on a production, the show would go on.

The broadcast began in black and white with the thundering chords of "Look in the Sky," then fading into a color shot of the band in front of a small audience. The band played on a sound stage decorated with a few flowered panels, and cameras weaved through a group that included fans, station personnel, and even a few children. At the sound check prior to the show, the volume of music in the studio had been so loud that it put a line across the video monitors. The tubes inside the cameras were picking up the heavy vibrations, and the studio technicians finally eliminated the problem by blanketing the amplifiers. But any problems behind the scenes were never evident to the viewers at home.

The broadcast was primitive by today's standards, but for the time it was a major undertaking. When Glass Harp would break between songs, there would be interview segments with Martin Perlich and the band that were put together by a local college film student. Those segments were choppy at times, even overexposed, and at one point Perlich had to stop the conversation to answer a ringing telephone, but years later it still stands as a slice of life in 1972 Cleveland. There were also scenes of the band sledding and horsing around in the winter snow along with canned performance footage. Unfortunately the sound of the group on stage in clubs was distorted and unusable, but it was its sound in the studio that was most important and that came off without a hitch. The program included segments of Keaggy, Sferra, and Pecchio talking about their Youngstown roots, their families, religion, and hopes for the future. One particularly poignant moment, which actually never made it to the final cut, featured Pecchio holding his baby son, Ted, and offering his advice on the child's possible career in music. Years later, Toto, as baby Ted was nicknamed, would prove to be a very accomplished bass player in the funk bands Funkomatic and Mr. Tibbs and later with Col. Bruce Hampton. Few people who watched that night realized that they had just seen the future of music, eight years before the advent of MTV.

The performance included a lot of the concert-hall staples: a tasteful drum solo by Sferra and Keaggy mugging for the camera during his guitar solo and trading licks with Pecchio on flute. It was a tribute to the engineers in the booth that the show came off as well as it did, because the sound both on television and the WMMS simulcast was crystal clear. Just before the show ended, Keaggy thanked the audience with, "Praise the Lord. Remember, God loves you," and launched into a new song, "Do Lord," as the credits rolled and the program faded to black. It was a triumph for the television and radio stations and especially for Glass Harp.

Drummer Sferra had a liking for some of the more ethereal musical lyrics of the day, so after the broadcast the band jumped into a car for a quick ride down Brookpark Road to the Yorktown Theater to catch the last part of T. Rex's second show. Despite his superstar status across Europe, Bolan was unable to impress the Cleveland audience. He seemed to walk through the set and didn't really ignite the crowd until "Bang a Gong (Get It On)." Glass Harp left the theater unimpressed as well.

Coffee's On

Despite the uncertainty of new ownership as well as the future of the progressive rock format, Cleveland's FM rock stations continued to develop new ways of reaching out to the public. WNCR solicited work from local artists for a People's Art Gallery at its Playhouse Square studios. They opened the space for a couple of hours in the morning and afternoon, and the response was just another indication that people were embracing the station's mission. WMMS also started a five-minute *Students' Rights Program* three days a weeks at 4:00 in the afternoon, focusing on civil liberties and fighting "the good fight" against the Establishment and even airing a show on veterans' rights. But WMMS was to begin another type of programming that would become a Cleveland tradition.

On March 22, 1972, a young songwriter named Carol Hall stopped by the studios at 11:00 A.M. for a live performance. She had written songs for Barbra Streisand and was getting a lot of well-deserved attention for her own self-titled album. Her performance proved to be a hit with the audience. Plans were made to continue the live shows, and while Hall may have since drifted off into relative obscurity, the concert series was only starting its long run. The station christened the shows *Coffeebreak Concerts*.

A singer from Cuyahoga Falls named Alex Bevan had been getting tremendous reviews for his work at local clubs as well as for his first LP, *No Truth to Sell*. He had tried to break out of Nashville, didn't like the way they did business, and came back to Northeast Ohio. Bevan was quickly booked to show the WMMS audience what he was all about. He arrived at the studios wearing a red jersey with "Bovine" emblazoned across the back, unpacked his guitar, and let loose with a sensational array of songs showing that his reputation was rightly deserved. A handful of people, including the *Plain Dealer*'s Jane Scott and three St. Augustine Academy students, stopped by to hear Bevan sing music he wrote in Greyhound Bus stations, folk standards, and gut-bucket blues. In a conversation between songs Billy Bass asked Bevan if, with his range of talent, he could also yodel. Bevan was seated next to David Spero and quickly tore into "You Are My Sunshine," complete with a yodeling refrain that drew applause from everyone in the studio. It was clear that Bevan loved what he did for a living and so did the audience. The phones lit up, and more *Coffeebreak Concerts* were on the way.

In the coming months the series hosted shows ranging from the electric blues of Robert Junior Lockwood, to a folk duet with Bevan and Hessler Street's John Bassette, to Jackie DeShannon. Bassette had gotten a lot of attention with a local hit from his privately distributed EP *Weed and Wine*, and his pairing with Bevan was a folk-lover's dream. The shows would often rebroadcast late-nights on Sunday, and even though it would go through a lot of tinkering over the years, the *Coffeebreak Concerts* became a longtime broadcasting staple.

Expanding the Horizon

The FM stations continued to experiment on a number of different levels. On the technical side WNCR had pioneered triphonic rock with its WGAR simulcast of the Damnation of Adam Blessing album *Second Damnation*, and in April 1972 added the much-hyped quadraphonic broadcasting to its portfolio. The station aired *Quad Hours* of music specially suited to that type of four-speaker sound, but that style of broadcasting was doomed nearly from the start. Like the wars that would later see Beta videotape fall to VHS, there were several formats for quad, and different record labels put out product that wasn't compatible with certain equipment. Still the concept intrigued a lot of the listening public, and the experiment continued for some time to come.

Vietnam, urban decay, and other social issues were also on the front burner. In the early 1960s Cleveland was still the sixth largest city in America, and its streets bustled with people, even drawing nearly full houses at Cleveland Municipal Stadium for weekend baseball games four or five times each summer. By the early 1970s white flight to the suburbs had taken its toll, Cleveland's nightlife was dying off, and the only things blowing down downtown streets after 6:00 P.M. were tumbleweeds. WNCR began a series of Renaissance Outlook shows, *Why Is Cleveland Dying?* Although they didn't provide any solid answers, the programs did allow listeners and guests to vent off steam about the state of the city. The station's *Job Bank* show helped employers match up with jobless listeners, many who found themselves that way because companies wouldn't accept their lifestyles and long hair.

Jeff Gelb started premiering *New Sides* of albums on his Monday night show of the same name. It was a wise move because as David Spero notes, "When you look at the music that was being created at that time, on any given day there were eight great albums coming out. It wasn't just Beatles and Stones and Who—Cat Stevens came along, Free, and Peter Frampton. Mott the Hoople, and Lou Reed." It was a golden era for fans of freeform rock.

WNCR took notice of the response to the WMMS *Coffeebreak Concert* and countered with a live broadcast of a folk concert at Case Western Reserve University featuring Mimi Farina and Tom Jans. Other live events were to follow on both stations. Meanwhile the emerging "Jesus freaks" found a show on WMMS that ran in the wee hours of Monday morning and 6:00 to 8:00 A.M. on Saturdays called *Goodtime Religion*. The show centered on religious lyrics in rock, and the Saturday spot was a relatively bold move considering that public service programming on other stations was relegated to times most people weren't near a radio.

WMMS and WNCR had pioneered programming concerning student rights, social issues, gay concerns, the environment, and women's liberation. John Lennon had been tweaking the nose of the Establishment for years with his comments about the popularity of the Beatles, Vietnam, and his political views. His "Ballad of John and Yoko" in 1969 found little time on AM stations because of the line, "Christ, you know it ain't easy," and a single Lennon released in April 1972 was about to revisit that situation all over again. It was titled "Woman Is the Nigger of the World," and its release to radio stations caused a firestorm of controversy. The title came to Lennon during a discussion with Yoko Ono, and while "Ballad" would sell over a million copies, this latest release would be a hard sell to get radio play—except on WNCR and WMMS.

WJW-AM called the record "offensive," WIXY hedged by saying it was "undecided" about adding it to its playlist, and WKYC-AM said it was unlikely to be added because "Lennon isn't often in our format." WJMO-AM, one of Cleveland's urban format stations, pointed out that Ike Turner had used the objectionable word in some of his work, but that Lennon's song wasn't really its style of music. WABQ-AM, Cleveland's other major urban outlet, said that some of its listeners might have some reservations.

WNCR's music director Seth Mason said there was very little reaction when his station played it. At WMMS Billy Bass downplayed the word the way Lennon used it, telling the *Plain Dealer,* "It's healthy for the business to make people wake up and think. Lennon could have said 'Woman Is the Polack of the World,' you know. He's saying that he takes for granted that free-thinking, liberal blacks and liberal whites will understand the usage of the word. If you saw Lennon on the *Mike Douglas Show,* you know he loves controversy and he's a master of getting publicity. Women's Lib is in vogue now, isn't it? Lennon could have treated that song a million different ways, but he took the one that would grab attention. Musically, I like it." While the double LP *Sometime in New York City* would sell reasonably well based on the Lennon name alone, its heavy political message didn't win many new fans.

One political message that drew a huge response concerned the upcoming presidential primaries. Because it occurred fairly late in the primary season, the May vote in Ohio could often be a "king maker" for aspiring candidates, and there was a big push among the youth movement for Senator George McGovern of South Dakota, who made it very clear that he wanted the United States out of Vietnam. To bolster his chances, a huge concert and rally were to be held on April 28, 1972, at the Cleveland Arena, with music from Paul Simon (in only his second solo outing since the split with Art Garfunkel), Joni Mitchell, and James Taylor. Jack Nicholson, Warren Beatty, Julie Christie, Goldie Hawn, and Peggy Lipton signed on to act as celebrity ushers. McGovern and Ted Kennedy would be there as well, and there was a major promotional push to get out the young vote. But the campaign of former Vice President Hubert Humphrey still had a lot of support in the Buckeye State, especially among organized labor, and it promised to be a hard-fought battle for Ohio's votes. It became very clear at the studios on East 50th, where Humphrey stopped in for a taped interview on WHK-AM and where Nicholson and Beatty appeared live on WMMS to promote the campaign and the show for McGovern. Only a wall separated them. McGovern won the primary but later suffered a big defeat in the general election. Richard Nixon and Spiro Agnew would never finish their second terms, and McGovern and the people who voted for him could hold their heads high knowing he was never hounded from office.

The Free Clinic had long been an ally of Cleveland's "freek" community, but it was a friend in need. It was about to celebrate its second anniversary and found itself needing funds to continue operations into the next year. At that time, the East and West Side locations were treating thousands of cases, many of them related to

WMMS Staff (1972). (Clockwise, starting from the lower left) Denny Sanders, David Spero, Joyce Halasa, Tom "Tree" Kelly, Billy Bass, Shauna Zurbrugg, and Martin Perlich. Collection of Joyce Halasa

drug use or sexually transmitted diseases. But revenue sources were few, and the Free Clinic needed money to keep its doors open. Plans were announced for a benefit concert, A Day to Remember, to try to raise as much as $275,000 to continue the Free Clinic's work. The date was May 20 and the location was Edgewater Park, with acts such as Eli Radish, Glass Harp, folk singer Ellen McIlwaine, and the all-female rock band Fanny. The Cleveland Transit System arranged to have shuttle buses transport fans to the site to avert the massive traffic jams seen in years past, the Ohio National Guard brought in two five-hundred-gallon water tanks to cool down the crowd, and some local grocery chains donated hot dogs and buns. WMMS started heavy on-air promotion and was on-site as well to handle any last-minute emergency. Early that Saturday, the crowds started pouring into the park.

The crowd was peaceful and extremely orderly and, despite a threat of showers, settled in for a long day of music. It swelled to as many as 35,000 by mid-afternoon, with only three drug overdoses seeking assistance at the Free Clinic tent, though a lot more needed help for drinking too much alcohol. It was a triumphant promotion for both the Clinic and local radio, but as afternoon turned to evening, and with some of the musicians offering inspired and sometimes lengthy performances, a number of acts still had to go on, and time was running out on the permit's 10:00 deadline. The headliners were standing by, and a few hours before the curfew, calls were hastily being made to city officials to keep the park open. Mayor Ralph Perk heard an appeal from Councilman Dennis Kucinich, and the two appeared at the show, telling the 15,000 remaining that the permit would be extended to 11:30 that night. A huge wave of cheers erupted from the crowd, and Perk also suggested that he'd be open to having a similar mega-concert from Cleveland Municipal Stadium. After the announcement, Perk also commented on how orderly the crowd had been and said that was a major influence on his decision to keep the show going. The Clinic fell short of its goal, with donations totaling about $20,000, but the concert was still a tremendous success in showing how the Greater Cleveland "heads" could gather toward a common goal, hear some incredible music, and even cooperate to help clean up the site afterward.

Not long after the concert, Martin Perlich had made a decision to test the media waters on the West Coast. He announced he would leave WMMS on June 1 for a job at Metromedia's Los Angeles outlet, KMET-FM. News of his impending departure shook the rock community, but Perlich clearly felt it was time to go. He wouldn't be the last to make that decision in the months to come. Meanwhile, Denny Sanders took Perlich's place as WMMS program director, Tom "Tree" Kelly became music director, and a salesman with a voice that started in the basement was brought on board to do overnights. His name was Len Goldberg, but soon everyone knew him as "Boom," for obvious reasons.

Goldberg remembers his audition,

The station was being sold, and Martin went to L.A. because he heard [the station] was going to be country, which did not happen. Billy called me—he was trying to think of someone with a deep voice—and I wasn't doing anything at the time. He said, "Would you come and take over the all night show?" I told him I didn't really know the music. I wasn't that hip to what was current, and he said, "You don't have to worry about that. You can play anything you want! Let's just hear what you sound like on the air." So, I came in around two o'clock one morning, sat down, and the first record I played was something I was currently listening to, Shawn Phillips. It was a cut that ran about twenty minutes, as one song segues into another. I never even gave it a thought that I was going on the radio, and the record would only play for so long and you had to talk or play

Len "the Boom" Goldberg (1972). With a voice that could shake the rafters, Len Goldberg had no problem getting his message across verbally or otherwise. Photo by Bob Ferrell

commercials, or whatever. This thing kept going and going and going. I think I only talked twice in the hour, but they called me the next morning and said, "You got the job." I started that night, midnight to six, six shows a week.

Goldberg was a man who made his feelings known, particularly when he spoke. Glass Harp was one of several Cleveland groups getting a major push from WMMS, and they did a *Coffeebreak Concert* the week Goldberg was hired. He was a fan, but as Dan Pecchio would later recall, the guys in the band were a little afraid of him because of his piercing eyes, thick hair, and of course that bellowing foghorn of a voice. They all remain fond of each other to this day, although Goldberg isn't any shorter on the decibels.

The summer of 1972 was a time for experimentation. WMMS forged an alliance with the Viking Saloon, a college bar on Chester Avenue across from Cleveland State University. It smelled from stale beer, but it was turning into an important location for live music. Artists such as Tiny Alice and other local favorites drew respectable crowds, while another WMMS-endorsed venue, the Smiling Dog Saloon on West 25th, was bringing in acts such as Odetta, Dave Van Ronk, and others. WNCR's live-music plans included a Summer Renaissance Fair, combining folk, blues, rock and jazz, along with artists and craftsmen peddling their wares. The station also continued its series of live concert broadcasts from Agency Recording.

But WMMS may have taken some of the most daring programming moves. For a time, the station started playing songs by artists such as Fred Astaire ("Something's Got to Give"), Marlene Dietrich ("Love"), Kitty Kaye's Kats ("Fish Truck Boogie"), and the Bo Jazz Carolina Serenaders ("Chicago"). Everyone from station manager Bass to Shauna Zurbrugg got in on the act, using 78s borrowed from Judy Sherill, who used the radio name Aunt Judy. The experiment was seen as a novelty, though Bass and Sanders also took to the airwaves of Case Western Reserve's WRUW-FM, under the names Dr. Soul and Mr. Funk, for a one-shot evening air shift to play some of their R & B favorites.

Rumors had circulated about Nationwide's plans for WNCR, and Walt Tiburski resigned from the staff with a note that included a handwritten addition stating, "I will be at WERE!" That station was now pursuing a controversial talk format called People Power that was getting a lot of heat from politicians, newspapers, and the public for topics that often bordered on the offensive, and for hosts that pushed the envelope whenever they could. There were also some firings at WNCR, but they would result in a showing of the true spirit of community being felt in Cleveland and with friendly rival WMMS. WNCR had taken a hit in the ratings, budget cuts were ordered, and Mason and the well-traveled Lee Andrews, who had been through a number of radio jobs in the past few years, were given their walking papers, although Carolyn Thomas was able to hang onto her overnight shift by negotiating with Jack Ambrozic. Gelb, Digby Welch, and Lynn Doyle became the music-selection committee.

Even though they were technically competitors, WMMS and WNCR had a bond in that they served the same audience, and personnel at both stations were very friendly toward each other. As a result, on June 23, 1972, staff members from

WMMS, along with a group of listeners, joined in front of WNCR's Playhouse Square studios to carry picket signs protesting the firings. No one was rehired, but it sent a loud message that WMMS practiced the same brotherhood it preached from its electric pulpit.

Both stations did some lineup shuffling that summer, with Spero pulling morning drive duties at WMMS. Bass tried his hand at a six-hour progressive jazz show on Sunday afternoons. FM radio was still pretty freewheeling and unburdened by tight formats, and disc jockeys were allowed to be as creative as possible on the air. Spero remembers, "If there was a big storm, we'd grab a sound effects record with rain segments, put it in the background, play a whole set of rain songs—maybe read a poem, or take a couple of calls about the weather—and create these atmospheres. In fact, Denny Sanders was the one who taught me you could do it the other way, too. If it was raining, play some sunshine songs. That works, too!" Formats were at the discretion of the deejay, although a coming new station would start to change that in fairly short order.

WNCR tried its hand at rock and roll television again in a simulcast of a canned concert with WKYC-TV featuring John Denver, Mort Sahl, Linda Ronstadt, and others broadcasted in quadraphonic sound. It was aimed at registering young voters, who hadn't come out quite in the numbers expected after the eighteen-year-olds got the right to cast their ballots. Quadraphonic sound continued to get a lot of press, even though a relatively small number of people had the equipment. Despite that WNCR aired music programs, and even Lenny Bruce material, in that format in specials sponsored by Carling Brewing Company, which limited the number of commercial spots and restricted the content to topics like the ecology and recycling. WMMS brought its *Coffeebreak Concert* series outdoors with a five-hour Sunday show at Chester Commons with Jesse Colin Young of the Youngbloods, Orville Normal, Alex Bevan, and John Bassette performing.

Another player entering the radio arena was entrepreneur Nick Mileti, who bought WKYC-AM and FM. In September 1972 he announced plans to change the call letters of the AM station to WWWE and the FM station to WWWM. But WERE-AM's general manager Paul Neuhoff objected, saying that the AM call letters were too close to WERE, a premise that Mileti simply brushed off. Neuhoff argued that there was too much potential for confusion among those reporting to the ratings services, but the FCC sided with Mileti. WKYC would soon be known as "3WE."

Some of the programming at WMMS and WNCR was starting to sound painfully familiar. Southern rock had taken off in a big way with the Allman Brothers Band's *At Fillmore East* live double-album, and other bands from south of the Mason-Dixon line like Wet Willie were getting a big push from the record honchos in New York. Bass had had his fill. WMMS built its reputation on being different, and he wasn't convinced that the next big band should be coming out of Alabama. But Sanders told Bass about an album by a British artist that intrigued him and might even break the spell the Dixie musicians were starting to have on radio. The new guy had an androgynous look and did songs by Biff Rose and his own music about people like Andy Warhol and life on other planets. Best of all, this guy was good—really good. Bass had a good feeling about this singer and put him in the rotation. The album was *Hunky Dory*, and the artist's name was David Bowie.

Ziggy Invades Cleveland

Phones began to ring when WMMS started playing David Bowie. Songs such as "Life on Mars?" "Andy Warhol," and "Oh! You Pretty Things" from *Hunky Dory* and "Width of a Circle" from the previous release, *The Man Who Sold the World*, were a welcome change from the Southern rut that a lot of stations were falling into. Quality aside, there was simply a glut of Confederate boogie, and a lot of the classic groups like Quicksilver Messenger Service and the folkies were starting to sound tired. Bowie was a fresh and exciting alternative. A new LP was due shortly, and RCA was aware of WMMS's interest. Bowie's first real U.S. tour, an upcoming eight-city swing across America, would certainly include Cleveland in the schedule.

Upstart WGCL-FM wasn't winning many fans playing taped music and announcers, so it accelerated its plans to put live deejays on the air. On August 21 it dumped automation for a new lineup, although morning drive time would still air tapes for a while. In the coming months, the talent roster would include Truckin' Lenny, who would later find fame in Cleveland radio under his real name, Lynn Tolliver.

The WMMS push behind Bowie paid off with a sold-out Cleveland Music Hall show weeks before his appearance. His new album was called *The Rise and Fall of Ziggy Stardust and the Spiders from Mars*, and rumor was that Bowie would be basing a good part of his on-stage persona on the title character, an imagined glam-rock idol doomed by stardom. Bowie had said in interviews that he expected to be the first rock star assassinated on stage and that he was also bisexual, an uneasy admission for the time. For many people it was one thing to accept gay rights, but it was an entirely new issue to pay to see a homosexual artist. It was anyone's guess how Bowie's stage show would be accepted.

WMMS didn't flinch. It had taken stands before and continued to push for what it thought was in its audience interest. One example was trying to encourage listeners not to watch the upcoming Miss America Pageant, calling it "insulting to the American liberated woman." That was a very bold move, considering the advertising revenue it stood to lose from WKYC-TV, which carried the contest. For a time the station also experimented with eliminating commercials, mentioning sponsors only at the beginning, middle, and end of each hour's set.

David Bowie (1972). The American debut of Ziggy Stardust in Cleveland was a smashing success thanks to support from WMMS.

Opposite: While initial response was lukewarm in cities like Chicago and Boston, Bowie played to more than 23,000 people in three sold-out shows in Cleveland. Photo by Fred Toedtman

The controversy over Bowie's sexuality was small potatoes, and the deejays simply stressed the obvious, that it was his music that really counted. *Ziggy Stardust* was getting attention everywhere, even rising to number four on the WIXY charts, but Bowie's camp was extremely grateful to WMMS, which broke *Hunky Dory* at the same time as Boston's WBCN. Bowie chose Cleveland for a distinct honor. His show here would be his first real concert date in the United States.

There was still a lot of publicity surrounding Bowie's openness about his private life. On the day of his debut in Cleveland, Disc Records ran an ad for his two latest LPs stating, "David Bowie—in town in drag." The show was christened *WMMS Night with David Bowie*, and Billy Bass was named to emcee the event. Concertgoers showed up dressed in everything from the modified hippie uniform of fringe and flairs, to *A Clockwork Orange*–esque chic, with even a few brave souls in drag. The house lights dimmed, the strains of Beethoven's *Ninth Symphony*, "Ode to Joy" filled the air, and Bowie and the Spiders raced on stage bathed in rapid fire strobes. They launched into "Hang on to Yourself" from the new album and were greeted with a huge roar of approval from the crowd. Bowie and the band even looked like aliens, with exaggerated costumes, brightly colored razor-cut hair, and onstage posturing that hinted more of musical theater than a rock show. Every song, from "Space Oddity" to "John, I'm Only Dancing," got a bigger and bigger response. At one point Bowie dropped to his knees simulating oral sex with Mick Ronson's guitar, drawing screams of approval. At the finale, when Bowie sang "gimme your hands" during "Rock and Roll Suicide," people in the audience thrust their arms above their heads creating an eerie sea of humanity. His American debut couldn't have been better. On this night in Cleveland, Bowie owned the world.

Holding court later that night at the Hollenden House Hotel, Bowie, surrounded by his bodyguards, met the press. He was shy, painfully thin, and humbled by the stunning ovation he received. Reporters from local papers and *Rolling Stone* magazine asked him about every aspect of his music, career, and expectations, and as Bowie sipped a glass of wine, he went on with a quiet enthusiasm about what the evening had meant to him. Rock and roll and Cleveland radio would never be the same again.

Strangely enough Bowie's tour was not receiving the same response in other U.S. cities as it did in Cleveland. Some venues, like in Boston and Detroit, did very well, while others drew fairly small crowds. But the response in Cleveland was so huge that Bowie and the Spiders were quickly booked for another show later in the tour and on a much bigger stage. WMMS announced it was bringing Bowie back for a show at Cleveland Public Hall, the city's largest concert venue, on November 25, 1972. Tickets sold out immediately, and there was so much demand for tickets that a third show was added for the very next night. No other artist had ever been brought back for an encore show on his first tour, and this time Bowie would sell nearly 20,000 seats. It was a stunning victory for an act that had been relatively unknown at the beginning of the year and for the station that put itself firmly behind him.

Despite its phenomenal success, there was good deal of risk associated with WMMS promoting Bowie as much as they did. The emerging "glam-rock" scene had not repeated in the United States the success it saw in Europe. This was true for other types of music and bands as well.

One example was the popularity of T. Rex. Its success in Europe was comparable to Beatlemania, but Marc Bolan and the band went flat in America. The first tour was greeted with some enthusiasm, but the second major crossing of the States was met with indifference. The band's appearance at the Allen Theater, with the Doobie Brothers as the opening act, was destined to draw just a few hundred people. Part of it was due to Bolan's apathy to the crowd, not to mention his rumored heavy drug use, and pretty much going through the motions just to get paid. But most likely the biggest reason was that the band's latest release, *The Slider*, didn't have the same intriguing hooks and melodies heard on *Electric Warrior*. There was also an issue with the way T. Rex was marketed by its record company at the time, Reprise. Like Cleveland's own Raspberries, the label pushed T. Rex as a teenybopper band appealing to young adolescents. A lot of it was based on looks and image, and articles about these groups were more likely to be seen in *16* or *Tiger Beat* magazines than in *Rolling Stone*. It didn't matter how good the music was on the albums or how huge the response was in Europe. In America, the "freeks" didn't have time for kids' bands, and that hurt record sales. In retrospect, both the Raspberries and T. Rex have been cited as groundbreaking acts for their time, with artists ranging from Courtney Love to Bruce Springsteen lauding the Raspberries, and everyone from Bowie himself to Guns N' Roses paying homage to Bolan. However, big success in the States at the time eluded them.

Bowie remained extremely grateful and loyal to Cleveland and WMMS. When he appeared on Broadway in *The Elephant Man*, he noticed the *Plain Dealer*'s rock critic Jane Scott in the audience and sent out an assistant to bring her backstage for an after-show dinner. When Bowie was profiled for the BBC production *Dancin' in the Streets,* he praised the city of Cleveland for his early and ongoing success. Whenever his concert tours sold out as soon as they were announced, Bowie would always ask WMMS if the station would like an interview and would personally call back at the appointed time. Years later, when Bowie brought his heavy metal incarnation of Tin Machine to the Agora on Euclid Avenue, he left word to allow backstage access to "Jane Scott, and anyone from WMMS."

Despite the victory with Bowie, there developed much uncertainty about the future of WMMS.

Sold!

A simple fact of life is that radio stations are sold and formats are changed according to the whims of the owners. But the early 1970s was a time of activism, with lessons learned from the massive Civil Rights March on Washington, the bloody riots at the Chicago Democratic Convention, and the slaughter of students at Kent

State University. Protest brought attention, if not immediate results, and the impending takeover of WMMS from Metromedia by Malrite Communications was unsettling to Cleveland's counterculture.

The future of the station was uncertain, even though Malrite's Milton Maltz assured the *Plain Dealer*'s Raymond Hart that the policy of presenting progressive rock would remain the same. Maltz told the paper, "An exhaustive study of the Cleveland broadcast spectrum has been made—with the conclusion that the existing format should not be changed." He went on to say, "We have filed notice of our intention of retaining the present format with the FCC. And we don't anticipate any changes in personnel, and we are going to retain the WMMS-FM call letters." In response, Billy Bass told the paper, "That sounds like good news. The people of Cleveland obviously have responded favorably to what we're doing." But some still had their doubts.

Henry Speeth had been politically aware since he was a child, the product of a family dedicated to public service, and his father was a Cuyahoga County Commissioner. In fact, he was running for State Representative and trying to mobilize the youth vote when he met Martin Perlich, who invited him in to the WMMS studios to review the station's logs. Speeth was an activist, very much a child of the '60s, and after that meeting at 50th and Euclid he launched a campaign to save the WMMS format. He recalls,

> We filed a petition [with the FCC] to deny that transfer of the license, which is a right that every American has who listens to the radio. It was easier than any political thing I ever did, because you don't have to be a registered voter to sign the petition. All you have to do is listen! We'd been trying to get the eighteen-year-old vote through in Ohio, and we were successful in that, but we had this wonderful group of people who just believed. It was a wonderful, wonderful time to be alive!
>
> We knew how to process this place, or that place, how to get in, how to get out. This was the easiest signature campaign of my life, because everyone loved WMMS. The "Zoo News," the astrology forecast. We wanted to preserve it, so we founded what we called The Committee to Deny the Transfer of WHK and WMMS to Malrite Media.

The group gathered tens of thousands of signatures, with hundreds of telegrams, and even the FCC admitted it had never seen anything like it. "The basis," Speeth continued, "was that this was the only hippie channel—the only place that gave alternative news. It was during Vietnam. Our people really don't read the newspaper, and you're denying access to a full segment of the population in Greater Cleveland. If you allow this transfer, you'll kill that station." Speeth also stresses that the concern wasn't with just one person but a dedicated group of people inspired to keep their music on the dial. The fight froze the transfer of the stations for a year. Then Speeth says the group received an invitation to dinner by the folks at Malrite. That group included Speeth; lawyer Michael Thal; the vice president of the county's Young Republicans, Sterling Wagner; the president of the Young Democrats, Mark Price; James Vaughn; Tom Gatz; and people from the *Great Swamp Erie da da Boom.*

Both sides sat down at the Keg and Quarter at Swingo's on Euclid Avenue, where it was stressed that Malrite owned the station, but they were open to suggestions from the petitioners on programming. Speeth says, "I told them they could take WHK and make it as hillbilly as [they] want, but, 'You keep the same format on WMMS, and in a year's time, if you're not making money, then you can change the format.'" Malrite agreed. But when Speeth suggested company officials sign a contract, he remembers them saying they couldn't, "I'd dealt with politicians all my life. I told them I really wanted to see it in writing, but I also promised it would be sealed in my lawyer's safe for a year, and no one would see it." Speeth says that Maltz agreed, and both sides left with an understanding of where the station was headed. Looking back Speeth recalls, "At that dinner, Maltz asked me if I'd like to be a salesman. He was a man of his word. He didn't touch the format. He left WMMS exactly the way it was for one year."

Until the FCC gave its final okay, life would go on as always at Cleveland's FM rock stations. David Spero and Alex Bevan presented *A Folk Special for a Sunday Afternoon,* and a rising local star named Peter Laughner brought his Wolverines to the *Coffeebreak Concert.* Laughner was a rock journalist as well and a contemporary of New York's Lou Reed, but he angered that songwriter with a particularly scathing review of one of his albums. While Cleveland wasn't in the same league as New York in a lot of different ways, it had a longstanding tradition for spotting musical trends, and Laughner was held in high regard as a critic. His comments could have an impact, not only on future sales, but on an artist's credibility as well. WNCR solicited stories from listeners about their strangest dreams, and ten winners would get complete Pink Floyd sets. They presented the stories in a special Renaissance Outlook program, with the disc jockeys reading the most surreal stories to a musical background of Pink Floyd and Amon Duull. Some of the tales were obviously concocted, while others may have been chemically induced, although no one asked about the source.

Billy Bass joined Phil Keaggy again on WVIZ-TV, but for a far different reason than their Glass Harp simulcast earlier that year. It was part of a special and telethon, co-hosted by WEWS-TV's Fred Griffith, called *Time for a Showdown* about the epidemic of sexually transmitted disease (STD) in Northeast Ohio. As many as one in eight Northeast Ohioans suffered from some form of STDs, including gonorrhea and syphilis, and this program helped fund local clinics offering treatment.

At the same time, a grassroots movement was coming to its final resolution. On the final day of October 1972 the FCC gave its approval for the sale of WMMS and WHK-AM from Metromedia of New York to Malrite of Ohio. A closing date would be set in the coming weeks, and Malrite was anxious to move in.

The sale of a station is always unsettling, although the deejays rallied to carry on as best they could. They did their shows like before and continued to make promotional appearances. Though Spero might have thought twice about his runaway ride on a Barnum and Bailey elephant. It weighed two tons, and while Spero tried to give her instructions in English and French, the elephant just kept rambling down Euclid Avenue taking in the sights. When the trainer finally caught up to the two, Spero found out that the elephant only understood German.

In the first week of December 1972 WMMS finally changed ownership. At WMMS Shauna Zurbrugg had made an important career choice and turned in her notice. She had decided to head west for music and writing work. Her final day was December 9. Len Goldberg replaced Joyce Halasa in the lineup, and Denny Sanders manned the afternoon drive time. The biggest change of all would occur in just a couple of weeks.

David Bowie's success in Cleveland opened the doors to a lot of artists who weren't getting widespread attention until Bowie acknowledged their influence. One of them was Lou Reed, the former member of the Velvet Underground, who had always looked at Cleveland as one of the few places outside of New York that understood what his music was all about. The Velvet Underground had enjoyed great success at La Cave, a Cleveland club, and the time had come for Reed to emerge on a much larger scale. WMMS joined with Belkin Productions to promote Reed and opening act Garland Jeffries at Music Hall on December 22, and concertgoers received a special program. Inside the program, the Belkins and WMMS offered holiday greetings. WMMS's read:

> 1972 was the weirdest year of all time for us here a WMMS. Until recently, we never knew whether we were or not—now we are. Thanks to the people of Cleveland, the Committee to Save WMMS, and special thanks to Henry Speeth.
> 1973 will no doubt be equally as weird, so . . . "Freak out on a moonage daydream" . . . Oh Yeah.

The back of the program was designed in part by Spero, and highlighted letters in the "Peace on Earth" greeting spelled out "End War." For some, a different war was about to begin.

Bass did not seem comfortable with the new owners, even though they increased his salary. Sanders suggested that they meet with the incoming management, and on December 28 he sat down with Maltz. Sanders had already told Maltz that KSAN-FM in San Francisco was a good indication of what an FM rock

Hopes for the Future. The program for the Lou Reed show (1972) at the Music Hall held hopes for world peace and that FM rock would continue in Cleveland. Collection of Joyce Halasa

station could do and, while he couldn't predict how soon, was sure it could be a profitable format. He also stressed that Bass was an important music person and Cleveland personality and that all sides should get together.

The meeting did not go well and at times was adversarial. It was clear that the two sides wanted to succeed, but each had different visions of what that success could be and how to achieve it. Bass gave Maltz a list of demands, including being given the title of vice president, and suggested that Maltz accept them or else. He angrily walked out of the meeting and, with Halasa, turned in his resignation. As Sanders recalls, Maltz turned to him and asked, "Is that the guy you think should be running this radio station? I've got a good mind to pull the plug right now!" It was evident that he might have done so with little more provocation.

There were other departures too but not as dramatic as Bass's. After just two weeks at WMMS Jeff Gelb decided to do the same, heading back to his old haunts at WNCR the following week.

Bass told the *Cleveland Press* why he resigned and didn't hold back. "Money isn't the most important thing to me," he said, "The new owners took away my power to make decisions. They took away our engineers and my salesman. We had the best engineers in town. Our production standards were high." Bass continued, "We worked hard all year building a radio station. We attracted a big audience. We made profits. What these people are into is a small town operation and we are a big city. You have to do professional radio." He went on to tell the *Plain Dealer*, "In the past year, WMMS-FM, with the combined efforts of the people of Cleveland and the existing staff, built a station that is not only important in the Cleveland market but important on the national level because music starts in Cleveland. . . . I don't feel the new owners have the amount of professional energy that the existing staff has. Consequently, the situation was unworkable."

Hal Fisher, Malrite's vice president and the new general manager at WMMS and WHK, countered those charges. He responded, "It is unfortunate that Bass lost sight of his goals. Upon acquisition of the station, we consulted with him, gave him a substantial increase, and put him on an incentive program. We instituted a sizable promotion budget, and enlarged the staff of WMMS. We did not impose any musical or creative restrictions on any member of the staff." Fisher continued, "Since our acquisition, Bass has been absent much of the time, and has had substitutes filling his air time. Other times, he has failed to appear at the station to do his program or carry on the day-to-day affairs of the station. When payroll expenditures had to be approved for WMMS earlier this week, he was supposedly in New York without notifying me he was out of town." Then Fisher made a pledge to the public, saying, "We are committed to making WMMS the best progressive rock station in the country," but he also noted, "We are appreciative of Mr. Bass for bringing WMMS into prominence in its format, as it probably is one of the best free-form radio formats in the country today." It was an ugly split. Other deejays were recruited to fill in the airtime with longer shifts.

Sanders witnessed the brief but very volatile exchange between Maltz and Bass, and he didn't want a similar confrontation. He was a student of radio, knew the mechanics of the business a little better than most, and also knew about the other markets where Malrite had stations. Maltz learned that Sanders was aware of radio in markets such as Rochester, St. Paul, and Milwaukee, and the tension eased a great deal after that. The two found common ground and had a long talk about formats and the future of WMMS. Sanders said, "Please just let me handle it, and I'll stabilize things. Just let me take care of it." As Sanders remembers, Maltz looked him in the eye and said, "Just don't lose my license! You got it?" It was at that meeting that the seeds of the future were planted and would bloom for a very long time.

There was some interesting programming going on up the dial at WNCR as the station aired specials showcasing "roots" artists like Bessie Smith, Big Mama Thornton, Lena Horne, Lizzie Miles, and Big Maybelle, with some Janis Joplin thrown in for good measure. It also programmed a seventy-two-hour marathon over the New Year's weekend, featuring jazz, blues, and classical recordings, interspersed with clips from *Howdy Doody*, *Milton Berle*, and *Your Hit Parade*. It was unique, but in truth their audience was not likely to be in any condition on New Year's Eve to truly appreciate that programming. As it turned out, there wouldn't be much more opportunity for experimentation in the near future.

Shake, Rattle—And Roll On

It came without warning—to the staff and the audience. On January 16, 1973, WNCR started a musical transition from progressive rock to a "contemporary" sound. Change was rapid, with a new format expected to be fully in place within just days.

Nationwide's Phil Sheridan made it clear that WNCR was not going the musical route of WIXY, saying the station would be "modified Top-40 ... with a broader reach and larger playlist than most stations." Bob North was named the new program director, and promised that 85 percent of the music played would be album cuts with tight, very structured production. The employment axe swung down on Jim Minard, Norman Moore, the newly married Jeff Gelb, and the rest of the airstaff. Fortunately Gelb bounced right back at WMMS in his old partner Shauna Zurbrugg's late-evening slot. Despite the early attempts to make WNCR different from its sister station WNCI-FM in Columbus, the decision was made to make it very similar, with the two stations sharing a news director and Sheridan taking the general manager's post at both.

News or a feature was now slotted at the same time every half hour, twenty-four hours a day, and music was no longer picked by the disc jockey on duty. Sheridan said it was an attempt to broaden the listening base to include more female listeners. Prior to this the station's strength had been with males aged eighteen to twenty-four.

An ad appeared that Friday in the *Plain Dealer*. It read, "And then there was One. WMMS continues its tradition of serving Cleveland with radio that is real creative stimulating free adult imaginative innovative meaningful unique intelligent progressive and above all entertaining. The One is WMMS 101 Stereo." The jocks at WMMS had gone through their own transition and took no great satisfaction in seeing their friends down the street out of work.

Malrite's John Chaffee came to love WMMS and its format, saying, "I was thirty-two at the time, so it was a thrill to see what you might call its 'incubation' period. Prior to Malrite taking over, WMMS never had much financial support, so it was definitely time for a positive change. At the time, Cleveland was Malrite's biggest market, and there was no way this company was going to lose here. No way." It was obvious that the AM and FM stations needed marketing muscle, and the top management set to work making them as visible as possible. "I got to know Denny, who also did mornings for a time. While he was a good knowledgeable jock, he wasn't a morning man. He was also not a programmer, because he was so easily rattled and sensitive to everything. But frankly, he might not have wanted to be a programmer at that time." Chaffee says that it was obvious there was a need to go in a different direction and find leadership there. The job would be filled in very short order.

Where Do We Turn?

The remaining staff at WMMS desperately wanted to survive while still embracing the basic principles of the station. Denny Sanders and Milton Maltz had agreed to stay the course with the present format, but the future depended on ratings and getting new listeners. One of the more interesting promotions at the time centered on a *Coffeebreak Concert* by a North Ridgeville native who did parody songs. He had released some music on a major label and was anxious to break out in the Cleveland area. The plan was to have listeners enter a drawing for A Dream Date with Martin Mull, where the comedian-singer would escort the winner to dinner at a Royal Castle, continue on to his old stomping grounds for a tour of North Ridgeville, and finish with a scenic drive up I-77. The only requirement was that an entrant had to specify their gender. For some reason Mull needed to know if he was going to escort a boy or a girl.

Despite the departure of Billy Bass, the WMMS staff seemed to appear stable, and while it might not have been evident at the time, a key hiring was about to occur. Another package came from that guy at Cleveland State University, the

Kid Leo Travagliante (1974). Leo trimmed his shoulder-length hair to a more manageable style within about a year of his arrival at WMMS. Photo by Fred Toedtman.

same one who approached Ginger Sutton some time ago, but this time it went to Denny Sanders, who decided to call him in for a chat. On the appointed day, Sanders came out to the lobby to greet a thin, longhaired student wearing a weird purple leather hat. He said his name was Leo.

Lawrence Travagliante got that name from a nun who mistakenly called him "Leo" at the school he attended on the city's southeast side, and it stuck with him. Since he was a sports fan he added the title "Kid" to his name because it made him sound like a boxer. His radio career started at Cleveland State, where he was pursuing a business major. He wasn't that crazy about the Top-40 format at WCSB-FM, the college radio station, and he joined with another student to try and change it into something a bit more credible. At the time, the station was only broadcasting over a PA system to the school cafeteria, but Travagliante thought the music could still use a lot of improvement. When he was comfortable there, he started sending his tapes to WMMS and finally got an audience with Sanders.

Sanders had been in the business for a while and recognized something he liked in Travagliante. The kid in the ugly hat had a lot of enthusiasm. He waved his hands as he spoke, and after the two picked up the stuff Travagliante knocked off Sanders's shelf, Sanders hired him for a weekend slot. On Valentine's Day 1973 Kid Leo's star began its rise at WMMS.

In March the prediction made in 1969 by *Plain Dealer* columnist Karl Burkhardt that FM would make its way into upper ratings within five years started to come true. The latest Arbitron ratings showed three FM stations poking their way into the top ten, led by the "lush sounds" of WDOK-FM, WQAL-FM, and the new kid on the block, WGCL-FM. Prior to this, AM ruled the airwaves, and AM station WGAR did finish second, followed by WHK-AM with its best showing in years just before the old staff was let go. It showed that people were willing to invest in better audio equipment for their cars, where most of the day's radio listening occurred. There were also inexpensive FM adapters, which gave sound in mono but it was a band with no static. It was estimated that 90 percent of Cleveland-area homes now had at least one FM receiver, and so did 40 percent of the cars. That ranked as some of the highest FM penetration in the country. The days of AM dominance were numbered.

Just a few months after walking out of WMMS, Bass and Joyce Halasa sat down with a writer from the free alternative paper *Scene* to explain what led to their departure. The article stated that there were a number of reasons for the split, including conflicts with Malrite over the way that WMMS would be operated and "the pressure of running the station with very little help from the staff." It also claimed that the final straw was a dispute over the promotional budget.

Bass and Halasa said that they wanted to do a billboard and television campaign but couldn't get the funding from the new owners. Bass went on to say, "They took authorization for promotion monies out of my hands. I was no more than a highly paid program director. We wanted billboards with a full color painting on them—nothing but a picture of a lady floating over the city. We wanted to leave them up all winter, and then, in spring, tag them with 'WMMS wishes you Spring,' or something to that effect. It would have been great because people would have been wondering about it all winter." They went on to say their decision to resign finally came after a confrontation over another ad, this one proposed for the *Plain Dealer*.

Scene also reported that there had been talks among staff members about a mass walkout as a means to get their concessions, but when the time came to exit the rest of the staff stayed put, prompting Bass to say, "It really hurt us, they were afraid of losing their jobs."

Commenting about his tenure at WMMS, Bass said, "I was sick of dull, pretentious, hippie bullshit. I wanted to prove that an FM station could make money and be interesting. You don't need to have a Madison Avenue image to exist." The article ended with the lyrics of Lou Reed's "Wagon Wheel": "Yeah, you got to live, yeah, your life / as though you're number one / Yeah, you got to live, yeah, your life / and make a point of having some fun."

WMMS did some fine-tuning on its frequently changing lineup. Those changes included Sanders on mornings, followed by Len Goldberg on middays, David Spero from 2:00 to 6:00, and some new faces rounding out the day. One of them, Steve Lushbaugh on the overnight shift, had previously launched his own pirate radio station on 1610 AM from his bedroom. Tom "Tree" Kelly was named production director and moved to a weekend shift. It wouldn't be long before he followed the others who left for the West Coast.

There was another noticeable change in mornings. As emphasis moved slowly toward the FM band, a lot of morning-drive motorists were still tuned in to stations like WJW-AM and WERE-AM to get traffic, news, and weather even though they had to sit through dreadful music to get that information. Sanders offered an alternative, and in a fairly bold move for a free-form station started giving time, traffic, and temperature along with the news reports. It was the start of a successful move to make WMMS a personality-driven, full-service station.

Spero welcomed some familiar Cleveland names to sit in on his show. The inimitable Doc Nemo had returned for a visit from California, hooked up with Barry Weingart, and the two took over Spero's program, with Nemo waxing nostalgic about Dorothy Fuldheim, the Warrensville Workhouse, and Ernie "Ghoulardi" Anderson. They also played some truly inspired roots music, ranging from folk to roadhouse boogie. It was a walk down memory lane that is still talked about by a very appreciative audience.

Dick "the Wilde Childe" Kemp was still hanging out in Cleveland. He had been replaced as the production director at WMMS a few months before and went to a different kind of station—a gas station in Oberlin that he managed for a time before heading back to WIXY. He took over the evening shift from Jeff McKee, who headed to the warmer Texas climate, and also helped with the station's promotions department.

Another returning "WIXY Superman" (as they were called) was Eric Stevens, the station's former music director. Stevens was just a sixteen-year-old Cleveland Heights High School student when he got his first radio job. He became an intern for Jim Stagg and Jim Runyon at KYW-AM. He would eventually produce *Runyon and Folks*, a show centering on the growing popularity of folk music. The position would lead him to an opportunity to interview the most important new voice on the folk scene, Bob Dylan. Stevens remembers, "I was a little scared. I had all my questions written out, and he would pause, and I would think he lost interest. But he'd start talking again!" His career took off rapidly from there. By the time he was seventeen, Stevens was both production and music director at WIXY. Now he was coming back as the station's program director, but his career would take a drastic turn in the coming years when he would focus again on FM rock.

WNCR was still a player, though what was left of its progressive audience was hesitant to stay. As the station was airing concerts by groups like Cold Blood from the Agora, it was also presenting live sets by middle-of-the-road acts like the Fifth Dimension. Speculation was running high about yet another possible format change.

And a shift in management was coming to WMMS. John Gorman, who had worked with Sanders, wasn't an air personality, but he did have a deep-rooted sense of what kind of programming would work. But stations in Boston weren't biting, and Gorman ended up delivering mail for Boston's Department of Public Works. Gorman recalls,

> I didn't want to be on the air. I wanted to be behind the scenes, and it was my whole goal in life. Well, I knocked on the door of every station in Boston trying to get a job, and it was always part time here and part time there. The time came for me to make a decision. I had an opportunity to take a course with the Commonwealth of Massachusetts, which would lead to this position with the state. I can't even think what the position was, though it had something to do with communications. Probably something with press releases and public relations. Anyway, I had to let them know and it was almost a no-brainer.
>
> At the time, I lived in a cool apartment building off of Harvard Square. Couple of the guys from the J. Geils band lived there, like Magic Dick—and I used to see Peter Wolf in the laundry all the time. I really wanted to be in radio, but I was getting nowhere.

Finally he rang up his old friend Sanders in Cleveland, who admitted that he couldn't have called at a better time.

Sanders had become close to a physical wreck handling the program director's and music director's jobs, along with his air shift, during a critical time for the station. He went to Maltz and stated, "Our music department is a mess! I need someone to handle the music for a station." But he did it in a clever way. They walked over to the record library, which was in a mess with records not filed and scattered about. "Look at all the records we have," announced Sanders, "and it's property that's not being taken care of!" As Sanders recalls,

> You could talk to Milt about aesthetics and ethereal things forever, and it may or may not be resolved. But when you show him goods—concrete goods—in disarray, now you're talking, and it was obvious that something had to be done. I mentioned John Gorman, and said he would be a terrific music director. In the back of my mind, I'm thinking if John works out as music director, I know he would accept the program director's spot.
>
> My feeling was that if I told Milt Maltz that I couldn't handle all this stuff, it was too much work and I'm more of a commodity on the air, something might happen. This is where I wish I had a crystal ball, because the way things would go in a few years, I wish in some ways I would have chucked the on-air gig and stayed as program director. But in those days, an off-air p.d. on an FM station was unusual, because at AM stations you had a budget for such things and we didn't. I was a commodity on air, and made a little more money as a jock than they would have paid if I was only the program director. It still wasn't really good money, because we didn't have the budget, and I wasn't about to take a pay cut! Plus, I wasn't about to leave the daily show because, frankly, I was needed to go back on at night, which was the most critical time period for the station at that time. Remember, in those days, there weren't as many FM tuners in cars, so the real music-oriented impact for stations like ours was when people listened on home stereos in the evening, in those pre-cable and VCR days. It wouldn't make sense to not take advantage of someone with my background in that shift. It seemed like a good test run for John.

This would turn out to be Gorman's first real major job in radio. He walked into the station on July 2, 1973, to become its full-time music director, and within a few weeks he had the department in good shape. It wouldn't be long before Gorman ascended to the program director's spot, and that was also a calculated move on the part of Sanders. He explains, "I eventually went to Milt Maltz and said, 'Here's your program director.' One reason I did it that way was if I went to Maltz and said, 'I don't want to be program director, get yourself a new p.d.,' he would have gotten a new program director, and who knows who he would have gotten! This was a way of trying to keep things together."

Gorman and Malrite's John Chaffee began working very closely and started making important programming decisions, defining WMMS as a voice for Cleveland with local people at the helm. Chaffee says the reasoning was simple:

Malrite was a company that grew out of small towns, and serving its communities. It grew by covering high school football, and local council meetings. The focus was always local. Always. So no matter how big a market was, we wanted to keep it local. We tried to be everywhere you could be locally, or in places you would never expect a station to be! We tried to do things you would never expect us to do. John Gorman and Denny thought that way, too, and even though they were from out of town, they saw Cleveland as a receptive city. It was almost like a "harmonic convergence"—a time and place where the audience, the music, the station, the company, the attitude, fads and changes all just sort of gelled.

Chaffee is among a growing number of people who doubt it could ever happen again like the way it started back in 1973.

Gorman remembers,

The FM was ignored. [Malrite] was putting a tremendous amount of money into WHK-AM. They had changed to a "cover hits" format, which played nothing but hit tunes, but not by the artists who made them famous. They would play "Unchained Melody" by Johnny Mathis, but if they played "Chances Are" it wouldn't be by Mathis, but rather someone like Perry Como or the Everly Brothers. It was a bizarre format that went through a lot of reincarnations in the short time it was in use. WMMS was a complete and total joke at the time.

Chaffee had hired me on the phone. I pictured him as this very straight, very conservative looking person. I was told on my first day that I was going to meet him, and Denny described him as looking like a hippie on *Dragnet*. Sure enough! John was very cordial, and we set about trying to make sense out of this mess on the FM.

I was told WMMS was seven months into a one-year agreement on this particular format, so I had five months to turn the station around. It was like, "Holy shit! What am I going to do?" I just came in from Boston, and I had everything I owned in a U-Haul truck, and this station was so freeform that you could play classical music to jazz to rock. Even though that's a charming way to do radio, it doesn't get ratings and it doesn't get very many listeners because you're all over the road.

In those days, the WMMS studio was the size of a closet, and we had all the albums lining the walls. It was a firetrap! So, WHK has this format where they're playing 300 cover tunes over and over, and we've got all this new music coming in. Why can't we switch studios with them? They don't need all this space, and we did. The WHK studios were three times as big as WMMS, so I thought we should swap studios so we could stretch out and even have enough space to bring in people for interviews. As it was, three people in the 'MMS studio and it was crowded! I went to the chief engineer, a guy who looked like George Gobel, and asked him if we could switch. Here I am, a twenty-three-year-old kid, and he looks at me and starts laughing! He says, "Don't you kids get it? You know what FM means? Find me!" And he laughed some more that I would even ask to switch.

I tried to explain to him that society was changing, and that this music couldn't be played on AM radio. He kept laughing, and said, "It's never going to be anything! The only reason you kids have this FM at all is because some idiot at the FCC came up with this cockamamie idea that we have to have different programming from our AM." I countered by saying, "We try to look at ourselves more as a classical station than a rock station. We're not playing jingles. We're not playing a tight format." He kept laughing. Any time I tried to bring up the cultural change in the world, he'd counter with something like, "Can you buy a car with an FM radio in it?" I started to tell him about FM converters, but he cut me off saying, "How many people are going to special order a $200 radio so they can get FM? How far does the FM signal go?" He wrote off stereo radio as nothing more than something for drug-induced people to listen to. I argued that people listened to this type of radio in their homes. It was part of their life, and besides, our songs were too long for AM radio. Some of them were twenty minutes long! He starts joking, saying, "Can you go into May Company, or Higbee's, or Halle's and buy an FM table radio? How about an FM clock radio?" I tried to explain that people were buying FM components, but he brushed it off as stuff for audiophiles. "You kids have got to realize that AM is where it's at, and neither me nor Mr. Maltz will waste a lot of time on this puny FM radio station!" I was so pissed off that I told Len Goldberg and everyone else about that conversation. I didn't say it to him, because you respected your elders—at least to their face—but I thought, "I'll show you, you bastard!"

The AM stations got priority, and if something broke down on the FM the engineers fixed it when they had the time because WHK was the moneymaker and more important. Most people remember the engineers always called it "the FM," rather than WMMS, and had just as little regard for the disc jockeys.

However, there was a shift under way on the air and on the streets. Sanders remembers,

In 1973, the counterculture was splintering. In hindsight, it was too late to be as eclectic as we'd been. The audience wanted a more rock-oriented service, and didn't get all the interesting little asides we'd been putting on. The folk music. The little pieces of jazz. A classic blues 78 from Bessie Smith. I used to play some Bessie Smith next to Janis Joplin, and the audience just didn't get it. It was too late to be that eclectic, and that's a lesson I learned from Billy [Bass].

He always spun it rock. He allowed Perlich to be eclectic, and he allowed me to do the same. Billy was very Mott the Hoople, David Bowie, and very trendy. When I became program director, that was my mistake. I opened it up too wide, too late.

Clearly the format would have to adapt, and it appeared that WMMS had the right people to do it.

WMMS continued its live broadcasts from the Smiling Dog and other venues, but efforts were made toward a different type of publicity, putting the mushroom logo on the backs of people on the streets, making them into walking billboards. For the growing counterculture, the t-shirt was a uniform for both sexes, and there could be a sense of camaraderie if one person recognized another who listened to their favorite station. The shirts hit the street with a $1.99 price tag, with proceeds going to the Free Clinic. The price was right, the cause was there, and phase one of a very astute marketing campaign was under way.

WMMS also offered a unique prize in its on-air marketing campaign: a "crash pad," a free place to live for nine months, offered by WMMS along with the fabled blues club The Kent Kove. The location was the Silver Meadows Apartments on the outskirts of Kent, and it was the perfect prize for a student, or anyone, looking to relocate for the school year. A Kent State University med student from Bedford who worked two jobs took the prize and eliminated the need to commute to school.

Two other personalities at WMMS got their first taste of publicity that September. Kid Leo's overnight show, airing the "best of the imports" Monday through Friday, and his Saturday 6:00-to-midnight shift got its premiere publicity in Jane Scott's column, as did a reggae show airing Sunday mornings at 1:00 A.M., hosted by the station's production director, Jeff Kinzbach. Both names would enjoy a long relationship with the local press.

Another new arrival made her debut on WMMS on October 29, and that was Debbie Ullman, who took over the morning drive slot from Sanders. Along with his jobs as music director and head of creative services, Sanders assumed the evening 6:00-to-10:00 slot, where he would remain for close to thirteen years. He would be the first on-air staff person to remain in place for an unprecedented run in Cleveland radio history.

Ullman came to WMMS from Hartford, Connecticut. The twenty-five year old seemed very much the "earth mother" type, with a penchant for acoustic music, granola, and a back-to-basics lifestyle. She would get up at 4:00 A.M. to do a half hour of yoga exercises in preparation for her show and said that it gave her energy to get through the morning.

Bass was still very much in the news, moving back into the record business on a much higher level. He was named national promotion director of albums at

RCA, just months after his angry departure from WMMS. It was turning into a fast ride for Bass, who began as a regional R & B promotion man in August 1973, was quickly moved to Dallas as head of regional promotion, and would find himself in a national spot by January 1974. It was becoming obvious that RCA thought a lot of Bass after Bowie's triumphant Cleveland premiere. In fact, Bass was offered a chance to manage Bowie's career. But as radio personality Bill Randle did when Elvis Presley gave him a similar offer, due to the success of Randle promoting Presley including a historic appearance at Brooklyn High School in Cleveland, Bass turned Bowie down in order to concentrate on his own career.

WMMS remained focused in its programming. In addition to continuing live broadcasts, both locally from the Agora, Smiling Dog, and other venues, and internationally from the BBC Concert Series, it also presented rare sides and unreleased music. Radio historian and personality Archie Rothman presented nostalgia shows on Sunday nights at 11:00, when most "freeks" were burned out from weekend partying and were cooling out for the workweek ahead.

Efforts were still being made to ease any continuing fears about WMMS, which could preoccupy the staff in a critical ratings period. In a personal letter to each of the airstaff dated April 12, 1973, Maltz wrote:

> As you know, the ARB Rating period begins today and continues through May 9.
> Although many of you were concerned about the Company's dedication to the format, I trust that is well behind us and that we can proceed forward as the winning team you are.
> Having worked for, built, and owned many stations, I have never met a staff that cares more about its product and its listeners than the people of WMMS.
> You are to be congratulated for your spirit and your growing professionalism.
> Thank you for a great station! The entire staff, John Chaffee, Hal Fisher and I wish you much success.

Donna Halper joined the WMMS team after Gorman heard her show on Cambridge's WCAS-AM, a daytime station near Boston. She did a folk-music show, and Gorman liked what he heard, extending an invitation to Halper to move to Cleveland. Halper recalls,

> I had never been away from the east, having worked my way through college while living at home, and was just getting on my feet financially [paying off those college loans]. I was thrilled to get an offer from WMMS! I had applied to many stations, but back then, you could still say you didn't hire women. The FCC had issued a ruling in 1970 forbidding that, but you can't legislate attitudes, and a lot of stations did not want women on the air.

> The job offer at WMMS paid less than I was making as a teacher in the Boston public schools, which was how I was supporting myself while seeking full time radio work. My WCAS air shift was part time, but I loved every minute of it. I gave up tenure in the Boston Public Schools, and embarked on a journey that would change my entire life. My family was puzzled. My father, God rest his soul, and I never got along, but my mother sort of understood why I had to take the job, although she wished it were not so far from Boston. I put my belongings in a car and drove to Ohio.

Gorman recalls bringing in Halper because the station needed some new blood,

> My philosophy of the station was for everyone to bring in new ideas, and we would combine them into a radical new package. David Spero liked a lot of mainstream stuff, like the Beatles, Eagles, and a lot of stuff that was safe and popular. Len Goldberg was very heavy into jazz and slightly avant garde, while Denny Sanders was into your basic progressive rock. Steve Lushbaugh was into fusion jazz, David Bowie and the glitter rock stuff. Out of all these types of music you would get a station that worked. Plus, WMMS had no budget, and anyone willing to work for a little over minimum wage or the AFTRA minimum was welcomed aboard. Donna came on to help forge all those styles into one cohesive unit.

The FM rock stations kept serving up their brand of music and other items of interest to the "freek" community, but a cold dose of reality hit in early 1974 when WMMS provided a public service by airing the numbers in the military draft lottery as soon as they were pulled. The government had decided to fill its military quota for Vietnam by randomly pulling birthdays from a drum, and that was the order in which young men would be called to service. If someone's birthday was among the first, chances were they would be called for service very soon. David Spero was not only the morning disc jockey reading those numbers as soon as they came off the news wire, he was also at risk for military duty. He recalls,

> I sat there as they were coming across the wire service, reading them, playing songs, and scared out of my wits about it! I was number 312, but I couldn't be happy because I was calling all these other numbers of kids that were going to go over there. I was very naive. I grew up in a wealthy suburb, Beachwood, and kids in Beachwood didn't go to Vietnam. I didn't know kids that really went there, except on the phone lines, and that's where it hit home about what was going on, what's really going on. There were sisters of brothers who were killed saying, "This was his favorite song. Could you play this?" And it happened every day. It was real tough to take. On the radio, we were living a non-reality.

WNCR, which had been limping along with a Top-40 contemporary format and was desperately trying to maintain its credibility with the youth movement, had sent up the white flag. In March 1974 the station abandoned rock for modern country. Looking at the market, it painfully made sense because there was a wide-open audience for that type of music. The staff was put on notice, a new general manager from an FM country outlet in Richmond was brought in, and the station targeted the eighteen-to-thirty-four-year-olds who weren't afraid to say they liked the music coming out of Nashville. The rock field was narrowing, but only for a very short time.

The Long Battle Ahead

WMMS was getting a lot of attention from its peers and the record industry, and that led to one of its greatest moments on air.

In April 1974 John Lennon was going to do some phone interviews with the RKO stations around the country to promote his new *Walls and Bridges* album. John Gorman heard about the plans and knew that many of those RKO outlets were Top-40 stations. Gorman had spoken to Lennon by phone before, so he quickly rang up Capitol Records, saying AOR stations were just as viable as the Top-40s, particularly WMMS. It was a thrill to speak with a former Beatle, but Gorman felt conversations like that belonged on the air. It worked. Cleveland would be the smallest market to get an interview with Lennon. He would be calling Los Angeles, New York, Chicago, San Francisco, Boston, and Cleveland, but there was a lot of behind-the-scenes wrangling even after the interview had been finalized.

Denny Sanders says that interview came down with little notice. He recalls,

> I heard that morning that I would be speaking with John Lennon at two that afternoon. We had one production studio at that time for both the AM and the FM. A few hours before the interview, as soon as I knew it was going to happen, I raced downtown feeling confident I could pull this off and I went to the production director to reserve the studio for that call. I was told in no uncertain terms that WHK production was far more important than whoever the hell I was going to be talking to.
>
> I went to Hal Fisher, and said, "I have John Lennon calling at two o'clock, and the WHK production guys won't give me the studio! I need that studio, Hal! It's John Lennon!"
>
> Fisher's eyes got wide, and he says, "You do? He's one of my favorite actors."
>
> I said, "Not Jack Lemmon. John Lennon."
>
> Fisher got a concerned look on his face and asked, "Who's John Lennon?"

Dumbfounded, Sanders told him, "He's one of the Beatles."

"Didn't they break up?"

Fisher promised to figure out a compromise with the production people. They dug through an old closet and pulled out a tube Magnacord tape recorder, which Sanders says, "looked like something they liberated during World War II when the Allies crossed the Rhine." The only phone patches were in the control rooms and the production studio, so they had to run makeshift lines down the hall, held together with alligator clips, and fed into a crude handheld phone receiver with a mute switch in the handle. The lines went down a well-traveled hallway at the station, but it was the only way to get what promised to be history on the air.

While Sanders was in the studio doing some last minute checks, Lennon rang up the WMMS switchboard. Verdell Warren took the call and started frantically paging Sanders, who couldn't hear the call in production. Gorman finally tracked him down, and Sanders started the interview on his makeshift taping system.

Despite the system used to record the interview, it came off well with a very animated Lennon describing his new LP and post-Beatles life and career. But there was also an occasional click, and Sanders explains, "Those were people stepping on the cords in the hallway. Plus, the handset I was using had a mute button in the handle with a dirty connection, so there was a crackle or squeal on the tape. At one point, right in the middle of the conversation, which I was able to edit out, a fill-in receptionist picked up the line and asked, 'Who are you holding for?' Luckily, it was a pause in a thought and wasn't difficult to splice out." Whoever that person was will likely never know that she interrupted one of the truly memorable interviews ever aired on WMMS.

David Spero continued his very busy life in and out of radio, and sometimes he would have to spend a few days away from WMMS. When that happened the station brought in high-profile guests to host his show. Some of the luminaries included up-and-coming recording artist Michael Stanley, a Hiram College graduate and member of the group The Silk whose name was starting to become a lot better known outside his Cleveland base; Linda Ronstadt; Maria Muldaur; and even former White Panther leader John Sinclair, who was celebrated in a Lennon song after getting a ten-year prison sentence for having two joints. That song helped Sinclair get a new trial and freedom, and he was no less vocal after getting out of prison.

Joe Walsh was having a conversation with Spero one day, when he told him about Stanley. Stanley's father was Stan Gee, a former Cleveland radio personality who was currently a salesman at WHK-AM. Gee would occasionally mention that his kid was doing a record with Walsh, but Spero didn't put two and two together and realize that they were talking about the same guy. Walsh thought Spero could play an important role for his friend.

102   Radio Daze

Within just a few weeks, in March 1974, Spero announced that he would take a leave of absence to manage Stanley's career. Spero told the press, "It's a friendly split. I'm not upset with anything, but I just felt I needed a change now." It would prove to be a very savvy move on his part because Stanley was about to join the Eagles tour as the opening act. It gave Spero management experience on a national level and introduced him to a headliner band whose influence would touch him for decades to come. True to his word, the split was a friendly one, with staff members presenting him with a set of the ten best WMMS memos of 1973 and Kid Leo delivering a cream pie square in Spero's face.

FM radio still didn't pay huge salaries, and Gorman and Sanders were limited in the money they could offer talent. As a result they went back to the station that produced Leo and hired two disc jockeys from Cleveland State's WCSB-FM. Their names were Betty Korvan, who was hired to do Sunday evenings, and Spero's replacement, "Matt the Cat" Lapczynski. Both would settle into long-term employment at the station.

Korvan didn't plan on ending up on the radio. She became interested in philosophy while attending Lourdes Academy, where some friends convinced her to try doing the news at the college station. Korvan was supporting herself by typing bills at a trucking company when she sent in a tape to WMMS.

Matt had a more prolific radio career, sort of. Born in Belgium and migrating to the Cleveland area when his father accepted a position as church organist at Immaculate Heart of Mary Church, Matt was always interested in the medium. While living in Maple Heights during high school, Matt hooked up with a neighbor who ran a highly publicized pirate station out of his home. The station broadcasted on a number of frequencies, did remotes and interviews from the Chanel High School Firebird Festival, operated a request line, held giveaways, and it was even rumored to have been selling commercial time. After the FCC shut the station down with a threat of prosecution, Matt was asked to audition for WNCR. The station declared him to be "too progressive sounding" for its needs.

Cleveland State University's radio station was a sort of farm team for WMMS. Just about everyone sent demo tapes to the commercial stations, but prior to Leo's break he and Matt had approached Lee Andrews at WLYT-FM with a package concept for a type of progressive Top 40, with Leo proposing he do a 6:00-to-midnight shift and Matt doing overnights to 6:00 A.M. Matt remembers it as "low impact programming. An experiment. Nothing that would tie up the middle of their day and very similar to what WIXY had done with Billy Bass, doing more adventurous programming at night aside from the regular Top 40 during the day." They paid a visit to the station and didn't get to see Andrews, but Leo finally did get through on the phone. Andrews seemed interested but was dismissed from the station before he could hire the two on. Leo was the first to get into WMMS, and Sanders reportedly asked him, "Any others like you over at Cleveland State?"

Soon after, Matt got the call and after reassuring Sanders that he wasn't out of control, his big chance finally came. Others followed. WHK's Ed Richards called over to Cleveland State, and Matt, who was still at the station for a time, picked up the phone. Richards needed someone to edit copy, write some news, cart up interviews, and so on. Matt politely declined the offer but asked Richards to stay on the line and called over another broadcast wannabe, Ed "Flash" Ferenc. "I watched him, and his eyes are bugging out and he's getting excited," Matt later recalls. "He took the job." Fate had played another important hand.

WMMS had won a lot of notice from the record industry when it broke David Bowie, and its influence on the Cleveland market was growing rapidly. Two artists in particular were to benefit from that attention in 1974. Of all the imports passing through the station, Leo took a shine to Roxy Music, the band of "new romantics" that had been getting rave reviews in the British press. He talked them up a lot on his *Import Hour* show and, as a result, was instrumental in getting the band signed to a stateside deal with Atlantic Records. Before he left, Spero had been a fan of a New Jersey singer's debut album and said he would play Roxy Music on his show if Leo gave a close listen to Bruce Springsteen. Both artists would have a profound effect on Leo's career.

Singer Bryan Ferry of Roxy Music paid a visit to the studios on May 4, 1974, to sit down for a chat with Leo, who at that time still had fashionably long hair past his shoulders. Leo had a reputation as something of a streetwise "greaser," and very long hair was part of the uniform of the pseudo-hippie community of the day, so he wore it that way for obvious reasons. Ferry got along famously with the WMMS staff, but his visit would have an impact on Leo in particular. Not long after Ferry's visit, Leo came in with a new hairstyle reminiscent of the Roxy Music front man. He realized he didn't have to have long hair to be cool, and it stayed right around that length right up to the present day.

Another artist that broke in the Cleveland area that year was Rush, thanks in great part to Donna Halper. Halper says it all started with a record promoter named Bob Roper:

> He always had a reputation for breaking Canadian music. I'm "spiritually" Canadian anyway, possibly born there in a past life. I'd also been friends with a number of people in the Canadian music industry, having a pretty good history as a music director during college, and I'd also written for ABC Radio for a couple of years and produced a series called *Retro Rock* for the American Contemporary Radio Network. So a lot of promoters knew me, and if I liked a record, I had the program director's ear and could get the record played. They even sent me imports, and one day I get this album on the Moon label, and I admit, it was really ugly. We're talking about a loving-hand-at-home production! But there was a note from Bob Roper at A&M Records Canada, and it

read, "My label passed on these guys because everybody up here is playing nothing but wimpy mellow Jello. I can't sign these guys, but somebody ought to. I hear Cleveland is a rock and roll town, so give them a listen."

I dropped the needle on "Working Man," and the moment I heard it I knew it was a Cleveland record. I took it downstairs to Denny Sanders, he gave it a listen and thought it was a new Led Zeppelin album. I told him it was Rush, and forget the ugly cover, what do you think of them? We listened to "Finding My Way," and "In the Mood"—which we thought was sappy and stupid, but that was okay—it was just a good album! We ended up playing four tracks, and the next thing I know, Peter Schliewen [from Record Revolution] gave me a call and said they were getting requests for the new Zeppelin album. I explained it was Rush and got the number for their managers Dick Wilson and Ray Daniels. At first they couldn't understand why I was calling them, and they couldn't figure out why the record was getting so many requests in Cleveland. They didn't even have a deal in the United States. They shipped twenty-five copies to Record Revolution, and they were sold out in a couple of days.

The next thing I know, Jules Belkin is on the phone and wants to know how to get Rush on a bill! It led to a spot with ZZ Top at the Allen Theater, who got annoyed with Rush because they were getting so many encores. On their first trip down here, they were terrified, because they'd never really been out of Toronto before. Just before the show, Dick Wilson came up to me backstage, put his hand on mine and said, "Don't worry, Donna. We won't let you down." That's how it all shook down.

Suddenly, all the record companies are calling, and it came down to Columbia, Polydor, and Mercury. I didn't want them to go to Columbia, because I thought they'd get lost there. At that point in time, it didn't seem like Polydor had any hits, but I thought the national record promoter for Mercury, Cliff Burnstein, was really sharp and really understood the band. He promised the label would make them a priority.

The rest is rock and roll history.

Yet another significant player moved into position that May, and that was Carl Hirsch, who assumed the general manager's post at WMMS and sister station WHK. He was a native Clevelander, a Kent State University graduate, who did time at WKNT-FM and WERE-AM before heading to Philadelphia and eventually back home. Carl Hirsch would later prove to have a major influence on Cleveland radio.

Popular music was all over the map in mid-1974 as stations and record companies all searched for the next major trend. Punk rock was starting to emerge in England, and heavy metal continued to show growth as well. However the big news at WMMS in June 1974 wasn't as much about music as it was about imaging. It came in the form of a twenty-foot cartoon buzzard in full color sitting atop a mushroom. It was painted on the side of the Music Grotto record store at 24th and Euclid, and WMMS asked listeners to get a look and let the station know what they thought of the new mascot before it went on t-shirts and bumper stickers. Just a few weeks later, the station sponsored a contest offering a listener use of the "Buzzard Car," described as a 1974 Renault, equipped with an electric rollback sunroof and a four-channel quad system in bright orange and decorated with the now-infamous WMMS Buzzard logo. How the bird would become infamous in such a short time is questionable, and it could be argued that Gorman used the concept of "perception is reality" to its fullest advantage. The reality was that the Buzzard would prove to be as important to WMMS as any of its deejays.

Jimmy Perdue, who had been a disc jockey at urban outlet WABQ-AM, got a break at WMMS doing overnights. Perdue had street sense and a good ear for jazz, though his tenure would be for a relatively short time. He would sometimes deliver stream-of-consciousness raps on the freedoms he was allowed on the air, occasionally wandering about the station far off the microphone, making bizarre sounds with ashtrays and anything at hand, talking a little bit to his dog, return to the console to thank the station for the great leverage it allowed him, and then cap off the speech with a very ordinary song that had been played to death in the weeks and months before.

Belkin Productions' World Series of Rock events at Cleveland Municipal Stadium proved to be a tremendous success, with the Crosby, Stills, Nash, and Young show setting a new record for a concert at an enclosed venue, and WMMS took notice. At the show featuring Emerson, Lake, and Palmer and Joe Walsh, the station managed to hang a huge black-on-yellow banner in the stadium. The goodwill and harmony among concertgoers, due in no small part to substances that many were able to smuggle in, and an on-air promotion by the station helped convince another promoter, Bill Graham, to give WMMS sponsorship of the George Harrison concert that would be coming to Cleveland in late fall. Sadly a snowstorm cancelled the show when Harrison's plane was grounded in Chicago, but it wouldn't be the last time that WMMS would swing that kind of clout with big shows coming through town.

A steady parade of important personalities made their return to Cleveland in the coming months. Lou Reed was part of a special record signing at Record Revolution on Coventry Road that drew close to a thousand fans. He was promoting his latest release, *Sally Can't Dance,* and while he was there Reed enjoyed a conversation with his Cleveland counterpart and sometimes rival Peter Laughner. The group moved on to lunch at Nighttown Restaurant on Cedar Road, where Reed schmoozed with members of the local press and WMMS. However, two other arrivals drew almost as much attention. Former WNCR deejay Lynn Doyle, now at KMET-FM in Los Angeles, paid a surprise visit, and soon after he was joined by the inimitable Billy Bass, now a record executive and back on his old stomping grounds to oversee Reed's promotional tour. Bass returned as the conquering hero, now a player on the national scene, and he was destined to return a lot sooner than people thought and at a most unlikely place.

There was some very innovative programming down the dial at WGCL-FM, and it originated with a group of high school kids. It was a Junior Achievement project, a public service show called *The Young Generation,* and one of the people behind it was a seventeen-year-old radio fan named Errol Dengler out of Willowick. It would prove to be one of the most talked-about shows of the year. Dengler says,

> We had on a guy named John Gray, who taught a human resources class at Kent State University. He came on with us and acted as if he was a personnel director at a major corporation, and we interviewed him about how to get a job with his company in the current business environment. I was talking to him on the air, and Gray started making some very snide comments. We had a pretty diverse panel, and Gray dropped things like, "Well, we all know that black people are really lazy." One of the girls who was sitting there was an African American. It started to get a little uneasy.
>
> Pretty soon, he's stirring up all this commotion, and the phones lit up like a Christmas tree. People were asking, "Who is this guy? Where does he get off saying this stuff?" In reality, Gray was trying to demonstrate the dangers of bigotry, and he did a very good job of pretending to be this very biased personnel manager. He insulted every single persuasion—white, black, female—"You have long hair, so you're a hippie. What do you know?" Then he started talking about discrimination and what it was all about.

The amazing thing was, despite his very European features, Gray revealed at the end of the show that he was in reality a black man from South Africa. This came as an incredible shock, especially to the participants in the studio, who could never have guessed his true racial background with the inflammatory comments Gray made to fire up the dialogue. He would later go on to infiltrate and expose the Ku Klux Klan and other hate groups, which could have meant death if they'd have learned his true origin.

It was a real feather in the cap of Dengler, demonstrating an understanding of how to use the media to its fullest potential and generating a huge response to the station's public affairs director, Bill Bailey. He was so impressed by the young broadcaster that he gave him additional work around the station. Dengler would do a lot more in the years to come.

Winds of Change, Winds of War

The year 1975 would prove to be a year of positioning and discovery for the FM rock stations. Len Goldberg resigned his on-air position at WMMS but stayed on to host the Wednesday night broadcasts from the Smiling Dog Saloon. But listeners were having a major effect on the station as well, and in a most disturbing way.

The station admitted that it was getting requests—for disco songs.

Dance music had always been popular, but the current club trend was a major departure from the ranks of tie-dyed hippies that peppered the landscape. Discos were starting to pop up a lot more frequently around Northeast Ohio, and club deejays were nudging some live bands into the unemployment line. Artists like La Belle, Rufus, and the Average White Band were starting to creep onto the WMMS surveys, and while there was some very well-produced dance music, this was still extremely unsettling.

It had taken over clubs across Europe and had a firm foothold in New York at places such as the Hippopotamus and Le Jardin. In Cleveland, it was the Mad Hatter, another club called Anna's Bananas, and She's a Freak, Too, which was in the basement of the Inner Circle Restaurant in Beachwood. It was only a matter of time before disco music infected radio as well. Billy Bass predicted it would have a huge impact on Cleveland. But Bass's own impact was still being felt as well.

During his visit in January 1975, Bass again walked through the doors at 50th and Euclid, this time in his role as RCA's national album promotions manager. Under his arm he held three Gold Records for WMMS, thanking the station for helping break David Bowie in the United States, for which Bass was a driving influence.

Meanwhile that same month, long-held suspicions were coming true as the new owners of WWWM-FM made a huge announcement. Sometime between mid-February and March 1 the station would drop its "easy listening" format in favor of a more contemporary sound, though it still wasn't certain what that programming would be. Automation would be dropped in favor of round-the-clock disc jockeys. Former WIXY publicity director Eric Stevens took that same position at WWWM, along with the operations manager post, so it seemed very likely the new station would air a rock format of some type. Cleveland radio was keeping a very close eye on the new WWWM. But a tragedy would soon occur to a WMMS personality that would have a major effect on that station and, for that matter, Northeast Ohio's radio scene for many years to come.

Morning drive deejay Debbie Ullman had only been on the job a short time when she was involved in a head-on car crash in January 1975. She was very fortunate to survive, but her injuries put her in Robinson Memorial Hospital in Ravenna for what was expected to be at least a couple of months, with a long recuperation

M-105 Bumper Sticker (1975). WWWM's programming would prove to have a far-reaching impact on its rival down the dial, WMMS. Collection of Bill Stallings

period after that. Ullman's jaws were wired shut, and the only way she could communicate for the next three months was by grunting. It gave her an opportunity to do a lot of writing, and thinking, and she was forced to leave WMMS. Looking back, Ullman calls it a period of personal growth that eventually led to an extended trip to Colombia, South America, to recuperate. She reportedly invested in a seventy-acre coffee plantation, with orange and banana trees, and contemplated her future.

The station was forced to find a replacement fast. It quickly recruited a current member of the staff, production director Jeff Kinzbach, who took over until a permanent replacement was named.

It was right around that time that Donna Halper was having doubts about her future at WMMS. She recalls,

> From day one, I knew I had disappointed John Gorman and everyone else at the station. I had short hair—I used to wear it long, but as a teacher had to cut it, which was a small compromise to look more "professional" and make a decent salary while I was seeking radio jobs—and I didn't dress like a hippie chick, and worst of all I was straight, which back then did not mean your sexual orientation. It meant whether you did drugs, and I did not. I never felt more alone in my life, and I was afraid I had come to the wrong place, but I didn't want to lose the chance to do radio, so I stayed.
>
> In that time, I encountered quite a bit of sexism, but mostly I encountered loneliness because I was so different from everyone else there. Back then, Cleveland was mainly a bar town—I don't drink either—and it had few bookstores, few places to go meet people. I've always said I was too hip for the straight people and too straight for the hippies! I loved rock and roll, I was always a nonconformist, a feminist, and also politically active, which made my elders uncomfortable. I didn't fit the drug culture, and I didn't drink and wasn't into

free love, which made the people my own age think I was strange—and maybe I was. I learned to keep to myself.

It was not an easy time for Halper, who remembers general manager Hal Fisher as being very conservative, having a photo of Richard Nixon decorate his office, and who likely didn't care for the FM programming, despite the money it was starting to pull in. When it was discovered that Gorman hired a female music director, Halper was told to type out the logs every day along with her other duties, even though her male predecessor was not required to do the same. They were most likely paid differently, too, and not to Halper's advantage.

WMMS had a dilemma on its hands, and that was how to handle its growing popularity in the key-money demographics, meaning young listeners with expendable cash. The station reserved the right for creative approval on the spots, sometimes rejecting the content and recutting the ads. Carl Hirsch was quoted as saying that as much as 20 percent of the advertising offered to the station was nixed, with 10 percent having to be turned down because of unsatisfactory content, and the other 10 percent because they just didn't have the availability. The station still had to accept some spots because, as Gorman told *Cleveland Magazine*, "There are also laws against refusing advertising, and we don't need a hassle with the FCC right now!" In reality, it was a tough fight turning down some agency spots because they held future revenues from other important advertisers.

One reason that WMMS was so strong in its growing audience niche is that it didn't have any real competition in the progressive format. WMMS was the only freeform game in town. That was to about to change drastically.

At 7:00 P.M. on Tuesday, March 4, 1975, WWWM officially changed formats. It kicked off the new M-105 with the original version of "There's No Business Like Show Business" followed by Emerson, Lake, and Palmer's "Hoedown." Stevens, who was only twenty-seven but already had eleven years in the business, headed the new station. The "wunderkind" had promised an all-album station, heavy on listener requests. It was set up in a "fish bowl"-type studio in the shopping mall at the Park Center complex on East 12th and Chester, so passersby could see the air personalities at work. M-105 went commercial-free in its first weekend, showcasing hour-long blocks of artists such as Bowie, Carly Simon, Elton John, and others on its playlists, and it planned to distribute 10,000 program guides throughout Northeast Ohio every week with concert info and profiles of the station's disc jockeys. It also debuted a weekly live event at the Agora called *Le Disco Electro* every Sunday night, promising "the most unusual show of its kind, featuring a completely new stage set and the ultimate in audio-visual stimulation."

Within a month of its debut, Tom "T. R." Rezny joined the staff. Like many of the others, he was a native of Greater Cleveland, growing up across the street from Doc Nemo in Warrensville Heights. Rezny graduated from Bedford High

School in 1971 and then headed to Bowling Green State University. One of his first orders of business was to string up a wire antenna so he could still pick up FM radio from Cleveland. Rezny got a shift at the campus station, which soon led to other positions around Bowling Green and Toledo before heading back home. He would prove to be an enduring voice on the local scene.

One of those listening very closely to the new station was Malrite Broadcasting's national program director John Chaffee, and what he heard sent up a red flag. He recalls,

> We were still playing Hawkwind and all that kind of stuff, and doing fairly well, but I freaked when I heard classic rock coming out of those speakers twenty-four hours a day. I got hold of John [Gorman], and I said, "I'm going to tell you something. This is dangerous! If we don't start playing some oldies, we're in deep shit! They'll love us, but a lot of listeners will go for the familiar, and we're going to lose some audience."
>
> I did a twenty-four-hour monitoring of M-105 and us. Every day part [disc jockey shift]. Six of them for an hour each, and I put the music list for those hours [on paper] and I labeled them on a sheet as station A and station B. We had a jock meeting, and I passed that list out to everybody. I said, "It's 6:30 in the morning, your alarm has just gone off, and you've got an hour or so to get ready for work. Which station would you want to hear at 6:30 in the morning, A or B?" They voted, and we did that for the entire list. Everyone turned their sheets in, and when we tallied it up 60 percent said M-105! I said, "You! The guys who are running the station would rather listen to your competitor! What does that tell you?" That afternoon, Kid Leo played the Eagles' "Hotel California," and we started playing some older rock too. From that point on, M-105 never got more than a 4 [share of the overall audience].

Chaffee recalls it as a defining moment for WMMS because even though it played basically album rock, it also allowed the disc jockeys to play Top-40-like tunes and put up a wall against anyone hoping to battle the station on those terms. The station had transformed itself from one that started out way on the edge, playing music that was really familiar only to true music aficionados, to clearly understood mainstream rock radio. And for the growing number of people getting FM tuners and converters in their cars, it was a lot more listener-friendly to a much wider audience.

After sifting through hundreds of audition tapes, WMMS was able to kick off its Sunday night *Local Color* show in early 1975, with performances by Peter Laughner and Rocket from the Tombs with Crocus Behemoth. The Rockets had been getting a lot of favorable attention as an art rock-punk outfit, and both these acts would win eventual acclaim as groundbreaking artists from Northeast Ohio.

Also in February, publicity director Gorman was able to speak in a conference call with John Lennon, who assured him that a Beatles reunion was out of the question, though he was back with Yoko Ono. He also had no plans to tour like George Harrison because of prior recording commitments, and he was thinking about producing an album for his old mate Ringo Starr. Lennon urged listeners to write to their congressmen to help keep him in the United States, a long battle from which he would soon emerge victorious. The conference call was an honor shared with eleven other stations but still a coup for WMMS, which aired segments of the talk for several days over the station.

There was another victory for the station, and it started as a rumble that would grow to enormous proportions in the coming years. The station got hold of an advance bootleg copy of a new song by Bruce Springsteen that was due to be on his forthcoming LP. It had been playing a copy of the single, "Born to Run," months before the release of the album of the same name, and perhaps no other song would become so closely associated with a radio station in years to come than that particular tune. In fact, M-105 finally had to threaten to pull all Columbia products from its airwaves unless they got a copy of "Born to Run," too—which they finally did—but WMMS had a big head start playing it. It would become synonymous with the weekend, Kid Leo, and the Buzzard.

The growing popularity of disco was still casting a long shadow over radio, and stations were still trying to decide what place, if any, it had in their programming. When M-105 tried its Sunday night El Disco Electro at the Agora, it died a quick death after just one outing. Space-age-clad dancers writhed to recorded music, with a drummer and two synthesizers adding a live beat. The music was described as "celestial," the club was papered with M-105 banners, and orange and green balls hung from the ceiling. The emcee wore white tie and tails, and most of the people who attended simply watched, and many left. Disco may have worked at the Mad Hatter and Anna's Bananas, but the Agora had a reputation as a club for live music, and disco was just too much a departure for that crowd. It also left M-105 a bit red-faced, having stumbled in its first live promotion. The Sunday night series went back to live music, with bands like Reign, Molkie Cole, and 15–60–75.

Crosstown rival WMMS continued to push bands such as the European techno-group Kraftwerk and a new Boston band, Aerosmith, at Cleveland State University for its first headline tour. But another event would show the true impact of WMMS. For dozens of years, the tiny town of Hinckley had held an annual daylong celebration to welcome back the turkey vultures that, according to legend, returned there every March 15. No one was sure why they came, or for that matter even cared, but it was a great way to get out and celebrate the impending arrival of spring. The turkey vulture is better known by a different name, the buzzard, and WMMS grabbed the opportunity to link its mascot with the event. The week before the event it urged listeners to show up and welcome the birds, not to mention the very popular Murray Saul, but no one—from the police, to the event sponsors, and even to the radio station—expected the kind of response it got.

Tens of thousands of people, maybe as many as 30,000 more than usual, flooded the tiny community, causing safety officials to scramble to contain the crowds. At that time, Hinckley had a two-lane highway heading into the center of town, and a handful of cops with flashlights directing traffic. Before WMMS started talking it up, that's all that was really needed. Hundreds of people left their cars on the side of the road and hiked in, rather than deal with traffic that was stalled for miles. If they were expecting a Woodstock-like free concert, it wasn't in the cards. It was pretty much local folk singers Alex Bevan and John Bassette sitting on a picnic table, strumming guitars but still some very fine music. All was peaceful but very hectic, and the annual sausage-and-pancake breakfast quickly ran out of food. Angry calls were made to the radio station; the deejays were just as surprised as anyone by the turnout, and requests were made by city officials not to promote the event the next year or anytime after that. But if a WMMS promotion could get this kind of response, it was clear that people were listening and willing to support the station, and the time had come to kick it up to the next level.

WMMS had debuted a staple of its promotional campaign, the Buzzard bumper sticker, putting them in stores as far away as Windsor, Ontario. Disco still posed a nagging dilemma for FM programmers. They couldn't deny the popularity of dance music, but they built their reputations on a different type of rock. It was a huge audience, and by March WMMS had its airstaff doing live shifts, Thursday through Sunday nights, at Studio B, the disco at the Picadilly Inn on Euclid Avenue. Saul and Denny Sanders were the first to kick it off on Wednesday nights, and the station also hosted an outdoor disco night hosted by Saul and Matt the Cat at a Record Theater parking lot in North Randall. Studio B would attract luminaries such as Hunter and Ronson; members of Chicago; Tony Defries, with Bowie's Main Man label; and Blood, Sweat, and Tears. It was a trend that many in the music business were keeping a very close eye on. Saul's live "Get Down" raps drew a lot of response at the club. Admission to the Penthouse, which showcased live music at that same location, would allow access into Studio B, and for quite a while long hairs sampling the new type of nightclub would show up expecting the WMMS disc jockeys to play the live version of Lynyrd Skynyrd's "Freebird." It was an interesting mix, especially in the early days.

While disco was gaining in audience, mainstream rock was still the major concern of the FM stations and their audiences. The Beach Boys and Chicago drew more than 35,000 fans to Cleveland Municipal Stadium in late spring, and despite a late afternoon shower, the show was a smashing success. WMMS late-night man Jimmy Perdue hosted it, although it was a greeting by another WMMS staff member that drew the most comments. Spotted at Gate A, quite hard to miss, was

Chaffee, who was decked out in knee-high silver boots, gray leggings, very short shorts, and a red jacket. The thirty-five-year-old Chaffee also had a long mane of silvery brown hair and stood six-foot-seven in those boots. Chaffee was known for his eye-catching apparel, especially at the station. While owner Milton Maltz would usually be seen in a gray or blue business suit, Chaffee's taste in colors and clothing ran to the opposite end of the spectrum.

As WMMS was scoring its victories, there were still some issues that had to be addressed. Production director Kinzbach had done a credible job filling in on mornings, as did some of the other deejays, but a permanent replacement had to be named for that position and the music director's post. Another voice from Boston, Charlie Kendall, came in to fill those spots that June, with Shelly Stile from WCUE-FM signing on to do part-time and weekend work.

Kendall, who had been working in the media since the age of fourteen, arrived from Boston's WVBF-FM and remembers WMMS as being "just an amazing place all the way around. I entered a very creative and productive environment. Oddly enough, at that point we were still the kids in the back room, and I didn't think management cared what we did as long as we didn't get arrested."

Gorman ran the station like a general under siege and assured Kendall that he could even play Iggy and the Stooges in morning drive as long as he had a great time while he was doing it. That was all Kendall needed to hear. "Hey, I took a pay cut from Boston to go to WMMS, but it was actual free-form radio. My old station in Boston started out that way, and then a consultant came in, and all of a sudden it turned to 'top-ten singles and up-album cuts only.' That's when I made the decision to head to Cleveland."

Kendall's arrival brought a new dimension to the station. He was named music director, and having come from a Top-10 station in Boston, tried to infuse that same type of high energy into the very laidback, progressive format. He also felt the rest of the staff tried to calm him down a bit when the energy levels got a bit too high, which seemed to average out nicely. It was a very loose format. "I just basically put sheets on the covers of the albums so that people could write down what they played, what they liked, and some comments about it. A jock might say, 'Hey! Leo likes this one. Let's check it out.' The way we checked it out was to put it on the air and see if we liked it." Kendall also notes, "As far as we were concerned, we had no real competition, and we had no risk. The audience was with us to the point that they would sit through a couple of bad songs because they knew you were going to get to something good. People also felt comfortable enough to call in to say things like, 'Who likes that song? That sucks! He doesn't know it from his ass!' and that was cool. That's how relaxed it was!"

WMMS may have seemed like an endless party on the air to the folks listening at home, but it was an extremely aggressive and tightly run operation. Disc jockey Steve Lushbaugh sums it up,

John Gorman was a general at heart, and it's probably good for the world that he went into radio instead of military school! He was just crazy! Every day there would be full-page memos about killing the competition, the Chimp swinging on his rubber tire when he heard us get a hot new album, and how we stole his bananas from under his nose when we got a concert promotion instead of M-105. When we went out to public places, if anyone from M-105 was there, we weren't allowed to talk to them, and John would shun them. Drinks would be spilled on people when we walked by, or you might be in crowd talking to some record guy, and you'd suddenly flick your cigarette on someone's jacket. It was very hostile to the competition!

Chaffee agrees that the spirit could best be described as "militaristic." "I have to say that John [Gorman] put his imprint on everything as a war. Maybe it got carried too far at times, but everything was a major battle. It was drilled into you every day—the jocks, the marketing people, the sales people, everybody—we were at war with the world, and we would pull out all stops to win."

Just a few weeks after the Chicago-Beach Boys show, the Rolling Stones made their first Cleveland appearance since the 1960s, this time at Cleveland Municipal Stadium. The band had played the Akron Rubber Bowl in 1972, with Stevie Wonder as the opening act, but had steered away from the city in previous tours in part because of the lukewarm reception it got during the British Invasion. But this time the biggest audience to date for that tour, 82,500 fans, jammed the Stadium, and WMMS made its presence known throughout the venue, with some fans even waving homemade WMMS signs on bed sheets. Another Stadium show, with Yes and Joe Walsh, would be greeted with similar enthusiasm.

August 1975 stands as an important landmark date for WMMS with the arrival of Dan Garfinkel, who took over as promotions director. The Pepper Pike native was a familiar face on the downtown scene, had done similar work for the Playhouse Square Association, and brought a new perspective to putting WMMS in the public eye. Garfinkel suggests, "I had some experience that I brought to the table that was probably a little broader [than most promotions people]. There are a lot of people in radio who are 'radio people,' and that's all they know. That's not such a bad thing, especially if you're a programmer, but in my area of marketing it was very valuable to have these other experiences." He also proposed in his first interview with the station's Carl Hirsch that WMMS produce a major free concert. Shortly after his hiring, the wheels started moving to accomplish that goal.

M-105 countered by hiring the former publicist for Joni Mitchell, the Beach Boys, and Glass Harp, Lee Helper, and Belkin Productions brought in Jim Marchysyn to help with their public relations work.

Garfinkel saw the station's marketing potential right from the start, saying everything it accomplished came from a true collaborative effort:

John Gorman was, obviously, a large part of the architecture of that, as well as Denny and Leo. Everybody that passed through the doors. Steve Lushbaugh and Kinzbach, everybody contributed something to the mix. Betty Korvan, Bill Freeman, all of them, and obviously, the ownership. Milt Maltz gave us free reign. There weren't a lot of barriers to doing whatever we thought needed to be done. There was also a level of authenticity. We put the station on the air that we wanted to listen to, and that came through the speakers. We cared about the music, and we cared about the artists, and that's what makes successful marketing. The greatest marketing in the world will not make a crap product successful—at least not for very long.

It was still tough keeping talent at the other stations. Management at WMMS had started a subliminal word-of-mouth campaign that it wasn't cool to listen to any other station and neither were the people who did. As a result, big voices such as J. R. Nelson, Charlie Tuna, Shotgun Tom Kelly, Skip O'Brien, and others who went on to widespread fame at other stations couldn't make a dent against WMMS. The station also became a lot more involved in the dance club scene around this time. By September 1975 Leo was taking requests and playing music like K. C. and the Sunshine Band's "Get Down Tonight" at Jicky's after Dark in the Flats. The building that housed Jicky's was a flatiron-type on Columbus Road and used to be D. Poo's Tool and Die Works, then later Otto's, before going the disco route. In addition WMMS asked listeners their opinions on a proposed two-hour Saturday night disco show. It resulted in a request show called *Disco 101*, hosted by Shelly Stile, that aired from ten to midnight. That genre was getting a major push from the industry, but there was something else happening in rock and roll that would prove to be even bigger than disco, and like with Bowie WMMS had been one of the first to jump on the Springsteen bandwagon. That station's support, especially from Leo, helped sell out Springsteen shows at Cleveland's Allen Theater and Akron's Civic Theater. That mutual loyalty would prove to be a key weapon in the coming years.

M-105 saw the tightening grip WMMS had on the key age groups and conducted an exhaustive research project gauging the musical tastes of Northeast Ohioans. It would be very tough to beat a station with such a head start image-wise, and one of the image problems M-105 had to counter was the depth of music knowledge shared by the WMMS disc jockeys. Every one of them seemed to have some sort of inside information about the artists they played, and Matt and Sanders in particular seemed to have their fingers firmly on the pulse of the industry. Sanders was also the Midwest head of the Progressive Radio Announcers professional policy group within the industry. He was among a group chosen to co-emcee the group's first rock awards, which also boasted critics from the *New York Times* and *Rolling Stone* magazine. When it came to rock and roll credibility, it was pretty tough to go up against competition like that.

Feeling Lucky, Punk?

While the energy of the emerging disco scene was running high, it seemed as though many of the hard rock stalwarts of the early 1970s were moving into a more relaxed style. For example, Eric Clapton produced some of the most memorable rock guitar performances ever, but after his heroin addiction and eventual rehab, he settled into a more laidback style, continuing with his version of Bob Dylan's "Knockin' on Heaven's Door," with which M-105 scored an exclusive premiere. There was still a lot of power and energy in rock and roll, but some key artists of the day were now leaning toward ballads. There was still a large segment of the audience that wasn't ready for disco and needed the release of high-octane rock.

Local bands like Circus, Lover's Lane, and Left End all had an audience niche, and the Michael Stanley Band had taken its version of Cleveland rock to a national stage. However, a band at the Viking Saloon on Chester Avenue would prove to have a major impact on the world rock scene, even though it never lasted more than a few months. At first people thought Rocket from the Tombs was a joke, and members such as Crocus Behemoth and Charlie Weiner were playing it for a laugh. But by the time the group broke up at the Viking Saloon on July 27, 1975, its lineup had proven to be very serious about their craft. Peter Laughner decided to pursue a songwriting career, and Behemoth still planned to write and maybe start his own group. Guitarist Gene O'Connor would join with vocalist Stiv Bator and drummer John Madansky for a new project as well. All would be in the music press in the very near future.

M-105 was changing personnel too, with Joyce Halasa, who did weekend duty there for a very short time, leaving because of health reasons, to be replaced by another WMMS alumni, Gary Edwards. Bill Stallings, the former drummer for longtime favorite Cleveland area club band Freeport, arrived to settle in for a lengthy stay.

WMMS was about to get another indication of its growing stature, when Murray Saul visited the Party in the Park at the Hanna Fountains Mall. These parties were usually Friday night affairs sponsored by the Greater Cleveland Growth Association. Radio stations would cosponsor the events by providing personalities and entertainment, and WMMS set up remote lines for Saul's rantings on-stage. When he delivered his kickoff salute, Saul was greeted by the roar of hundreds of well-oiled fans that came across like a "mini-Woodstock" on the air. The station was definitely doing something right.

Within weeks of their departure from Rockets, O'Connor and Bator added a new rhythm guitarist named Jimmy Zero. Name changes were in store too with O'Connor taking the name "Cheetah Chrome," and Bator briefly flirting with the name "Stivan Scarlett," though he quickly went back to Bator. They were still searching for a name for the new group, with Blitzkrieg a very strong possibility. Eventually, the group debuted as Frankenstein, with "Johnny Blitz" on drums, and "Magnus, the Robot Fighter" on bass.

Mainstream rock was still a very hot commodity, and WMMS scored big with new music from Bruce Springsteen, La Belle, and Greg Allman, though the U.S. premiere of Pink Floyd's *Wish You Were Here* LP on a Sunday morning drew a huge response. The juggernaut continued to roll, but it had yet to grab the top prize in the Arbitron ratings race. By October the station announced it would broadcast in quadraphonic sound around the clock. It kicked off the promotion with a Seven Days of Quad promotion, inviting listeners to hear that format at local audio stores.

The promotion had some obvious drawbacks. There were multiple quad formats, with Discreet Quad needing a separate cartridge to play the specially made and higher-priced albums. The version used by WMMS was called SQ Quad, which was a synthesized format, and the station traded out the equipment from a local audio store. The signal was encoded before it went to the transmitter, and if a listener had the proper receiver or decoder at home, they would get a simulated, four-channel sound. But they also had to have four speakers, and that was an added expense for serious audiophiles. That format created added distortion known as "multipath," which caused a "picket fence" effect to seem as if the station was quickly fading in and out.

While WMMS was continuing to tie in to the very shaky quad craze, M-105 took a page from WMMS's book by getting its deejays out in the streets. On October 18, 1975, the very resilient Tom Rezny was pressed into service for a promotion that only a young, hungry, and extremely fit disc jockey might attempt. It was billed as the Incredible Skate Day, and he was set to roller skate all the way from the State 8 Rollerdome in Cuyahoga Falls to M-105's studios at Park Centre, collecting pledges for the American Cancer Society.

Weeks before the event, the station tied in with J. B. Robinson Jewelers and local schools to collect pledges for the trip. Listeners were invited to join in anywhere along the route at designated stopping points, provided they could bring pledge money with them. Television and newspaper staff showed up for a press conference held at Burke Lakefront Airport a couple of weeks before the event. Since M-105 billed itself as "The Greatest Air Show on Earth," it was only appropriate that Rezny arrive in a helicopter, dressed in a specially designed World War I pilot's gear. Local dignitaries such as prominent African American business leader Arnold Pinkney and Cleveland Indians owner Ted Bonda, all associated with the Cancer Society, showed up to laud Rezny's efforts and cheer him on to success. After receiving proclamations from the cities of Cleveland and Akron, Rezny took the mike to assure the crowd that his initials stood for "terrific roller skater" and that he had been training by eating a special high-protein diet and doing five miles on his wheels every day. The station promised to keep the crowds abreast of Rezny's progress and vowed that he would stay on the road as long as the pledges continued.

Rezny remembers the events leading up to his marathon, "It was Eric's idea, and I did some training beforehand. Even so, there were times that I had to wonder what I'd gotten myself into, and why I was doing it in the first place, especially as I skated through some of the more unsavory neighborhoods in Cleveland." It took a full day, and by the end of the trip he had worn the wheels down to the rims—the rubber had been completely worn away. "The lady from the American Cancer Society took the laces to have them bronzed, and they gave them to me on a plaque. I still have the skates, too!"

It was a forty-mile haul and took about ten hours to complete. It brought in about $4,500 for the Cancer Society, which in 1975 dollars was a pretty good piece of change. For Rezny, it was a high-profile event that would help keep him in the public eye for many years to come.

M-105's promotions department wanted an advertising campaign that would stop traffic, and that's just what they got for pennies on the dollar. The station's bumper stickers started popping up on stop signs around Northeast Ohio, and some were even printed to match the red paint used on those signs. It's not likely that that promotion was sanctioned by upper management, because it still came down to defacing public property, but if a station wanted its call letters seen, stop signs were a prime place to put them.

In the waning days of hippiedom and the continuing drug scene on the streets, new marijuana laws were a big concern among FM rock fans. When the new laws went into effect that November 22, 1975, Saul took to the airwaves of WMMS at midnight with a special report on how they would affect fans of the weed.

Another report by a WMMS personality brought an official end to a special era in radio. At the end of November, Kid Leo wrote a full-page article for the *Plain Dealer*'s "Rock Reverberations" page in the paper's entertainment supplement, *Action Tab*, covering a wide range of topics. They included Cleveland music ("We play national material—if it's good music, it belongs any place in the format."), the 1960s, ("I think the 60s are ending about now. Now we are really starting the 70s. The emphasis is shifting back to entertainment instead of being 'relevant.'"), and Springsteen ("He's the best actor since James Dean. He feels stories and he tells stories. I hope he lasts."). He also expressed concern about Mayor Ralph Perk's leadership and finally said that he would no longer consider WMMS's format progressive rock. Leo wrote, "That's outdated. I call it radio. But I heard a good word in the trades, AOR. That's Album-Oriented Rock. That's a name for the 70s." Progressive rock was dead. Long live AOR—and in radio terms, it would live a relatively long time.

Leo was quickly becoming one of the most recognizable media personalities in Northeast Ohio. He had developed a street-smart, aggressive air-style that appealed to blue collar Cleveland and was expanding from there. But the music industry

The Who (1975). Support for them was intensified by FM radio in Cleveland. Photo by Janet Macoska

can be a volatile, emotional place. That became evident at a post-show party at the Keg and Quarter Hotel honoring the Who after the group's sold-out Richfield Coliseum gig. Betty Korvan had been a longtime fan of the band and did her air shift after seeing their performance in December 1975. Although other stations were invited, WMMS always made it a point to establish its presence and often dominated an event. The Who party was no different. The bar was busy that night as members of the group mingled with local music folks, including original James Gang guitarist Glenn Schwartz and drummer Jim Fox, and Korvan's show was blasting through speakers in the room. Unfortunately Korvan made a critical remark about guitarist Pete Townshend's performance that night, suggesting that he wasn't at his best, which infuriated Townshend. A frantic John Gorman called Korvan on the hotline to advise her to cool the criticism.

Meanwhile, the irrepressible drummer Keith Moon added to his reputation as a lovable lunatic by showing up at the party dressed in a Cook County sheriff's outfit, complete with handcuffs. He drank and talked and generally entertained everyone at the party. Finally, he ended up in a conversation with Leo, who eventually asked about the cuffs. Moon quickly snapped them on Leo, shackling him to a nearby blonde. Then Moon ran out the door with the key, if there was a key, never to return.

Leo was the one person there who was very likely the least entertained by Moon's madcap behavior. But his station was heading into a very busy and, ultimately, prosperous time, and Leo would have plenty to smile about in the weeks, months, and years to follow.

PART TWO
FM ROCK COMES OF AGE
1975–1983

The Pot Begins to Boil

While the air staffs of the FM outlets were living out their rock and roll fantasies, it was still a business and a chess game among station programmers. The popularity of WMMS continued to spread quickly and to a great degree by word of mouth. The station even started getting fan letters from as far away as Alaska, where a pipeline worker received five hours a week of WMMS programming sent on cassettes by his brother back in Cleveland. His fellow crewmembers became fans as well and started inquiring about WMMS merchandise.

But listeners in Northeast Ohio were the lifeblood of local radio, and the ratings were showing continued strength for the FM band. WGAR-AM's morning man John Lanigan got a lot of attention very soon after his arrival in Cleveland, and in early 1976 he ruled the morning drive race, though WHK-AM's Gary Dee and WERE-AM's new "all news" format were posing strong challenges. Dee, in particular, was a major player who was a pioneer in "shock talk" radio, making the comments heard on *All in the Family* sound like a Saturday morning cartoon. He insulted listeners, used ethnic stereotypes, and even staged his own death. Plus, his headline-grabbing marriage to television personality Liz Richards gave him the kind of publicity other shows could only dream about. Even still, WMMS came in a respectable fourth, showing its growing influence on the ratings.

WMMS started doing some very innovative programming with its *Instant Radio Spectaculars,* which included recorded major concerts and unreleased recorded work from leading artists such as David Bowie and the Spiders from Mars, in their last concert together where they were joined on stage by Jeff Beck; Bob Dylan and Mike Bloomfield at the Newport Folk Festival from 1965; and a rough mix of John Lennon's appearance at Madison Square Garden, circa 1972. It was a prime opportunity to hear historic recordings that weren't likely to ever appear on record, though plenty of tape machines were capturing those sounds around Northeast Ohio. The station had such pull that it could solicit suggestions for what it might play in the future, and if they didn't have it, chances were that a few strings could be pulled to get that music on the air. The station was even able to use its promotional influence to get a facsimile of the statue used on the cover of the Led Zeppelin LP *Presence* for an on-air giveaway. It was numbered 999 out of the 1,000 available, and even though the album drew lukewarm response, it proved to be a very popular promotion for WMMS.

WMMS offered promotions of other kinds too, staging "Smokestock," a free show at Cleveland's Edgewater Park on Lake Erie. This was the concert Dan Garfinkel earlier had proposed to Carl Hirsch in his first interview the year before. Hoping to relive the success of similar live shows, including a WNCR birthday party at the same site, WMMS offered a slate of acts featuring the Pure Prairie League, the Michael Stanley Band, Angel, former Montrose vocalist Sammy Hagar, and a special appearance by one of the shining lights of Woodstock, Joe Cocker.

The funny thing about this concert is that the city of Cleveland granted a permit for the show with the understanding that it would remain orderly and fans would mobilize to clean the park afterwards. All went well until Cocker took the stage, did a couple of songs, and then projectile-vomited on the stage. Garfinkel recalls a rider in Cocker's performance contract stipulating placement of a fifty-gallon garbage can behind the speaker stacks, based on Cocker's similar behavior at many of his shows. Cocker left the stage and the concert ended early, but true to their words, a group of concertgoers cleaned up after the crowd.

Cleveland radio was getting a lot of national attention for breaking new artists at that time. After the success of Bowie and Bruce Springsteen, among many others, bands were chomping at the bit to get a break in Northeast Ohio. Cleveland may have been getting its share of bad press, but the music scene had fans everywhere such as Artful Dodger. The Washington, D.C.–based band played Cleveland like a second home and even thanked Kid Leo on one of its LPs. WMMS rival M-105 gave the band a substantial push as well. That station was making significant inroads against WMMS, but both stations were content to push ahead creatively and, for the most part, ignored each other on the air, at least for a time. But that would change.

There were also changes happening in the music itself. The emerging punk scene was getting a firm foothold with relatively little airplay on commercial radio. Rock poet Patti Smith's appearance at Cleveland's Agora rock club in January 1976 won over David Spero, Leo, and John Gorman, with musician and journalist Peter Laughner calling her the "logical consequence of the verbosity and possibilities that have been dormant since the 60s." Smith became a fan of the city, telling Murray Saul on his WMMS *Jabberwocky Show* that, "[I] could feel Cleveland before I got here. I felt like I was home." She even sat in to do a disc jockey's shift during Leo's afternoon drive show and later paid tribute to her two sold-out Agora shows by releasing a live version of "My Generation," recorded on stage with the Velvet Underground's John Cale.

Even with that reaction and the relative freedom WMMS deejays had in picking songs for their shows, punk was so different from much of the music of the day that it posed a problem for programmers. Disco was still showing plenty of muscle, but so was punk. Cleveland band Frankenstein had changed its name to Dead Boys and became a favorite at the New York club CBGB's, riding the wave of popularity generated by a furious stage show and a new single called "Sonic Reducer," destined to become a classic of the punk genre. Dead Boys shared stages with such darlings of the New York punk scene as Television and the Ramones and were quickly becoming one of Cleveland's favorite musical exports. Dead

Boy Jimmy Zero would later say, "Outside of New York and L.A. and London, Cleveland was one of the very first cities to embrace punk. We had a lot of good musicians here, and the scene was hot, too." Despite the notoriety of the Dead Boys and the growing importance of punk, that type of music was still having a hard time getting a foothold on Cleveland radio outside of the college stations. A good part of it had to do with radio formats, but it was also an angry form of art, and such a radical departure from other '70s music that it scared a lot of programmers. It was very difficult to find a proper niche for punk.

But the local concert scene always seemed to have a great influence on Cleveland radio and vice versa. A few weeks after Smith's concert, Paul McCartney and Wings announced an international concert tour for the upcoming spring that had M-105 sponsoring a show at the Richfield Coliseum in May. It was McCartney's first U.S. concert tour since 1967 and a definite coup for M-105.

And just eighteen months after debuting at the Agora, Springsteen's career had skyrocketed, thanks in no small part to the early support he received from WMMS and especially Leo. Jim Marchysyn with Belkin Productions, knowing the history between Springsteen and WMMS, came to the station with a gift-wrapped package of concert tickets addressed to Leo, who was sworn not to mention anything about a concert until his show went on the air that afternoon. Leo read the announcement cold; Springsteen was booked to do two shows at Cleveland's Allen Theater. The first show sold out in just thirty minutes—a tribute to the power of "The Boss" (Springsteen's nickname) but also of WMMS, which wielded a huge influence on its audience. The second show sold out in about ninety minutes, setting a small-hall record for ticket sales.

After those shows, Springsteen paid a visit to WMMS for his first studio meeting with the man who helped break him in Cleveland. Leo greeted Springsteen by saying, "You have given the greaser back his respect, and I thank you for it." Springsteen had a lot to thank Leo for as well, joking with band member "Miami Steve" Van Zandt about his success in Northeast Ohio, saying, "You didn't double park my Rolls Royce in front of the station, did you?" It was Leo's first interview with Springsteen, and topics ranged from the formation of the E Street Band, to his early days, to his favorite music. At one point Leo asked, "Did you ever want to chuck the whole thing?"

Springsteen replied, "I've never had that choice. After thirteen, I never had the ability to do anything else. I think if you can quit, you should. I never had that option." It was a surprising answer, but for his many fans it was fortunate he never made that choice. It was a memorable visit, and after he was off the air Springsteen hung out with the staff. Deejay Matt the Cat, who had an eye for video and was often seen lurking around the station with his Super 8 movie camera, captured Springsteen at the studios that day, even footage of him just leaning against a car outside the building as people passed by without a second look.

But big changes were in store at WMMS, although at the time no one realized how much of an impact they would have on the station and the audience. On September 2, 1976, WMMS morning man and music director Charlie Kendall was supposed to be one of the prize catches at a promotion at the rock club Agora that was raffling off well-known Cleveland men to its female clientele, but it didn't happen. Kendall had other things on his mind as he bid an emotional goodbye to his morning audience, announcing that he would be leaving WMMS for KZEW-FM in Dallas.

It was a great move to a bigger market for Kendall, but it left a major gap in the station's air assault team. Steve Lushbaugh, who had been with the station for some time, asked for a shot at the morning slot and had the support of many of the staff. As he recalls, "We made that pitch, but I think the company was kind of against it. They wanted somebody with a big voice, and said as much when I spoke with them about that job. But we convinced them I could do the job, so they gave me a crack at morning drive."

It would prove to be a lot tougher than Lushbaugh originally thought:

John Chaffee gave me a three-page memo with formatic rules, saying in no uncertain terms that I would say the temperature three times every half hour, the call letters, and so on—and it was like they put me in a tight little box. I almost felt like I was set up to fail. Sure, I was green and never did mornings before, but I had some comedic talents and did some voices. I did the show with Flash for awhile, and as luck would have it, one day I started having problems with the equipment or started talking about the money being bad; I'm not sure what it was. But I remember saying if we could invent something good, we could make a lot of money and retire. It started an ongoing series of bits about things we were inventing in the studio.

Now by the third week, my dad had told me this joke, and he was a really straight guy. Never told me a dirty joke in his life, except this one time. He said, "You know what a pussy stretcher is?"

I couldn't believe he said that, and I just kind of stammered a little and said, "Uh, no." He told me, "That's what they use at the animal hospital to wheel the cats around."

So, the next day, "Hey, I got another idea, Flash," and told him the joke. Well, Flash gets all flustered, and I asked, "Wait a minute. What are you thinking?" I did the punch line, and I figured it was over. The next week, it's, "I got an idea, Flash. This didn't work, that didn't work, the pussy stretcher didn't work. How about this?" Turns out the g.m., Gil Rosenwald, heard it and wrote me an angry memo. He was not happy. He writes, "How dare you say that? That's a violation of the National Association of Broadcasters Code, and not in keeping with the Malrite tradition!" Gil was hot.

WMMS Staff (1976). (Clockwise from lower left) Betty Korvan, Jimmy Perdue, John Gorman, Kid Leo, Jeff Kinzbach (center), "Matt the Cat" Lapczynski, Steve Lushbaugh, Ed Ferenc, and Denny Sanders. Photo by Fred Toedtman

Looking back, what I should have done is gone into his office right away, face to face, and said, "Gil. Come on. We got Murray on the air talking about drugs and ass and tits and nipples every Friday night for eight minutes. I do a double entendre thing, and you're ripping me." But I didn't. I put it in a memo and wrote, "I'm surprised you would use the NAB code as justification for trying to censor me. What I did was just a quick joke!" Next morning, there's a message in my mailbox from John Chaffee, and it wasn't pretty. He said I shouldn't address my general manager in those terms and even suggested I was damn lucky to still be on the air. Well, guess what? A couple of days later my luck ran out. They made me production director, and Jeff Kinzbach—who was probably their first choice—was named the new morning man. So, basically we switched jobs, which in the end, turned out to be the best thing that ever happened to me.

But it would be some time before the switch would prove to be beneficial to the station.

Transitions

Whoever succeeded in the Cleveland radio market had to understand the audience and the Midwest, blue-collar mindset held by most of the listeners. WMMS was looking to fill the overnight slot, and luckily it didn't have to look far. Program director John Gorman's Boston connection was still very much alive, and he recalled an incident some years before as he was driving down Boston's Storrow Drive. As he told *Record World* magazine:

> Storrow Drive . . . it's like four miles of dead man's curve all the way, and I punched up 'BCN, but somehow I hit the wrong button and I went past 'BCN and landed on this station WVBF. It had just come on the air. I hear this guy coming on. Well, I nearly drove off the road. I was hearing "Funhouse" by the Stooges at like 6:15 on Friday night. I ended up listening to that station the entire night, and I heard rock and roll that I hadn't heard for a long time. . . . I was saying, "Goddammit, it can be done! This guy is doing it!"

Gorman was listening to Bill Freeman, who called his show *The BLF Bash*. Gorman put the call out to Freeman and then anxiously awaited his arrival. There are plenty of radio legends about Freeman's coming to Cleveland, but this is the way Gorman continues the story:

> I had an FM converter on my radio with a dial on it, and if I went over a bump I would lose the signal. All of a sudden, I'm hearing Crosby, Stills, Nash, and Young's live version of "Southern Man" from *Four Way Street*. I'm blown away! The album had just come out, and Steven Stills and Neil Young are trading

Bill Freeman (1976). Direct, brutally honest, and with a wry sense of humor, Freeman called his *B.L.F. Bash* show "the American listening sensation."

these guitar solos on one of the greatest live recordings ever. Hey, maybe WBCN has a clue here! Then I heard Bash's voice. That's when I realized it wasn't WBCN, but it wasn't like anything I'd heard. Right after the Stooges, he plays "Ramblin' Rose" and "Kick Out the Jams" by the MC5! I'm hearing this great music and this guy with a gravelly voice calling himself "BLF Bash," and I couldn't figure out what "BLF Bash" meant! They were a Top-40 station in Framingham, but they broke every rule. The station had a Top-40, high-energy feel and still played some great music. Unfortunately, it was a short-lived format.

By the time I got to downtown Boston, which from Harvard Square to Beacon Hill is a pretty good drive, I had made up my mind. Somehow, some way, it can happen. It took another year, but I ended up going to Cleveland. One of the things I did when we had openings was to try to hire this guy. I couldn't get through to him at the station, but I did get hold of Charlie Kendall, so that's how he ended up coming on board before Bash. But then I heard Bash was available, too. We were able to get hold of Bash, and we brought him for a concert. Bash didn't like to fly, so he came in on a Greyhound bus, and when he arrived at the station he had a small nylon bag. I asked, "Where's your luggage?" and he said, "This is it!" It was a six-pack.

Freeman was quite a unique character. He was a heavy reader, shunning away television, and loved trains so much that he would spend his vacations at the Union Pacific rail yards in North Platte, Nebraska. He would wear the same style of clothes year in and year out: jeans, sneakers, a blue work shirt, a baseball cap, and a toothpick. Freeman's reputation thrived on such eccentricities. Gorman has called him "the greatest all-night talent in the history of the world."

Freeman took the mike in September 1976, initially working the 2:00-to-6:00-A.M. shift. The *Cleveland Plain Dealer* noted that he had worked in Mobile, Fargo, Tucson, and Indianapolis, and he liked sports and beer. He told the *Cleveland Press* that he intended to play the songs "people want to hear," saying, "I've often wondered if the reason for the CB [citizens band radio] boom is because the people can't get what they want on the air." Bash went on to say, "I graduated high school in 1960, and went from the surfing scene to hiding out in the mountains until I got a draft notice." When he went to the service, he recalled, "The barracks upstairs was into rock and roll, and the downstairs was country-western. We were all into radio.... I think being an R&R deejay is second only to having a rock and roll band. I am still going to be the world's oldest rock and roll drummer."

He then took aim at commercial radio of the 1970s, saying that stations had fallen into a "Let's Make a Deal" syndrome, with lots of giveaways to attract listeners, and non-rock-and-roll programmers had turned their stations into "Ford [Motor]-type production[s]." Freeman first heard WMMS as he was passing through Cleveland en route to the Dakotas and Montana from a Rolling Stones concert. He liked the station's attitude, far more than his subsequent job at WVBF-FM, which the paper said he compared to a "top 30 pig." "If you want to hear good rock and roll," he said, "you are not going to hear it on a Top 40 station." Freeman would prove to be a truly distinctive performer, and each evening he would carefully arrange a series of worn scraps of paper along the console. No one knew what was written on those scraps, and no one cared. Freeman was an artist, and the airwaves were his canvas. A new masterpiece was created every night.

Throughout his tenure at the station, Freeman was always trying new things; sometimes controversial, but always creative. The Associated Press once reported a hoax from the Russian news bureau that a UFO had landed in Red Square. That night Freeman went on the air to "interview" the aliens telepathically. He asked, "Hey! If you guys are really aliens, why don't you make the Terminal Tower disappear! [pause] Whoa! We'll be right back!"

He was also known to editorialize the news he reported. Freeman was asked to do a brief headline segment at the top of each hour, ideally not lasting more than two or three minutes, but they would occasionally stretch to more than fifteen.

He would announce, "Alright, it's time now for headlines from the WMMS newsroom, and then we'll throw them in the 'Bash Bucket' where they belong." One broadcast had Freeman report, "Oh, look, BLF fans. They caught General Manuel Noriega, and they're gonna put him on trial. Just wait until he gets on the stand! Then let's see who runs for cover [long pause] George Bush!"

Later, during the war in the Persian Gulf, Freeman discussed plans to go there and launch a radio show called *Ali Babla Bash*. He said that when he had their confidence, he was going to show the country how to really win the war with him as their leader. He went on to say he would next set his sights on Northeast Ohio, where he would carry out his political agenda of trying the Cleveland Indians for treason.

His opinions didn't just extend to world events. Once after a McDonald's spot by the Texas Tornadoes had aired, Freeman opened the mike to say, "You know, BLF fans, it's a damn shame that the best music on the station is in a commercial!" Such was the stuff he would come out with on his show.

Freeman's freewheeling attitude didn't just exist on the air. When the official staff photos were being taken, Freeman wanted his shot at the Greyhound bus station. When that idea was nixed, he told the photographer to meet him a half-hour before his air shift in an alley behind the Statler Office Tower, where there was an old couch. He told the photographer to work quickly, especially since he didn't want his picture taken in the first place. That night, a fundraiser was being held for Democratic presidential candidate Michael Dukakis, and Freeman noticed men with rifles on the Statler roof. When he was told why they were there, he launched into a political diatribe that even the Secret Service men were hesitant to interrupt. Freeman wouldn't stop talking for the photographer, so the photo shows his toothpick as a blur.

Freeman didn't want any special treatment, never went to the staff parties, and didn't mind working holidays. All he expected was a paycheck. When WMMS was moving from the Statler to the Skylight Office Tower, the morning show started giving away stuff on air that wasn't going to make the move. Freeman opened his locker and unloaded every pin and giveaway he had received from the station over the past fifteen years.

The addition of Freeman wasn't the only WMMS programming move. By year's end Steve Lushbaugh was officially named as the new production director and Jeff Kinzbach was put in the hot seat in morning drive, and he would stay in that slot for many years to come. Dan Garfinkel did his movie reviews, Murray Saul popped in with an occasional comment, and Ed "Flash" Ferenc, who was getting more airtime since his arrival from Cleveland State radio, did news. Meanwhile, music director Shelly Stile was waging a campaign as chairperson of a group called GIPAL, which stood for "Get Iggy Pop a Label." Detroit rocker Pop had a European release but didn't have a deal here in the United States. Stile was a fan and would soon see her efforts pay off in a big way.

Kinzbach was perfectly happy doing production duties and admits that the switch to being on-air was a bit uncomfortable because he and Lushbaugh were so close. He recalls,

> There were no rules, and I could play just about anything I wanted. The way I approached the show was to look for the kind of stuff I'd want to wake up to, with everything from the Doobie Brothers to Santana and everything in between. The bottom line was to have a good time. Basically, we were like the first or second cup of coffee for people. We wanted to wake them up easily, and give them a soundtrack for a typical day in Northeast Ohio. We were doing the same thing our listeners were. We partied and had a good time, and tried to reflect the average Cleveland person of those days. The whole basic concept of WMMS was to create a radio station that we would want to listen to, and hopefully, everybody would want to listen to a station like that.

By the beginning of 1977 Saul and his listeners were starting to get weary of the "Get Down" routine, and he decided to do his final weekend salute at a station promotion from the Cleveland Convention Center. It was broadcasted live via phone lines, and the assembled crowd gave a huge ovation. Saul continued *Jabberwocky* on weekends, as well as the *We the People* segments three times a week. That same winter, WMMS started to promote Kinzbach and Ferenc more extensively as a team, sending the morning show with Garfinkel on highly visible promotions and getting their names in print too.

It was near this time that the WMMS Buzzard made its television debut in a thirty-second cartoon television commercial by the artist who created the bird, David Helton, with a soundtrack produced by Kinzbach, and the clip was an instant attention-getter. Television would prove to be a very important weapon for WMMS as it waged war on its competitors. But it wasn't all competition, as FM rock radio had come a long way since the Martin Perlich days on classical station WCLV-FM, and there still remained a great spirit of cooperation between WMMS and WCLV, even though they were competitors. For example, broadcasts of the program *New World of Jazz* from the Agora expanded to include Saturdays on WCLV and the same show on WMMS the following night. It wouldn't be the last time the two stations would show this type of cooperation.

By this time it was clear to everyone that WMMS and its sister station, WHK-AM, had outgrown their facilities on 50th and Euclid. There was more action farther downtown, and it was a lot more accessible to news events and promotions. Plans were made to pick up stakes and move the base of operations to the old Statler Hotel, now the Statler Office Tower, at 12th and Euclid, and it happened the morning of February 13, 1977, but not without incident. Relocating a radio station can be a lot like moving a graveyard. It can be done, but it's quite a job. Tons and tons of equipment, records, and history have to be moved while not affecting the on-air sound. That Sunday morning police cars raced to the WMMS studios after a frantic listener reported that the station was playing Gary Wright's "Dream Weaver" over and over again. It was quickly explained by the on-air deejay that engineers had put the station on maintenance control for three hours so they could test the equipment. Incidentally, there was a way the deejays could signal if there was trouble in the studio. If someone from the station heard the Beatles song "Help" on the air, it meant that there was an emergency of some sort and that someone should call the authorities quick. It was the one song a listener could never request to be played on the air.

The spacious new offices were far more fitting of a corporate headquarters, covering nearly the whole twelfth floor, with eleven of the forty-one rooms devoted to WMMS. It was a two-million-dollar renovation, including new equipment, furniture, and custom-carpeted walls. It was a bit sad to move from a historic location like the WHK building, but the new offices had plenty of history too. The Statler Hotel housed hundreds of dignitaries over the years and was even mentioned in the Marx Brothers' film *Animal Crackers*. Baseball great Ty Cobb had stabbed a doorman there, and just one flight up from WMMS had been the longtime home of WGAR radio. When the Orson Welles version of *War of the Worlds* was first aired in 1938 on WGAR, it was a young board operator named Jack Paar who was on duty at the Statler Hotel location. He took the show off the air when panicked listeners called in but was ordered to put it back on by an angry program director. Coincidentally, just on the previous Halloween night, the *Buzzard Theater of the Air* staged a parody version of H. G. Wells's *War of the Worlds*, followed by Orson Welles's radio production from 1938.

The Statler had a rock-and-roll history too as the home of WGAR's sister station, WNCR, and Jimi Hendrix stopped in its basement club, Otto's Grotto, during his one appearance in Cleveland. A lot more history would be made at that location.

It was a Herculean effort, moving more than fifteen thousand albums and box upon box of promotional material and FCC documents. It took the whole staff the better part of a weekend to get it together, but the move was a success, and a lot of other folks benefited from it too. The staff uncovered a few surprises, such as a rare photo of Lou Reed smiling, and hundreds of posters, photos, and other memorabilia that had been lost for years. All of that plus a framed Buzzard animation cel from the recent cartoon commercial were donated to the Radio and Television Council of Greater Cleveland. The station offered a clearance sale of sorts at Daffy Dan's t-shirt shop to get rid of outdated shirts, windbreakers, and halter-tops.

The WMMS morning show inaugurated the station's new home by playing Southside Johnny's "We're Having a Party."

Bowie Goes Pop

Even before his triumphant North American debut in Cleveland, David Bowie had made it known that he was greatly influenced by Lou Reed and Iggy Pop, and rumors began to circulate that Bowie was about to thank Pop in a big way. "The Thin White Duke" (Bowie) had achieved superstardom practically overnight, working with everyone from John Lennon to Bing Crosby, and word was that he would be touring with Pop's band but only as a sideman. Bowie had met up with Pop at New York's legendary rock bar Max's Kansas City and agreed to produce an LP that would eventually be titled *The Idiot*.

The date for Pop's Cleveland show was set for Monday, March 21, 1977, at the Agora, and tickets were snapped up in minutes. A second show was hastily added for the next night, and after another quick sell out, fans anxiously awaited the return of Bowie in an intimate concert setting, even if he was just going to sit behind a keyboard.

Pop was just as anxious to return to the Cleveland area. He'd played the city a number of times, with varying degrees of success, including a date at the Allen Theater in June 1974. A maid found him passed out in a nearby hotel room; he was revived and did three songs before stumbling off the stage. This tour was supposed to reestablish his musical credibility, as well as win new friends.

One of his best friends, David White, was the son of a Shaker Heights insurance man. Pop met White while a student at the University of Michigan at Ann Arbor. White went on to study for his doctorate at Harvard. There was another Shaker connection too, and her name was Wendy Weisberg. They had met through friends, and when she visited Pop after a concert visit to Ohio, they quickly got married. But it soon became apparent to them both that at only nineteen Pop was too young and way too ambitious to make the marriage work, so it was annulled after only two weeks, but Pop retained fond memories of Weisberg.

Electricity filled the Agora at the first show on Monday, as celebrities such as Todd Rundgren and various members of the Runaways mingled with fans who didn't know what to expect but were anxious for it to start. Blondie opened the show with a taste of music from New York's Bowery scene, and soon after it was time to see what Pop could do. An explosion of light filled the stage to the opening chords of "TV Eye," and true to his word Bowie sat behind a small keyboard, chain smoking, and happily playing along with his friend and mentor. The group also featured Hunt Sales on drums and his brother Tony on bass (their father was slapstick comedian Soupy Sales, and the brothers would later show up in Bowie's hard rock outfit, Tin Machine). The crowd was extremely receptive, with only a few calls for Bowie to play his own hits, and by show's end Pop had won a lot of new fans in Cleveland.

What a lot of people don't know is that Bowie had a keen interest in breaking the Pop band in Northeast Ohio, and part of it centered on the weekly broadcast of Agora shows on WMMS. The Agora's Hank LoConti remembers the concert well:

> When the show was over, Bowie went upstairs to the studio, and that must have been about 11:30, 12 o'clock at night when he got there. He actually engineered the show himself. That was on a Monday night, and the show was going to air on WMMS on Wednesday. He had two security guards—one was asleep on the sofa as you walked in the door, and the other one was in the studio with him.
>
> I left there about one o'clock in the morning and came back the next day around nine. They were still working. He worked until about ten A.M. that day.

What many didn't realize was that Bowie himself produced the show that aired on WMMS.

Early the following Wednesday, Pop, Bowie, Blondie, and the rest of their traveling road show piled into station wagons and vans and headed to their next gig at the Columbus Agora. Years later, Bowie would fondly remember everyone mooning each other from their cars as they drove down I-271.

Buzzards, Penguins, and All Kinds of Animals

The move to the Statler Office Tower put WMMS a lot closer to concert venues, advertisers, and the city's political base at City Hall, which was a break for the news department it shared with sister station WHK-AM. Public demand brought Murray Saul back to kick off the weekend but not with his famous "Get Down" tag, which he thought had become passé.

It was right around the same time as the move that Jeff Kinzbach started to introduce into his morning show some things that would become and remain very popular. Kinzbach says, "All you ever heard from other jocks was joke after joke after joke, and most of them were stupid. I came up with an idea for the 'Token Jokes of the Morning,' and the name, of course, could be taken two ways. That was the whole idea anyway. Listeners would send in jokes, and we'd send them a t-shirt if we read them on the air." A WMMS t-shirt was a very hot commodity in those days, almost like currency. Kinzbach adds, "Listeners were hearing some of the best jokes and loved hearing them on the air. Plus the audience benefited, so it was like the listeners doing part of my show for me."

For another feature, Kinzbach was inspired by a lot of the mob-related bombings happening around Northeast Ohio in 1977, recalling, "[Mobster] Danny Greene, and so many others. One morning I came on the air and said [to the listeners], 'Jeez!

They're blowing up this, blowing up that. If you had a chance, what would you blow up?' Man! I was getting calls for weeks and that kicked off, 'Blow Something Up!' Everybody gets ripped off. Everybody gets stabbed in the back, and raked over the coals, and this was their chance to vent! The worst story of the day got to press the button." The popular syndicated *SCTV* television series at that time featured a recurring skit where two hillbilly brothers, played by John Candy and Joe Flaherty, would blow up celebrities—punctuated with the phrase "blow 'em up real good." Explosions seemed to have been a normal part of pop culture in the late 1970s, a stark contrast to the sensitivity that developed in the 1990s after the Oklahoma City bombing and in even more recent years following the World Trade Center attack. Even *The Muppet Show* featured a "mad bomber" character named Crazy Harry, played by puppeteer Jerry Nelson, who would pop in at any moment with TNT plunger in hand, and Ernie "Ghoulardi" Anderson was well remembered for decades for blowing up various items while hosting movies on WJW-TV's *Shock Theater*. But Kinzbach says he did not copy such bits and that his came directly from newspaper headlines.

Then there were the Morning Mind Exercises, in which Kinzbach would ask daily trivia questions about any number of topics, with promotions director Dan Garfinkel offering his own trivia contest at 8:45 every Monday. All of these new segments turned into cornerstones of the morning show that would remain until Kinzbach's very last day.

Kinzbach and company were gathering strength, ratings-wise, but the spring 1977 ratings showed that only one FM station was in the top five in the all-important morning-drive race, and that was Tom Armstrong at WDOK-FM, who came in third. Malrite's talker Gary Dee on WHK was still king of the hill.

In other developments, WMMS had scored big by aligning itself with the Belkin Productions' World Series of Rock shows at Cleveland Stadium. The concerts had embraced the spirit of the giant Woodstock festival of 1969 with good vibes and a true sense of community among the tens of thousands who would come out to the shows.

However this was not the case on June 25, 1977, when the British powerhouse Pink Floyd played to more than 81,000 fans. The problems began outside the stadium, where a reported four thousand angry people, who couldn't get into the sold out concert, clashed with Cleveland police and security guards for nearly two hours. It started when the gates were locked shortly before the concert, and a crowd started throwing rocks and bottles near the Memorial Shoreway. Two cops were injured, more than sixty people were arrested, and the resulting publicity brought the stadium shows under intense scrutiny. More than a hundred patrolmen and thirteen mounted police responded to the melee. Meanwhile, Pink Floyd's performance inside the stadium that night was described as "inspired" and drew dazzling reviews from fans and critics alike. The show set two records: one for the largest crowd ever at the stadium and an international record for the highest number of tickets sold for one act. Still the clash with angry music fans was a black mark on the day. Their behavior was inexcusable, and it was fortunate that no one was more seriously injured. Ironically Pink Floyd was promoting its *Animals* LP, a title that might have seemed appropriate considering the incident.

Pink Floyd wasn't the only highlight that summer. Fleetwood Mac was riding a huge crest of popularity with its *Rumours* LP, which was shattering sales records in the United States and around the globe, and WMMS got the brass ring by sponsoring the group's two concerts at the Richfield Coliseum. Both shows offered tickets by mail order only and sold out quickly. Fleetwood Mac had made its mark as one of the most popular recording acts in the world, and wheels were turning at WMMS to capitalize promotionally on the band's visit to Northeast Ohio.

But there was a lot of activity in the halls of WMMS in the weeks prior to the Fleetwood Mac show, with some folks saying goodbye and some new arrivals. Saul left the station, saying it was like, "the closing of a chapter. They were the most wonderful four years of my life." His voice was missed, but Saul remained a popular fixture on the local music scene and went on to do publicity for the Numbers Band. Longtime Cleveland favorite Alex Bevan and the Buzzard Band, with deejay Matt the Cat on guitar, premiered a new song titled "Ho for the Weekend," which the station described as "Murray Saul with strings." Bevan had also penned a song that would become his signature song for decades to come, singing about a "Skinny Little Boy (from Cleveland Ohio)." It became a staple of the WMMS playlist. And twenty-year-old Al "the Bear" Koski made his debut as a weekend and fill-in disc jockey for WMMS.

As the Fleetwood Mac shows approached, the folks at WMMS were struck by inspiration concerning the band's logo and how the city, and station, could ultimately benefit. On September 26, the day of its second Richfield show, WMMS, Fleetwood Mac, and its label, Warner Brothers, joined to announce that a penguin was being purchased by the band to arrive at the Metroparks Zoo the following March. The group's John and Christine McVie—along with Henry, a penguin shipped in for the event from Sea World in nearby Aurora—made the announcement. It got plenty of print and television coverage for everyone involved, and the radio station didn't have to spend a dime.

Another promotional stunt involving the animal kingdom had far more sinister overtones. Boomtown Rats had released their debut album on Mercury Records, and the head of album rock promotion contacted WMMS's John Gorman about a promotion, that might catch the eye of a station's music director, involving large laboratory rats packed in formaldehyde and sealed in transparent plastic bags. Gorman recalls:

> [Mercury Records's head of rock promotion] almost lost his job. "You're going to send what to these program directors to play Boomtown Rats?" He asked

me to write a letter to his boss and say I wouldn't be offended if that happened, which I did. Then I asked what he was going to do with the rats. He said, "I don't know. I think we're going to pitch them," and I asked him to send me a box of them.

When we got the box of rats in, Bash walked by and asked, "What's that? Hey, can I have one of those?" Somewhere along the line—and I have to be honest—it wasn't my idea, and I can't think who came up with it, but somebody said, "You ought to send some of those over to M-105." There was a rumor that Eric [Stevens, M-105's program director] was very germ conscious.

The trick was thought to be so devious that Gorman was a bit hesitant to follow through on the suggestion, but it was finally decided to pack up the box to look like an early, exclusive release from a major artist and record label. It is still not known what Stevens's reaction was, or if it even got to him. Gorman continues,

> These were the days we were both battling. Eric was part of it in the Sixties when stations fought to get the new Beach Boys or Beatles. This was the same thing, but instead of the Beach Boys or the Beatles, it was Queen and Led Zeppelin. We knew there would be something coming out, like a new Eagles or Linda Ronstadt, and we put the return address of the distributor, and rushed it over there through Bonnie Speed or something. To be honest with you, I have no idea what the story was once it got there. I heard all the legends about people screaming, but I don't believe any of them. I'm sure the music director or someone else opened it before it got to Eric Stevens.
>
> Keep in mind that that kind of thing happened only a few times. I've heard the stories that we were doing it for years, but only two or three times tops.

By November, Shelly Stile had resigned her post as music director at WMMS, and Kid Leo assumed the role. There was a wide variety of music to choose from, and the WMMS playlist ran from Elvis Costello to Blue Oyster Cult and just about everything in between. WMMS was also starting to play a song that would become as much a standard for the station as Springsteen's "Born to Run." Bill Freeman hosted a program called *After Hours Joint* that ran from Saturday midnight to 6:00 A.M. Sunday. It was designed as a soundtrack for the ride home from the bars, and he started playing a song by Funkadelic at the same time every week. It was "Maggot Brain," and it became a longstanding tradition of the WMMS weekend fare.

Exciting things were happening on the local music scene too. New wave and punk continued to make inroads. Dead Boys had been signed by Sire Records and were still playing Cleveland-area gigs, like the Garfield Heights Rec Center, the Phantasy club in Lakewood, and the Rock Factory in Akron. If one was to walk into Hideo's DiscoDrome on Cedar Hill, there was a very good chance that a band like Devo would be playing to a handful of customers. Disco was still getting more airplay than punk, but the latter was gaining influence.

Sadly, a very influential personality on the local scene had passed into legend. Peter Laughner had been hospitalized frequently for liver problems and was only twenty-four years old when he died in his sleep at his parents' home in Bay Village. With a knack for spotting music trends, he had helped start bands such as Rocket from the Tombs (later to be known as Pere Ubu) and had helped establish Cleveland's punk scene at a time when few cities embraced that culture. Laughner's influence would impact the local scene for years to come.

The Start of a Dynasty

Despite the competition, it was becoming clear that WMMS was claiming a huge cross section of listeners in the Greater Cleveland market. Ratings continued to increase, and the morning show anchored by Jeff Kinzbach and Ed "Flash" Ferenc was becoming increasingly popular. T-shirts were spotted all over Northeast Ohio saying, I WAKE UP WITH JEFF & FLASH . . . AND GO TO BED WITH BETTY KORVAN! a subliminal suggestion to turn one's clock radio to the station. It was showing results, with the morning show posting a significant ratings push in the fall Arbitron ratings, putting it just behind the number-one-rated Gary Dee program on WHK-AM just across the hall. There was good reason for that.

Besides the popularity of the format and music mix, WMMS had capitalized on its reputation as a groundbreaking station and a music insider. There was also a good mix of personalities in the morning and a number of daily features that would become morning drive traditions, such as the Free Ticket Window and Mr. R.'s Astrology Report. Ferenc had stayed on to do news after Steve Lushbaugh's switch to production, and he and Kinzbach proved to be a popular pair.

WMMS wisely moved the show into a "full service" direction, realizing that frequent time, traffic, and weather updates would attract a bigger audience that might otherwise look for that information on others stations. The station's jocks were getting television attention too, with Kinzbach and Korvan exchanging cream pies on the *Hoolihan and Big Chuck* Friday night horror show after Kinzbach reportedly made certain "disparaging" remarks about Korvan's athletic abilities. This chapter of the Korvan-Kinzbach rivalry ended with Korvan winning the televised pie fight.

Perhaps Denny Sanders, the station's early evening host, had best summed up the station's success when he told the *Plain Dealer Sunday Magazine,* "Let's face it, we're a commercial station. But we're commercial art, and we attempt to put some flair into what we do." The paper also pointed out the heavy emphasis on personality and the emergence of the so-called Buzzard Nuclear Army, which banded

I Wake Up with Jeff & Flash (1977). This t-shirt graphic urged listeners to put the Buzzard on their clock radios as well as in their cars. David Helton

listeners together with the common goal of promoting the city and the station. WMMS was ranked number one with listeners twelve to sixty-four, with more than half of the eighteen-to-thirty-four male market favoring the station. Thirty-six percent of the female audience also rated the station as its favorite in that same age bracket, and despite the trend toward stricter programming, WMMS continued to do otherwise. The paper quoted program director John Gorman taking aim at M-105 and similar stations, saying that they had become "nothing more than automated juke boxes. There's as much soul in that kind of format, as much flesh and blood as there is in a piece of tin."

Sanders went on to say, "Not only do you have to play good music, but you have to express some kind of positive life force. As a person broadcasting, you must make an impression. That conveyance of distinction is imperative, or else you haven't left your mark on the world, and what was your life for?"

In conclusion, Gorman summed it up: "We're trying to appeal to a very large demographic area—anywhere from teens to people in their mid-to-late thirties.

Buzzard Nuclear Army Logo (1976). The station's icon took on many forms over the years, including this Buck Rogers–type symbol, for the way it planned to symbolically "zap" its competitors. David Helton

And we're trying to make everything as palatable as we can but still remain creative. We're exposing everybody to everything."

Plain Dealer writer Joe Crea pointed out an extremely important fact in the article: Because of the numbers and influence WMMS had, its deejays set trends and created a "Cleveland style," unique to the city and its people. The station set the musical tone of Northeast Ohio and, as a result, tightened its grip on the Cleveland market. The station's influence was so great, and growing, that the national trade paper *Radio and Records* did a huge spread on it in early 1978, interviewing Gorman, Kid Leo, and some of the management.

The other stations didn't just throw their hands up in the air and give up. M-105 had a respectable following and brought back a familiar name to strengthen

Betty Korvan (1977). Pulling in impressive numbers for her nighttime show, Korvan would become one of the most popular female disc jockeys in the country. Photo by Janet Macoska

its lineup, David Spero. Spero had recently been married and had just left the Michael Stanley Band, figuring that life on the road was not something he needed at this point in his life. He says, "Michael was playing hundreds of dates a year, and we were gone all the time. Plus, I really missed radio."

After a brief stint on middays, Spero kicked off a new morning show on M-105 in February 1978 with a load of celebrity interviews, including Peter Frampton, Eric Carmen, and Eddie Money, and plenty of bits too. Among his initial promotions Spero offered to wake listeners up in person, broadcast his 6:00-to-10:00 show live from their houses, bring them breakfast, and have a chauffeur standing by to whisk them off to work. The first winner in the Good Morning, David contest was a suburban Middleburg Heights housewife, and public television station WVIZ-TV stopped by to tape part of the visit since M-105 offered a breakfast remote for the station's annual fundraising auction. Spero also featured Cleveland finance director Joe Tegreene doing concert and music reviews. Tegreene was a very visible member of Mayor Dennis Kucinich's inner circle, and it didn't hurt the station to have a friend in City Hall. Plus, Tegreene truly enjoyed music. Doug Johnson, a former Solon resident and Dartmouth College graduate, was brought in from a Tampa, Florida, station to replace Spero on the abandoned midday shift.

But album oriented rock was still king in Northeast Ohio, and WMMS and M-105 battled for that huge audience. In fact, at WMMS Korvan pulled huge numbers at night, not an easy feat considering that radio traditionally doesn't put the same promotional push behind evening shows, and with an 11.7 rating in the winter Arbitron ratings she could boast of being the top-rated female disc jockey in America. WMMS itself was having great success raising funds for the Cleveland Ballet through a special commemorative t-shirt, showing that people were happy to embrace just about anything WMMS promoted.

The station retained its major impact on the local concert scene and, as noted in the *Cleveland Press,* probably staged more rock shows than any other station in the world. From smaller venues like the Agoras and Cleveland Music Hall, all the way through Public Hall, Blossom Music Center, and the Richfield Coliseum, WMMS had its hand in helping bring some of the biggest acts in the business to Cleveland. It had forged strong partnerships with Belkin Productions, the Agora's Hank Lo Conti, and record companies in helping break major acts in the city. As far as WMMS was concerned, every show it was linked to could boast what Leo called the "Cleveland seal of approval . . . our call letters," and plenty of those shows carried that seal.

Obviously M-105 was up against a ruthless competitor in WMMS and especially Gorman. Today Gorman laughs off much of that war, saying that it was never really directed at anyone in particular and was simply an aggressive business atmosphere. In fact the two men thought to be "blood enemies" in those days have since met socially on several occasions and are now great friends. But back in 1978, the staff at WMMS lived by the slogan, "We don't go to work. We go to war!" and a look at memos passed down to the staff in those years would definitely point to a bunker mentality.

Just about anyone who meets Gorman now would like the man. He has a quick sense of humor, knows music and radio, and is an affable and outgoing sort of person. At the same time many people who worked with him, and especially those at competing stations, describe him as a fierce and unrelenting perfectionist who wasn't about to let anyone stand in the way of his mission to dominate the Northeast Ohio airwaves. And his style of programming left countless victims in its wake.

Gorman's memos to the staff at the time outline a super-aggressive and very often mean-spirited approach to competition. They were often several pages long, and sometimes came within just a day or two of each other. M-105 as a station, and program director Eric Stevens in particular, were always referred to on the air and in the halls as "the Chimp," and Gorman seemed to know every move they made. A WMMS staff memo dated February 3, 1978, was addressed to "The Buzzard Nuclear Army . . . at War & Winning!" and titled "Chopped Up Chimps, Pulverized Pigs & Blown Out Brillo Pads! Mass Murder on the Streets of Smokestack City." "Brillo pad" was a reference to Stevens's hairstyle, which Gorman just loved to mock. Gorman writes, in part, "Have you monitored 105? They are sounding better now than any other time. I've heard some excellent forgotten oldies, and I think it would be worth your while to give them a good serious listen. Never underestimate the enemy. Always be on guard for attack!" The memo ends with "Kill! Kill! Kill!" and if there's anything that could be said about Gorman's style of management, it certainly was inspiring—and extremely graphic.

Another memo came just a few days later, titled "Buzzards Conquer the Universe! Chimps Die and Brillo Pads are on the Endangered Scum List! Shoot to Kill! Could Anyone Withstand the Buzzard Nuclear Barrage?? No!!! Another WMMS Chimp Crushing Weekend!!" Under the heading of "Propaganda," Gorman writes,

> The Buzzard Nuclear Army [BNA] successfully gained control of the press in the city. The shock waves from last Sunday's *PD* magazine article are quickly rupturing the banana cream foundation of that eyesore at 40th and Euclid. . . . The Chimp would have broken his neck (if he had one) from falling off the tire upon seeing the special Super Buzzard *PD Sunday* story. In fact, every other radio station in Smokestack City is on the rag about the current BNA takeover of the papers. Today's *Friday* magazine is again filled with more BNA propaganda. . . . The BNA pulverization is just beginning. Wait 'til we release our biggest, most deadly BNA nuclear attack: The Tenth Anniversary!! Special bumper stickers, decals, buttons, t-shirts, and lighters will be flooding the already Buzzard flooded Smokestack City market shortly! David Helton's 10th anniversary logo is another BNA masterpiece!

Under the heading of "Buzzard Investigative Squad Reports," WMMS spies reported,

> The Chimp is working on what it calls "a gala anniversary special" for 105 in March, which it claims will include free concerts. . . . At yesterday's Radio & TV Council luncheon, Brillo Pad was spotted walking out the door (like a slobbering German Shepherd) with [Cleveland mayor] Dennis Kucinich!!!! We've gotta speed up our takeover of City Hall. The Buzzard Investigative Squad is on the case. The next attack: Death to anyone who tries to get to City Hall before we do!!!! Die, Chimp, Die!!!

It was seen as critical to Gorman that WMMS debut new music by established artists before any other station in town, and in many cases WMMS beat out stations all across the country. Later on in that same memo, Gorman writes about the station's premiere of Jefferson Starship's *Earth* LP:

> As it turns out, WMMS was the first station in the US of A with the album (even Bonnie Simmons of KSAN didn't get it 'til late yesterday.) We had a 24 hour exclusive on it. The Chimp finally got a hold of a copy late yesterday afternoon. By the way, that so called exclusive Springsteen cut Spero had yesterday turned out to be nothing more than a bootlegged live cut of only fair quality. It was a tactical error on the part of the Chimp. We are now guaranteed an exclusive on the new Springsteen LP, which according to the CBS branch, could be out as early as mid-March!! Die, Chimp, Die!!

There were many who thought Machiavelli could have gotten pointers from some of the WMMS management, if he'd had the opportunity.

Gorman says about those memos, "It was just the humor of the time. I've kept every memo throughout my entire history at WMMS, and it was that bizarre type of *National Lampoon* humor. Remember, those were the days of 'Buy this magazine, or we'll kill this dog.' They were never meant to be taken literally."

The WMMS staff shared in every victory and pored over every detail in Gorman's memos. One came under the heading of "Die, Chimp, Die! You Had to See the Look on that Fat Pig's Face," stating,

> Tuesday night, Joe Tegreene, sitting in for the Florida vacationing Dennis Kucinich, gave the Babys the keys to the city at the Bond Court hotel. Although no one remembers exactly what the line said in Kucinich's proclamation, read by Tegreene . . . it had to do with WMMS being the best station of its kind in the world. Not bad! Hizzoner is listening to the Buzzard these days . . . in fact, his younger sister (17) was there, and she is a Grade A WMMS fan all the way!!!!! You can fight city hall . . . but you can't fuck with the Buzzard!

It went on to promise that, "David Helton is working his fingers down to the bone [on] the Buzzard News . . . a double one with a complete full-length story on the history of the Chimp." Helton's *Buzzard News* was an in-house publication, similar to an underground comic book. His drawings were reminiscent of some of that genre's better-known counterculture cartoonists, and usually every bit as graphic.

Later that month WMMS was poised to strike again, this time with an exclusive concert announcement that would stun its competitors. A confidential memo dated March 10, 1978, states,

> This is top secret (in other words . . . don't mention on air or in the company of those other than family)! As you know, Aerosmith managed to make Grade A

assholes out of themselves this past year. First we had all [the] cancellations, then when they finally hit town, their show was so bad that they ended up with "Worst Concert of the Year" in our WMMS–*PD* Music Poll.

Well, they blew it across the country, too! The Kings & Queens LP didn't come close to projected sales & the kids on the street are starting to call them "has beens!"

Lieber–Krebs, their management, is about to attempt to change their image via having the band play a number of small hall dates in key cities . . . and Cleveland is one of them. Originally, no one station was to be involved in this promotion, but thanks to Leo, Walt, Diana & Belkin . . . the show is now 101% ours!!!! Never underestimate the power of threatening one's life!

The date is April 5th (the day before the book begins) [the "book" is industry jargon for the published results of an Arbitron ratings period] and the place is Music Hall. We will announce the show on Friday, March 17th, between 9 and 10 PM. The deal is that we must do an Aerosmith "special," then announce the show. What we'll do is run the old King Biscuit Aerosmith Concert, then make the Music Hall concert announcement . . . DON'T BREATHE A WORD OF THIS ON OR OFF THE AIR 'TILL THEN!!!!

Such was the power of the Buzzard.

The never-ending battle for Northeast Ohio raged on, with Gorman preparing his "troops" in a memo, IT'S TUNE UP & BLOW THE BASTARDS OUT OF THE MARKET WEEKEND! SHOW NO MERCY! SHOOT TO KILL! LET'S CLEAN UP CLEVELAND! REMOVE THE SCUM: CHIMP, PIG, BRILLO PAD, SLIME TWINS & BLOW RAINES!!!!! THEY DON'T DESERVE TO LIVE!!!!! KILL! KILL! KILL! THE BUZZARD NUKES RULE! THE APRIL-MAY ARBATTLE IS CREEPING UP FAST. THREE WEEKS BEFORE THE FULL SCALE WAR ERUPTS! WE WILL LIVE WITH THE RESULTS OF THIS WAR THROUGH THE END OF THE YEAR. . . . IF YOU THOUGHT LAST APRIL-MAY WAS A DIRTY FIGHT . . . JUST WAIT 'TILL THIS ONE! It stated, "We may laugh at that no-necked basket case Chimpoid's showing in the Jan. Feb . . . but those assholes at 105 are fighting for their survival, and after spending a lot of time monitoring 'em this week, a word to the wise is DON'T WRITE 'EM OFF YET! Their sound has improved a great deal this week (especially at night!) . . . we cannot afford to slack off! . . . After giving 'em a good long serious monitor, I'm inclined to say that ZIP should be considered a serious threat!"

Under the topic "Personalities," Gorman goes on to say,

Both stations are encouraging raps between jocks. Sometimes they fall flat on their faces . . . and sometimes they come off! We need more on-air interaction. Don't get carried away or get silly . . . but it would be nice (not to mention . . . human) to hear occasional raps before the shift changes. . . . Bring out the best in each other's personality! ... Also, if you're coming out of a spot cluster & the last spot features an WMMS voice ... acknowledge it! ... Show your personality on-the-air. You know who you are. Let those listeners know, too. ... Booth announcers are a dime a dozen ... but there is only one Buzzard Nuclear Army! ... We've made Cleveland a major rock & roll force ... don't be ashamed to admit it! If you've got it ... FLAUNT IT! Now is the time to fine tune ourselves into being a total way of life with our listeners! ... Our image has to be the best! Project it!

Days later, another memo to the staff suggested double-number ratings are again within their reach, reminding that,

Boom Boom (Goldberg) first said this in '73 or '74 and it's an excellent description of Cleveland: A number of "islands," cults, lifestyles which somehow, manage to mesh together (he was talking about music, not ethnics.) We are most successful when we're able to embrace this total audience. We got that 10.6 in the Oct.-Nov. war by locking our hard cores in, plus stealing additional bodies from the Chimpo, ZZP, G98, and (probably more of a factor than originally believed) WGAR. ... Our "progressive" audience alone ... (people attracted to WMMS because it's a progressive rock station) account for a total audience share of around 5.0–5.5. If we attracted strictly a progressive rock audience and nothing else, we'd probably get anywhere from a 5.0 to a 5.5. ... That is a compliment! This is an audience which recognizes the uniqueness and progressiveness of WMMS! They know everyone's name and airshift. They closely follow the further adventures of the Buzzard Nuclear Army. They know each & every one of you even if they've never met you. They discuss WMMS. They may not like all the giveaways and contests, but they would probably turn the radio off and put on a record before putting on 105. (They will occasionally listen to M105 out of curiosity or just for the hell of it ... but they have no respect for it. ... Some of 'em have been listening to progressive radio since the days of Bass, WNCR, and Doc Nemo. Others are younger. They started listening to WMMS four, three, or two years ago. But the one thing in common this audience has is the ability to recognize our uniqueness, progressiveness, and so on. They are our most cherished listeners. Their heads are in the same place as ours ... or close to it. There is no other radio station in Cleveland (and certainly no other progressive-AOR in the country) with a hard core, loyal audience like ours. (Gary Dee may have incredible listener loyalty, but it is one person as opposed to an entire radio station!)

He went on to say,

the non-loyalists. They'll listen to the station offering 'em the most. And it's this transient audience we need to hit the double figures. They would never give you a creative answer as to why they listen to a particular station. Maybe they like the Eagles and 105 just played 'em so, right now, at this moment, 105 is their favorite station. If we play the Eagles and the same kid hears it, suddenly WMMS is his or her favorite station.... They know WMMS is 101 and the Chimp is 105. The buttons in their cars are set to G98, WMMS, 105 and ZZP ... and they use those buttons when they hear an unfamiliar tune or a song they don't like.

All the promotion in the world ain't gonna help you if you sound like shit. (WIXY, WERE, M105, WGCL, WZZP ... and even WMMS have all proved that at one time or another. Luckily, for us, it was very brief and corrected. Others haven't been so lucky.) ... The Chimp knows that he needs at least a 5.0 to survive the war and he's shooting for a 6.0.... The Slime Twins at G98 need a 5.5 ... and even then they may get booted out the door. (It seems to be a tradition over there.) ... Everyone is fighting for their fucking lives. Only one or two at most will survive this war.... It's our biggest propaganda attack ever! ... that crazy fantasy of ours and turning it into reality!

Gorman then warned his "troops," "Be aware of what is happening around you ... besides WMMS. Identify with your audience. What is hitting their hot button besides music. Find out what movies are hot (although the hottest ones need no introduction), TV shows ... mention Saturday Night Live. Who are the hot faces ... actors-actresses.... Dope (marijuana) ... it is a big part of our audience's lifestyle ... recreation ... etc, etc. Learn as much as you can about our audience. Know their language. Refer to things in the language of the street.... WAR BEGINS: DIE, CHIMP, DIE!" These memos helped unify an already-strong WMMS staff and struck fear in anyone outside the station who might ever have a chance to read one.

With that kind of fiery emotion at all levels of the station, it's not hard to see how the WMMS juggernaut was starting to roll toward complete domination of Northeast Ohio.

Station politics aside, the most important product was still what was on the air. In May 1978 WMMS had exclusive premieres on new releases by the Rolling Stones ("Miss You"), Tom Petty and the Heartbreakers ("You're Gonna Get It"), Joe Walsh (the *But Seriously Folks* LP), Bob Seger (the *Stranger in Town* LP), and up-and-coming new wave act Mink De Ville, all in the space of ten days. Shortly after that in early June, Kid Leo had the premiere of *Darkness on the Edge of Town*, the first Springsteen LP since 1975. Clearly the station had solid connections in the industry and the kind of clout to get product "diverted" its way.

Radio, like any successful business, thrives on relationships, and it made sense for WMMS to use the growing power of its ratings and influence to barter for additional success. That was the topic of a confidential memo from Gorman on May 16, 1978, concerning the Michael Stanley Band. It read,

Jeff and I went over to Belkin Productions to meet with Jules, Mike, Jimi Fox, Carl Maduri, Wendy Stein and Michael Stanley. We listened to the new MSB album and were encouraged for our comments and opinions.

Bear in mind that a few members of MSB are still close to the Brillo Pad... including Michael himself. Despite their poor choice of friends, Michael Stanley will be a guest on Jeff's show (opposite the Brillo Pad) to premiere and discuss the soon-to-be-released MSB album.

While at Belkin's, Jules discussed the upcoming WMMS summer concerts. We've got some hot ones lined up under our call letters, including (not for publication): Fleetwood Mac, ELO, CSN, Boston, Heart, etc. We could end up with a new WMMS World Series of Rock show every 2 weeks in July and August!

All Belkin asks for in return for these shows is a heavy push on the Michael Stanley Band over the next few days to help support the upcoming Coliseum show. Please do. The Chimp has been politicking his fucking ass off, trying to steal whatever he can for M105. Jules is sticking with WMMS, asking us for just one favor, which is in our best interest to carry out for him.

Stay hot on the MSB show and play. They'll be listening.

A lot of people did listen to WMMS, and Stanley did get a lot of play on the station. But Stanley would never have had the success he did without his raw talent and savvy management, because all the promotion in the world cannot force people to accept a product if they don't like or want it.

Just a few weeks later, the first World Series of Rock was announced for July 1 at Cleveland Municipal Stadium, and the station announced that specific information on tickets would be made exclusively on WMMS in the near future.

When the Rolling Stones show arrived, WMMS put its wheels in motion to ensure that everyone at the stadium knew they were the station of choice. A staff memo outlined plans, saying, "We will present another 'Rolling Stones Orgy' in honor of their appearance... and it will be commercial free and piped into the stadium between acts. Can you imagine a Metro quarter-hour of 88,000????" Just a few days later the spring Arbitron ratings were released, showing WMMS doubling and in some cases tripling the margin held by M-105. On June 22 Gorman sent a memo to the staff announcing, "We've sent the Chimp a fitting present for his showing in the April-May war: A box of dead rats and a picture of a Chimp in crutches! Kick 'em while he's down!"

At M-105, Doug "Dawg" Johnson, who had only been with the station a few months, announced that he would be leaving the station for a time. He was commissioned to write an off-Broadway play, described as "medieval entertainment," which was a project he'd wanted to do all his life though he wasn't expected to be gone for long. Also at M-105, Spero and his producer Joanne "Froggy" De Pompei had been airing the "Not Ready for Drive Time Players," a take-off on *Saturday Night Live* featuring humorous bits and parodies, which proved to be a very popular feature. Even though she didn't have a regular shift, Spero's constant mentions of De Pompei made her almost as popular as anyone at the station.

At WMMS in the last week of July, Walt Tiburski was named station manager, which meant he oversaw all day-to-day operations at WMMS and put his neck on the line if something went wrong. Tiburski presided over a station that introduced a "Nuclear Navy" (a twenty-four-foot speedboat named *Thunder Buzzard*), a luxury "Buzzard Bus" that escorted listeners and staff to special events, and briefly considered even a "Buzzard Jet." Also, Gorman announced to a select number of staff members the formation of an editorial committee, which included Leo, Tiburski, Matt the Cat, Ed Ferenc, and two others, stating, "Our first attack—THE KUCINICH RECALL! Think of ways & ideas you would want WMMS to approach the recall (obviously, we will take a pro-recall stand . . . it's how strong we want to get). Since four Buzzard Nukes are Cleveland residents, it gives us more credibility to take this stand." A historical note: Mayor Dennis Kucinich and his cabinet had upset the political and business communities with what many perceived as an "in your face" attitude about serious matters concerning the city. Some chalked it up to their youth, and others inexperience, but whatever the cause, there was a push to have Kucinich recalled as the city teetered on financial ruin. While Kucinich may have had his hands full with the recall drive, there wasn't much likely to go wrong at WMMS in the coming months.

Happy Birthday—From the Boss!

A tenth anniversary is a special occasion in any business's history and a hallmark to celebrate in grand style, especially in the constantly changing world of radio. Prior to August 1978, WMMS had been talking about its ten years on the air, its meteoric rise, and its pulling the city together in the common spirit of rock and roll. But Cleveland needed a celebration, an all-out love-fest, something to embrace as its own in a very difficult time in the city's history. For the better part of a decade, WMMS had been that source of comfort, energy, and pride, and if the station was going to celebrate its tenth anniversary that month, it had better be a real celebration.

The station did not disappoint its fans. Just a few days before the anniversary there was an announcement about a concert, a *free* concert at the Agora, and what a show it promised to be. Bruce Springsteen, one of the biggest concert draws of the 1970s, in the middle of his eighty-five-date, sold-out Badlands Tour, was go-

Hank Lo Conti (1977). His original vision for a college-age club turned into the Agora, one of the most respected rock venues in the United States. Cleveland Press Collection/ Cleveland State University

ing to honor his friend Kid Leo and the city of Cleveland with a free performance at the Cleveland Agora. The show would be broadcasted to Cleveland and seven other cities, but the big news was that it was happening here, and WMMS was making it happen.

Even though admission was free, chances of getting tickets were slim. Listeners were asked to bring stamped, self-addressed envelopes to a Record Theater store, and 375 of those entries would get two tickets apiece. The record chain collected thousands of envelopes for what would prove to be one of the hottest tickets of the decade. The concert was to be broadcasted live on WMMS, but the real prize was to be at the show.

Agora owner Hank Lo Conti says that concert came together in an odd way, and part of it concerned his club's thirteenth anniversary:

> There are so many stories of how it happened, and the more I look at it the more I think that people at Columbia Records might have played both ends! I was sure WMMS was looking for somebody big, but I was in Los Angeles at the time that Columbia had their annual party to give away the gold and platinum albums, and all the accolades. I was invited in '78 because we broke Meatloaf in 1977, and Steve Popovich insisted that I be invited to the party.
>
> At the cocktail party before the dinner, I was having a drink with a fellow from Cleveland who was a promotion man for Columbia in the early seventies and worked his way up to New York. We were talking about the Agora, and some of the stunts that he had done there in his younger years. I said, "We're going to be celebrating our thirteenth anniversary," and he said, "You should have somebody special! What about a—Springsteen!?"

Lo Conti didn't think he was serious and said, "Yeah, and I'd like to go to the moon, too. I didn't believe he really meant it, but the next thing you know we're planning a Springsteen show! I thought I did all that at the cocktail party." But he later found out that WMMS had been working on its tenth anniversary, so he "would say we probably got Springsteen more for WMMS than the Agora [anniversary]." But today Lo Conti says everybody won, so it didn't make any difference how it happened.

The show wasn't mixed at Agency Recording like other shows broadcasted live from the Agora. The label brought in a mobile mixing truck, which resembled an old bus, from Star Fleet Studios. The sound would be mixed in the truck, sent up to Agency Recording, on to WMMS, and then out to the network of other radio stations.

Anticipation was running high as the night of the show approached, and when the network switched to the Agora stage that August 9, the show opened with the sounds of screaming fans and Leo triumphantly announcing, "Welcome to the WMMS tenth anniversary concert. Ladies and gentlemen, the main event. Round for round, pound for pound, there ain't no finer band around. Bruce Springsteen and the E Street Band!"

It sounded exciting on the radio, but the sound inside the Agora was absolutely deafening as Springsteen and his band bounded onto the stage. Springsteen called for some lights, grabbed the mike, and opened the show by saying, "Leo—he must have memorized that at home! I know you did. Cleveland—how you doing? You ready to shake them summertime blues?" A blistering set played to an audience that never sat down during the entire show. Springsteen joked with the crowd, even mentioning Leo in a song praying for more watts for the station. It could only be described as a triumph for WMMS and the entire city.

During the show, Denny Sanders stayed in the mixing booth at Agency Recording, which acted as network control for the event. As he recalls, "The [sound production] mix was fed to the Agency board, and I was added live to do the network open, intermission, return, and encore bridge. Leo then came upstairs and did the intermission cutaway and show close." What many may not realize is that after the broadcast signed off Springsteen came back to do one more song, "Twist and Shout." Two different bootleg recordings of the show circulate, one taken from the air that ends with Leo signing off, and the other came out of the truck and has Springsteen doing the final song.

It was a night that no other station could match, and one of the defining moments in Cleveland's rock and roll history, right up there with the Moondog Coronation Ball, the Beatles' shows, and Otis Redding's final performances on the local *Upbeat* television show and at club Leo's Casino. Every station in town, every rock and roll fan, and even the harshest critics of WMMS had to sit down, catch their breath, and just say, "Wow!"

Television, Togas, Playboy, and the Buzzard's Enemies List

The increasing power of WMMS and rock and roll radio was growing at a fever pitch across Northeast Ohio, and other media took notice. There had always been some type of newspaper coverage, but now television was hoping to lure a crossover audience into its grasp. There had been experiments over the years, starting with the popularity of Herman Spero's locally produced dance party series *Big Five Show*, which was later renamed *Upbeat*, and *Ed Sullivan, Hullaballoo, Shindig, Music Scene,* and *American Bandstand* on the networks, plus the *Live at the Agora* series aired locally, all of which got terrific response. WKYC-TV news reporter Royal Kennedy did a piece on David Bowie's *Diamond Dogs* tour at Public Hall in the summer of 1973, which raised eyebrows among the station's older viewers. Music got a big response and Cleveland television honchos took notice and looked to radio people for help.

WEWS-TV featured M-105's David Spero on its *Afternoon Exchange* talk show, offering on-scene interviews with personalities ranging from Gregg Allman (from Quail Hollow Golf Course) to the Beach Boys (caught poolside at a local hotel). Spero scored a coup with a one-on-one talk with members of Fleetwood Mac, who made a special trip to Cleveland to explain guitarist Lindsey Buckingham's

Fred Griffith, Meatloaf, and David Spero on *Afternoon Exchange* (1978). Now with M-105, Spero joins Griffith to interview Cleveland International recording artist Meatloaf in the studios of WEWS-TV. Photo by Janet Macoska

illness that forced the postponement of their World Series of Rock show at Cleveland Municipal Stadium. That interview didn't sit well with the competition.

The interview was done at the Bond Court Hotel, which is where Fleetwood Mac held its press conference. Although WMMS was there and was able to send it back to the studio via microwave antenna, getting it on the radio first, Spero still had scoop of the exclusive chat. Spero also did music reviews for the television station, and he knew the layout well. After all, he spent his teenage years helping his dad with the *Upbeat* show at that same address.

John Gorman was livid, writing in a lengthy memo on August 11,

> Unfortunately, we did get caught off guard on one incident that I personally felt great pain about: Spero's F. Mac interviews. Here's the story: When word leaked out about Saturday's F. Mac press conference, David Spero (with the help of Dave Lucas and Ted Cohen) contacted F. Mac's PR person, Sharon Weisz, and, under the call letters WEWS, Channel 5, offered TV coverage of the event if he could land a quickie interview with the band. Since F. Mac wanted to get as much media exposure as possible to explain the cancellation, Weisz said OK

(she had no knowledge of Spero's association with 105).... Believe me, I'm out to kill that motherfucker for fucking with that band.... On the positive side: F. Mac know who we are, (they've had enough contact with us), and we will win & rule on August 26th.... No one can withstand a Buzzard Nuclear Curse!

Denny Sanders got some television time on WEWS within just a few months, both as a guest and interviewing acts like Suzi Quatro. WJKW-TV also cozied up to WMMS, featuring Kid Leo for a time on its *PM Magazine* talk show, and Jeff Kinzbach was a frequent guest on that station's *Noontime* program with Mike Keen, aired in front of a live audience. All the programs had a following but were fairly short lived as more homes in Northeast Ohio got linked to cable and had more options from which to choose.

WEWS presented a Michael Stanley special in July 1979 titled *Michael Stanley: Stage Pass*, with Mike Belkin of Belkin Productions credited as executive director and Spero as production supervisor. But rock and roll television was in for a huge change, one as big as the radio switch from AM to FM. There was a monster still on the drawing boards called MTV that would change everything, including radio. It debuted to a small cable audience in 1981 but grew rapidly from there. By spring 1983 rock and roll television was well-established, thanks in most part to MTV.

The "radio as war" mentality continued at WMMS, with program director John Gorman often called out of town for industry functions and meetings. Gorman memoed his "troops" before one such road trip, reminding them that Kid Leo would be acting program director during his absence and that they should not create any extra problems for him.

He wrote, "Keep the boola & booze out of the building.... As soon as Labor Day weekend hits, we begin Chapter 6 in the Further Adventures of the Buzzard Nuclear Army. Shit, it's hard to believe that most of us lived through five years of this craziness. Anyway, we're getting real close to the time when we'll have to gear up for another season of WAR! Let's get all the leftover problems & shit cleaned up now so when September hits we'll all be clear to wage war against all those assholes." He warned the staff to "Monitor the competition. I ain't lyin' when I say that 105 is sounding good." Gorman told them to give the listeners the music they wanted, saying, "The one thing they don't want to hear is what you could call 'hippie era' music. Stuff that sounds dated. We're in a period of time when, with a few exceptions (the survivors who are still around today: Who, Neil Young, electric CSNY ... and there ain't too many more), no one wants to be reminded of '68, '69, and '70. The peak hippie years."

As the summer of 1978 was drawing to a close, WMMS had scored decisive victories with the Bruce Springsteen show, lots of media coverage, and plenty of exclusives. That's not to say the competing stations had given up the fight. But there's an old saying that "perception is reality," and even though M-105 would get

The Buzzard Song (1978). David Helton

visiting artists, interviews, and premieres, people would still assume that WMMS had them first, even if that wasn't the case. It's hard to fight that kind of publicity.

WMMS launched a new volley in the radio wars with its Buzzard Record and Filmworks label. Dan Garfinkel says, "We had to put something on the label, and David Helton, Denny Sanders, John [Chaffee] and I came up with that. It was obviously tongue in cheek." The first official releases from the station's label were the *Pride of Cleveland* compilation LP and Alex Bevan's single, "The Buzzard Song," backed with "Ho for the Weekend." It was basically a promotional 45, naming all of the disc jockeys and featuring a Helton Buzzard image on the picture sleeve. The record sold for $1.01, and while no one expected it to make the *Billboard* Hot 100, it did show that the station was starting to flex its promotional muscle in new and creative ways. Garfinkel also says that the total output of the Buzzard's film production was a couple of animated television spots produced by Helton.

Meanwhile, M-105 welcomed back Doug Johnson to its midday lineup. The writing work on his play was finished, it was due to open in November, and as M-105 headed into the fall it needed Johnson's strong presence. He was still up against

WMMS's monster ratings and the very popular Matt the Cat, and Malrite wasn't in this race to lose.

In a memo to the "Buzzard Nukes," dated September 15 and titled, "Weekend Attack! It's one month before the war! Shoot to Kill! Torture, Maim and Destroy!!!" Gorman warned that,

> Stations are loading up their guns. Both WGCL and WZZP are planning major campaigns . . . aimed right directly at us. . . . The chimp ain't sitting on his fat ass right now, that's for sure! The chimp is working on getting on the right side of the record companies. Instead of adding his usual two albums per week, the fat fucker's getting aggressive! This past week, he added six albums! And now he's hitting all the record companies for promotions [contests and prizes to lure listeners] during the book . . . The chimp was quoted as saying: "I'll get WMMS while they're sleeping. Everyone thinks they're stars and immortal over there. I'll get WMMS because they're too sure of themselves. . . ." War is on!! Here's what we're planning for war: We'll be doing an exclusive TV simulcast (probably two of 'em within the book) of Southside Johnny with special guests Bruce Springsteen, Miami Steve and Clarence Clemons for the new "Live at the Agora" TV show. Negotiations are still being worked on, with Channels 5, 8, and 43.
>
>   If you checked this week's *PD Friday* magazine, you'll see the chimp is suddenly trying to build his airstaff into personalities! Spero is out to get Jeff. Doug Johnson is out to get Matt, and so on. The battle has TWO faces. It's not really Spero vs. Kinzbach, or Johnson vs. Matt. The theme of the war is OUR KIND OF RADIO VS. THEIR KIND OF RADIO . . . LET US NEVER FORGET THE OCT/NOV OF TWO YEARS AGO WHEN WE ALL DID BECOME TOO SURE OF OURSELVES. . . . It's going to be a dirty, dragged-out Oct/Nov!! It will make the April/May look like Sesame Street! . . . THAT BASTARD MUST DIE!!!! . . . We are going in for the KILL NOW!! . . . Keep your breaks tight! BE READY WHEN YOU OPEN THAT MIKE! Have your rap ready! Know what you're gonna say! It's a dirty fucking war, and no fucking Chimp or Spero is gonna out-do us on any fucking thing!! Go for the jugular vein. Every weekend is important!!! DIES, CHIMP, DIE!!! WE WILL WIN AND RULE!!!

As the fall 1978 ratings campaign got under way, WMMS got the jump on the *Who Are You* LP, which was one of the most anticipated new releases of the year. However, a couple of weeks before the premiere, word came from London that Who drummer Keith Moon had died. Rock radio was buzzing with the details, and Jeff Kinzbach commented on Moon's passing in his regular spot on WJKW-TV's *Noontime* show. At M-105, Johnson was playing the Who's "Slip Kid" when the report first crossed the wire. In a sad, dark way, Moon's untimely death provided a promotional boost for rock stations as listeners stuck by their radios for details.

On a different note, WMMS had firmly positioned itself as "pure Cleveland," even though M-105 had an entire air staff of Northeast Ohio natives. M-105 deejay Bill Stallings tried to correct that perception by playing a daily segment featuring artists with local connections like Joe Walsh, Eric Carmen, the Outsiders, the James Gang, and the Euclid Beach Band, but others stations also played those artists. Although music was the driving force, stations were always looking for a hook to get new listeners and keep the audience they had.

The WMMS morning show took its campaign to the clubs, capitalizing on the popularity of the film *National Lampoon's Animal House,* throwing toga parties hosted by Kinzbach and Ed Ferenc at the Cleveland Agora. The "kings and queens of togaland" were chosen at midnight, and winners were judged on the originality of their togas. There were even jello-gulping contests. As silly as they may seem, the contests offered some pretty good prizes, as winners received tickets to five shows of their choice at the club. These would prove to be very long nights for Kinzbach and Ferenc, who still had to do their morning drive shifts the next day, but youth was on their side, and they could shrug off the lack of sleep. Kinzbach also made an appearance on behalf of W.E.L.C.O.M.E., which stood for "West Side, East Side, Let's Come Together." The group hoped to ease the rising tension of court-ordered busing, which was due to begin soon in the Cleveland school system.

WMMS helped *Playboy* magazine in its search for a "Girls of the Midwest" pictorial and later hosted the winners, and local news media, when the magazine was released. Among those interviewed on the morning show was Nina Blackwood, who had been in some local bands and would win fame a couple of years later as one of MTV's original on-air personalities, as well as appearances in other television and film.

WMMS was marketing everything from clothing to chocolate Buzzards and was all over television with Kinzbach, commercials, and its *Live from the Agora* series, simulcasted with WJKW. Even stores that advertised stereo equipment had "100.7" across the dials in their print ads (The station started referring to itself as 100.7 rather than 101 FM when digital read-outs became more common on stereo dials.), and M-105 had a semiweekly magazine in local record outlets. By the end of 1978 the WMMS morning show had climbed to sixth overall, ahead of M-105 at number nine, and Denny Sanders and Betty Korvan ranked number one in their respective slots.

Gateway to the Eighties

WMMS entered the final year of the 1970s on a huge promotional push. Kid Leo was profiled in the pages of *Rolling Stone* in January, naming him as one of the most influential radio voices in America. And in a very ambitious programming move, WMMS offered a twenty-four-hour long "Buzzard Beatles Blitz," with demos, outtakes, interviews, and a lot of little-known information about the band. It was an impressive production and drove home the scope of knowledge the disc jockeys had about the artists and music they played and of the unique material they were able to bring to the air. To top it off, John Gorman used his influence in the industry to secure an authentic autographed copy of *Abbey Road* from a former employee of Apple Records. The station also tried its hand at jazz and aired Cleveland Browns safety Thom Darden's *All Pro Jazz Show* late nights on Sundays, produced by Leo. By midsummer, Len "Boom" Goldberg was hosting the very popular all-request *Solid Gold Sunday* morning show, heavy on music from 1964 to 1970, showing that WMMS was able to embrace a wide-range of the rock music spectrum without alienating its core audience of modern rockers.

The air staff performed as *The Buzzard Theater of the Air* in a series of vignettes offering clues for one of the more imaginative contest promotions to date. It was called the Buzzard Mystery Contest, and each day various members of the staff would act out a scene with some type of error for the listener to spot. For example, "Lenny Danders" might stop the auction of the original music to "Roll Over, Beethoven," signed by John Lennon and Paul McCartney, claiming it was a fake. The first caller to correctly tell why it was a fake (in this case because the song was written by Chuck Berry) would qualify for some impressive prizes, including cash, trips, and electronics, among other big-ticket items. It was reminiscent of the "Encyclopedia Brown" magazine and book stories about a boy detective who would solve mysteries based on the clues established throughout. That promotion ranks as one of the cleverest bits presented on WMMS.

Meanwhile, there were changes at M-105, which by year's end would introduce its Lo-Dough and No-Dough concerts, which showcased up-and-coming bands for a very small ticket price, if one was even charged at all. And in the morning drive, David Spero ended his shift on a Friday, and within a few hours ended his affiliation with the station.

Spero says he missed radio after moving into music management, but it was a different business when he returned, especially when it came to playlists: "At WMMS, that never existed! My initial deal with Eric Stevens, the program director, was basically, 'Yeah, there's a playlist that I could play around with.' It didn't take long before we had our first little battles over that. Strangely enough, it was over a song

Buzzard Beatle Blitz Ad (1979). The staff at WMMS put long hours into its tour de force tribute to the continuing appeal of the Fab Four. David Helton

that I thought was going to be a huge hit that he had no intention of playing by a brand new band called the Police. It was a song called, 'Roxanne!' That was our first little battle, and I realized at that point that it wasn't the way it used to be."

Spero says that radio had a different feel that time around, "It was a job—and you can never go home [to how things used to be]." The split was described as "amicable," following the differences with Stevens. Spero explains, "It wasn't any one thing, just an overall difference of the way things should go." Some of it was linked to M-105's tightening playlist in light of intense competition from other stations. Spero was a radio veteran and didn't harbor any bad feelings, though he did tell the *Plain Dealer*, "I suppose I should change my phone number." Its last three digits were 105. Doug Johnson was then given the early morning shift.

What hadn't been revealed until now were the intense negotiations to bring Jeff Kinzbach and Ed Ferenc over to M-105 for the mornings. Stevens doesn't remember who approached whom first but says those talks were very serious: "It was after David was gone. We basically had a deal if we could agree on the dollars. We came down to the one yard line." The talks were unsuccessful, but if they would have been, the history of Cleveland radio could have gone in a very different direction.

The ratings and promotional battles among the FM stations raged on, beginning with the January-February ratings, and the powers that were at WMMS were already looking several months down the road. In a memo dated February 23 titled "The First April-May War Roll-Up Attack Memo!!! Prepare Now!!! The Biggest Bloodbath in the History of Radio is about a Month and a half away!!! Believe it!! The eyes of the Entire Radio-Record Industry are on Cleveland!!!!! (Whew!)," Gorman told the staff,

> The industry is asking the question: "With the changing trends in music, lifestyle and general priorities of the masses, can a WMMS still survive?" The industry looks at WMMS in amazement! WMMS is considered a one-of-a-kind! It turned Cleveland into the legitimately and nationally recognized "Rock and Roll Capital of the World." BUT, they ask, could WMMS have peaked? They ask the question based on the fact that WMMS has downtrended in the last couple of books. It's still the #1 rocker, without a doubt, but WMMS' ratings for this past Oct/Nov weren't as high as the April/May '78 or the Oct/Nov of a year ago . . . and at the same time, M105, although by no means setting the world on fire, has been slowly uptrending book-by-book.
>
> The industry is important to us from the standpoint that they do supply necessary arms needed for war. As long as we're hot, we're in a position to bargain. But fuck the industry for now. Our war is fought on the streets of Cleveland.
>
> We'll have the first of many BNA war counsel meetings. Collect your ideas, plans and concepts now. The Buzzard Investigative Squad has already uncovered some of the Chimp's April/May plans . . . and, no shit, they're all pretty heavy and a lot more powerful than anything they've tried on us in the past. . . . Don't take anything for granted! More on this at the BNA war counsel meeting. In the meantime, start fine tuning and planning for the WAR!
>
> The reason WMMS has survived (and went on to become the single strongest influence to our listeners) is directly due to our ability to tap right into the street and audience we serve. We've been successful in constantly changing with the times. Now, we find ourselves, a decade later, in a world totally opposite from the one which created us.

But M-105 had some promotional business of its own on the streets of Cleveland early in the year. Johnson was enlisted to pull a contest winner in a rickshaw all the way from the station's studios at 39th and Euclid to the Cheap Trick show at the Palace Theater. It was a rigorous ordeal, and Johnson would later comment, "I couldn't stop because I was afraid I would collapse." Spero had chosen that same location for his "sock hop and drop" promotion, dropping hundreds of ankle-socks full of prizes to help promote the movie *Grease*. It took a lot of guts to hold those events just a few hundred yards from the WMMS studios, and as expected WMMS retaliated. As Gorman described it in one of his "weekend war reports," his station crashed their party with predictable results. "It was bad enough," he wrote,

> When the WZZP van (shit, compared to ours) parked at Chester Commons during M105's promotion . . . but when we sent our high school intern help

Ruler of the Airwaves (1977). David Helton

down there to not just pass out WMMS stickers . . . in trade for M105 stickers ("wouldn't you rather have one of these than one of those?") . . . our "party crash" enraged the Chimp. One of our interns unknowingly went up to the Chimp himself and asked that no necked bastard if he wanted a WMMS sticker. Our interns didn't stop there. They started harassing (non-violently) M105's help by offering them $20 for their stacks of 105 stickers. . . . Our listeners even positioned themselves and dropped WMMS stickers on Dawg Johnson. He didn't know how to react when he suddenly spotted WMMS stickers sailing down around him. We got rid of hundreds of WMMS stickers and destroyed an excellent number of M105 stickers in the swap. The M105 interns finally attacked our party with water pistols . . . but it was too late. The damage was done!

What you don't know is that we received a report from someone who was at 105 when the Chimp and his troupe returned from Chester Commons. Here's that report:

The Chimp . . . walked into 105 fuming over the "party crashing" incident. He immediately ordered Dawg Johnson on the air. He told Dawg to "go crazy" on the air and complain about "the other station's" actions. Every word he said came from the Chimp.

The Chimp kept calling, calling and calling their control room, telling Dawg to break after just about every record to complain! A sample group: "John (making reference to me), if you wanted an M105 bumper sticker, all you had to do was ask"; "Somebody should clip that vulture's wings" (in later repeats of the same lines the word vulture was substituted with bird). The Chimp assumed that we monitor him the same way he monitors us. He turned the whole station into his sounding board! . . . Yes, you could say we once again successfully ruined his day!

Sometimes the mean-spirited WMMS station memos carried a dark humor. One jokingly made mention of a vacationing Ferenc getting a groin transplant from a chihuahua, and the missive on February 20 outlined Gorman's trip to the *Radio and Records* Convention in Los Angeles. It read, "Leo will be shaking down all the record company heavies for exclusives, concerts and more military aid. While we're away, the Sword [Denny Sanders] will be wearing three hats: acting PD, MD, and Minister of Propaganda. Be around to give the Sword any help he may (and will) need. He can't afford to lose any more hair. It's bad enough that people are stopping him on the street and asking if he's one of the Bee Gees."

And while the wartime memos were often fall-down funny, Kinzbach talks about at least one incident on the street that borders on the sinister that revolved around station WGCL-FM, better known as "G98."

Taking its cue from WMMS, the WGCL promotional machine would sometimes show up at WMMS-sponsored events. Kinzbach says,

> They would pull up in their van and start passing out t-shirts. I knew some guys who were, well, let's just say they were the local "muscle." I mentioned it at dinner to one of them that we were having this problem, and I'd done some favors for them. He said, "We'll come by and take a look." They came to the Party in the Park, everybody is having a few beers and enjoying themselves, and all of a sudden, the 'GCL van pulls up. They start passing out their t-shirts, and these guys went over there. All I remember is that it was a short conversation. Next thing I know, the doors are closing up and the van is gone!

Just about every move at every radio station was closely monitored by WMMS, and Gorman found himself acting as coach, cheerleader, quarterback, and every other conceivable type of leadership role it might take to inspire his "troops." In yet another lengthy message to the staff, Gorman wrote on March 30:

> War is picking up (and they're all aiming at us.) . . . We are vulnerable to sneak attacks by the Chimp and every other station in town. You can have a brand new car with four Michelin tires . . . but if one of 'em is flat, you're standing still. . . . Be enthused. Be Proud. . . . We're not out to hurt, maim, or torture. We are out to K-I-L-L!!!!

Ian Hunter (1978). His "Cleveland Rocks" would prove to be an early anthem for Northeast Ohio's musical pride as well as the traditional starting gun for the weekend. Photo by Janet Macoska

Later in that same memo Gorman mentions "Future Heavies"—soon-to-be-released songs expected to make a huge impact, including Ian Hunter's "Cleveland Rocks." He ended that call to arms with, "WAR IS ONLY TWO WEEKS AWAY!!! KILL THOSE BASTARDS!!! SHOW NO MERCY!!! WE WILL WIN AND RULE!!!" Gorman often looked at promotional blitzes and rating periods as "times of war."

Inside memos show that WMMS was very concerned about M-105's programming and promotional plans. In a staff letter titled, "Chimp Hunting Season is About to Begin!!! We Gotta Have 'Em This Week or We're in Trouble Next Week!!! Slaughter Those Bastards!!!, etc.," Gorman wrote to the sales department in particular,

> Our programming is firebombing the shit outta everyone else on the dial. You really think you're sitting pretty . . . don't cha! Well, guess again. Get outta that dream world and get back into reality! . . . The Chimp really does have the biggest promotional budget ever! . . . It's minute-to minute! It's dirty!! It's bloody!! Every shit ass station is fighting for its life and every one of 'em is aiming their guns right directly smack dab at us! . . . Don't rile us up! We're under a lot of pressure. More pressure than any other book in the history of the BNA. Have respect for us. We'll help you as long as you don't con job us!
>
> The fact of the matter is . . . for the next four weeks your careers are in our hands. We'll all win with a hot book. We'll all lose with a bad one. Remember, hot tubs can be repossessed!!!

Later in his "Declaration of War" Gorman advised,

> We gotta lock ourselves into the central nervous system of our audience. Dive bombers don't operate on automatic pilot. . . . In the Jan.-Feb. war, we kicked the shit outta that Chimp. In the April-May, we've gotta finish him off and burn his cancerous diseased no-necked bastard body beyond recognition!!! We must pull a 3 share higher than our next nearest enemy . . . and hope it's not the Chimp. The overall sound of the station has improved a great deal since the weekend. . . . We can't afford one "automatic pilot sounding" minute! . . . We will win and rule . . . but we're gonna have to fight dirty to do it! Our prime concern is that audience. We want it all!!!

M-105 and its staff were well aware of what was being said and written about them. The record promotion crews worked every station, and gossip spread quickly, but most of the disc jockeys at M-105 say they really didn't care. It was a game, and they played it, but it didn't keep them up at night staring at the ceiling. For example, Eric Stevens might be at his desk having a cup of coffee and reading the trade magazines when his secretary buzzed in to say someone like Murray Saul was waiting to see him. Looking back, he recalls, "I knew he was friends with Gorman, and I would do things like stand on my desk, and start screaming and yelling. Calling Gorman names, things like that, knowing it would get back to him. But it was all an act!" Saul would leave, the secretary would come in, and Stevens would calmly ask, "How did that sound? Real?" She would rate the performance, and Stevens would smile, pick up the paper, and go back to where he left off.

Down the street at East 12th and Euclid, a spy was found lurking at WMMS. In his bulletin of April 20 titled "Espionage," Gorman tells the staff,

> The music director of WLYT, it turns out, was also working at WHK part-time and stealing WMMS memos and info for Rock 92. . . . He was fired as soon as this discovery was made yesterday afternoon. Word is on the street that his days are numbered if the Buzzard Nukes get their claws on him. It's possible that this clown leaked WMMS material out to others as well. He hasn't long to live.

That same day, in a memo titled "Slaughter Those Suckheads!!! This is the Real Thing!!! Kill! Kill! Kill! Buzzard Nuclear Energy Must Be in Overdrive!!!" Gorman recounted an exchange with WGCL's highest-rated disc jockey, Tim "Birdman" Byrd, at the Stark Records–Cleveland International label industry party at Akron's Tangier restaurant. Gorman wrote,

> Three people (including WZZP's Birdman) asked me where we got the new Fleetwood Mac single. Birdman was brazen enough to ask if he could get a dub of our copy if he doesn't get his by Monday. I told him he could, providing he said "courtesy of WMMS" over the intro. He walked away, still thinking we have a new, exclusive Fleetwood Mac single (which is actually two heavily Nicksed cuts from the new John Stewart with Lindsey Buckingham and Stevie Nicks album. It's still a WMMS exclusive and should recieve [sic] heavy airplay over the weekend. (Word is, the Chimp is f-u-r-i-o-u-s and once again called up Warner Bros. in L.A. screaming about our Fleetwood Mac exclusive). If Birdman, the Chimp and Stark Records executives believe it's the new Fleetwood Mac single, imagine what the street—our audience—thinks! Play the living shit out of those two cuts this weekend, every show. . . . A large amount of our audience will lock their dials on WMMS just to hear what they feel is the new Fleetwood Mac–WMMS Exclusive! . . . Our enemies are monitoring us to counter-program us at every turn. Is it coincidence that both Birdman and the Chimp heard the Stewart–Nicks–Buckingham tracks, hailed as WMMS World Exclusives?

Gorman also gave additional insight as to who the station saw as its chief rivals and how to counter them, writing,

> M-105 has a hefty TV promotional schedule and, although I haven't seen their spot, it was being raved about at the Stark/Cleveland International dinner last night. The word on the street is (and rightly so) . . . "WMMS has got a fight on

M-105 Staff (1978). Chuck Collins, "Benson," Doug Johnson, Bill Stallings, Tom "T. R." Rezny, and Suzy Peters. Collection of Bill Stallings

its hands this time," and that statement couldn't be closer to the truth. . . . If you monitored M105 and G98 last night, you heard how hot it really is. M105 is sounding the best they've sounded in a year. . . . We've gotta sound even hotter than ever before!

Gorman added,

M105 has Suzi Peters (who has improved a great deal) . . . calling herself "the new lady of the night." G98 has Caroline (I never hear a last name) calling herself "the foxy lady of the night." WHK has Carolyn Carr doing her S&M version of country radio. . . . It's about time someone coins the phrase "jiggle radio" because that's what the night-time on Cleveland radio has turned into (but only WMMS has Jenny [sic] Cheeks!) One of the Camelot store managers who was at last night's dinner/party at Tangier's was telling a group of people that night time radio in Cleveland (he's originally from Columbus) "is like *Playboy* vs. *Penthouse*." This guy is 23 years old. He spends most of his time listening to WMMS . . . but he does sample M105 and G98, and spent a good amount of his conversation (which was around five people) . . . including Ellen Foley . . . comparing Babz [Betty Korvan] with Suzi Peters, G98's Caroline . . . and Jenny Cheeks (who he wants to meet because of her name). I asked him, without identifying myself, who he liked the best. His reply . . . "I haven't made up my mind yet." Also overheard at the dinner . . . an argument between two invited (in other words, they were not in the business themselves) guests over what station plays the best and most rock and roll. The WMMS person called us a "legend." The M105 supporter said, "WMMS is resting on its image." I did break in to say something, to the effect of a polite "fuck you" to the M105 supporter. I said, "If 105 is such a great station for rock . . . why didn't anyone from that station show up at this party?" Then I turned around and blinds them with my WMMS jacket and walked away . . . but not before saying, "Tell me M105 broke Ian Hunter, Meatloaf, Springsteen, and Roxy Music and I'll call you a liar!" Anyway, as you can see . . . we don't own the audience. We gotta win 'em!

Success in the radio and entertainment fields is often measured by the influence wielded by one's friends. Recognizing that, WMMS was hugely influential, and it usually didn't matter how someone might have gained disfavor—to cross the station was like placing a noose around your neck. For example, Gorman wrote on April 27,

> Add these following names and locations to the BNA enemies list: Mayor Dennis Kucinich and fellow cronies (for proclaiming today as M105/Bruce Springsteen Day in Cleveland), the Pirate's Cove (for selling out to the Chimp. . . . They sleep with the Chimps), the Flatbush band (for composing a song for M105. Let them try to sell records off of M105 play), Bruno Bornino of the *Press* (for four straight weeks of M105 headline stories in *Showtime*.) Wait and see. One of these days, they'll come crawling to us, asking for the support the Chimp could never give them. When they do, we'll spit in their faces. The BNA does not forgive! Fuck us over once and you're in the shithouse forever!

Few dared to test the Buzzard's patience.

WMMS continued to monitor its enemies and report to the "troops" from its spies at the front. Gorman's memo titled "The Enemies" reported,

> Bob Travis, PD at G98, is sending out resumes to major market radio stations and record companies. He wants out at G98, and things are "grim" at WZZP. The pressure is on Birdman (Tim Byrd). His staff is divided into political groups, all hoping to be on the side that's winning. One of Birdman's buscards or billboards was defaced by one of his own jocks. He's in tight and not worried about his job . . . but he's openly blaming his jocks for the poor Mediatrend to the GM and Sales Manager. More spy reports as we receive them.

John Gorman (1978). There were those who compared him to Machiavelli. Gorman took competition to serious new levels. Photo by Fred Toedtman

And those reports continued to pour in. In a "Confidential—Classified" memo titled, "Slaughter Those Shitheads! Dirty Tricks Galore! Kill! Kill! Kill!" Gorman wrote on May 4,

> We landed an exclusive on the new ELO single. We were the first in the world to play it. The second the Chimp heard it . . . he called everyone regionally and nationally, at both CBS and Jet (ELO's Label) . . . The heads of CBS and Jet exploded! The single was not scheduled for release until Monday and its release was to be accompanied by a major promotional campaign (it was to be the first new ELO release to be distributed by CBS). Due to the fact that WMMS had the exclusive, Jet and CBS decided to rush release the single to avoid any additional damage, thus blowing their original planned campaign.

This was indicative of the power of WMMS; it could have disregarded the intentions of a major record company, which represented many important artists on the WMMS playlist and debuted product on the station, in favor of its own ratings campaign. They may have been livid, but they didn't mess with WMMS or else their artists could have taken a hit at one of the most important radio stations in America. That doesn't mean that the companies sat back and let WMMS do what it liked, but it would have been futile to cut off access to their artists and music. WMMS could always get the music, but the companies wouldn't always be able to break new acts or releases. They had to pick their battles, and if they battled with the station, it would have been full-scale nuclear war.

Evidence of that line of thinking can be found later in the May 4 memo in which Gorman commented on Stevens, saying that he would,

> Regret his current war activities. What goes around comes around ... and if that fucker wants to play those kinds of games, he better prepare himself by shopping for a coffin, because, as of now, we are out for revenge! If he wants to play dirty, he's come to the right place because we're gonna teach that daddy's boy what fighting dirty is all about! When we get through with that diseased scum bag, he will regret ever being born! It is time to go in for the KILL!
>
> This war has shaped up to be everything we predicted it would be. Fuck ethics! Fuck fair play! It's just plain fuck fuck fuck the competition! Every format, every demographic ... everybody is shooting their biggest guns in this one. ... SHOOT TO KILL!! DIE, CHIMP, DIE!!! YOU WILL ROT IN HELL!!! WE WILL WIN AND RULE!!!

Today the parties on both sides of those memos—Gorman, Stevens, Spero, and the rest—laugh off the content of those letters as being in the "spirit of competition" and hold no ill feelings toward each other, though one gets the feeling that a good amount of time was needed to reach that resolution.

Even though FM continued its march toward the top, the winter Arbitron ratings books showed AM was still king in the morning, at least for the time being. In mid-March, Gary Dee at WHK-AM was barely edged out of the number one spot by all-news WERE-AM, which benefited from some serious weather and its battles with Cleveland mayor Dennis Kucinich's administration. While WQAL-FM was number one overall, it was followed closely by WMMS, and WGCL was also gaining strength.

The record community was quick to congratulate WMMS on its strong showing, and a sampling of mailgrams from that time might have caused serious friction with other stations if they had been leaked. Capitol Records offices in Madison Heights, Michigan, wrote to Gorman saying, "Congratulations you bum—that's the way to hang that chimp by the balls," and, "This ARB should make you very happy. Congratulations to you and you[r] entire staff. You guys are the greatest rock & roll station in the States. Boy, the Chimp must be bleeding over this one!" In another message, credited to Columbia Records' Beachwood offices, and distributed to the staff under the heading "Dumbshit Telegram of the Year ... aka The Worst Stroke of the Year," Gorman is told,

> The most recent Arbitron is a statistical testimonial of the greatness of WMMS and the brilliance of your programming direction. In terms of being totally in touch with the market and as a reflection of the needs and desires of your vast audience, WMMS is a mirror image of Cleveland ... and we both know the importance of mirrors.

The book is a reconfirmation of my belief that WMMS is the greatest and most important radio station in the United States if not the world.

I believe in the Buzzards and I believe in you. Long live the Taurean imperative in Cleveland. . . . Long live WMMS.

Clearly, gaining the favor of WMMS could prove to be very beneficial to a record company trying to break new artists and product.

But all of the success enjoyed by WMMS couldn't ward off tragedy. Headlines concerning a former WMMS deejay sent shockwaves through the local radio scene. Tom "Tree" Kelly, one of the best liked and most respected people in local radio, headed out to Los Angeles in 1974 when he was just nineteen years old. He accepted an offer from another former WMMS personality, Martin Perlich, to become a partner in Aural Traditions, a company that produced syndicated radio specials. They joined with Rachel Donohue, whose husband Tom developed the progressive rock format at KSAN-FM in San Francisco and was later an inductee into the Rock and Roll Hall of Fame. The trio had done some impressive work together, including a highly regarded special featuring Perlich's old friend, Judy Collins.

But in March 1979 Kelly felt it was time to head home to Cleveland. He decided to sell off some of his property before picking up stakes for the move north and placed an ad in the *Recycler*, a free classified paper in southern California. Among the items listed was an unset diamond worth about $2,000 and fifty-five opals valued at $3,200.

On March 23 a man in his twenties and a female companion knocked on the door of Kelly's Venice, California, studio. When Kelly allowed them in, the man drew a .45 Colt Commander. There was a struggle, and the gunman was wounded in the scuffle, but Kelly was shot three times and critically injured. The man and woman fled the scene, and Kelly was left to lie bleeding on his studio floor for more than an hour.

Kelly was going to be part of a combination farewell party and WMMS reunion at Billy Bass's home in Los Angeles. Among some of the other folks planning to attend was Jeff Gelb, now on the staff of *Radio and Records*. Wanting to confirm plans, Perlich decided to call Kelly but was horrified by what he heard: "He managed to get the phone to his face. He said he had been shot and couldn't breathe! I called the police." Kelly was rushed to Marina Mercy Hospital in Marina Del Rey and underwent a series of operations. He lingered for fifteen days, but his wounds were too serious, and he died in the hospital's intensive care ward.

A suspect in the shooting was caught, and a benefit was held in Los Angeles to help defray more than a hundred thousand dollars in hospital costs. His loss was deeply felt in Cleveland radio, and an April 14 memorial service was held at Lakewood Park Cemetery in Rocky River. He was eulogized as a consummate professional with a keen ear for music and production, and Jeff Kinzbach paid tribute to him on his show that following Monday.

Kelly was one of Cleveland's most creative production people, using sound to convey emotion and impact, and setting the quality bar very high for those who aspired to achieve those same levels. In his time, Kelly had raised the level of FM radio, especially in the way it was marketed, and embraced by the listeners. Kelly may have been a memory to much of Cleveland's radio audience, but his influence as a broadcaster and especially as a dear and trusted friend would live on to this day.

The year ended with WMMS scoring a major coup by working out a deal to broadcast Todd Rundgren's New Year's Eve show from the Richfield Coliseum live, but it turned into a major headache for engineer Frank Foti. As he remembers it, Foti says,

> It wasn't like our live shows from the Agora, which was attached to Agency Recording. We had to bring in a remote mixing truck, and Agency's Arnie Rosenberg—God rest his soul—drove it in to the Coliseum, which presented another problem. Richfield wasn't part of Ohio Bell, which meant we couldn't get the special audio circuits from the phone company linking the Coliseum to our studios. We had to set up a special microwave connection from the hall to our transmitter.
>
> We went on the roof of the Coliseum and set up antennas, and dropped microphone cables from the rafters to the stage area. That particular night, for whatever reason, Todd Rundgren's people would not let anyone from the station in the truck to assist with the mix. Sure enough, somebody patched something wrong so we ended up with a buzz on the air for the whole concert! I fought tooth and nail with his people—they even had roadies and bouncers around the truck—while we were live on the air, and they wouldn't let me tell them we had a serious technical problem! The whole show had the right channel out of phase and a hum on the air. I wanted to kill Rundgren's people!

That wasn't his only problem.

To put that show on the air, Foti had to throw a switch on cue, stopping the signal from downtown and picking up the concert feed sending it to the transmitter. That meant if the musicians decided to screw around on stage and start swearing, the deejay had no way of taking it off the air. To do that, Foti had to stand by at the transmitter and take the cue from the disc jockey, switch it over, and stay there for the show until the final "Thank you and good night!" Then he threw the switch back to the studio. That New Year's Eve, while most young guys kissed their girlfriends at midnight, Foti's date was the transmitter.

At the dawn of the 1980s it was clear that WMMS was the station to beat, and that would be very difficult if not impossible. The stability of its longtime staff, savvy programming, and a very talented promotions department had made it as much a part of Cleveland to its core audience as the Terminal Tower, and that

Todd Rundgren (1977). He established an early fan base at clubs like the Agora, but his entourage was anything but cooperative during his live show on New Year's Eve broadcast on WMMS. Photo by Janet Macoska

audience was still growing. To top it off, *Rolling Stone* had named WMMS 1979's "Radio Station of the Year" in its annual readers' poll, and Mayor George Voinovich marked the occasion by designating January 30 "WMMS Buzzard Day in Cleveland." In addition the city issued a proclamation to the station for its fundraising drives benefiting the Public Library, the Cleveland Ballet, the Museum of Natural History, and the Free Clinic, as well as Cleveland State University's scholarship program. That honor also won the station a feature segment on WJKW-TV's *PM Magazine* and put faces to the voices heard on the radio. With press like that, programmers around the city had their heads in their hands trying to counter the monster that lived at the Statler Office Tower.

WMMS was attracting talent from other stations, including M-105's Walt Masky, who brought his *Homegrown* program to WMMS and renamed it *Ohio Homegrown*. It was heard twice a month in the 10:00 P.M. Sunday slot, and the first show featured Peter Panic and Flatbush, who were very hot on the Cleveland club scene. Masky told the *Plain Dealer* that he moved the show because WMMS supported local music in its regular programming, along with the *Coffeebreak Concerts*. He also revealed plans for an *Ohio Homegrown* LP to be produced in collaboration with WMMS.

But it was college radio that was breaking some of the more eclectic artists and definitely playing more local music than any other station. Diane DeFrasia of the Phantasy nightclub says that her club, along with college radio, "Helped the music scene in Cleveland by getting in the underground bands. Otherwise, it was strictly Bruce Springsteen." There's a lot of truth in that statement. While many of the new wave and punk bands got little airplay, they could still sell out clubs and larger venues based on word of mouth and college airplay. When the bands were too big to ignore and if they didn't stray too far from the musical boundaries at the commercial stations, they would start getting greater exposure.

It became evident to the other stations that WMMS was out-promoting them at just about every turn, and Dan Garfinkel was rewarded for his efforts by being promoted to director of advertising and marketing for WMMS and WHK-AM. Station M-105 tried to counter by naming a former record industry manager to its promotions post. It also rechristened itself as "Cleveland's Classic Rock," a term that would prove to be a thorn in the side of WMMS in years to come. Even so, it was very tough to battle WMMS when the station was getting sponsorship of most big-name shows coming through the city, including the huge World Series of Rock concerts at Cleveland Municipal Stadium. The station was able to make those 80,000-plus-patron shows even bigger by putting Matt the Cat and Al "The Bear" Koski on scene for live broadcasts through the day, interviewing headliners like Bob Seger, Eddie Money, Def Leppard, and the J. Geils Band.

At this point, every station was scrambling for radio turf, grabbing any opportunity that might get them extra publicity. M-105's Doug Johnson even did a fire-eating act on WEWS-TV's *Morning Exchange,* and the man who took t-shirt art to a new level, WMMS's Buzzard creator David Helton, was getting national recognition once again doing cover art for two radio station promo compilations from Epic Records featuring Cheap Trick's "Daytripper" and New Musik's "Straight Lines."

Kid Leo's swagger, style, and attitude had become as much a part of Cleveland as Lake Erie, and he gave the competition a multimedia whammy for the important fall 1980 ratings race. First, he made his big-screen debut in Paul Simon's *One Trick Pony,* and soon afterward he was seen as a member of the "Buzzards" street gang in Second City's *From Cleveland* television special aired on CBS and locally on WJKW-TV. And Jeff Kinzbach was getting press over reported "feelers" from

New York powerhouse WABC. Cleveland may have been weathering bad times, but its radio was proving to be among the best in the nation.

Football fans were enjoying ABC's *Monday Night Football* on the evening of December 8 when a distraught Howard Cosell announced that John Lennon had been gunned down outside the Dakota in New York City. Lennon had appeared with Cosell on the NFL show, and his death had an impact that could be likened to the killing of John F. Kennedy. Like their counterparts across the globe, Cleveland radio went on disaster alert, immediately suspending regular programming to play Lennon's music and taking calls from stunned listeners. WMMS aired Denny Sanders's 1974 phone interview with Lennon, and the murder sparked massive international mourning, drawn together in great part by radio. WMMS sponsored a candlelight vigil the next evening on Chester Commons. On December 10, Yoko Ono issued a statement asking for a ten minute vigil of silence to be held at 2:00 P.M. EST the following Sunday. M-105 and WMMS complied, as did a number of other stations, by going off the air for that time so people could reflect on the impact Lennon had on their lives. It also helped to show the continuing impact of rock and roll radio.

There were pivotal moves under way at some of the stations. Walt Tiburski had been named general manager of WMMS, and Jon Anderson left his communication and promotion director's job at the Blossom Music Center concert venue to take over the promotions job at WGCL. Anderson was well schooled in the record industry and in dealing with artists, and he would earn his stripes in a relatively short time.

In December a new voice was added to WMMS. Dia Stein, who came up from Florida, was very aware of the legend of the station and especially of Leo.

Born and raised in Albany, New York, Stein admits she was a bit naive about the medium, thinking AM and FM stood for "American music" and "foreign music." She worked at a number of stations around the country during her rather young college years, pursuing a journalism degree at Syracuse University at age sixteen. Ending up in Tampa, Stein was invited to a reception at the city's Agora Ballroom franchise by chain owner and Clevelander Hank Lo Conti. It was there that she caught sight of Leo and his very pregnant wife, Jackie. Lo Conti encouraged her to introduce herself. Mustering up her courage, Stein walked over to say hello. Leo was familiar with her, having heard her show the previous day, and he offered her a job on the spot.

Stein remembers her first day as being very intimidating. "It was a Friday," she recalls, "and I thought I'd go in the production studio to do some whiz bang stuff and make them marvel at my creativity! I grabbed some records out of the studio, but little did I know that on Friday afternoons Leo played 'Born to Run,' and 'Friday on My Mind,' and 'Cleveland Rocks.' As it happened, I borrowed 'Born to Run' for a montage I was trying to make, and the door flew open and Leo was furious! He yelled at me, yelled at me, yelled at me—my first day! I went into the bathroom and just cried! I thought, 'Oh my God! My career is over!' That was pretty much my first day."

Better days were ahead for both Stein and the station.

Dominance and Decay

M-105 was seeing a lot of change in the early months of 1981, most notably veteran Eric Stevens leaving the program director's post, and radio altogether, to form his own creative consulting firm. Another vet, disc jockey Bill Stallings announced his resignation, and Doug Johnson moved to overnights, doing his *Midnight Express* from midnight to 6:00.

It would be a herculean, if not near impossible, task to derail the WMMS machine, and if someone worked mornings at that time at any other station, they might as well have kept their bags packed. More bad news for the other stations came in February 1981 when it was announced that Jeff Kinzbach had decided to stay in Cleveland, despite offers from New York's WABC-FM. Mayor George Voinovich honored the station by proclaiming February 18 "Buzzard Day in Cleveland" after WMMS was named for the second time in a row "Radio Station of the Year" by *Rolling Stone* magazine. The proclamation was delivered to Matt the Cat during a *Coffeebreak Concert* broadcast with the band Wild Horses.

Just a couple notches up the dial an upheaval was underway that would change the radio landscape. WZAK-FM had become the city's only full ethnic broadcast outlet after WXEN-FM switched to dance music and became WZZP, Zip 106. The Zapis family owned the station and decided to pull the plug on WZAK's ethnic format in March, switching to an adult contemporary sound. The final ethnic show on the station was Duane Dobies's *Polka Convoy*, and with the fading strains of an accordion the full-time nationality shows on WZAK passed into history. The next morning at 6:00 A.M. the new WZAK format was inaugurated with Lenny Williams's "Fancy Dancer."

Owner Lee Zapis recalls,

> WXEN had dropped the ethnic format to switch to WZZP. In doing so, they gave all of their producers a thirty-day notice. They used that month-long period to organize rallies to fight the format switch. Ultimately, they failed, but the FCC still had the authority at that time to dictate a station's format. When it came time for us to change formats, instead of giving a thirty-day notice, we just informed them the day of the switch and just did it. By the time we had switched formats, the FCC was out of the business of dictating for-

mats. Also, as stupid as this sounds, we still kept the ethnic programs on Sunday in an effort to maintain some revenue. My father finally agreed after a few months to dump those shows altogether.

By spring 1981 Kinzbach and Ed Ferenc had decided to spread their wings and market themselves more extensively outside the confines of their radio show. The result was the club Jeff and Flash's Monopolies, which opened in May in the old Starz Nightclub in Lorain. Together with veteran club owners Gary Bauer and Geoff George, they offered live rock Tuesdays through Sundays. For the first couple of weeks, the two hosted evenings at the remodeled club every night, but they still had to maintain their show in the morning. After a while, they did weekend duty at Monopolies or the occasional special event. The *Plain Dealer*'s Mary Strassmeyer speculated in her daily column that the opportunity with Monopolies might have been a deciding factor in Kinzbach's decision to stay in Northeast Ohio.

The club design was based on Parker Brother's Monopoly game board and highlighted Cleveland-area landmarks. George, who had remodeled clubs around the country, oversaw the design. Bauer had a lot of club experience over the years as owner of Akron's Urban Cowboy Saloon, and he had managed the Cleveland Agora. He served as operations manager for Monopolies. It opened to great fanfare with searchlights and Kinzbach and Ferenc arriving by limousine, wearing top hat and tails, walking down a red carpet. It would not be the morning-show personnel's last venture into the nightclub business.

The club got a lot of attention, and the January through March ratings period showed a huge gain for the WMMS audience. The station ended up dumping easy listening WQAL-FM from the number one spot, and the WMMS morning show beat the top-rated Gary Dee on sister station WHK-AM. This was no small feat, considering that young listeners eighteen to thirty-four were generally harder to track because they usually didn't fill out ratings surveys. Even though this Arbitron ratings period was extended by a month, the surge in youth numbers was a surprise. Nonetheless, WMMS had the numbers, and those were big-money demographics to advertisers.

Radio was still getting a lot of attention from television, with WKYC-TV's Ed Miller profiling disc jockeys such as Kinbach and Ferenc and WGAR-AM's John Lanigan on his *On the Radio* feature reports, and Kid Leo continuing his on-air work with WJKW FM's *PM Magazine*. Leo also won national attention with an interview in the May 22 issue of industry trade paper *Radio and Records*. He told them, "People live rock and roll in Cleveland. It's more than recreation; it seems to be a political statement." Leo said the city's obsession with the music had led many to hear a "Cleveland sound." Noting artists such as Bruce Springsteen, David Bowie, Bob Seger, Roxy Music, and Michael Stanley—all with firm footholds on Northeast Ohio airwaves—Leo said, "I couldn't tell you to put a hook here, a chorus there, use this tempo or that beat and you'd come up with a Cleveland sound. But I can tell you it has something to do with rock and roll with soul. They're not following formulas; they're always taking chances."

Former WMMS staff member Jeff Gelb, who had been watching the station's climb for years from the West Coast, wrote the article. It was a much different station from the one he left, and it was far more aggressive. As John Gorman recalls, despite the lofty ratings, "The thing about being number one is that everybody's out to get us. Our numbers put us in a position where we have to fight harder than ever." WMMS was perfectly willing to take on all comers.

Leo was also pushing new music in a taped Sunday night program called *The Grand Jury of Rock and Roll*. It was an interactive show—or as interactive as it could be at that time—with Leo presenting five new releases and asking listeners to phone in their "verdict." The show featured a good number of imports and "out of the box" artists and showcased some very promising bands, like Franke and the Knockouts.

Competing stations in the market were hard pressed to battle WMMS on its home turf. The station was just as aggressive on the street as it was on the air, often humiliating deejays from other stations at joint appearances. M-105 and its management had long been labeled by WMMS as "the Chimp," and the same thing went for WGCL-FM's "Baboons." There was an *esprit de corps* at WMMS as it rode the tidal wave of ratings, and the embarrassing moments struck hard at the employees of their competitors. WMMS artist David Helton was commissioned to do a series of vicious memos, comic books, and other artwork that often "found their way" to the opposing stations. For the most part, managers at other stations claimed that level of competitiveness was petty and immature and brushed off the continuing assault from the station. Whether it was true or not, WMMS still had a huge war chest, a very aggressive promotional department, and management that would have made General George Patton proud. They struck hard at their competition, beat them on the air, and tried to break their spirit at every turn.

One of the measurements of that success was something called "exclusive cume," the cumulative audience that would note only WMMS in their Arbitron surveys. WMMS had extremely high exclusive cume, with some rating periods showing its exclusive cume was higher than the total cume of other radio stations. As Dan Garfinkel puts it, "Those were people who were afraid they would miss something if they didn't listen to WMMS. We were breaking exclusives, insane contests, and there were so many people coming by the station. There was always something going on."

Things weren't as rosy for the city of Cleveland, though, on the national scene. Comedians took mean-spirited jabs at the city, and anyone looking for a cheap laugh could usually throw the word "Cleveland" into the mix and get the results they wanted. The worst part was that pride in the city continued to slip, despite some serious gains in trying to reverse the civic malaise that had settled in years before. The rest of the nation took shots at the city, and for the most part, people in Cleveland were willing to accept them—to a point.

Randy Youngman was a sportswriter for the *Dallas–Times Herald,* and on April 25 he decided to take the low road by launching a volley against the city. Relying on worn-out phrases like "the mistake by the lake," comments about the weather, and even the city's nightlife, Youngman proceeded to rip apart anything even remotely connected to Cleveland. He even took on a Cleveland civic campaign that likened Northeast Ohio to a plum, writing, "If New York is the Big Apple, then Cleveland is the Big Prune (and you know what happens when you eat too many prunes)." He claimed, "The only thing worse than the city's air pollution is the water pollution in Lake Erie (no swimming allowed) and the Cuyahoga River."

There were other snide, and inaccurate, comments, and that didn't sit well with Kinzbach. Soon after, Kinzbach went on the air to rail against the writer, saying, "I think Mr. Youngman needs his head straightened out." Cleveland radio was one of the few consistent high points for the city, and it quickly became a crusade, with Kinzbach urging listeners to write Youngman and his newspaper to retract the column. If that didn't work, WMMS would fill the stands during the next Indians-Rangers game for a very vocal protest, but that didn't stop Youngman.

He wrote a second column, mocking the campaign and taking even more shots at Cleveland. That prompted Kinzbach to call for a listener boycott of the Indians-Rangers game on July 7. He told the *Plain Dealer* that Clevelanders could "support the Indians other days, but boycott the Rangers game. That will hurt them in their pockets. I don't want to carry on a feud with an irresponsible sports writer. Clevelanders don't have to lower themselves to that."

In truth, Cleveland fans didn't support the team the way they should have in those days anyway. As the Tribe slid down the standings, the few thousand that showed up for games by July would make the cavernous Municipal Stadium look nearly empty. As for hurting them in the pocketbooks, it was probably the vendors that took the biggest hit. In those days, attendance could be so sparse that a fan could have had their own personal hot dog and beer vendor. If a player hit a ball into the stands, a fan could take a relaxing walk to retrieve it without fear of having to fight another fan for the souvenir.

Perhaps the boycott made a little difference, but it also made for some great radio, and on a higher level helped to show that Clevelanders were tired of the jokes and had allies on the FM dial. Youngman later spoke with Ferenc about the incident, claiming he got more than eight hundred letters of complaint, but he refused to apologize. He also admitted that he really wasn't that "down" on the city and said that people in Cleveland should lighten up and show they have a better sense of humor.

It was during this same time that the major league baseball strike had hit a nerve with Cleveland fans. A prolonged walkout could ultimately mean cancellation of the season, and while the Indians were far from contenders, a good number of people still enjoyed a day at the ballpark. Not only that, but the midsummer classic, the All Star Game, was scheduled for Municipal Stadium and would be put on hold until the strike was settled. Angered by the walkout, Kinzbach took to the air to gather hundreds of Clevelanders for a "Big Boo" to protest the strike. The crowds gathered at a Party in the Park to vent their anger in a 130-decibel scream that drew the attention of the *Guinness Book of World Records.*

And there was another fight brewing, this one off the air, that would affect the entire radio market. It centered on a very successful local businessman who for years had recognition from his frequent radio advertisements that identified him as the Diamond Man. In July 1981 Cleveland jewelry merchant Larry "J. B." Robinson agreed to purchase M-105 and sister station WBBG-AM from the Cleveland-based Embrescia Broadcasting Corporation for a reported four million dollars. He established Robinson Communications and asked the station's management, including president Thomas Embrescia and vice president James Embrescia, to stay on. While WBBG was switching over to big band sounds, Robinson stressed that M-105's format would remain intact, and the only hurdle left was to win FCC approval.

That would prove to be a bit more difficult than originally thought. Former Cleveland mayor Carl Stokes said that he had a verbal agreement to buy those stations, telling the *Cleveland Press,* "I didn't know that it happened! I'm not only surprised, but I would anticipate filing a lawsuit in federal court against the Embrescias within a week for dealing in bad faith. We had agreed on all the essential terms except the actual signature." He went on to say, "There will be no transfer of the [station's] license for the next three years until Chief Justice Warren Burger has spoken about it!"

Robinson would only say he had "a very high opinion of Carl Stokes," while Embrescia said, "I tried to call Carl, to alert him about the sale. I left a message, and he didn't return my call." The fight was far from over.

Baseball Buzzards

One of the more notorious bands to emerge from the Los Angeles music scene was the Plasmatics, led by former porno film queen Wendy O. Williams. She was now sporting a mohawk, destroying televisions, and even riding speeding cars into walls of fire. Williams was also prone to disrobing at concerts, sending up red flags for the vice squad, and resulting in two heavily publicized decency trials—

including one for her Cleveland Agora show. The furor was just starting to subside when Belkin Productions announced that the Plasmatics would be making another assault on Cleveland at a Music Hall show in July.

Williams had gained a lot of notoriety with a stage act heavy on nudity and destruction. After court battles in Cleveland and Milwaukee, it wasn't likely that Williams would be showing as much skin at a Cleveland Music Hall show in July 1981. But the onstage vandalism was a whole different issue, and the band joined with WMMS to use it to their advantage. Williams had gained a reputation for doing death-defying stunts, like riding atop a speeding car that smashed into a wall of televisions. Since the average listener probably didn't have a pile of televisions ready to be scrapped, and almost everyone had that one problem set that just wouldn't die, WMMS offered the Weekend of Lunacy, a contest in which the grand-prize winner would get a visit from Williams and her sledge hammer, and the television would soon be on its way to Boot Hill. In return, the station provided the victim with a brand new, state-of-the-art set, and four runners-up won mohawk haircuts—or one that more fit their personalities—from a local stylist. The band also got a lot of publicity at a Michigan show where it blew up a car on stage, and the promoters began searching for a 1969 Chevy Nova to do the same in Cleveland.

It wasn't long before city officials pulled the plug on the upcoming Plasmatics concert. Fearing another wave of negative publicity, Cleveland fire officials denied the band the needed permits for its pyrotechnic stage show, the first city on the tour to do so. In response the Plasmatics claimed they didn't want to "compromise their artistic integrity" by not giving a complete performance and refunded the ticket money. But that didn't mean that Williams and friends wouldn't be appearing in Cleveland.

The good news centered on the settlement of the baseball strike, and plenty of sports stars visited the city. Cleveland had just hosted baseball's All-Star classic at Municipal Stadium, and WMMS capitalized on the renewed interest following the players strike to stage its own game with high profile players. The station's Baseball Buzzards met a group of local athletes and folks from the music community calling themselves the Will Rogers Celebrity All Stars in a benefit game at the Byron Junior High field in Shaker Heights. The $1.01 admission went to the Will Rogers Memorial Fund, which aided those in the entertainment industry, but it's unlikely the fund's namesake could have ever envisioned the lineup that took the field that Sunday in July.

The Buzzards team included staff members Denny Sanders, Jeff Kinzbach, and Len Goldberg, with Mike Stanton, Joe Charboneau, and Duane Kuiper of the Indians acting as "ringers," ready to show a little muscle on the field if the WMMS team needed runs fast. The Will Rogers team included the Tribe's Mike Hargrove and Len Barker; announcer Mudcat Grant; Paul Warfield of Cleveland Browns fame; Cavaliers basketball team favorite Bingo Smith; promoter Mike Belkin; David Spero; and Michael Stanley, Tommy Dobeck, and other members of the Michael Stanley Band. Rounding out the rosters for both teams were Fee Waybill of the Tubes, Franke from Franke and the Knockouts (who had opened for the Tubes the night before at the Richfield Coliseum), Alex Bevan, the Tribe's Wayne Garland, and wearing a customized Baseball Buzzards t-shirt, in her debut as a pinch runner, the infamous Williams herself.

It was a great day for baseball, and the 1,200 or so fans who showed up got more than their money's worth. Barker took the mound for the All-Stars and threw a lot of hits in the early innings, and teammate Charboneau knocked the hide off the ball for the Buzzards time and again. Williams proved to be far more valuable on stage than on the field, though no less reckless. She and All-Stars catcher Dobeck bonded more than once at the plate, plowing into the drummer as she slid home, and another time Dobeck tagged her in the butt as she raced in to score. A photo of that fateful meeting later showed up in the pages of *Billboard Magazine*.

The rules went out the window early in the game, with Sanders having to be carried off the mound, and fans swarming the field for autographs before the end. When the smoke cleared, the Buzzards were the clear winners, 26 to 8. Plus, the event raise more than $1,200 for the charity, and in the end Williams was able to show she wasn't nearly as dangerous as people might have thought.

The Long March Continues

WMMS continued to strengthen its grip on the Cleveland market through the summer of 1981, ranking number one overall, with Jeff Kinzbach in a dog fight with Gary Dee, who barely edged out WMMS's drive time voice as the city's top morning personality.

The station also continued to showcase local talent, though more sparingly. The *Coffeebreak Concerts* continued and listener Jerry Bush got airtime on WMMS after sending in a demo tape, as did Wild Horses, who even played a Geauga Lake Park concert for the station. M-105 was also playing Wild Horses and other local acts, though the major part of the programming day went to international acts. There were a number of promising bands in Northeast Ohio, and one in particular had an impact on WMMS.

By August 1981 Dan Garfinkel was starting to feel a bit frustrated and was thinking about a move. He says, "Back in those days, broadcasters were limited to owning seven AM and seven FM stations. Within a broadcasting company like that, there weren't a lot of opportunities to progress. Part of it was that I was sort of

The Michael Stanley Band (1983). From left, Michael Gismondi, Danny Powers, Kevin Raleigh, Michael Stanley, Bob Pelander, and Tommy Dobeck.

starting to repeat myself. It wasn't as challenging as it had been in the past six years, and I had gotten involved with a band called the Generators, whom I thought—and still think—had tremendous potential. A friend of mine was an attorney, and was also interested, so we decided we would start a managing company to see if we could get the Generators a record deal."

With the Indians again winding down a dismal, albeit strike-shortened, season, attention was diverted to Cleveland Browns football. The "winningest team in football" at the time, the Browns had come tantalizingly close to a playoff berth the winter before, and fans were hoping for an even more impressive finish in 1981. WMMS gathered the staff again in September for another loud showing at a Party in the Park, but this time it was a "Big Cheer" welcoming the Browns match-up against the San Diego Chargers.

The fans gathered to cheer for some other hometown heroes as well, buying out three appearances by the Michael Stanley Band at Blossom Music Center. It was a major event for Northeast Ohio rock fans, and WMMS hoped to capitalize on it by broadcasting one of the shows live from the venue. The problem was that it had never been done from Blossom before, so engineer Frank Foti, a veteran of the Todd Rundgren–New Year's Eve–Richfield Coliseum concert fiasco a few years earlier, was again called in to do the impossible. He recalls the meetings leading up to that broadcast:

> Blossom Music Center had the same problem as the Coliseum. We couldn't get the special audio lines. I said, "You know guys, maybe there's a way we can get a microwave signal from Cuyahoga Falls to the transmitter in Seven Hills, and then we might be able to do it." I went out on the top of the pavilion with a couple of stage hands, and we did a test to see if we could do what would prove to be the first broadcast of its kind from Blossom. Denny [Sanders] and John [Gorman] were all excited about it, and again, we had to have somebody out at the transmitter to switch the feed on. Denny gave us the cue to go live, and it sounded great—but this was also a time when wireless microphones were a new concept. Michael and another guitar player were on wireless mikes, and no one knew that one of the wireless mikes was on the same frequency as the State Highway Patrol! Right in the middle of one of the songs, we hear something like, "One Adam Twelve"—a police call that went out over the air and into the audience! It didn't happen again that night, but to this day, anyone who remembers that concert always brings up the cop in the middle of the song.

The concert has shown up in bootleg tapes over the years, and the call can be heard as plain as day. But Stanley recalls that in the early days of wireless stage gear, Cleveland was one of the few places that a musician could use that type of equipment. He says, "There were too many radio stations. Chicago was worse than New York, because you have to have a certain amount of free band space to pick up the mike on the guitar. If you didn't have enough bandwidth, you were sunk. We were playing some venue near Chicago, and we were in the middle of 'He Can't Love You,' or 'Lover,' one of those—and all of a sudden, we were overtaken by some kind of meringue music from a Spanish station!"

Support from Cleveland radio gave the Michael Stanley Band a huge home base, drawing some critical attention from other parts of the country, including *Rolling Stone* magazine. But while Stanley was able to make inroads in some other markets, most didn't offer the same type of "pride of Cleveland" airplay that made the band superstars in their own hometown. Stanley, Kevin Raleigh, and their bandmates returned the favor by filling in as disc jockeys at M-105, WMMS, and WGCL-FM.

WMMS got a bit of a jolt in the summer Arbitron ratings when the soft sounds of WDOK-FM took over the number one spot, dropping WMMS to number two, and M-105 ranked eleventh. Like in any war, there is continued fighting highlighted by major battles, and that was the case in Cleveland radio in 1981. WMMS continued its close association with the Michael Stanley Band by announcing a two-concert run set for New Year's Eve and New Year's Day that would allow the listeners to pick the songs the band performed. And M-105 capitalized on a storyline in local artist Tom Batiuk's *Funky Winkerbean* comic strip when Bill Stallings hosted the Air Guitar Night at the Cleveland Agora. The strip's storyline featured Crazy Harry visiting the Agora to compete in the contest in a "national air guitar" championship, and Batiuk himself was on hand to help judge the proceedings. It made sense for M-105 to get close to Batiuk after he had provided rival WMMS with publicity in the comic strip, which was seen in over three hundred newspapers. Batiuk had been a fan of WMMS even before the Buzzard was born and continued a warm relationship with the staff. M-105 apparently wanted in on the publicity as well.

But another concert battle would prove to be one that got the most headline space, and that was with the Rolling Stones at the Richfield Coliseum. WMMS had been laying the groundwork to own that show before the concerts were even announced. In October the station was announced as the "official Northeast Ohio Rolling Stones radio station" at a Chicago news conference by Jovan, the perfume company that was sponsoring the tour. That helped guarantee the station access to exclusive information about the upcoming Cleveland dates, as well as other shows around the country and a planned surprise at the end of the tour.

Concert dates for the Coliseum were soon announced for November 16 and 17, and the shows promised to be the hot tickets of the year. That was quickly followed by a contest that WMMS labeled Stoned Alive, a forced-listening affair that challenged listeners to write down the title of every Stones song played over a forty-eight-hour, late October weekend, with a pair of tickets going to each of twenty-five winners. The contest drew such a huge response that a second phase was mounted the very next weekend. WMMS came back with yet another wave of tickets in phase three of "Stoned Alive" and with a mail-in contest for a ten-seat loge to the show.

M-105 aired special Rolling Stones programming and a contest consisting of reciting hundreds of Stones tunes in alphabetical order, with chances for free tickets to the shows becoming greater if a contestant didn't miss a title. The station also had tickets for the listener who correctly identified the number of times a special Stones–M-105 jingle was played in a set period of time. Despite such gimmicks, WMMS's own promotional war chest, along with some very clever programming and a full head of steam from the ratings, made the station an almost unbeatable opponent in the ratings war.

The importance of the shows in Cleveland was obvious. The Rolling Stones tour was the biggest ever in music history, getting some kind of national publicity on an almost daily basis, and the *Plain Dealer* announced plans for a special supplement to showcase the Richfield Coliseum concerts. When the Stones finally arrived in Cleveland, it was nearly impossible to escape their influence. Bassist Bill

Wyman phoned Kid Leo for a one-hour talk on his show and met with the press at Swingo's (the group's Cleveland base) to promote his solo record, "Je Suis un Rock Star." WMMS's Betty Korvan presented Wyman with three Cleveland Browns jerseys brought in by listeners. Plans were announced for the *Rolling Stones Christmas Concert* pay-per-view special, to be simulcasted on Cleveland television station WCLQ-TV's Preview service and WMMS.

The concerts drew widespread praise, even if Wyman thought the Monday show was mediocre and plagued with equipment problems. At the Tuesday show, it just didn't matter. For the final songs, Mick Jagger did a quick costume change and bounded onto the stage wearing one of the Browns jerseys that Korvan passed on to Wyman. The station had scored a decisive promotional win over its competition and within days issued a new shirt of its own. Sales benefited Cuyahoga County's Council of the Disabled American Veterans and read "Ruler of the Airwaves." These were very heady times for WMMS.

In October, former Cleveland mayor Carl Stokes filed a petition with the FCC to block the sale of M-105 and sister station WBBG-AM to Larry Robinson and his Robinson Communications. Stokes told the FCC that he had a verbal agreement to purchase the stations but cited "a situation riddled with fraud and deceit" and claimed that the current owners used his offer to leverage more money from Robinson. The petition claimed Stokes made a written offer totaling $2.2 million to buy the stations in January 1981, though the offer was later reduced to $1.95 million. Stokes said he was surprised to read in the July 20 edition of the *Cleveland Press* that the stations would be sold to Robinson for about $4 million. He also claimed that FCC records would show Robinson actually paid $2.6 million for those properties.

Stokes noted that even though Cleveland's population was 42 percent African American, there were no black station owners in the city. He stated that the group that owned the station—developers Nick Mileti and Joseph Zingale and broadcasters Thomas and James Embrescia—along with their lawyer, persuaded him to take over the stations, even offering to help him raise the money to buy them. The petition claimed that WBBG was losing $20,000 a month and that the Embrescias thought a black-oriented news and talk station could be a moneymaker. Thinking that his offer had been accepted, Stokes said he spent more than $36,000 in legal fees preparing to take over the stations. Stokes wanted the FCC to stop the sale. Robinson was said to be surprised by the move and was quoted as saying, "There is absolutely no way I can understand why Carl Stokes would sue me." Responding to claims of a verbal agreement, Zingale told the *Plain Dealer*, "I don't think that was the case at all."

Another One Bites the Dust

Some very significant changes were put into effect at various spots on the dial early in 1982 in the effort to tap into the huge audience that WMMS continued to gather. WZAK-FM's management put some serious money into commercials starring Tim Reid, aka Venus Flytrap, from the television series *WKRP in Cincinnati*. WZAK placed ten- and thirty-second spots featuring Reid in primetime programming to help stake its claim to Cleveland's underserved urban audience. And in December 1981 M-105 brought back a familiar voice by adding former WMMS personality Al Koski to the lineup. The former "Bear" was using a different name now, Jeff, and held court overnights and Sundays.

But WMMS continued to reap the rewards of local and national press and its own energetic promotions and marketing staff. One rule of thumb from that era is credited to Dan Garfinkel, who first pushed the Buzzard as a symbol for the station and the city. The goal was for motorists heading into Cleveland for work every day to see the call letters and mascot at least three times before they reached their jobs. Strategic placement of billboards, print and television coverage; carefully planned promotional events; and excellent presentation and graphics made the Buzzard a very attractive package and at the same time highly visible. Jim Marchyshyn, Garfinkel's successor at WMMS, tips his hat to a system that was in place when he came on board. "Dan Garfinkel was the godfather of the whole promotions movement in FM rock radio," according to Marchyshyn, "because he made WMMS a big part of the community. I built on what was already in place."

WMMS linked itself to the Preview Television system with the Rolling Stones pay-per-view concert finale by simulcasting the stereo sound. That was a promotion that was up in the air until almost the very last minute, as stereo lines for the show weren't connected to the station until an hour before the Stones took the stage. Even so it was a great success, and another big audience tuned in to the station for the full audio experience. At a Wednesday-afternoon *Coffeebreak Concert*, Matt the Cat announced that *Rolling Stone* magazine's readers' poll had named WMMS its station of the year for the third consecutive time. On February 17, the day after the readers' poll issue hit the stands, Cleveland Mayor George Voinovich once again declared it "Buzzard Day" in the city and flew the station's flag in front of city hall.

That sparked a media blitz, with coverage in the *Plain Dealer, Cleveland Press*, and local and national magazines. The *Wooster Daily Record*'s Bob Dyer wrote that some Wooster residents refused to move to a certain neighborhood because they couldn't pick up the station's signal as well in that part of town. He also headlined stories like "Kid Leo: Godfather of Rock," and "Sex & Drugs & Rock and Roll" (about how they don't really mix). The *Cleveland Press* did an equally flattering piece in its business section, focusing on the station's marketing strategies. Despite its continuing success, WMMS still felt some stiff competition from WGCL-FM in a number of critical areas.

Denny Sanders explains,

> By the early eighties, AOR radio, like it or not, had deteriorated to essentially a nostalgia service, and when they did play current music it was by established acts. I remember as late as '81 or '82 looking at the AOR playlist adds in *Radio and Records,* and it was Aerosmith, Eric Clapton, Moody Blues, J. Geils Band. The damned list could have been printed seven years before! WMMS was always an unusual station, because we had a pop sensibility that a lot of AORs didn't have, which could be traced all the way back to David Bowie. When we were first playing Bowie in Cleveland, the height of playing *Ziggy Stardust* and *Hunky Dory,* and about the time he played the Music Hall show, I went back to Boston to visit some friends. They never heard of Bowie, WBCN was not playing Bowie, but instead played a lot of hard rock, Emerson Lake and Palmer, Yes, Crosby, Stills, Nash, and Young, and the Band. A lot of AOR programmers thought David Bowie was too pop. That later changed. We were just ahead of the curve.
>
> We tended to be ahead of the curve on a lot of things that had a pop sensibility. We had a sort of "Cleveland pop" feel, so we never had a problem playing some pop oriented material. I remember one of Leo's favorite tracks was "Dancing Queen" by Abba! He used to play it in the afternoon, and it was wonderfully produced and infectious record. If you listen to it, it's not too far off from the Ronettes! We played music from Jimmy Cliff's *The Harder They Come,* Madonna, and the station was sort of to radio what *Rolling Stone* was to rock journalism. There wasn't always a rock star on the cover. Sometimes it could be Bob Hope, or a movie star, or a political figure. But it was written with a rock and roll attitude for a rock and roll audience. We played selective Madonna, and certain material like that, but we played it with a rock and roll attitude and it was very heavily day-parted, so it was downplayed at certain times of the broadcast day. We were also very careful how it rolled out on the air. You would never hear Madonna go into a Michael Jackson, the same way you would never hear a Led Zeppelin into a Deep Purple.
>
> By the time we started to facelift the station in the eighties, we sort of had three segments of music. We had hard rock like Aerosmith, Zeppelin, and so on. Then you had the other side, which was Psychedelic Furs, OMD, Tears for Fears, U2, INXS, and the like. Then there were the middle acts, which had credibility with both crowds. The Pretenders, the Kinks, the Rolling Stones—and those were always the link acts, so a typical music flow at WMMS was Deep Purple

into the Pretenders into INXS. Or Tears for Fears into the Kinks into Black Sabbath. It was a very successful and smart way to go. But there was a problem.

We were the first in the market to play Prince's "Little Red Corvette." I remember a conversation with Leo about a show on the USA Network called *Night Flight*. It was on overnights, and was a lot hipper show than a lot of the MTV stuff. They had a live Prince concert clip, and I was riveted by it. Leo saw it, too, and we both said we were missing the boat! We were doing these Foreigner block parties, and John Gorman agreed. He liked Prince, too, and we played "Corvette." The audience went for it, and now it's time to do some giveaways. Warner Brothers told us, "You can't give away Prince LPs! He's a CHR [Contemporary Hit Record] act, not AOR! You have to go to the CHR division, and he doesn't deal with you!" Meanwhile, WGCL is giving away the Prince album, and decide to add "Corvette" five or six weeks after we did! Then it's time to do the show, and we can't co-present it! Warner Brothers wants to give it to the CHR station. Same thing with INXS. They call it a CHR act! Psychedelic Furs? CHR! All the new acts are being called CHR! They offered us the new Moody Blues, but by this time the call letters are getting old, and you can go down the crapper real fast with an old set of call letters and the image that you're behind the times. But we were all still young and vibrant, and listening to new music that we wanted to play—and we did.

John wanted to put that to a stop, called up *R & R*, and said, "Okay. We're a CHR station." We didn't change a thing on the air, but just changed our designation. Some of that hit the press, and they said, "WMMS is going Top-40!" Bullshit! We merely changed the *R & R* configuration so we could credit for the acts we were breaking and playing!

WMMS would find itself fighting that battle on a much larger scale in the weeks and months to come with what turned out to be a very controversial advertisement.

Lynn Tolliver had distinguished himself at a number of Cleveland stations, and in the spring of 1982 he signed on again at WZAK. Looking back, station owner Lee Zapis says, "Ray Calabrese, who was managing the Dazz Band at the time, kept telling us we should hire Tolliver. We finally did in May of 1982, and he was with us ever since."

M-105 was still pushing ahead, with promotions like the "$5000 Rock Guarantee" if someone heard less than five songs in a row before a commercial and was the first to point it out, they would get a check for that amount of money. It was a forced-listening contest that had proven to be very effective in the past for other stations and usually sparked a good amount of interest for people tuning in, hoping to make a quick buck and at the same time liking what they heard so they would keep it on their dial.

Lynn Tolliver (1977). Known as "Truckin' Lenny" in the early part of his career, Tolliver would rise to become one of the most prominent R&B programmers and air personalities in the country. Cleveland Press Collection/Cleveland State University

The "$5000 Guarantee" was eventually awarded to someone who caught a disc jockey playing less than five songs in a row. Public service spots, deejay talk, and comments from listeners were not included in the time allotment. It's anyone's guess whether this was the case at M-105, but many radio stations plan an exact time to award their grand prize to show that people really do win by listening to them.

But a second winner was definitely not planned, and there was another $5000 giveaway. It happened during a promotion with a local florist that gave listeners bouquets for Mother's Day. The deejay announced that the tenth caller would receive a prize package and roses from the florist, whom he mentioned by name. The phone quickly rang, and a voice said that he'd won the $5000! The disc jockey said that what was going on was a contest, but the guy on the other end said that by mentioning the name of the florist, it became a commercial, so the station ended up paying off a second winner.

Businessman Larry Robinson's influence as an owner was evident right from the start. He'd been a familiar voice on radio and seemed to truly love the medium. He had a lot of influence and plenty of stories to tell about his own experiences. Robinson had a keen interest in M-105 and would sometimes pop by the studio when morning host Marty Sobol was doing his show or visit deejays in other timeslots. That didn't sit well with some of the air staff, but Robinson was part owner of the station, and no one asked a person of his stature to leave.

Another voice that started to be heard in 1982 was producer Dave Sharp, who worked at M-105's AM sister station, WBBG, doing overnights and then did extra duty helping to produce the M-105 morning show. M-105 had a reputation for its relaxed atmosphere, but it was also a tightly run ship. In fact, the disc jockeys could be held financially responsible for their mistakes, with the money collected going into a fund for a year-end party. That could be seen in a staff memo from program director Phil DeMarne titled "Party Fund," and it read,

> Just a reminder of what your contributions to our party fund will be for errors:
> $25.00 Break format for whatever reason.
> $20.00 Miss announcer meeting (unless excused by me in advance)
> $10.00 Unprepared, miscue, or . . .
> $5.00 For failing to fill in transmitter reading logs.
> Let's hope the party fund is very small so the "Boys" will have to pay for it.

Sharp recalls how different the FM deejays were from most of the people associated with Robinson and his partners:

> I think it was at the Palace, and there was a big party there for all the agencies and buyers in Cleveland, and the jocks from both stations. We were all supposed to wear tuxedoes, and there was the AM staff—with Carl Reese, Ted Alexander, and Bill Randle—some of the great names in radio were there. Then there was the FM side, with guys like TR, and Bill Stallings, Jeff Koski, Jennie Cheeks, and Alan Sells, and I was kind of in with those guys. There was a free bar, and there was partying as the night went on, and the group kind of split

up into two factions. There were the people who were enjoying the food, the hors d'ouevres—the higher class people, and then there were the jocks. At one point, I saw one guy with his shirt off, and a couple of guys actually had their own bottles in their hands. It was hilarious. Bad behavior, but fun to see, and I think that was the nail in the coffin.

Sharp also recalls people anticipating a "Black Friday," with a major overhauling of the station and format. They wouldn't have to wait long.

In June 1982 the local radio scene was shaken by news that had been rumored for months. A staff meeting was called at M-105, and DeMarne started spreading the word that a big change was about to occur. M-105 would cease to exist in its current form at midnight, June 13, to be replaced by a soft-rock format, a new staff, and eventually the call letters WMJI-FM. That move came after a reported $50,000 research study on the musical tastes of Cleveland listeners, even though M-105 was showing strength in the latest Arbitron ratings.

Mike McVay replaced De Marne as program director, and he kept Chuck Collins on as production director and Jennie Cheeks, to be known as Jennifer Anderson, on the overnight shift. McVay had a long history programming radio and was last in Cleveland at WWWE-AM. But the exiting disc jockeys were not happy and still had plenty to say about the change. "We had really only found out about two weeks before," according to Bill Stallings, "and during that final week it started to really settle in. People were stunned, and obviously, not too happy. You just wonder what the hell you're going to do when that happens."

McVay explains that the decision to flip formats was not a hasty one, saying a lot of research was done prior to the change. It proved to be the right move, though Cheeks recalls the deep depression that settled over the whole staff: "Basically, we had nothing to say about it, and very soon after that, the new staff came in. It was only for a day or two, because they wanted to make it as clean a break as they could. I thought it was done very sloppily."

Oddly enough, M-105 had a fan in one of its bitterest enemies. Gorman says,

> I have nothing but the highest compliments for Eric Stevens, who's also a friend. Would I have spent all that time over M105 if they weren't a major threat? WMMS wouldn't have happened without M105, because they were a damned good competitor. Eric was a very good programmer, and when he signed on M105 it didn't sound anything like WMMS. I'll go on record saying he was actually doing a better job of being WMMS than we were! The first few weeks M105 was on the air showed me they were a serious threat, and I always treated them as a serious threat. I also always had the highest respect for Eric Stevens because he was taught by the best at WIXY, and his father was a record promoter, so he grew up in the business. So did David Spero, and that gives you a great advantage.

At 9:00 A.M. on the day of the announcement Sharp was walking through the newsroom, which had once been part of the old WIXY studios. He remembers, "It was either Dick or Tom Embrescia—one of the two—came up to me and said, 'Could I speak to you for a second in my office? It's private.' We sat down and he said, 'Listen. I want you on the air'—during the change—'and just basically babysit the station for about five or six hours. Just keep the format on. The other guys are going to be out of here, because we're going to have a transition from one station to the other.'" It was an unfortunate turn of events, but Sharp reported to the studio at the appointed hour. "There was no format anymore," he recalls, and some disc jockeys were said to be leaving a lasting impression, forcing Sharp to pull records to get him through the next five hours. "People were just coming in and grabbing albums by the armfuls and taking them out to their cars."

There was a pretty big party in the station parking lot, with some listeners stopping by to toast the staff. No one was certain what the new format was going to be, and Sharp started getting calls on the studio hotline from other deejays asking what was likely to happen. Later in the shift McVay stopped by the studio to introduce himself, and that Sunday it was McVay who went on the air to announce the station's plans. To kick off the new format, the very first song played was the Lovin' Spoonful's "Do You Believe in Magic?"

It's often said that the only constant in radio is change, and transitions like these are far too frequent. Radio is a passion, and when people work as closely as they do with a fairly small group, a deep sense of loyalty and friendship tends to develop, even though it seems that there are always goodbyes in the industry. Sharp says, "I was a young kid and really looked up to everyone on M-105. They were my heroes, and my friends, and I loved them all. I was real lucky to be part of it. At the same time, my loyalties seemed to be to myself, and having to do what I was asked to do. In honesty, I wasn't there much longer. For my career at the time, it was probably really stupid, and I probably did the wrong thing. But I was only 18, and it just wasn't my scene at the time."

The expected change had been the focus of water-cooler talk and the press for some time. After it finally occurred, Cleveland's weekly entertainment paper, *Scene*, sent writer Keith Rathbun to speak with Tom Embrescia, Robinson, Larry Pollock, and lawyer Wilton Sogg, who had all put the new station on the air. Robinson said, " We went through many months of research; we've now spent close to $80,000 in talking with Greater Clevelanders—asking them what they want in radio. That research will never stop, but we've (already) done enough that we have a pretty good idea of what's wanted, and that's what's being played right now on 105. All we're doing is giving the people back what they asked for."

Embrescia added, "What we started with in 1974–75 with Eric Stevens was one thing musically, and we've always shifted and evolved through the years since. There's no question that what we had back then was not necessarily what we had on Friday (June 11—"Classic Rock's" final day) nor necessarily what we had today (June 14—the kickoff of "Magic Radio"). Music—that's the key point—keeps shifting and changing. But the idea for the station was the same then as it is now: that's to be in the rock music scale, and rock is anywhere from 'GAR to 'GCL to WMMS to 'ZZP today depending what age group you're in."

The article went on to say that the enormous amount of research money went to The Research Group, headed by the respected Bill Moyes, a California consultant who had built one of the premier research organizations in the United States. As Robinson described it, "What they look for first is 'Where is the hole?' By that they mean, 'Where is the unmet need in radio in Cleveland?' What they've come up with is that the M105 listener has grown, and that the station should grow accordingly." Embrescia backed that up, saying,

> Basically, they came back to us and said that the (biggest potential) audience was growing older. The Baby Boom is over. When AOR started there was a big drive towards teens; now those teens are older. We're all part of that Baby Boom and now the biggest age group is right where we are so we've got to continue to play them on that basis. WMMS can't move (because they're so firmly positioned on top of the ratings). Their hard core listeners are 18–24, that's it. Who they used to appeal to five or ten years ago has grown older.
>
> We want to get out of that 5.9 share (a reflection of M105's last ratings book) and get into double digits. That's really where it's at for us. We don't want to be second to anybody any longer. We want to beat WMMS; beat WGCL; beat WZZP. Before we were always positioned 'Us against WMMS.' Today, we're saying we're in competition with everybody. We want to be #1 in the market. We're committed to do that. We're going to spend in the neighborhood of one half million dollars to support the station, to get the feel of what we are out on the street.
>
> We're going to be a hybrid, there's no question about it. We think that we'll take audience from WMMS; that 30-year-old who just can't hang in for a heavier rock sound. And we'll definitely take audience from people being lulled to sleep on 'ZZP. We're a hybrid. I guess that's a good term to use.

Embrescia also told *Scene* that research did support some of M-105's programming but not enough to keep that format. He cited the *Tasty Nuggets* show as one program the public really seemed to like, and it evolved into WMJI's *Lunchtime with the Oldies*. He also predicted a surge in the ratings, and he would eventually be proven correct.

The station unveiled its new lineup with Mike Ivers doing mornings, McVay for a two-hour shift until noon, followed by Ron Foster, Chris Eliot in afternoon drive, and WIXY and WGCL veteran Tom Kent overnights. The station made its programming statement perfectly clear, promising "your favorite songs from yesterday and today" and saying it would "encompass a wide spectrum of contemporary music." McVay was quoted in the *Plain Dealer* as saying he wanted to compete with WMMS but that "such a diversity of competition is much like going after General Mills, General Dynamics, and General Motors with a product called General Custer." In radio, it's wise to pick the fights, especially with a new format. But another station wouldn't follow that advice and took aim directly at WMMS.

WLYT-FM management decided to capitalize on the death of M-105 and WMMS's slow shift to more pop-oriented music. As a result in the final days of August 1982 the station began dumping its Top-40 format, changing to album-oriented rock. The "drop dead" date to have the new format in place was Labor Day, and new program director Tim Spencer said that while there would be some shifts in personnel, it wouldn't be anything near the mass firings at M-105.

WLYT was owned by United Broadcasting Company out of Bethesda, Maryland, and its president, Jerry Hroblak, told the *Plain Dealer* that rival WMMS had "broadened beyond AOR. They're trying to do something for everybody." His idea was to switch WLYT to pure AOR and force WMMS to back up and fight his station on its own turf. Hroblak also downplayed the importance of the WMMS promotional machine, saying, "I think people listen based on what they hear, not on what they see on billboards and TV."

WMMS didn't seem worried. It had already built up such a huge image and commanded so much attention that it only needed to go about its business and do radio the way the station thought best. WLYT, suffering from pockets of poor reception and fewer assets than WMMS, just wasn't much of a concern.

Meanwhile, Jeff Kinzbach and Ed Ferenc were announcing the closing of their Monopolies club in Lorain to look for a spot in downtown Cleveland. The two were minority owners in the club and were paid for the use of their names. The future of Monopolies was put in the hands of principal owners Geoff George and Gary Bauer, though the unemployment rate in the Lorain area probably sealed its fate some time before the official announcement.

WMMS scored a major coup with two of the most hotly anticipated concert events of 1982. The Who had been longtime favorites of the Cleveland audience, but the death of Keith Moon and later the horror of the Cincinnati show—where nine people were trampled to death as crowds pushed into the concert hall—had taken an emotional toll on the band. It had announced that its 1982 tour would be its last, and its final U.S. shows would be at the Richfield Coliseum. They were set for December 12 and 13, and WMMS knew it had to dominate those shows. They ended up guaranteeing two sell outs, and for that promise were allowed to have their call letters on the tickets. Immediately the other stations cried "foul." After being assured that there would be tickets set aside for their own giveaways, they had to deal with their prizes having their main competitor's call letters written

From left, Jules Belkin, Dean Thacker, Carl Hirsch, Milton Maltz, John Gorman, and Walt Tiburski (1982). The concert promoter and Malrite management salute the Who's tour sponsor after WMMS wins the right to get its call letters on every ticket for the two upcoming shows at the Coliseum. Photo by Janet Macoska

across the top. A promotions person at WGCL decided to cross out the letters with a black marker, and that turned into a public relations disaster.

The nights of both shows were bitterly cold, and when people who had the tickets with the crossed-out call letters tried to get into the Coliseum, ushers refused them entry because the tickets had been altered. There were a number of angry confrontations, and most of the people were able to eventually get to their seats, but a very vocal group swamped the studio lines and showed up in the lobby the next day to vent their anger. Kinzbach likens it to "buying stock on margin," and says the success of that maneuver opened the door for the station to do the same with a number of other major shows that came through the area.

By January 1983 Robinson was being touted as the "wunderkind" of Cleveland radio, having taken WBBG and WMJI from the ratings cellar to near the top of the heap. Robinson had spent a lot of time around the broadcasting field, but the bottom line was that he was a very smart businessman and knew the value of putting the best people in the right positions. Robinson's choice of McVay to program WMJI reaped big dividends fast, especially when they bought into the "Magic Radio" soft rock format. In the fall 1982 Arbitron ratings period, WMJI sailed past WZZP-FM, with the upstart coming in at number three overall, and WZZP ranked number seven. WMJI's morning host Ron Foster also nudged past WZZP's Ken

Morgan, with Foster at number eight and Morgan tied for ninth. But the big story continued to be WMMS, with Kinzbach staying as king of the morning, though WGAR's John Lanigan continued to make his presence known at number six. Lanigan may not have had an FM signal, but he sure had the audience and, given the opportunity, could do some serious damage on that end of the dial.

On January 17 WMMS general manager Walt Tiburski got the call that the station had again been chosen as "Station of the Year" in *Rolling Stone*'s annual readers' poll. It was the first time a station had won that distinction four years in a row. Other stations around the country had taken notice of that honor, and one, KLOL-FM in Houston, even mounted a "Beat the Buzzards" campaign. Wheels began to turn immediately for another highly publicized, free listener appreciation event featuring someone who seemed to like Clevelanders a lot, Little Steven (E-Street Band's Steve Van Zandt) and his Disciples of Soul.

When the day of the show arrived, a WMMS flag reading "Thank You Cleveland" flew over Cleveland City Hall, and Mayor George Voinovich declared that Wednesday yet another "Buzzard Day," along with a special proclamation honoring the station. Not to be outdone by the mayor, WMMS presented Kid Leo a pizza with ten candles, marking his first decade at the station. Van Zandt took the stage to a huge roar of approval, and the Disciples of Soul covered everything from Marvin Gaye to Duke Ellington. They even dedicated a song to "the princess of Little Italy," Pam Popovich, the Lake High student and daughter of Van Zandt's friend, Steve Popovich from Cleveland International Records. The concert and live simulcast caused a buzz on the streets that continued for weeks.

The Northeast Ohio–Jersey connection grew when Leo and Popovich attended Van Zandt's wedding on New Year's Eve. It was held at New York's Harkness School of Ballet, and it was an event peppered with big-time celebrities. Bruce Springsteen was best man, with Gary (U.S.) Bonds, the E Street Band's Max Weinberg, and Richie "La Bamba" Rosenberg and Jean Bevoir of the Disciples of Soul all in the wedding party. Blues legend Little Milton played during the ceremony, and the "Bronze Liberace," Little Richard, was on hand to read the service, with the E Street Band's Roy Bittan and Clarence Clemons, Southside Johnny Lyons, and filmmaker Martin Scorsese all in attendance. The wedding band from the film *The Godfather* played at the reception but gave way to Springsteen and the band, who played "Hungry Heart" and then kicked into a high-energy performance well into the night. It had become painfully evident that with this kind of inside connection, no other station in town would be able to lay claim to one of the most popular acts in the 1980s like WMMS had with Springsteen and his camp.

With that kind of promotion and connection, it was becoming increasingly difficult for other stations around the city to get any type of attention, but that didn't stop them from trying. The biggest change came at WGCL with the addition of "Dancin' Danny" Wright, and he came in with a bang.

Wright was originally from Michigan, and worked a number of markets before coming to WGCL from Boston. Just before his arrival, Wright had won the title "Fastest Talking Disc Jockey in the Country" on NBC's *Games People Play* and was hired on to fill the afternoon drive job. Wright says, "My first thoughts about coming to Cleveland were understandably not all that great! After all the jokes and put-downs, I was prepared for the worst. I can still recall my first morning at the Holiday Inn–Lakeside. Looking over the city from the top of the building, I thought, 'Wow, this city looks pretty good!' Then I had breakfast in the hotel restaurant, read the paper—the people were friendly, the paper was filled with pretty positive stuff. As I got to know the people of the city, I was even more impressed—strong, resilient, no pretense, straight ahead all-American city!" As it turned out, it was another prime example of being in the right place at the right time, for both the station management and Wright. "In all honesty," he recalls, "I came to Cleveland because my Boston gig had soured big time, and G98 had offered me a gig before Boston. I called to see if it was open, and voila! I was off!" He was on the job only a few days when he got word that *Billboard* magazine had named him its "Disc Jockey of the Year." Wright would prove to be a formidable opponent, but the big question remained: Did he have the stuff to beat Leo at WMMS?

At WMMS Denny Sanders hosted a series of Cleveland Orchestra concerts for young people, and this would continue a relationship with the "best band in the land" for some time to come. The ratings game started to take an odd twist in April 1983 when the winter Arbitron ratings were published. There were some changes at the top, although those numbers might have been deceiving. There was a definite tilt toward easy listening overall, with WDOK-FM and WQAL-FM ranking first and second, showing WDOK's "Tall Ted" Hallaman as the top morning host. But Kinzbach and company were just two tenths of a point behind Hallaman, and then a noticeable drop to number three with WQAL and WGAR's Lanigan at fourth. The ratings showed a distinct line being drawn between the growing rock audience and the older audience that preferred beautiful music, news, and talk. WGAR-AM was getting a lot of listener support, and morning host Lanigan continued to gain strength despite the increasing number of radios tuned to the FM dial. Clearly, bigger things were on the horizon for some of the staff at WGAR.

Most of the stations could see the importance of reaching audiences beyond the radio with some physical presence for them to focus on. WMMS expanded into publishing with *Best Seats*, featuring articles, interviews, and venue layouts for local concert halls. This proved to be a handy publication for concertgoers because they could pinpoint the location of their seats as soon as they received their tickets. It also revealed some interesting facts, such as the Richfield Coliseum numbering its main floor seats from left to right, and Blossom Music Center numbered them right to left. WMMS also premiered a late-night syndicated show on Monday nights called *Rockline*. While it was technically a call-in talk show, it featured some of the biggest artists of the day and offered listeners the opportunity to call toll-free for a chance to question people like Robert Plant, George Harrison, and many others. It also played a lot of music and seemed to fit well with the station's image as "Northeast Ohio's music authority."

April 1983 was a good time to be at WMMS. Ratings were steadily climbing, despite some very tough competition across the dial, and the longtime staff was starting to feel very comfortable. There was an aggressive spirit at the station, but it was also a festive one, as WMMS carved a niche the size of the Grand Canyon in its target audience. As a result, morning host Kinzbach was treated to a surprise thirtieth birthday party at Stouffer's Inn on the Square, though surprise turned to shock when Ferenc jumped out of the birthday cake.

The year 1983 also marked the station's fifteenth birthday, and with the success it had seen over the years it was an event to be celebrated. A commemorative t-shirt aimed at raising money for charity was issued, but the audience expected and wanted something on a grand scale. Springsteen came in to celebrate the station's tenth birthday, which set the bar pretty high as far as what the audience had come to expect. Promotions head Jim Marchysyn, art director David Helton, and the programming and air staffs were all put on "red alert" to come up with a memorable series of events to mark the occasion.

At the same time, the power of WMMS would give the station an advantage its competition just wouldn't be able to match. In Cleveland, in far too many cases for the competition, artists wanting to win favor from WMMS went there first and fit in the other stations as their schedules allowed. But WMJI continued its rise, getting praise from the industry paper *Radio and Records* as the seventh-fastest rising station on its "top mover" list, and twelfth in growth among the nation's adult contemporary outlets. Despite the different formats and audiences, WMJI's growth was a warning of battles to come.

Another hint of the future could be seen in some of the local clubs. A group called Exotic Birds had been garnering lots of positive attention and drew the interest of Sanders, who was having a hard time finding the band's music on tape. Fortunately, he met up with Exotic Birds manager Dan Miller while hosting a Cleveland Orchestra concert at Severance Hall. Miller was a member of the orchestra and forwarded to Sanders the latest single. Among those who would graduate from the Exotic Birds was a struggling music store worker named Trent Reznor, who later found international fame with his band Nine Inch Nails.

The influence of MTV was expanding, not only airing videos by established stars but also giving time to bands that were mostly college radio fare. The college stations had the freedom to air pretty much whatever they liked, and artists like U2 were especially grateful for that early exposure. That also proved to be a critical outlet for new music in the city of Cleveland, which was the last city in Northeast Ohio to get cable television.

WLYT had changed call letters to WRQC some time before, and had gone through a number of formats searching for its niche. Sensing MTV's influence, the station's management joined with the Washington-based Rantel Market Research for a hard look at the Cleveland area audience and switched over to a "Rock of the 80s" format. That didn't sit too well with the station's AOR fans, but the disc jockeys told them that so much of the 1960s and 1970s material had been overplayed that they were leaving it behind. Prior to that change, the station's challenge to WMMS for the AOR audience lasted just nine months. WRQC's Linda Jackson says, "They actually could sell it [the format to advertisers], because the community that you were appealing to, the advertising community, was very much into that music. But the ratings were not all that good, and we had a very bad signal."

Although the city of Cleveland had shown its enthusiasm for WMMS after it had won national attention, it was still eager to work with other stations. But it was difficult for other outlets to battle the still very intimidating WMMS promotion machine. One example was the effort to mobilize listeners in a "spring cleaning" campaign in Cleveland during "Clean Up Ohio Day." WDMT-FM was getting industry attention by premiering songs such as Grandmaster Flash and the Furious Five's "New York, New York," but its "wrap up the trash party" on Chester Avenue paled in comparison to the Buzzard Nuclear Army's event on West Boulevard. WDMT shelled out for t-shirts, while Kinzbach and Ferenc handed out shirts, drinks, and music to just about everyone who showed up, and they drew quite a crowd. The "Clean Cleveland Day" that followed a few weeks later centered on the area near the Agora, with WMMS, Daffy Dan's, and the club offering free t-shirts, music, and a barbecue afterwards.

Cleveland's music scene had been drawing attention for years, and that May at his Richfield Coliseum show, Huey Lewis showcased a new song he had written in honor of Northeast Ohio, "The Heart of Rock and Roll." The song mentions a variety of major U.S. cities, but Cleveland was given special attention in a standalone lyric during the bridge. Listeners in the Greater Cleveland area embraced it like an anthem, which may have been in large part due to an urban legend that the song was originally written to focus solely on Cleveland while dumping on the other cities. This rumor may or may not have been sparked by a WCLQ-TV station ID that edits the chorus to say not "The heart of rock and roll is still beating," but instead "The heart of rock and roll is in Cleveland." Lewis explains that although he was inspired by Cleveland and wrote the song on the tour bus during a stop at Blossom Music Center, he denies that it was written exclusively for the city. In any event the song got a big push on WMMS, WGCL, WRQC, and MTV and gave Huey Lewis and the News a serious promotional boost.

Kid Leo and Huey Lewis (1983). The power of WMMS to break acts and Leo's stature in the music community helped inspire Huey to write, "The heart of rock and roll is still beating—in Cleveland." Photo by Janet Macoska

An artist who saw a great deal of regional success by aligning himself with WMMS was Michael Stanley. The Michael Stanley Band worked with just about every station in town, but it was evident that WMMS had been the greatest influence in establishing its huge Cleveland fan base, though some other parts of the country were slow to catch on. While setting attendance records in Northeast Ohio and getting good response in other select markets, Michael and the band were still looking for a hit that would break them nationwide. When the band's LP *You Can't Fight Fashion* was completed in the summer of 1983, a special listening party was held with an invited audience at the Los Angeles home of Gary Gersh, EMI vice president for A&R (artists and repertoire). The Stanley release was said to have gotten generally good response, and that was no small feat considering the guest list included luminaries such as the Kinks' Ray Davies and David Bowie.

When the release date approached, Belkin Productions brought in a good-sized sound system to Higbee's department store downtown auditorium for a special listening event for eight hundred people, hoping word of mouth and heavy airplay might push the boys over that final hurdle. The LP had a definite Cleveland-Midwest feel and used one of Higbee's street-side display windows for the album cover. Northeast Ohio loved Stanley. It was hoped that, this time around, the rest of the country would see what it was missing. The band filmed a video for the single "My Town" and gave the MTV audience the opportunity to see them. MTV's video jockey Martha Quinn especially seemed to like the clip, though it failed to break the band to the nationwide video audience as it had hoped.

And while their music certainly wasn't new, the Beatles continued to get a lot of airplay and interest on Cleveland radio. In fact WMMS offered special blocks of music, airing the U.S. premiere of the band's British Broadcasting Company (BBC) performances. The tapes had been found by an engineer for the BBC while searching its vaults for material to air on a twentieth-anniversary show. It featured never-before-heard recordings from the United Kingdom's *Teenage Turn* show and other BBC programs, plus interviews and anecdotes about the Beatles' early years.

This was an extraordinary time for local music fans, because someone could see new acts, like the Damned, playing on the West Side, rising stars like Exotic Birds playing at fashion shows, and old war horses like the Turtles playing in the Flats, all within a few days of each other. Plus radio stations were playing a wide range of music on their different formats, and college radio was helping to break future superstars.

It was right around this time that Malrite's New York station WHTZ-FM was having an effect on the company's other properties, particularly WMMS. WHTZ's Scott Shannon's *Morning Zoo* had caught on like wildfire in the Big Apple. Shannon acted as the "zookeeper" and host of a fast-paced show featuring zany characters from the studio and the street. The station also had an improved signal, upbeat contemporary music that appealed to a wide range of listeners, and marketing graphics by Buzzard creator Helton. Its rise to the top in the space of just a few weeks was seen as a creative and marketing coup.

A similar concept was tried at WMMS with the addition of "Spaceman Scott" Hughes and "Captain Kenny" Clean to the morning lineup of Kinzbach, Ferenc, and the rest of the voices on the show. Both shows were fast-paced mixes of comedy, music, and features, and they focused on water-cooler topics that people tended to talk about during the day. There was outside advertising as well, but the best promotion is having potential listeners ask, "Where did you hear that?" Both stations had their fingers on the pulse of the audience.

Perhaps the biggest change on the local radio scene came that summer. Michael Jackson's "Beat It" had just been released, and WMMS put it on the air. Jackson was an artist that would have been considered out of the WMMS format only a short time before this. The record industry was really pushing Jackson, and while rival stations like WGCL, WKDD-FM, and television's MTV didn't hesitate to air it, WMMS still had the image of a hard-rocking station with a Cleveland attitude. Was the Buzzard selling out?

It was April 1983 and as then–general manager Lonnie Gronek tells it, "The world stood still the day it happened! It's all anyone was talking about at the station, but it

An early version of the *Buzzard Morning Zoo* (1983). (Clockwise from lower left) Jeff Kinzbach, "Captain Kenny" Clean, "Spaceman Scott" Hughes, and Ed "Flash" Ferenc. Photo by Janet Macoska

was absolutely the right thing to do. The station had to do something, and at the time, it was the smartest—and maybe, musically—the most conservative thing to do. We couldn't hold our position as it was for long, and frankly, it may have extended the life of WMMS, and very likely improved it. The world was changing musically, and stations were very niche oriented, and it would have been extremely difficult to maintain our high ground in light of all that outside pressure."

John Gorman said that it was simply a case of taking risks and keeping up with the musical tastes of the audience, though with the power the station held over its audience, WMMS was instrumental in defining those trends. That power was tested with some very controversial additions to the playlist, most notably Jackson's *Thriller* LP. Gorman says,

> We'd been playing black music for years. You had Earth Wind and Fire in the seventies, you'd hear the Isley Brothers in the eighties. All the other album rock stations ignored New Wave, and we jumped right on it. Culture Club, the Eurythmics, Thompson Twins. These were acts that weren't getting played on AOR stations. We were a pop culture station, and we should reflect what was happening out there. *Rolling Stone* reported on acts that weren't part of the rock genre.
>
> Keep in mind that before we played Michael Jackson, we played Prince. But the fact is that Prince had a strong guitar, and people accepted him. The first Michael Jackson we played was "Beat It," with Eddie Van Halen on guitar. What's the problem? When Prince, and Time, and Sheila E all started happening, we played them as a part of pop culture. Prince's *Purple Rain* was being called the rock movie of the year! It fit, and that's what music is all about.

John Chaffee also remembers the day the station played Jackson: "It was an absolute brilliant move! When I heard it was going to happen, I thought, 'My God! That's crazy!' But the station had huge ratings at the time, and we were the new music station. And here you had what would prove to be one of the biggest selling albums of all time. How could you ignore it? Now, if it was David Gates and Bread, then I would have had reservations." It might have also been an attempt to stop WGCL and some of the other pop-oriented stations from taking credit for breaking the album locally. Whatever the reason, it worked, and the WMMS numbers remained as strong as ever.

Key members of the staff applauded that decision. Sanders asked,

> Did anyone criticize Eddie Van Halen for appearing with Michael Jackson, and doing the hottest guitar break that he had done in years? Does anyone criticize Mick Jagger for doing "State of Shock" with the Jacksons? Nobody went to those guys and said, "What's wrong with you?" There was a crowd out there that wanted to hear nothing but traditional seventies-style dinosaur bands. The WMMS trademark had always been breaking new acts, and staying as trendy as any commercial station could. If somebody would come back and say, "Well, you didn't do a hell of a lot with Throbbing Gristle," well—true for some of that stuff, but that was pretty extreme for commercial radio. This stuff wasn't! This was pretty commercial material, cutting edge and modern sounding, but by no means underground. This was all hit stuff!

But Sanders heard some of the strongest criticism from high school audiences:

> There were a lot of young kids back then who dressed like they were ten years behind the times. I used to do a lot of high school talks in those days, and I used to catch some flak in '83, '84, and '85—which, by the way, is when WMMS

had its highest ratings. WMMS was always very strong [with people ages] eighteen to thirty-year, plus 20-something shares, 30-something shares—and we were competitive [for the demographic of] eighteen-to-thirty-four [year old] women, but we didn't own it. That was WGCL. We had pretty good numbers [with] twenty-five-to-forty-four [year old] men, but didn't own it. WGAR had that, but we were competitive. We didn't throw the dinosaur bands off the air. We just expanded our attitude, and didn't lose any eighteen-to-thirty-four numbers, and while we heard some chatter on the streets, the ratings showed no real deterioration. In fact, we went right to first [with] eighteen-to-thirty-four women, twenty-five-to-forty-four men solidified, and we actually got competitive twenty-five-to-forty-four women, because we didn't sound like the dirty, grungy radio station of the seventies. We still retained some of that sound because we felt the need to. You're still going to run into people who think that we should have just kept doing Foreigner, Journey, and Moody Blues block parties. Ridiculous! We couldn't have possibly done that!

Sanders was equally amazed by that age group's musical tastes, as he often asked, "Who are your favorite acts?"

Some of them replied, "You played Michael Jackson and Madonna. We don't want to hear that stuff!" Then the eighteen-year-olds would run down their preferred list, including the J. Geils Band, Deep Purple, Moody Blues, Aerosmith, and other artists from that era.

"Like any new acts? Anyone who's emerged over the past year?"

Not a hand.

"Two years?"

Not a hand.

"Three? Four? Five?"

Nothing. "Especially out in the deep 'burbs," he recalls, "These white kids—they were still waiting for Led Zeppelin to get back together! Think about that. When, in the history of popular music, did we ever have teenagers embracing the youth music of twenty years ago?! It was as ridiculous as if people my age would ask when Frankie Laine would be playing Cleveland! I honestly found that disturbing that high school students and college age kids only liked music that was popular ten or fifteen years earlier, and I felt that we were part of the disservice! We weren't exposing them to newer material."

There may have been widespread approval within the station to play Jackson, but it wasn't unanimous. Kinzbach was against the move from the start:

> I felt that when we started playing Michael Jackson and Madonna, we had totally compromised ourselves. The station never, ever recovered from that. Ten years later, it was still showing up in ratings diaries, saying, "I don't listen

to WMMS because they play Michael Jackson!" That basically, in my estimation, came from greed. It was just pure greediness and ego on management's part, and believe me, it made me sick! That was just not what WMMS was all about. Michael Jackson and Madonna were not WMMS!"

He went on to say,

> You know what it was? More than anything, it was the fact that they were trying to stretch WMMS as far as it possibly could, to the boundaries of being a Parallel One reporting station to *Radio and Records* [signifying the most important, highest-rated stations of a particular format in the most important markets], which gave them extra status with the record companies. In turn, they thought it may help with the ratings if they could migrate a little toward Top-40. They compromised the product, and the audience never forgave them for it. There are a couple of people who I'd like to take their heads and bang together for pulling that kind of crap, because WMMS—with basically the original staff—should still be in existence today. It would have been metamorphosized into something a little different, but the truth of the matter is, there were some really greedy people that compromised that station. I've never forgotten that.

But Sanders counters that Kinzbach was the least musically oriented of any of the disc jockeys, saying, "Jeff never came from the music side. He came from production and the humorous radio personality side of it all. His musical opinion was coming from a different place."

Hughes remembers the morning that song debuted and says, "People were calling up and asking why, and the slug line we used was 'Hey! Eddie Van Halen is on it!' But those were very critical times. Obviously, growing up and listening to an AOR station playing that was strange, but I'm pretty sure that was John Gorman's vision. It was tough! I'm not sure a lot of the jocks were hip to it. In fact, at least one jock went on record saying that was 'the day the music died!'"

WMMS had a long way to fall if it didn't work, but the audience eventually grew to accept it. The deejays doing club work were getting complaints about the change, but people were still listening, most likely because there was still plenty of hard rock for the old fans, and the air staff was intact. Because the disc jockeys had all been there for such a long time, they added credibility to the music by including Jackson, Madonna, and the more pop stuff in their shows.

Summer is traditionally the big ratings period for rock radio, and while WMMS was still doing a lot on the streets, like "Christmas in July" parties and other warm-weather events, the main product was still the programming. The cornerstone for most radio stations is the morning show, which often defines the tone and attitude of the station, and keeps people listening through the day to sample more programming, and hear the commercials. With more people than ever tuning into FM on their car radios, the station's management was always on the lookout for ways to make the very popular morning show even more attractive to the Northeast Ohio audience. WMMS always had a news department and did traffic reports, but that summer the station decided to add sports to the morning show, and the honor went to Dan Coughlin. The irascible Irishman had been a sports fixture in Cleveland for years and was one of the *Plain Dealer*'s most respected and popular writers before he switched over to the *Cleveland Press*. Coughlin had an irreverent humor, knew how to speak to the average person, and would prove to be a very solid addition to Kinzbach's morning show.

For its fifteenth year anniversary, all eyes and ears seemed tuned to WMMS for some type of street celebration, and the station delivered on Labor Day weekend with a free concert to kick off the Cleveland National Air Show. A little over ten years before, Three Dog Night had filled the Akron Rubber Bowl and rose to be one of the most popular groups in the world. They were now on the comeback trail and agreed to do a free show at the Party in the Park at Burke Lakefront Airport to help get their careers back on the winning track. The concert drew a massive crowd, thanks to the power of WMMS and the memories of an aging baby-boom audience. That was followed two days later by a special free Appreciation Day concert at the Cleveland Agora, with a group that had been getting some interest called Fastway. The band was a hard-rock outfit headed by a young singer named David King, but it was the drummer who would have a bigger impact on the Cleveland scene just a few years later. His name was Jerry Shirley, former drummer of Humble Pie, and he would soon be a familiar face and voice around Northeast Ohio.

PART THREE
BENCHMARKS
1983–1988

Drugs

It would be naive to think that drugs didn't permeate the entire radio scene. In the 1960s, '70s, and '80s, drugs were evident at just about every station in town, including the AM dial. In fact, at least one well-known newsman had a drug bust in his past, and another freely admitted to dealing before he cleaned up and went straight. Drugs were a major part of the music industry, radio, and the lifestyles of many listeners.

At WMMS, drug references peppered a simulated on-air Christmas party in the late 1970s, and earlier in that same decade (about 1973), a radio spot offered a free pack of rolling papers just for mentioning the station at a local head shop. The station had its call letters on its own custom-made roach clip. In 1971, in a backhanded slap at the FCC, Ginger Sutton at WMMS could sometimes be heard inhaling something and grunting out the station's legal identification as he held his breath. Some of Murray Saul's Friday afternoon "get down" sessions catered to the mind-altered segment of the audience. The early versions of the Buzzard character were pictured with a joint, and a bootleg product—a bank not produced or authorized by the station—had a suspicious looking cigarette in its beak as well. In the early to mid-70s, there were concerns that a poison called paraquat was being sprayed on marijuana fields to cut down on production and, ultimately, its use on the street. It was said the paraquat would cause lung damage if ingested. Capitalizing on that scare, the Cleveland police offered a promotion through WMMS in which they would offer to test samples of the listeners' marijuana. People could send their dope to a post office box with a code name or number and then could get the results by calling a certain exchange. If it were tainted with formaldehyde or some other poison, the police would advise the listener in a taped message. If it were clean, they'd say so but advise the caller that "rolling your own" was still illegal. That public service ended when too many of the listeners started sending their samples to the radio station instead of to the post office. One trusting audience member even sent in a whole ounce and asked the police to send back what they didn't use. Others sent in a wide selection of pills and powder. But WMMS still had their license to worry about it, and the risk was too great to keep the campaign going. It came to an end soon after it started.

It was often rumored that many of the deejays "altered" themselves before going on the air. Jeff Gelb recalls listening to WNCR as he was driving back from a concert and wasn't exactly sure what the deejay on duty was up to, but it was pretty strange. His music selection was all over the place, and as Gelb recalls, "He was playing a record, and at the end of the song you could hear the needle being picked up. He put it back in the middle of the song, over and over, about four times. Finally, he came back on mike and said, 'I just had to do that. It sounded so good I had to hear it again.'" When Gelb arrived at the station, he asked, "What the heck were you doing there?" The jock just smiled back, stared through him, and said, "Hey! I'm tripping! It sounded great!"

Another incident happened in summer 1970, when David Spero was manning the evening shift at WNCR and fellow disc jockey Martin Perlich stopped by the studio to say hello. Some other station personnel came by to enjoy the conversation, and soon the air was heavy with a thick layer of smoke. Spero's father, promoter Herman Spero, had been escorting his old friend, Henny Youngman, around town after a gig, and he told the comedian, "Henny, I'm going to show you my son, the disc jockey." They were accompanied by another friend from a local agency and soon arrived at the Statler Hilton Hotel, found their way up the stairs, and walked into the studio unannounced. Herman Spero's eyes grew wide with disbelief, and a good bit of anger, when he recognized the smell of marijuana hanging in the air. David's eyes widened as well, but more out of surprise and fear, because he was still living with his parents at the time.

Youngman let out a wide grin and peered through the smoke to see Perlich in the corner, with his thick beard and shoulder-length mane of hair. "Oh my God," he announced in true Youngman fashion, "It's Jesus Christ." Perlich smiled back, and said, "You must be Henny Youngman. I never forget a tie." The two traded good-natured jabs back and forth for some time, and both greatly enjoyed the exchange, although the two members of the Spero family were obviously not at ease. David remembers, "My father is in horror! 'This is what he does when he goes to work?'" He grabbed Youngman, who was clearly having a good time, and left in a huff. David's mother, however, was a little more understanding of these things and, at that point, a lot easier to talk to. He explained the situation to her as best he could, but it was clear that David would not be going home that night, or the next night, or the night after that. Lucky for them, WNCR employees could get a room for about seven dollars a night at the hotel, and that's where David stayed for a week or two—"until the air had cleared." Figuratively, of course.

FM disc jockeys often found themselves on the receiving end of gifts from listeners, and one pair of deejays (who prefer to remain anonymous) was offered a joint rolled with what was described as, "African spearmint leaves." In reality, it was likely some kind of vegetation sprayed with PCP, or "angel dust" as it was better known, and they and a couple of friends found themselves staring at their shoes. One of them, who was the deejay who was scheduled to work an air shift later that evening, was walking around in circles on his front lawn. The whole group somehow made it down to the WMMS studios, but when the mike opened the air talent couldn't lift his head off the console and couldn't put two words together. He finally determined that he could do the breaks but only if he impersonated Bob Dylan, and that's just what he did for the next four hours.

The Buzzard and the "Buzz" (1975). The station's mascot holds what looks like a joint in acknowledgment of the lifestyle of its core audience. David Helton

A few years later, a disc jockey who was scheduled to work at WMMS on New Year's Eve decided to have some folks over for an early celebration. Like a lot of people at that party, he soon fell under the spell of LSD, when the phone rang. It was John Gorman asking the deejay to fill in at the last minute for Betty Korvan, who'd been in some sort of accident. The deejay recalls stammering into the phone for what seemed like a very long time, but was really just a few seconds, and finally admitted, "I can't—I'm tripping!" Gorman quickly called Len Goldberg to fill in for Korvan.

Much of the music of the time reflected the drug subculture and the lifestyle that followed it. There had been an ongoing debate about the real meaning behind songs like "Puff, the Magic Dragon," Donovan's "Mellow Yellow," and "Lucy in the Sky with Diamonds." WNCR general manager Jack Thayer told the *Plain Dealer* that progressive rock outlets saw more drug references in the music than any other format but stressed that his station had nothing on its heavy list that glorified substance abuse, although they would keep a close eye on the secondary list. In early 1971 David Moorhead at WMMS called it "a false issue" and claimed that he hadn't heard a song in two years that encouraged listeners toward any illegal behavior.

Many of the bands, though, openly admitted their drug use. The group Cactus was en route to Cleveland from Minneapolis in August 1971 when members of the entourage lit up on a Northwest Orient flight. The pilot radioed ahead, and police met the band as they disembarked. The incident was immortalized on the next Cactus LP *Restrictions* in a song titled, "Mean Night in Cleveland."

The drug culture was in full blossom on the streets, and in the mid-1970s a couple of well-known voices at WMMS had the chance to interview Paul McCartney during the first Wings tour when it stopped at the Richfield Coliseum. One of the disc jockeys took a hit of LSD and, after what seemed like an awful long wait for the drug to kick in, decided to do more. He immediately started having problems talking, so he relegated himself to holding the mike and running the equipment, while Shelly Stile conducted the interview with McCartney. The deejay was still feeling the effects the next day when he came in but set the scene for each question and made it sound as though he'd done the interview as was originally planned.

The reputation of sex, drugs, and rock and roll was a tough one to fight, especially when stars like McCartney, Keith Richards, and others were stung in well-publicized busts as they traveled to other countries. The radio personalities had to fight it too. The *Cleveland Press* ran a story in October 1980 titled, JEFF SOUNDS CRAZY, BUT HE'S NOT GONE TO POT about WMMS morning man Jeff Kinzbach. The article stressed Kinzbach's drug-free stand, quoting him as saying, "I guess a lot of people think we take drugs in order to be funny and loose on the air. We are crazy, and do crazy routines . . . but I'm not going to get up at 4:30 A.M. every day and risk 12 years in the business by taking drugs. It is important to keep a clear mind."

Kinzbach went on to say there was too much pressure and confusion during his shift and that everything had to run smoothly. "There is no way you could do that if you were intoxicated," he said, adding, "To dispel any rumors, I do not take drugs on the air, or at the station. Work is work and it is important. My career is going to be for the rest of my life. I can't risk being out of work."

The article continued, saying that despite Kinzbach's serious nature about his work, he could still enjoy himself "in the proper place at the proper time" but then discussed his concern over the failure rate in high schools. He was quoted as saying, "I don't care if people are into drugs or alcohol, but excess is detrimental to their health. Education and drugs don't mix too well. It is sad to see teenage alcoholism.... Younger and younger people are getting involved in drugs. When that happens at ages 10 and 11, it is an alarming situation."

Kinzbach also defended the music industry, saying that it couldn't be blamed if stations played for kids experimenting with drugs. No responsible station would ever endorse or glorify drug use, but songs like Sly and the Family Stone's "I Want to Take You Higher" (1968), Brewer and Shipley's "One Toke Over the Line" (1971), Bob Dylan's "Rainy Day Women #12 & 35" (1966), Tom Petty's "You Don't Know How It Feels" (1994), and many other songs had all gotten air time on the FM band, and they stated the artists' positions very clearly.

Kinzbach also defended the local concert scene, saying, "Kids aren't going to do anything at a concert that they haven't done already. In my 12 years of concert going, I have never heard a musician advocate the use of drugs. A person can't go to a concert and be introduced to drugs for the first time. We are lucky we are in Cleveland. We have a reputation for good behavior at concerts."

But not necessarily in the studio.

In the late 1970s, disc jockey Larry Bole went to visit a fellow WMMS deejay who was spinning records at a bar before their shift. The owner gave the deejay a bar tab, and since the Code of the Media is: "If it's free, I'll take three," the deejay was soon roaring drunk. By the time Bole helped that person to the studio, the deejay couldn't stand up and curled up on the floor in a fetal position. Bole spent the rest of the night cueing records and ads and propping the visibly shaken deejay by the mike for a quick break to show that they were on the job. After a minute or two in a vertical position, the deejay vomited in a trash can, took a couple of puffs from a joint to calm down, and curled back up on the floor until the next break.

Bole also recalls getting ready to head into work to run public-service shows on WHK-AM on the overnight-to-Sunday-morning shift. Matt the Cat was his roommate at the time and called Bole over to the radio to warn him of what he would face when he got there. The disc jockey on duty was obviously not in any shape to do his show, banging pots, leaving the mike open to walk down the hall for an ashtray, and generally making a fool of himself. It would be another long night for Bole.

When David Lee Roth visited the station to broadcast the nationally syndicated *Rockline* talk show, his record company put out a nice spread of food and drinks in the lobby. But Kid Leo stopped Roth at the studio door as he tried to walk in with a Corona: "No alcohol in the studio." There would be an occasional after-hours staff meeting–cocktail party in the conference room, but that didn't happen often, and most of the bottles that some staff members had in their offices just sat there. The staff always had an open bar at the *Coffeebreak Concerts,* but word quickly got around that restraint was the key to future employment. If a person couldn't control themselves, then they didn't keep their job.

Cleveland Magazine reported in its December 1986 issue that the federal government was investigating drug use at the station, specifically cocaine. There might have been a lot of stories about what happened on the Statler's twelfth floor, but they were probably just that—stories, possibly spread by competitors or former employees with problems of their own.

Gorman admits that drugs threatened the well being of the entire station at one time or another. "Most of the people doing drugs, especially cocaine, were doing it to keep up with everything that was happening at the station," he now says. "Not everybody did it, though a lot did, and everybody on the street did it, too. A couple of people did have problems, and their careers were hurt, but most of them, thank God, didn't. It was an intense place to work, and there was a lot on everyone's plate. Everyone put in long days at that station, and it was live hard, play hard. But was anyone playing music for drugs? No. That didn't happen."

Payola

One of the nagging realities of radio, even after the payola scandals of the 1950s, was that buying favors was still the best way to get a record played on a top station. The label rep had to show that his songs got airplay, and that started by getting chummy with the music director. Sometimes it simply involved dinner, a bunch of drinks, and a backstage pass, but in some of the more extreme cases, there were mortgages paid and kids sent to top rated colleges—all on the record label's tab.

Here's what former music director Donna Halper observed during her tenure at WMMS:

> Record promoters have been my friends for years. They came to my wedding, for heaven's sake, and they've been very, very good to me—but all very, very legally. The first day that I got there, the record promoters all lined up carrying—are you ready for this—bags of dope! I have to be honest. I didn't know what it was. I come from a background where we just didn't do stuff like that! Now, I'm no dummy. I'd been in broadcasting for a while at that point and heard about it, but I'd never really seen it. I'd seen the end result, but I'd never been given a bag of it to roll your own. So the promoters come in to pay homage to the new music director, and they bring their best offerings. One of them hands me this bag, and I gave a look like, "What's this?" What he saw in that look was, "This isn't good enough!"

"It's the best stuff I've got! You want better than that?"

I put two and two together, and realized, "Oh my God! He just gave me a bag of dope."

In those days, you got to the program director through the music director, and when a station is number one, what you want is what you get. If you liked chocolate cake, splendid. You got chocolate cake. You can call it payola, but it was also the cost of doing business. In most cases it was pretty harmless, and programmers didn't ask for anything like gold coins. In most cases, it was small favors that the record guys would do for you.

Here's how a promoter would get paid: In the Top-40 days, a promoter would get "spifs," which was a bonus that was based on how many reporting stations added their record. If a big-time music director didn't like your record, it didn't get reported, and you didn't get a spif. If your region had X number of reporting stations, and none of them reported your record, you not only didn't get a spif, you didn't get to keep your job. The honest record promoters found creative and amusing ways to get to your heart. In the case of people like me, or [WIXY's] Marge Bush, they knew that we were into social action, and they might be telling us about all the charity work their organization is involved with. In other cases, they paid a person's mortgage. Can you say "conflict of interest?"

In some cases, a look at a station's top ten might indicate something fishy was going on, with several artists from the same label coming out of left field to place on the list. But the bottom line seemed to be if a record just wasn't popular or plainly didn't sound like a hit, the person in charge of that list might come under additional scrutiny and often didn't stay at the job much longer.

News

Before the FCC relaxed its rules, every station in town was required to air news. Some stations had their disc jockeys "rip and read" off the news wire, but most had news departments and the FM rock outlets were no exception. Granted, some of the early broadcasts swung a bit to the left, reflecting their audience, although longtime and respected WGAR newsman Charles Day brought credibility to sister station, WNCR, as did his equally well respected FM counterpart Norman Moore. Both delivered straight-ahead serious newscasts for those looking for that type of information. WNCR prided itself on a strong news presence, aided by commentary from Martin Perlich during his *Perlich Project* show. There was a great deal of creativity in its presentation as well. The advantage was that most FM stations shared the AM newsroom.

WNCR bolstered its news product with topical tunes from a folk group, Bacha. In 1972 the seven-member band raised eyebrows when it was recording a calypso song for later use about a judge who jumped out of a window. The spirit of the music moved the musicians to march out of the studio, playing all the way into the elevator and down to the lobby of the Statler Hilton Hotel. The *Great Swamp Erie da da Boom* reported, "Members of the Ohio Library Association, checking in for their convention at the time, were somewhat startled to see three guitars, one bass, one trumpet, one maraca, and a saxophone emerge from the elevator ... but how can you argue with happy musicians?" For a time, the WNCR news sketches and commentaries were even syndicated to other Nationwide outlets.

The multifaceted David Spero traveled to Toronto to cover the Keith Richards drug trial for M-105. While M-105 never really tried to establish a news presence, the Richards trial had enough crossover interest to the music audience to warrant that kind of on-scene coverage.

As the years progressed and the need for more straightforward, unbiased, information became more evident, Shelly Stile and Ed Ferenc were able to establish a very credible FM news presence on WMMS. Some time after Stile left, a fine journalist named Jim Butler joined the staff, strengthening a newsroom shared with WHK-AM. Butler established solid relationships with newsmakers around the listening area, covered spot news, and brought a street presence to the department. He was also a fierce competitor. At the conclusion of a Bond Court Hotel news conference, the radio reporters in the room raced to pay phones in the lobby to call in their stories, but the microphone discs in the hand pieces of the phones had mysteriously disappeared. Butler casually walked to one of the phones, and he had one of them in his pocket. It became evident that WMMS was more than music and entertainment but becoming a full-service radio operation.

WMMS put out a solid, credible news product, though there were a few gaffes, one involving a botched bank job that turned into a hostage situation. Butler was on the scene, but Ferenc put in a call to the bank from the station—and the gunman answered. They had a taped conversation covering topics like "why not give up?" and even a song request, but that didn't sit well with the authorities. An angry FBI agent confronted Butler, demanding to know what the station was doing, and telling him to stop it immediately, but at the same time WERE-AM's news department also put in a similar call, for which the station took some heat. Fortunately the hostage situation was resolved peacefully.

Another mistake involved WMMS taping an exclusive WKYC-TV interview by reporter Tom Beres with an executive of the local Pick-n-Pay grocery-store chain. WMMS aired part of it the next morning without permission or attribution. It caused a lot of tension between the two stations.

But the little hiccups along the way were balanced by honors. The WMMS news department, in particular, distinguished itself with a number of prestigious awards, including first place honors from the Ohio Associated Press, United Press International, the Radio and Television Council of Greater Cleveland, Women in Communications, Sigma Delta Chi, and the Press Club of Cleveland. The department also took a national first-place award from the Associated Press for a documentary, *Fifty Years of a Never-ending Battle*, which detailed the popularity of Superman while his Cleveland-raised creators lived in relative obscurity. The station also won a Silver Medal in the 1991 New York Radio Festivals international competition for its *Source Report* special about Kent State University, *Four Dead in Ohio*.

Perhaps one of the high points of WMMS news coverage came when John Demjanjuk, the suburban Seven Hills man who had been accused of serving as Nazi death-camp guard "Ivan the Terrible," was released from an Israeli prison after the charges were dropped to return home to the United States. All of the Cleveland media anxiously awaited his return, hoping to interview him either at Hopkins International or Burke Lakefront airports. But Cliff Baechle and Carmen Tedesco, who also reported for sister station WHK, didn't think Demjanjuk would make himself accessible to the media. After looking at a number of regional airports, they determined that Demjanjuk would do better by flying into a smaller county airport. Sure enough, the two guessed correctly and Tedesco was the only reporter on hand when Demjanjuk emerged from a small private plane with Ohio Congressman James Traficant, who had long championed his cause. Appropriately enough, it was the name of the airport that caught their attention. The plane landed at Freedom Field.

Exclusives

The FM stations, and especially WMMS, prided themselves on debuting "world premieres," new music by established stars that gave the stations bragging rights that the music could only be heard on their spot on the dial. The label representatives often had preview copies to give program and music directors a hint of an upcoming release, but they weren't authorized to give one station a jump on another. They also didn't want to build up too much demand ahead of time for a product that wasn't yet available. But in the ratings wars, a hot new release would keep listeners riveted to one station. This type of warfare had been going on for decades.

Back in the 1950s, WERE-AM's Bill Randle was the undisputed King of Cleveland Radio. His keen ear had established him as a hitmaker, as he could identify coming musical trends with unerring accuracy. His acknowledging a certain artist could be a major career boost. Don Everly, of the Everly Brothers, recalled that his first trip north was to get Randle to break his record in Cleveland. Randle's numbers were through the roof, so record companies knew he could make or break a song or an artist.

The late Bud Wendell, who was a Cleveland radio legend in his own right, said that when he was playing hits on WELW-AM, his station and others would fume over the exclusives that Randle would air, so they devised a plan to get them at the same time. The stations would tune in to Randle's show on WERE and record a disc of their own off the airwaves when he played a new song. Minutes later they'd play the same cut, even though Randle had the only official copy. Randle quickly figured out what they were up to and countered by opening his mike during the instrumental breaks to announce, "You're hearing it first on WERE." The other stations resigned themselves to fighting for second place.

In 1973 David Spero struck up a friendship with David Bowie after WMMS broke him in the United States with his sold-out Cleveland shows. In a show of gratitude, Bowie made good on a promise to get an advance copy of his *Aladdin Sane* LP to Spero, who had the unofficial world premiere on his program and then locked the record away in his home. But friendships don't always bring home an exclusive.

Spero had worked with Michael Stanley for years, but when Stanley's *Cabin Fever* was released, he delivered it to Kid Leo first, and Spero waited along with everyone else. Stanley recalls, "It was an unwritten rule, but that's how the game was played. In Cleveland, WMMS had the ratings, so they got most of the new stuff first. WMMS and the Michael Stanley band 'grew up' together, and since there wasn't a whole lot happening in Cleveland at the time—the sports teams, the politics, the river catching on fire—the people who stayed were grabbing on to anything positive, and WMMS was sort of the focal point at that time. That was their town square, and that's where people got entertainment. That's where they expected to hear new music as soon or before it was released."

WMMS had the kind of power to move tens of thousands of albums, so a label representative might "accidentally" leave a copy of an album's preview tape, or a copy might anonymously appear in the mail. Matt the Cat recalls an anonymous call suggesting that a cassette of the new Rod Stewart LP might have been misplaced in a manila envelope at a Cleveland phone booth, and Buzzard artist David Helton remembers an album they retrieved hidden at a construction site.

Donna Halper gets right to the point:

You bet record promoters started bringing us exclusives. As soon as our ratings started going up, as soon as we became the "WIXY," as soon as we became the "WNCR," all of a sudden it was like, "Gee! I just happen to have a test pressing of the Yaha Band." Let's face it. They're professionals, and they don't want to alienate the competition, so they just happened to be in the neighborhood, and we just happened to get the exclusive. It goes on in every market, and anyone who says it doesn't is a liar.

As Halper correctly stated, those representatives still had to work with other stations, so after a time they became very guarded of some of their biggest artists before the official release date. It took competitiveness to a new level. Occasionally, friends at another station might get a copy and networked the tape to WMMS, but most often the station pulled its own strings.

John Gorman once had a wall papered with cease-and-desist orders. He states,

M-105 was a good station; they had their connections, and we had ours. There was a Carly Simon album, the one with "Attitude Dancing." I was at a *Radio and Records* convention, and I stole the acetate. It was playing in a suite, I saw it and.... We got to play the Boston record, too. We had it before anyone else in the country. Charlie Kendall had a connection at the pressing plant in Terre Haute, Indiana. Occasionally, we'd get something coming out of there.

In the early days of FM, the stations were still like love and peace, but we said, "Oh no! We're 'anti-this and that.'" It's rock and roll, and we took an aggressive attitude toward these things. At times, the station might get a pre-release copy of a hot song, but it might also come with the understanding that WMMS would very closely consider another artist. But, frankly, sometimes they were stolen.

What few people know is the ingenious way they were able to get the music without it leaving a representative's sight. Leo, Denny Sanders, or Gorman might meet with a label representative at the station and listen to an upcoming release in the production studio. What the label didn't know was that the production studio was wired to tape machines in other offices, and they rolled the entire time the representative was there. According to Gorman, it started when the station moved its operations to the Statler Office Tower in 1975. He asked the engineers about setting up a switch that would set off tape machines in other rooms, and that was the system they installed.

But Warner Brothers Records promoter Larry Bole, a former WMMS disc jockey, soon figured out what was going on and brought a pre-release copy of the hotly anticipated *Fleetwood Mac Live* to the station. Bole kept stressing about how excited he was about the record, and he would stop a song after a minute or so and say, "Now listen to this. There's something you've got to hear." As a result, the station couldn't record a complete song, and Bole noticed some very sheepish looking faces as he passed the offices where the tapes had been rolling. Even so, just a few days later, Bole heard Leo introduce the album as a world exclusive and nearly veered off the road when Leo promised to play the album straight through. "I don't know where he got the album, but I had all the pre-release copies in my car, and I put a call in to the main office. I told them about Leo, and said we had to service all the stations now." He remembers bringing a copy to the studio just as the final tracks were played and handing it to a very surprised Leo.

Another label representative soon caught wind of what was happening, and when he came to pitch a song, he offered them an exclusive listen—through Walkman headphones. But at least one representative wasn't aware of the taping system and brought in new music by the Rolling Stones for a prerelease listen. WMMS also ended up getting Queen's *The Game* album and a Jackson Browne album that same day. Gorman remembers hearing the Stones, knowing it would be a major coup to break it on the air. He recalls:

> This was a great album. I asked if we could hear the whole thing. There's nothing better than to say to a label rep that I want to hear some product. He said sure, and I started the tape machine.
>
> That night, as always, we waited until the branches [local record label offices] closed at like seven o'clock, and Leo starts talking about three big exclusives. I made sure I didn't answer the phone, because I'd be the first one they called. The record guys might call up the hot line, but the jocks would say, "I can't honor that. It's gotta come from John."
>
> So we're going to play three brand new songs, and this is a major coup. I didn't know it, but the head of WEA [Warner-Elektra-Atlantic] is in town, and he's having dinner at Swingo's with three of the top local guys. All of a sudden, the waiter comes over and says there's a phone call. One of the local guys gets up, takes the call, and comes back with his face very pale. He whispers in another guy's ear, but nothing's wrong. "Don't worry." Then the waiter comes back.
>
> Same guy gets the call, returns to the table, and "Don't worry. Nothing's wrong." Then he's called away a third time. The WEA guy finally says, "I can read a table. Will somebody please tell me what's going on?" and hears that WMMS just broke three of his songs. He starts getting red in the face and asks which three. He could take the Jackson Browne scoop, but then he hears we played Queen, and he's not happy. "Alright! What's the third one?" The local guy mumbles, "The Rolling Stones."
>
> The label paid a lot of money for the Stones, in part because they were a prestige act that could draw other talent to the company. When he heard the Stones were on the air, he explodes. "How could this have possibly happened?"

Three local record guys sat quietly and took the wrath of their out-of-town guest.

By the time some of the local reps figured out what was happening, Gorman had the switch moved. The old switch would trigger the production studio tape

Freddie Mercury (1978). WMMS used its pull to debut new records by top acts like Queen days and sometimes weeks before their official release. Photo by Janet Macoska

machine, and a similar situation was set up in the auxiliary studio. Eventually that feed went into the main studio, and once a disc jockey took Gorman's cue and actually broke the unsuspecting representative's song live on the air at the same time he thought he was playing it privately to Gorman.

M-105 had its scoops too, most notably songs by Paul McCartney and even Bruce Springsteen, but that could have strained relationships with record companies. Unless another station had ratings numbers like WMMS, which the labels couldn't turn their back on, it could be playing with fire. On one occasion, the program director at M-105 decided not to fight it and simply told a label representative to just give Foghat's new release to WMMS.

Sometimes a record would get networked to other stations before making its debut on WMMS just to throw the record labels off the scent. For example, Leo had a close relationship with Springsteen and his manager Jon Landau and was able to get Springsteen's new release before anyone. But Leo didn't want to jeopardize that valuable friendship by putting it on the air, so a deal was struck with sister station WHTZ-FM, the tape was flown to the Newark airport. That station's Scott Shannon, host of the *Z100 Morning Zoo*, picked it up from there. He took it back to his station, put the songs on individual tape cartridges, and every thirty seconds a voice would say, "Z100 world exclusive." Shannon left word that if anyone at WHTZ answered the back door over the weekend, they would be fired. Cleveland native Frank Foti was employed at the New York station doing engineering work at that time, and he stopped in that Saturday night to do maintenance. He counted ten "cease and desist" orders taped to the door. WHTZ, and soon after WMMS, got the jump on Springsteen's "Dancing in the Dark" weeks before any other station.

Bole had worked weekend shifts at WMMS for some time, while getting his foot in the door at Warner Bros. Records. He was eventually offered a fulltime position with the label, but he did one last show at WMMS. For that final outing he used the name "Be Bop Kirby," in part because he thought it might be fun to work under a different name, and also out of fear that other stations might charge WMMS with a conflict of interest by employing a record company rep. He very likely debuted Dire Straits' "Sultans of Swing," and he left a note for Leo saying he should give it a listen. Leo wrote back, "Play it at home," obviously stressing the separation of station and record label.

The world premieres would most often debut on a Friday evening after business hours. Deejays were told not to answer the phones or knocks on the door. Just play the music, and the station would deal with the consequences on Monday, usually a cease-and-desist order. It didn't matter. WMMS and the listeners got what they wanted.

Some exclusives even came from the labels themselves. Gorman says, "We'd take a chance on an artist, and then it was, 'Come on and do us a favor. We played Alex Harvey, and nobody else would, or some other artist. You can slip it to us.'"

But the labels didn't always cooperate with WMMS. When the station aired a new Rod Stewart cut well ahead of its release date, Warner Bros. cut off service in retaliation. Gorman says, "It was three months before it was due for release. It turned out that our sister station in Milwaukee, which was a Top-40 station, had a guy on the air with a brother-cousin-friend—it doesn't matter—who played with Rod Stewart's band and just happened to give him a copy of the album."

That issue was quickly resolved, but not long afterward Warner Bros. offered a special syndicated show featuring the Pretenders to M-105 instead of WMMS. An angry memo soon made its way to the WMMS staff announcing that "Bugs Bunny," a nickname used for Bole due to his affiliation with Warner Bros. had been banned from the station. The matter was eventually settled as well, and life went on for all involved.

There was also a network of disc jockeys around the country who passed along releases to their friends. WMMS got the jump on the Who's yet-to-be-released *Face Dances* LP when deejay Dia Stein gathered up engineer Foti, now at WMMS, to accompany her to the Greyhound station, where a dubbed tape of the album was coming in from a Buffalo station. But not all of the exclusives came from commercial product. Years later, an embarrassing tape from an isolated mike caught Linda McCartney's Knebworth Festival vocals, and it was broadcast coast-to-coast within weeks. The McCartneys were well aware of the tape. Linda cancelled an entire satellite radio–interview tour to publicize her cookbook when one of the stations let it leak to a publicist that they wanted to ask her about that performance.

With the advent of Internet music sites, digital audio, and endless sources for unreleased music, however, it's now a lot easier to get exclusives, and a lot more difficult for record companies to keep new stuff under wraps.

Get Down!

The *Cleveland Press* labeled him "Our Man Friday," and no words better expressed Murray Saul's role to an audience that couldn't start a rock and roll weekend without him. Every Friday around 6:00 P.M. Saul, who had been in the communications industry since 1951 and was selling time for WMMS, would enter the studio screaming out a rush of ideas and social commentary that commanded immediate attention, and added to the station's very distinct identity. He told a *Cleveland Press* writer that the whole salute idea started when program director John Gorman overheard him talking on the phone to his eighty-eight-year-old aunt, who was partially deaf. Saul remembers, "She would call me at the station and ask when I would come over to see her. I would tell her, 'I'll be there on Tuesday.' She would answer back, 'Wednesday?' 'No! Tuesday,' I would say a little bit louder. And she would say, 'Thursday?' and I would answer back still louder, 'Tuesday.'"

It may have been frustrating for Saul, but it was that comical exchange between the former Ohio State Fairgrounds barker and his aunt that helped give birth to a Cleveland radio phenomenon.

Gorman is said to have snuck a tape into the studio to open Denny Sanders's show on the night before Thanksgiving 1974, when Saul gave a rambling description of a turkey and introduced Sanders. Soon after, Saul's taped commentaries were moved to Friday nights.

Murray Saul (1976). The bellowing voice that welcomed the weekend became a fixture at several stations across the country as bootleg tapes of Murray's "Get Down" rant made their way to their airwaves. Cleveland Press Collection/Cleveland State University

Gorman recalls seeing fellow WMMS staff member Walt Tiburski dancing at a party, and Saul reportedly walked into the living room and cried out, "Walt, get down, dammit!" The "Get Down" line was used by Saul to end his commentary and start the weekend the very next Friday, and a legend was born. The station filmed a spot with a "Get Down" theme at a Party in the Park, featuring Doctor Hook and the Medicine Show, and it helped fan the flames of recognition for

Saul. (As one WMMS staff member observed in 1975, "He's getting laid more than any forty-seven-year-old man in Cleveland.") M-105 tried to capitalize on Saul's popularity with its Friday feature, *Greetings from Robere,* although Eric Stevens now tries to block that whole experience from his memory. "He was just somebody that walked in to the station on a Friday," but Stevens adds with a smile, "That's part of my resume that I would rather forget! It's just somebody I talked to, and I thought he was kind-of unique. So we thought we'd give it a crack." Robere didn't last long.

The Friday salute was Saul's first on air job, a collaboration between himself and Gorman, and his weekend rants became so popular that stations in Cincinnati and other markets would run bootleg tapes of his Friday ramblings. Gorman and Saul would meet prior to the salute, and occasionally Gorman would put stage direction at the top of the script, like "This is a real motherfucker. Scream all the way through!" At 6:00 P.M., a sonic boom would emanate over the airwaves. They struck hard at the psyche of blue collar Cleveland, sympathizing with the pain of the working class and always stressing the attitude of WMMS. There were veiled references to any number of "diversions," but Saul instilled the same energy and devotion in his listeners that he put into his weekly moment in the spotlight. No one who listened to Saul's time on the "electric podium" could be expected to do anything but welcome the weekend with outstretched arms and GET DOWN! A small sampling:

Gaaaaaaah! Freedom! We are free! Freer than hell! We have bought back our souls from the pawnbroker of the weekday grind! Those pens and pencils! The typewriters! The erasers! The memos! Ledgers! The drill press and lathes! He notifies us only of the bad! And he knows and defies us with a scowl on his face! We're discarded merchandise. He sends us memos with our numbers on them! Tell the boss to go to hell! Fahh-get it!

They've spent all that week trading our souls like baseball cards. Flooding our minds with their crap, and shiiahhhhht! We're going to drain our minds of that mental paralysis of the weekday grind! We're gonna solve those problems and get good and high. We have our souls! We bought them back! We're gonna release the full force of whatever exists, because it's Friday! Friday!! Friday!!! Friday!!! Now give me that almighty, mighty, mighty weekend salute! And give it to me loud! We've gotta, gotta, gotta, gotta get down to it all! Gotta, gotta, gotta, gotta get down! Gotta, gotta, gotta get down, please! Gotta, gotta, gotta, gotta get down—and longo, longo, longo weekend! Just 'cause I'm taking Yom Kippur doesn't mean I can't get Christmas! Fa-la-la-la-la-la-la-la! Oh, shiiiaaaht! What is all of that? Now, let's all speak Spanish the way it should be spoken! La-la-lowowowow! Yay-zah! Yow-zah, bow-zah, wow-zah! La-la-lolololoo! And roll them tunas! Ooh blah doo, ooh blah dee, oo blablablablablablabla dah! Shiii-aaaaaht!

The lungs, the lungs. It comes from the smoke! Well, shut off the smoke! The smoke that's coming out of the stacks! I have the proclamation. We are officially welcoming in the autumn! And we're gonna ball all through the fall! Eh—think about that! Other seasons may say it better, but the autumn does it the best! The best clothes come from the colors of autumn! The colors of the richest oranges, yellows, reds, the changing green–and let's not forget the gold! The blonde! The blonde gold, and the hay! The hay that's on the way! Just hearin' and nearin' and the harvest. We'll all be sliding down that cornucopia, a million colors around our brain.

The autumn is something to be proud of! Don't chill the buns! Carry a sweater! That way you won't have to worry about being cold! Snuggle up! Now's the time to wrap around the quilt! Northern Ohio! We will celebrate! We are the capital of autumn! The pumpkin festival is coming! The arboretum is in the autumn. We've got it all!

I've seen more bees this week than ever—ever! More bees are buzzin' now than in the summer! Beekeepers—on your mark! Get ready. Go! There's plenty of honey to eat and slurp around in. Autumn is the season to really roll around in it. Really get good and sticky! Taste it with your fingers! Your tongues! Get that tuna salad ready for the weekend! Keep ro-o-o-lling that tuna! Get ready to really roll around in those leaves! Get ready to really enjoy being in the back with your honey when the rains come!

Now's the time to really be an explorer! Play doctor even! We have so much to learn about the autumn, and it's gonna take all fall to learn! The easy life. The boulevard. Chairman Stick! We're gonna start right now, cause we gotta, gotta, gotta, gotta, gotta, gotta, gotta, gotta, gotta, gotta, gotta—gottagotta gottagottagottagottagottagottagottagottagottagottagottagottagotta gotta—GET DOWN, DAMMIT!

The Friday Get Downs were done live, but one exception was when Beach Boy Dennis Wilson came to the station to plug his solo album. Unfortunately, Wilson wasn't able to be there at six in the afternoon, so the station decided to tape it. Wilson had been a great fan of WMMS and made it a point to wear his World Series of Rock t-shirt when doing high-profile interviews. He was the real California kid, the only Beach Boy to surf, and a somewhat reckless soul, having had to deal with Charles Manson before the infamous Tate–La Bianca murders. Wilson really wanted to be Saul's guest that week, and a group gathered in the production studio to enjoy the exchange.

Steve Lushbaugh came in to engineer that session. He recalls, "It was kind of a loose party atmosphere, and Murray decided it was time to do the salute. I was rolling tape, and he starts this screaming tirade about California. 'Those people in California with their tops down, and their blonde hair, and their sunshine!' Murray just started ripping on California, and we're wondering, 'Where's this going? This isn't good.' Then he stopped at the end, and Murray leans right into Dennis's face, and yells, 'You know what!? I love it!! Aaaah!"

Wilson stared back at Saul, slowly got up from his chair until they were face to face, and said, "Murray?" and then his voice started to crack. "My father, Murray, used to put us in the bathroom and beat us until we were bloody!" At that point, Wilson broke down into sobs, loudly crying about the abuse he and his brothers suffered at the hands of their father. An eerie chill settled over the room as Wilson went on incoherently about the incessant beatings and indignities they endured, and everyone stood by dumbstruck as the tape rolled on.

Several minutes later, the people traveling with Wilson helped him through the door to escort him from the station, and one angrily grabbed the reel of tape off the deck. It didn't matter, because Lushbaugh was rolling tape in another studio. It was meant to snag a couple of cuts from Wilson's yet-to-be released album but instead was capturing a side of the Beach Boys legend that wouldn't be known to the general public for years to come.

Lushbaugh ran to the other room, and his heart sank when an engineer said he didn't realize they were taping and removed the patch cords. All that was left was a few minutes of a song, before the tape cut out. Perhaps it was best that it turned out that way, but it stands as one of the most bizarre moments with a visiting rock star.

Mr. Leonard

Despite all of the very real personalities who have graced the microphone at WMMS, the most bizarre is perhaps the station's only major fictional character: Mr. Leonard, a high-pitched jokester who billed himself as the "assistant public service director."

John Rio was just a teenager in his hometown of Salinas, California, when he got his first fulltime radio job of changing tapes at an automated station. "Your quiet island for beautiful hits," he recalls, "one after another after another." The year was 1974, and it wasn't long before Rio escaped the "quiet island" and headed to El Paso for a 7:00-to-midnight shift, and then moved on again to St. Louis. It was a drastic change going from a country to an urban format, but Rio took it in stride. He eventually landed a job at 93Q in Houston. It was there that the Mr. Leonard character started to take form.

Rio was doing overnights, but he stuck around to try out his Mr. Leonard material on the morning show. In 1982, after about six months, he became a permanent fixture on mornings. A year later he started doing Mr. Leonard in New York with Scott Shannon at Malrite's WHTZ-FM. It was in 1984 that Mr. Leonard

Mr. Leonard's Parking Space. John Rio's Mr. Leonard, the mysterious morning sidekick, even had a place reserved for the infamous lime green Pinto, but his appearances were few and far between. Photo by Brian Chalmers

finally made his debut on WMMS, while still doing shows in New York, Houston, and Philadelphia.

The character was loosely based on a salesman at the St. Louis station, whom Rio fondly refers to as a "scam artist." He recalls, "He wasn't a bad guy, but he would take you for everything he had. It was a nickel and dime thing. 'Can you slip me twenty 'til next Friday?' That sort of thing. Then you wouldn't see him for a month! You ask for the money, and he'd say something like, 'Your twenty?! You're not going to believe what happened! After you gave me that twenty-dollar bill, I went to put it in my wallet and a big old summer twister came roaring by! Snatched it right out of my hand! That was your twenty dollars, and I didn't even get to spend it! Why should I pay you back? It was an act of God! He's the guy you have to talk to!'"

The common assumption was that Mr. Leonard was a skinny little guy, maybe African American, but Rio never discussed his looks on the air, unless they were outrageously exaggerated in his favor. Whatever he was supposed to look like, Mr.

WMMS had printed business cards outlining Mr. Leonard's many duties, but very few of them made it outside the Statler Office Tower.

Leonard proved to be a huge hit. The response in New York made Malrite wonder if it would work in Cleveland. They didn't have to wonder for long. It was the same reaction in other cities, too, and Mr. Leonard even made it to national syndication as part of *Scott Shannon's Top 30 Countdown,* which was heard on two hundred stations.

Oddly enough the management at WMMS hesitated bringing in a character like Mr. Leonard at that time, because they already had the local Kenny Clean on the *Buzzard Morning Zoo.* But Clean soon left the show, opening the door for his logical successor.

Mr. Leonard changed and evolved over the years, but the basic lovable, fumbling character pretty much remained the same. He had a long list of excuses for not showing up to work, inventions that always seemed to go awry, and bizarre commentary meant to impress the *Morning Zoo*'s guests and listeners. But all it did was show his actual talent for kissing the Blarney Stone.

Rio thinks the character was an immediate success because, "Mr. Leonard was never out to get anybody. His scams always came back on himself, and Mr. Leonard would become a victim of his own misguided efforts." He says there were three basic types of Mr. Leonard listeners: "The first knew it was a bit and enjoyed it anyway. Then there was the group that wasn't really sure if it was a gag because it often sounded so real, and the third had no doubt that he was a real person and could not understand how he kept his job!"

But the success of Mr. Leonard had its downside as well, because his popularity often overshadowed the rest of the show. Looking back, Rio says with a laugh, "I dealt with it fine! It was the other people that had the problem, and I would have, too, if I was the host of a show and all of a sudden this guy comes in doing this character and starts getting all the fan mail and attention. 'Go do your own show, Bud!'"

Mr. Leonard's title at the station was an inside joke to those in radio, because while the "public service director" is really an important job, it's also boring and thankless, and Mr. Leonard was only an assistant! Nobody ever wants to be the public service director at a radio station, but Mr. Leonard always came off as being happy he just had a title. Rio points out that Mr. Leonard never did anything in service to the public anyway.

Surprisingly, despite his "mysteriousness," Mr. Leonard actually did quite a few personal appearances, just usually not publicized. But there were lots of people who didn't want to know what Mr. Leonard looked like. At a gasoline giveaway remote broadcast in the 1980s, a guy walked up to Rio and saw him do the character on the air. He watched for a minute, turned and started walking away, but looked back to say, "I don't know if I can listen to you anymore, now that I know what you look like!" That sent up a red flag to Rio, but still Cleveland saw more of him than any other city.

There was a lot of pressure to come up with funny bits on a daily basis, and many of them centered on what was in that day's headlines. If there was a union strike, Mr. Leonard couldn't come to work because he wouldn't cross the picket line, even if the line were miles away from the radio station. Once he couldn't come in to work because he was overwrought with grief over the death of Theodor Geisel, aka Dr. Seuss, so he eulogized him live on the air with one of the writer's supposed off-color rhymes that somehow never made it to print. Yet another time Mr. Leonard called in to say he was in mourning for his uncle, who "died suddenly from a fart in his liver. It went right to his brain!"

But still some of Mr. Leonard's sillier moments came from wacky songs such as "Pop Was a Weasel" or by his mangling the public service announcements at the end of the *Buzzard Morning Zoo* ("There's going to be a peter pull at St. Taffy's tonight. Er, eh, excuse me. A taffy pull at St. Peter's."). The bits may have relied on tired jokes and clichés, but Mr. Leonard's daffiness brought a certain charm to them.

Rio abandoned the Mr. Leonard guise for a time to do a morning-drive talk show in Tucson, but the character was in such huge demand that he returned to the airwaves soon after. Proving that he spans all cultures and formats, Mr. Leonard followed Rio all the way to a Dallas country station.

At one point Rio considered killing off the character. One possibility was having Mr. Leonard shot in the head as an innocent bystander at a liquor store robbery, and another was having him describe ways to clean his new gun live on the air, assuring the host that the safety was on.

Luckily, that never happened. Still, no doubt Mr. Leonard would have been absent for his own funeral.

The Buzzard

The WMMS Buzzard is a little like the Mickey Mouse of Cleveland. It's a character that is recognized around the world and came to symbolize the station and the city. Like the mouse, it had a humble beginning and grew to monstrous proportions, seen on everything from t-shirts and television spots to major motion pictures.

In 1974 program director John Gorman and Denny Sanders decided it was time to replace the station's old symbol, the mushroom. Because the mushroom didn't move, they felt that the station needed something that reflected the energy of WMMS and was as animated as its listeners. People listened to the station in their homes and in their cars, out on the street and in the air, so it made sense that a bird would be an appropriate character, but what kind of bird?

There's been a lot of debate over the years by past station employees why a buzzard was chosen, but whatever the reason it certainly made sense. There was a turkey buzzard around which a legend had formed that it returned, like the swallows to Capistrano, to Northeast Ohio, specifically Hinckley, every March 15. And the station was looking for something to convey "attitude." At that time WMMS's Sunday night programming included the nationally syndicated *National Lampoon Radio Hour*. The *National Lampoon* producers wanted to cut the show down to a half hour, and they did it with a sketch done in the show's typical fashion. They announced that some stations were cutting the show down to thirty minutes, and that the producers didn't like that. The show advised listeners that if that's what was happening in their city, they should—(dead air). The joke was that even though the show was going to be a half an hour long anyway, they wanted to give the impression that the local stations had cut it.

A local fan of the *National Lampoon* show was twenty-one-year-old David Helton, an artist at Cleveland-based American Greetings. Not realizing the real situation, he sent a three-panel cartoon to WMMS, protesting the change. Gorman and Sanders loved it and decided to give him a shot at designing the Buzzard, but there was one problem. Helton didn't include a return address. They were finally able to track him down by phone, and his girlfriend passed on the message that, "A John Gorman called and wants you to do some work for him." He returned the call, but since Helton couldn't come down to the station until after work, a meeting was set up with Sanders during his air shift.

When he arrived at the East 50th studios, Helton was brought into the station, and Sanders talked about designing some kind of bird, a predator, that would be the new symbol for WMMS: "We want a buzzard! Have him perched on a mushroom—with pieces of the competition in his mouth!" Slightly confused, Helton asked, "Well, how am I supposed to put other stations in his mouth?" and even Sanders wasn't sure. But Helton wasn't about to let his first freelance opportunity

David Helton's strip was meant as a lighthearted protest against the half hour version of the syndicated *National Lampoon Radio Hour* (1977). Cleveland Press Collection/Cleveland State University

Bird of Prey (1974). This is Helton's very first Buzzard as it appeared in the pages of *Zeppelin* magazine. David Helton

slip from his grasp, so he simply asked, "How soon do you need it?" When Sanders said, "Tomorrow morning," Helton stayed up all night to finish the job.

Helton had grown up being influenced by the work of legendary cartoon makers Walter Lantz and Tex Avery, among others, and although there's a hint of their styles in that first drawing, the finished product was an original creation. He used a lot of line work, similar to the style of another influence, Robert Crumb (who had officially left American Greetings the year before to work on his own under-

APRIL 1974   JULY 1975   1976-78   1979-PRESENT

The Changing Face of the Buzzard. The Buzzard had gone through an evolution over the years, becoming less ominous and more family friendly while still standing as a symbol of the WMMS attitude. David Helton

ground creations), but the long hair worn by the Buzzard was modeled after the style worn by Helton when he was in college.

At the time Sanders lived on the East Side of town, and Helton lived on the West Side. In true Cleveland fashion, famous for its East-West rift, neither knew much about the other side of town. They agreed to meet at a West 50th Street, though neither was aware that the street didn't exist.

After a confusing walk near where their proposed meeting place was supposed to be, a bleary eyed Sanders found his way to the next nearest intersection in Lakewood, the corner of West 48th and Detroit, to pick up the art. It turns out that Helton made that same confused tour, and they met, by accident, at the time they had planned. The drawing was of a scraggly looking bird, perched on a mushroom, but it worked. Helton got paid in albums and concert tickets, and the Buzzard had its print debut in *Zeppelin*, a weekly newspaper, on April 16, 1974.

The Buzzard has changed and evolved a lot over the years, much like any major drawn character. At first it appeared with other characters and was often seen with a suspicious-looking cigarette. But the crooked neck, stubbly beard, and scraggly hair soon gave way to a more-streamlined character. Some of the changes were due to the limitations of the printing process. Helton says, "The guy who did the first

David Helton (1986). Helton's graphics defined WMMS in a visual sense. He's pictured here in a scene from a television spot for the station. Photo by Brian Chalmers

Buzzard bumper sticker, he traced our artwork because I had so much line work in it, that it wasn't a hard edged line. I realized he had traced my artwork, and I was mad about it, but there wasn't much I could do. I was just some hippie kid from American Greetings." Helton's character became so popular that bootleg shirts and bumper stickers started to pop up here and across the country. Even Gorman was seen wearing one of the bootleg shirts that he bought for a mere three bucks. As Helton points out, "Most stations had no budget, and it wasn't long after we created our Buzzard that we got ripped off."

While the Buzzard was Helton's creation, it also had an aggressive side that could be attributed to Gorman, who had a reputation as a fierce competitor. Looking back, Helton says, "At times I had to tone it down, because I had a certain amount of good taste—and John had none. To him, showing Buzzards screwing under bed covers was hilarious. I found that a little offensive, however, so I toned it down. I didn't make it as blatant, or as obvious, as John's original intention, which for years made us a great combination." It wasn't long before marketing pro Dan Garfinkel was involved.

Helton recalls that there was something of a controversy over the copyright for the drawing, especially when bootleg items, like banks, started popping up: "I had it copyrighted under my name, just by simply drawing a copyright on there and putting my name next to it. Corporate never took it seriously for years, and never bothered copyrighting it. So people started ripping us off right and left. They didn't care."

One of the most important and recognizable promotional items ever put on the street was the WMMS Buzzard t-shirt. Garfinkel saw the marketing value of the Buzzard, saying, "When I got there [in 1975], they were manufacturing three thousand t-shirts a year, on which they were losing money. Shirts were not the number one priority on their agenda. I really tried to build a program out of it, and by the time I left the station, we were doing 30,000 t-shirts a year. 15,000 in the spring, and 15,000 in the fall. Plus, we were having the shirts custom made, with David Helton and myself designing those shirts from scratch. We had some great retailers that worked with us, and we generated tens of thousands of dollars for charities."

The bird turned up in the pages of *Playboy,* on NBC's *Saturday Night Live,* during rock-star interviews (the Beach Boys' Brian and Dennis Wilson and Bad Company's Paul Rodgers being just three of many), and even on a Haitian refugee featured in a PBS documentary. There were so many made for so many reasons, often in very limited runs, that no one, not even Helton, has a complete set. The Buzzard t-shirt became a world-known symbol of Cleveland, and photos of people wearing the shirts were sent to the station from outside the Eiffel Tower in Paris, the front lines of Desert Storm, and just about anywhere people from Cleveland would travel. There were also some very grim stories associated with those shirts. In the 1980s, a mother called the station to request that Lynyrd Skynyrd's "Free Bird" be played at a certain time when her son's coffin was being lowered into the grave. It was his favorite song, and he was being buried in his Buzzard t-shirt. Another time, in the early 1990s, a distraught woman, whose two young children and boyfriend were found dead in a car in Lake Erie, called to request a song in their memory. Her call went on the air, and she was extremely emotional through the conversation, gasping out her feelings between the sobs. As the deejays said goodbye, she sent her regards to *Buzzard Morning Zoo* mascot Mr. Leonard, and asked for a t-shirt.

The Buzzard symbolized the radio station, and Helton—and later his assistant and successor, Brian Chalmers—brought it to life, but that doesn't mean they had artistic carte blanche with the character. Station management didn't always understand the Buzzard or what it stood for, and the artists fought frequent battles over its use. For example, one manager didn't like Chalmers's t-shirt–design proposals,

The Buzzard's T-Shirt Debut (1974). This artwork graced the front of the station's long-sleeved t-shirt, which also carried the old logo on its back during the transition. David Helton

so the program director had his wife come up with a few concepts, including tiny buzzard heads popping out of flowers and a Buzzard Godzilla. Without saying where the designs came from, he unveiled the sketches at a staff meeting, where they were met with criticism and, sometimes, silence. He later decided to make his mark on the character by ordering a new haircut for the bird, which was quickly reversed by Gorman. Gorman was surprised as anyone to see that incarnation, saying it resembled Sonny the Cocoa Puffs bird. When OmniAmerica assumed control of WMMS in April 1994, the company scrapped the pony-tailed bird for, what Gorman called, "The son of the original, who was killed by bad programming, poor promotion, and neglect." The rebirth came on the first day of Omni's ownership, with a picture of the classic Buzzard breaking out of an egg.

The Buzzard has long been a point of contention: A general manager, who relocated from out of town, told Chalmers, "The Buzzard is an icon. It should never move. It should remain inanimate." It didn't matter to him that the Buzzard had been walking, flying, swimming, and doing every conceivable type of movement—even in actual animation—for more than twenty years before he arrived. Another general manager questioned why the Buzzard sometimes wore clothes, and other times he didn't. Shouldn't he at least wear a pair of pants? In perhaps the most outrageous comment, when Helton designed the Buzzard beach towel in the late-80s, another management-type asked if it was wise to have the Buzzard "dancing with a black chick." This is odd since no one cares whom the Buzzard dances with, and he was dancing with a female Buzzard who was just as black as he was. In most cases, cooler heads prevailed, though Chalmers had a unique talent of wearing people down with words, which often worked to his advantage.

As Mickey Mouse is for Disney, Superman for D.C. Comics, Alfred E. Neuman for *Mad Magazine,* and even Chief Wahoo for the Cleveland Indians, the Buzzard is more than just a corporate logo. He is a character, one that symbolizes the very essence of WMMS: its attitude, its style, and its freewheeling spirit. He has become an icon; a visual representation of all that rock radio is about.

That's quite a feat for a character that has never said a word.

How Did That Happen?

No matter how hard a station is programmed and promoted, the ratings game is a crapshoot, and like the NFL on a Sunday in autumn, anything can happen from week to week. Disc jockey Gary Dee, who held court on WHK-AM for many years, joked that someone would turn on the radio to keep their dog company, and that dog represented fifteen hundred homes. WMMS was the dominant FM station in town for a long time, despite some very stiff competition from the rest of the fragmented radio dial. The station's audience was a loyal one, but something happened in fall 1983 that had even the most seasoned radio columnists scratching their heads.

The summer Arbitron ratings were released that showed WGCL-FM was beating WMMS. WGCL had been showing strength for some time, even though it didn't get as many concerts, didn't have a mascot, and rarely had a television spot. WMMS, however, had a huge war chest and a great in-house marketing department. Billboards, bumper stickers, newspaper ads and articles, t-shirts, and even specially painted Volkswagen Beetles peppered the Northeast Ohio landscape, and a person couldn't move more than a few hundred feet before the call letters were staring them in the face again. So how was WGCL able to stun the Buzzard?

First, WGCL was not known for the longevity of its staff; some of the best voices in radio passed through the station's doors and walked right back out again, so it was tough to attract talent. Meanwhile, the core WMMS staff had been pretty much intact for almost ten years, proving that familiarity breeds ratings. WGCL had an image of a Top-40 station, ranking number one with the teenage audience, more than twice that of WMMS. Even so, it didn't have the kind of pull in the music industry that got the exclusives and premieres that WMMS had prided itself on. Then-WGCL general manager Kim Colebrook says the time was just right for his station to make its move, "It was a period of time where the station was at its best musically, because there was a lot of great product out there, and our rotation was tight. We worked hard to get there. Plus, we had Danny Wright on the air."

No doubt about it, Wright was a key element in WGCL's surge to the top, and he makes it perfectly clear that this ratings book was something he wanted dearly. Wright recalls, "I was totally psyched to whup the legendary Buzzards. One reason was because they disappointed me—and a lot of people. I came into town really looking forward to being in the city with the legendary AOR, and after I met a lot of the people who worked there, mostly in programming, I was surprised they were about as rock and roll as Maudie Frickert. And their tactics were really nasty, too."

But there is something about Wright, even back then, and that is that he never worried about ratings, saying, "I'm happy if they're good, and I like to look them over, but I'm more worried about my next on-air break than the statistical stuff. Once I saw what WMMS would attempt in order to stay on top, then I wanted to beat them, because they made it personal."

Sometimes the Arbitron didn't really reflect the Northeast Ohio audience. Summer was always a big book for music stations, but there were gaps in the process that could sway the ratings. It depended on who was filling out the ratings book, how they did it, and what influenced them. WMMS obviously benefited from having its call letters in so many places. One program director quipped that WMMS could turn off its transmitter for three months and still pull killer numbers. Then there were people who took their passion for a station to a different level and ended up giving their share of the ratings to the competition. Most of the stations would send a representative to Arbitron's Maryland headquarters to inspect the books, and they would sometimes return dumbfounded by what they saw.

The ratings basically depended on people accurately listing in a book what they listened to, and what time, and sending it back to Arbitron, which then compiled the stations' rankings. They could also be wildly inaccurate. For example, one person might have hated a particular station and wrote something like, "Smokin' Joe sucks." The problem with that is that "Smokin' Joe" was the name of the disc jockey that was written in that time slot, so his station got the point. Others might give a station the entire day when it was obvious that no one could listen for that entire length of time.

WGCL had always been competitive, despite its drawbacks, but it was still battling the WMMS machine. The WGCL strategy had been to concentrate on its strengths, which were teenage listeners and females in the eighteen-to-thirty-four age group, often similar to the younger demographics. Still it took a long, long time to beat WMMS.

The two stations had distinctly different sounds, and their personalities were like night and day. WMMS's morning show with Jeff Kinzbach and Ed Ferenc was a lot more focused and moved a lot quicker than Jim King's program at WGCL. There was also a matter of consistency. King's performance was often erratic, while Kinzbach and company were able to remain at a constant level of professionalism. WGCL's booth announcer, Bob Gott, had a cutting-edge sense of humor, but that only came through during his newsbreaks, and a lot of people didn't feel like waiting a half hour between reports.

Wright was WGCL's big gun, a consummate professional with a national reputation, who could have worked anywhere in the country and turned down plenty of offers because he really liked it in Cleveland. Wright proved to be the station's most popular disc jockey, but he still had to face the already legendary Kid Leo on a daily basis.

WGCL had another major player in its corner, promotion director Jon Anderson. Prior to Anderson's arrival, the promotion director's job had seen a lot of turnover, and WMMS had been the hallmark of consistency. But Anderson had a strong background in record retailing and was in close touch with the local and national music scene. Anderson was also focused, energetic, and had a distinct vision and idea on how his station could distinguish and promote itself. He was smart enough to know that, at that time, WMMS couldn't be beaten where it was strongest, so he pushed to get his deejays in clubs just about every night. Prior to his arrival, WGCL had to depend on record-company promotions, along with anything that the sales department could cook up. Anderson changed that too. All promotions from the time of his arrival had to have a distinct programming payoff for the listeners, otherwise they weren't even considered, much to the chagrin of the sales people.

Anderson wisely pointed out that no mascot could match the strength of the WMMS Buzzard, so he stayed away from that type of marketing. Instead, the station pushed its "G-Team," its on-air "G-Men," and traveled in its "G-Van" to give away "G-shirts." The ratings books started to show growth too. Over the course of five Arbitron books, WGCL climbed from a 5.2 ratings (compared to WMMS at 9.6) in summer 1982 to a 9.8 in summer 1983, against WMMS's 8.9. WMMS had only remained consistent during those books, while WGCL steadily edged toward them. And although WGCL might have been number one overall, the WMMS morning show still ranked far ahead of WGCL at the same hour.

Denny Sanders recalls it well, "They beat us in that one book," he says, "and they beat us in women and 25-plus men. But even if it was an odd book, and off by a few thousand people, it was close enough. We looked at that book and said, 'We are missing the goddamned boat.'" So was all of AOR radio.

WMMS wasn't sitting on its laurels. The station just kept giving the audience what it was known for, and it wasn't going to panic over a down book. The accuracy of that summer book had also been heavily debated, because a lot of advertisers just didn't believe the numbers. WGCL might have been popular, but there were many skeptics who didn't agree it was that popular. Some big-money sponsors decided to wait out the results of the fall ratings period, which was one of the most important for local radio. That's where WGCL stumbled. Instead of spending serious cash for a television and print campaign, they spent very conservatively and failed to capitalize on their summer numbers.

As Wright puts it, "I was pretty sure we couldn't sustain the lead, because we were a mom-and-pop station, and the owner wasn't all that interested in rock or CHR. He was more into our sister station, WERE. He should have poured a bucket of money our way to keep the war financed but actually pulled back with the attitude of 'Hey! We're number one—so we can relax.'"

Following that book, WMMS staged a full out media assault, and when the smoke cleared, the Buzzard had regained its perch. Although still showing some strength, WGCL would never take the top spot again. However, its influence would still be felt at WMMS.

The Battle Rages

The buzz on the streets was about a new technology that was about to change the face of radio and the listening habits of music fans everywhere. Compact discs had been talked about since the early 1970s, when the Grateful Dead hoped to introduce its music in the new form—although the group had thought it could be packaged in a pyramid or square shape. Years later, digital reproduction promised crystal-clear sound in more durable disc form, even though the public still had to be convinced it was worth replacing their record collections.

In October 1983 WMMS joined with WEA, Polygram, and High Fidelity for a public demonstration of compact discs (CDs) at the Agora. As always, it turned into a party, and those who could remember the night came away convinced that CDs were destined to kill vinyl records. That same month the station continued its series of free fifteenth-anniversary shows at the Agora, with a live broadcast of

Aldo Nova, an artist who was getting a lot of airplay across the rock dial. This was the era when a shot on WMMS could be a springboard to superstardom, not only for newcomers, but for veterans striking out on their own. Matt the Cat hosted the Joe Perry Project, fronted by the former Aerosmith axe man (Perry). Many of those who saw that show live suggested that coffee might have been a better beverage than what the band had obviously been enjoying before the show. Engineer Frank Foti recalls some other Perry appearances that underscore the importance of breaking out of Cleveland at that time:

> We had a permanent high-fidelity audio connection between the production studio at 'MMS and Agency Recording. At any moment, we could go live to the Agora, because we used to do the Monday night concerts. On occasion, we'd get a big act—like, Dire Straits, the Pretenders, the Police—and once we got clearance from the band, the jock would just bring it up on the board.
>
> When we did the Joe Perry Project, he was doing a Sunday night, Monday night thing. For some reason, the Sunday night sold out, but the Monday night show didn't, and they were going to let us broadcast Sunday night with the hope that it would help ticket sales for Monday. No one knows to this day why the regular engineer at Agency Recording wasn't there. Someone did something wrong at Agency Recording, and when they went on-air it definitely didn't sound right. It was distorted, out of phase, and you couldn't hear the vocals. Betty Korvan called me up at home, saying, "Frank! We've got problems! You gotta get out here!" I turned the radio on and—oh my God. This is horrible. She wanted to know what to do to fix it, and I just said, "Hate to tell you this, but just turn it off." Betty dumped the broadcast, and the next day, Joe Perry's people were pretty upset. Our people at 'MMS were upset, and the record label was upset, too.
>
> They wanted us to do the Monday night show live, but John Gorman would not do it. "After what you guys did to us, I'm not putting that trash on the air." John said the only way we would do it is if we had one of our people in the studios and to make sure "someone of substance" was there to make sure the mix is right. Joe's producer at the time was Jack Douglas, the legendary producer of Aerosmith, John Lennon, and so many others, and they flew him in. The only way that the show would go on the air was if Jack Douglas and I were in the room. He and I mixed the Joe Perry Project that night.

Douglas was one of the hottest producers in the industry, and one of the busiest as well. It was an indication of the power and importance of WMMS to have him drop what he was doing at a moment's notice for an impromptu trip to Cleveland, only to rush home and pick up where he left off as soon as the show was over.

Beat It

By late 1983 Michael Jackson was well on his way to becoming the biggest star in the world. Powered by the phenomenal *Thriller* LP and heavy rotation on MTV, the cable music channel that had been gaining strength since its inception in 1981 as more and more homes around the country were linked to cable. MTV exposed up and coming acts, like Prince and Madonna, as well as established artists trying to break new ground, and Jackson was using the video channel to his utmost advantage. MTV's popularity seemed to grow by the day, and as a result, a lot of radio stations started getting an increasing number of requests for artists getting that kind of attention. The reality was if radio stations wanted to keep listeners, they started programming music similar to MTV.

Jackson could guarantee sold-out arenas anywhere he chose to play. A long-rumored tour with his five brothers was finally announced at a New York press conference that December, and although they couldn't give a date, Cleveland was assured it would be the site of at least one show. In fact, Cleveland's Don King was signed by family patriarch Joe Jackson to promote the thirty-city *Victory* Tour, promoting the brothers' first album together in eight years. The Cleveland show promised to be a sellout and a huge promotional boost for the station chosen as its sponsor.

Meanwhile, WMMS was taking advantage of its strength, not to mention the offer of free television exposure, by getting its morning team as fill-in weather reporters on a Sunday night newscast. But not everything WMMS touched turned to gold. The British group Big Country had been getting a fair amount of attention on MTV and agreed to do a *Coffeebreak Concert* for WMMS that December. But just a few minutes into the set, lead singer Stuart Adamson stopped the show, saying, "We can't do this," and complained that his voice was gone. A stunned Matt the Cat was heard asking, "What? Is that it?" as the band exited the Agora stage, and when he returned the broadcast back to the studio, there was a period of dead air. Backstage, bassist Tony Butler said the band had "really partied last night," resulting in the first time a group had walked off in the series' eleven-year history. Whatever the reason, it was not a wise move to anger a station as powerful as WMMS. It was an embarrassing situation, and not much, if any, Big Country was played on one of the most powerful radio stations in the country after that. It made sense to stay on the good side of WMMS, because it was watched by the entire industry. Rising star Cyndi Lauper was a far more reliable choice the very next week.

Meanwhile, a very strong radio voice was rumored to be leaving Cleveland. WHK-AM's Gary Dee had exited the station earlier that year for a job in the Washington market, and that December, WGAR's popular morning man John Lanigan was said to be heading to a very lucrative position at Metroplex Communications' Tampa, Florida, station. Norman Wain and Bob Weiss, part of the team that made WIXY a legendary and very profitable station in the 1960s and early 1970s, owned that company. Lanigan's expected departure would give all of the stations a little more breathing room, as they hoped to siphon listeners away from his replacement, Paul Tapie, who was being moved from the afternoon shift. Lanigan and Tapie even did the show together in the days just before Lanigan's move, but the numbers started to fade after Tapie went solo. Even so, Tapie had other high-profile gigs in his very near future.

The Line in the Sand

At the beginning of 1984 WMMS's John Gorman moved up from his position as program director to operations director, and a Cleveland State University student, Gina Iorillo, became the programming and promotion assistant. Within a very short time this group would be making headlines in and out of the industry. WMMS and WHK-AM news director John Webster, who had a long and distinguished history in radio and a reputation for being extremely creative and a fierce competitor, left the station a short time later. "The Buffalo," as friends knew Webster, would soon come back to bite the Buzzard in a big way. At WMJI a new program director, Dave Popovich, was settling in to his new office. Popovich was another figure whose presence would be felt on a much larger scale.

There was an announcement being made from WMMS that seemed to have become an annual tradition but no less welcome. *Rolling Stone* magazine had named WMMS its "Radio Station of the Year" for the fifth straight year, despite a campaign by a Dallas station to "Beat the Buzzard," which brought them in only a poor fourth. The WMMS star continued to soar with the new Arbitron ratings book that put the station firmly at number one, despite the decline of similar AOR stations nationwide and the surprise book that had WGCL-FM beating the station the summer before. In reality, WMMS was so big that it was starting to feel the burden of its own weight trying to appeal to such a large cross-section of the audience. It was a publicity magnet, and every move was closely scrutinized by the industry, not to mention its huge audience.

There was a lot of pressure from within the company to maintain WMMS's ratings, reputation, and position, despite the changing musical trends and other diversions drawing people from the radio dial. As former general manager Lonnie Gronek recalls, "There was a great deal of fear and almost paranoia, as to what the station was doing. The more fear that was brought into it, the more people came into the mix, and as soon as eight or ten people have input on every decision, you have a ten-headed monster that couldn't possibly make the right decision every time. It was a very difficult way to operate a business."

Michael Jackson took top honors for best album, best male vocalist, and rock personality of the year in the annual WMMS–*Plain Dealer* Rock Music Poll, and the station's Kid Leo ranked number five in the personality category. Gorman heralded Jackson's recent sweep of the *American Music Awards,* his expected victories at the upcoming *Grammy Awards* ceremony, and his strong showing in the station's poll as harkening back to a time when "there wasn't any division between types of music." Even so, a good-sized segment of its longtime audience was wondering why WMMS was so supportive of a Top-40 act like Jackson.

The Beatles still drew a huge amount of interest on Cleveland radio, twenty years after their American debut, and stations across the dial paid tribute in a variety of ways. WMMS co-sponsored an event put on by rock photographer Janet Macoska and local record promoter John Awarski—both avowed Beatles fans—called Beatles-Faire '84, and the station's Denny Sanders appeared on WEWS-TV's *Morning Exchange* to promote the show and talk about Beatle collectibles. Plus, WMMS and WMJI both aired special Beatles programming, along with local television crews inserting related reports in their evening newscasts.

WMMS continued helping charities here and across the United States, with Leo jetting out to Los Angeles as national radio chairman for the T. J. Martell Bowling Benefit to fund research for cancer and leukemia. His efforts brought Rod Stewart; Britt Ekland; the Stray Cats' Slim Jim Phantom; Little Steven Van Zandt; and members of Berlin, the Go-Gos, and Motley Crue together to raise more than $60,000. Meanwhile, WGCL took a serious hit when Danny Wright picked up stakes to go to Washington's WRQX-FM radio. And it was now official: WGAR's John Lanigan was packing for his big move to Tampa's WMGG-FM to debut there on February 13.

Malrite was seeing a huge success at its New York station WHTZ-FM, which soared to the number-one spot in the nation's toughest radio market in just seventy-four days. Part of its success came from a new contemporary hits format, part from moving its antenna to the top of the Empire State Building, and mostly from word on the street. It would prove to have a ripple effect on the company's Cleveland station. Jeff Kinzbach remembers,

> The way that station took off so well may have led to management's decisions to try to make 'MMS even bigger by emulating a lot of the things they did there. Of course, we became the *Morning Zoo,* and that was from Scott Shannon at 'HTZ. Their whole thing was, "We just want you to be the *Morning Zoo* so no one else in Cleveland will take it." There were some things that we applied from WHTZ that really did work well for us. For example, the birthday

contest, which to this day is being used by radio stations. It was brilliant—a super promotion. But I think part of the WHTZ influence pushed management to add Michael Jackson and Madonna and try to steer us toward Top-40. This is Cleveland, not New York City. Our listeners wouldn't put up with it.

But WMMS had much to celebrate as well and welcomed spring with a free appreciation concert for its listeners with the Eurythmics. Among those attending were Cleveland city officials Jeff Comfort and Vince Lombardi, who presented a special proclamation honoring WMMS for its "Best Station" designation from *Rolling Stone* magazine. The Buzzard flag again flew proudly over City Hall.

Just a few days after the Eurythmics show, WMMS vice president and general manager Walt Tiburski announced his resignation to start WIN Communications with former Cleveland Browns player Thom Darden and other investors. WMMS was pursuing a prize it hoped would bring it even more visibility, and that was sponsorship of the Jackson *Victory* Tour. It was reported in the *Plain Dealer* that WMMS wouldn't have considered playing Jackson just a few years before, but now it was actively trying to attach itself to his Cleveland show. It appeared that WMMS had the upper hand following a live telephone interview on Leo's show with Jackson's personal manager, Frank Di Leo. He was a former Clevelander who left a vice president's position at Epic Records to work with Jackson and knew Leo from many years back.

On April 19 Di Leo gave first word on the tour on Leo's show, demonstrating just how serious the station wanted to be part of the Jackson concert. WMMS also continued to break new music, getting the jump on other stations with Rush's *Grace under Pressure*, the Eric Clapton–Roger Waters collaboration from Waters's *The Pros and Cons of Hitchhiking*, and the Michael and Jermaine Jackson duet, "Tell Me I'm Not Dreaming." The best promotion was still a solid air product.

Promotionally, it promised to be a very busy summer ahead for Cleveland radio, especially WMMS, and the station's marketing machine swung into high gear putting its people in the public eye. After their debut as Sunday-night weather forecasters, WKYC-TV's *AM Cleveland* cameras traveled to the Statler Office Tower to catch *Jeff Kinzbach and Friends* in the friendly confines of the WMMS studios. It gave viewers a rare look not only of the crew at work but also of the air studio that had hosted dozens of the biggest names in rock and debuted just as many hits for the Cleveland audience.

WGCL still had some steam behind it and hoped to increase its visibility by sponsoring a free show with the 1970s glam rock band Slade, whose previous hits, "Cum on Feel the Noize" and "Mama Weer All Crazee Now," had found new life with the mega-watt screamers Quiet Riot. The show was planned for May 25 at the Greater Cleveland Growth Association's Friday night Party in the Park at Mall C downtown. Slade's bass player had been hit by a case of hepatitis, forcing the cancellation of its American tour, but the band had already committed to this show. There had even been talk that WGCL was hoping to film segments of the show for a television spot, but there was an unforeseen roadblock. Prior to the WGCL show, WMMS had been presenting bands such as Blue Oyster Cult at that same location, and when WGCL announced the evening show that Friday morning, WMMS caught wind of it.

The war between WMMS and WGCL had gone to new heights, with WMMS labeling WGCL and its disc jockeys the "Baboons" and doing parody songs targeting that station. For example, David Bowie's "This Is Not America" was transformed into "This Is Not Baboonland," and Sanders joined "Spaceman Scott" Hughes to parody WGCL's Bob Travis and Tom Jeffries in a series of on-air skits. While hardly of the caliber of "Weird Al" Yankovic, the effort put into the on-air parodies spoke volumes of how serious WMMS was at undermining the confidence of its competitors.

Word got back to WMMS that Slade was going to lip-synch its WGCL show. Gorman believes someone connected with the city, and obviously a WMMS fan, tipped the station off about the permits and sound equipment the band would be using. The morning show broke the news immediately, and Gorman told the deejays to hit it hard through the day.

Slade's new release was being handled by CBS Records, and there began a disagreement between the label's Joe Carroll and Gorman that would turn into a clash of titans. At about 9:00 A.M. the station's studio hot line started ringing, and Hughes told Kinzbach that he might want to roll tape on the call. It was Carroll. Anyone who's acquainted with Carroll knows that he's one of the best record guys in Northeast Ohio, with a solid reputation. He also doesn't take things lying down. In an angry call, rife with very graphic expletives, Carroll exploded at the morning show, the station management, and just about everyone in between. Kinzbach continued to roll tape and asked, "Anything else?" to which Carroll shot back with even more angry words. Then he slammed down the phone.

The morning crew called Gorman on his car phone. He listened to the tape, and said, "Run it." They beeped out most of the offensive words, put instrumental music under it, and rushed the tape on the air. Leo was on the golf course as this was unfolding, heard what was happening, and raced back to the station. Adding fuel to that fire, the morning show gave out the business phone number for Columbia subsidiary Epic Records, as well as the district manager's name, saying if anyone wanted to complain, they should feel free to ring up the label. That didn't sit well with Epic's national office. Leo and Carroll had come from similar backgrounds and been friends for a long time, so this put them in a very uncomfortable position. Leo was a loyal friend, but he was also loyal to his place of business, so he would hear about it from both sides. It eventually resulted in Carroll being banned from the station for a couple of years.

WMMS War Chest Marketing Cover (1984). The "radio as war" mentality even made it to the some of the station's promotional material. David Helton

But that battle was far from over. WMMS put in a rush order for hundreds of "Baboon Buster" t-shirts and announced that Hughes would be handing them out free that afternoon near Mall C. Before the ink was dry, the shirts were being handed out, and people put them on. A sea of white t-shirts could be seen heading to the concert site, which effectively stopped the video crew hired by WGCL from panning the crowd for their television commercial. Adding serious insult to injury, once the band hit the stage, it wasn't playing in synch with the taped music.

The decision to scuttle the Slade show came, in part, out of frustration. WMMS was playing the band long before WGCL did, but as with a lot of other artists, WGCL would get the show sponsorship instead of WMMS, not to mention the revenue from spots advertising the show. This was the last straw, and WMMS started mentioning the enemy by name.

The power of WMMS could not be understated. Gorman, Leo, and the rest of the station had the ratings clout to wield huge influence in the music industry, and they wisely took advantage of just about every opportunity that came their way. Upcoming Bruce Springsteen shows at the Richfield Coliseum turned into "Buzzard love fests," and Leo was one of the few deejays nationwide allowed to host the concerts. WMMS had access to the artist and his band and developed a long history with the Springsteen camp. The other stations had to swallow hard when it came to linking up. If a station like WGCL gave away tickets to the Springsteen show, they knew they would be sending their listeners to an event that would bombard them with all things WMMS and be greeted on stage by its most popular, and powerful, air personality. Still the Springsteen shows were so big that they couldn't be ignored, so the competitors just weathered the storm and let WMMS have yet another moment in the sun. The station just did things on a larger scale to make themselves and the event larger than life.

The big prize that summer was still the Jacksons' *Victory* Tour, and a lot of stations were vying to sponsor it. But hope seemed to be waning for a Jackson show in Cleveland, as dates filled up around the country, and the only venue thought big enough to hold the concert was Cleveland Municipal Stadium. Word was getting around that Browns owner, and Cleveland Stadium Corporation operator, Art Modell didn't think all sides involved could agree on a date outside of the baseball and football seasons, and weather was a key factor as well. People were starting to make travel plans to see the tour in other cities.

WMMS was still expanding its scope of influence and exposure in other media. Kinzbach appeared in Tom Batiuk's nationally syndicated *Funky Winkerbean* comic strip. Batiuk also featured Michael Stanley's *You Can't Fight Fashion* LP on a character's t-shirt, as well as appearances by the Buzzard. WMMS reciprocated by having Batiuk in as a guest deejay. Finally, as the rating season made it to the halfway point, just after the Fourth of July, an announcement came out of Kansas City. At the opening show of their North American tour, the Jacksons had announced they would play Cleveland Municipal Stadium on October 19 and add more dates if necessary. But the fight was far from over among radio stations hoping to sponsor the show. Classical station WCLV-FM claimed that it had been named the "official station for the Jackson concert," jokingly referring to the upcoming Cleveland Orchestra concert at Blossom Music Center with Isaiah Jackson as guest conductor. It was a good indication of the frenzy that Cleveland radio found itself in as stations positioned themselves to land that big promotional prize.

Stages

Local radio and television personalities have always been a pretty good draw for live theater, and people like Linn Sheldon (local television kids' host Barnaby), John Lanigan, and others all tested the waters to great success. Certainly Kid Leo was no stranger to the performance stage, and for a week in mid-August 1984 he got a great response playing a disc jockey in the Kenley Players production of *Grease* at the State Theater. The show starred Christopher Atkins, then star of the television show *Dallas*, and it was baptism by fire for Leo because there was no rehearsal time. Even so, he talked it up on the air, the fans turned out, and it was a win-win situation for all involved.

A different stage performance proved to be a high-water mark for WMMS, as it announced a free, late-summer Appreciation Concert at Blossom Music Center, starring the Thompson Twins and Berlin. Besides honoring Cleveland listeners for putting the station at the top of the *Rolling Stone* readers' poll for a fifth straight year, WMMS was also celebrating its third straight ratings book with double digit numbers, which was a first in Northeast Ohio radio.

WMMS's dominance was very evident to the owner of WMJI and its AM sister station WBBG, Larry Robinson. Proved to be a very smart businessman, he decided it was a good time to sell those stations. Initial estimates put the sale price at $13.2 million, a sizable increase over the $6.2 million he paid for those stations three years earlier. Down the street at WGCL-FM, talks were under way to bring back Danny Wright, who had posed the only serious challenge the station could muster against cross-town rival Leo at WMMS. Wright had left Cleveland six months earlier to take a job in Washington. His return had been a very poorly kept secret, as he was booked on WKYC-TV's *AM Cleveland* two weeks before the official announcement. Even with Wright's marquee value, it would be very difficult to derail WMMS. As it turned out, Wright decided to stay put for the time being.

In fall 1984 rumors were already starting to surface that John Lanigan as well was being courted to come back to Cleveland from Tampa Bay, and that this time he might find a home on the FM band. If those rumors were true, whoever picked him up would have a big gun in the morning wars.

Only a little over two weeks before their appearance, the official announcement finally came down that the Jacksons would be playing two nights at Cleveland Municipal Stadium on October 19 and 20. What may have been even more surprising was that WMMS radio had been named the official radio station for the Cleveland shows, reportedly guaranteeing both concerts with a $2.7 million check to the promoters. The tickets weren't cheap at thirty dollars each, and 45,000 seats were available for each show. They went on sale at noon on October 2, the same day the concerts were announced. There was a pretty good chance the shows would be sellouts because these shows would be the Jacksons' only Ohio appearances, and the nearby Pittsburgh dates never materialized. That was a whole other market within a relatively short driving distance.

Even so, some eyebrows were raised when twenty ticket windows opened up at Municipal Stadium that day, and a relatively small crowd showed up for the first shot at seats. By day's end, only 26,000 tickets had been sold for a show that many thought would be sold out almost immediately. The price of tickets and time of year were obvious factors, as well as questions about the choice of WMMS to sponsor the shows. In fact, the Jacksons choosing WMMS as the host incensed Cleveland ward eight councilman Jeffrey Johnson. In an angry letter to the station, he charged that the station had a history of ignoring black and Hispanic audiences, and that "WMMS hadn't paid its dues. It wasn't with the Jacksons at the beginning. It came in only at the top." Johnson also claimed that WMMS had a monopoly with that sponsorship, shortchanging the minority community where most Jacksons fans could be found. In short, Johnson claimed, "The deal disgusts me and should be changed."

However, WMMS wasn't about to let those charges go unanswered, with Malrite Communications president Carl Hirsch saying, "Black artists have been played on WMMS for years. WMMS is not a white station, nor a black station, but instead a Cleveland station. I know of no other station or company willing to put up a guarantee of 2.7 million dollars. Furthermore, our business arrangements are such that not one cent of the proceeds go to WMMS." It was pretty hard to dispute those claims, but even so it was becoming clear that the Jacksons would have to depend on a big promotional push from WMMS to sell out their shows, and the station had to make sure it could recoup its investment. It would prove to be a very hectic two weeks until show time.

By October 12, WMMS was running full-page ads for the Jacksons' shows, stating how proud they were to be the host station and listing the box-office sites at theaters, shops, malls, and the stadium, where tickets could still be purchased. Mail order soon followed, and while no one seemed to be panicking, it was still a very anxious time for everyone connected to the shows.

On the day of the first show, there were more than 22,000 unsold tickets, and the Jacksons themselves (minus Michael) made the rounds at local children's hospitals, with news cameras rolling the entire time. The *Plain Dealer* ran a special supplement, covering everything from the song list to concert etiquette. That Friday night, more than 34,000 fans, many of them well-dressed and middle-aged, showed up to see Leo from WMMS welcome a highly produced and energetic show by Michael and his brothers, but it was still far short of a sell out. In fact, the Cleveland shows would be the first of several on the tour to come up short, but WMMS was still able to make an impression on its competitors and the listening audience that it could get what it wanted and no one had better get in its way.

There occurred, unfortunately, a major setback on the local scene. A fire at the Cleveland Agora on October 21 caused the roof to cave in and did extensive damage to the club. It had to shut its doors at East 21st near Chester and forced the postponement of future WMMS *Coffeebreak Concerts* at that site, although owner Hank LoConti vowed he would soldier on at a new location.

Meanwhile, the Michael Stanley Band released its tenth album, *Fourth and Ten*, that same November. Michael Stanley told Leo, during its world premiere on WMMS, that the LP's title referred not only to football but also to very serious considerations that might have to be made with the band's record label if this latest release didn't take off the way they had hoped it would.

WMMS also premiered its *Classic Rock Saturday Night* on November 24, with Betty Korvan hosting four hours of all-request favorites in response to a format that was drawing a huge following on the West Coast and was moving quickly toward the Cleveland area.

On November 27, some of the biggest names in British rock gathered in a London studio for an event that would have an impact on rock music as a whole as well as on the local music scene. In response to the suffering shown by starving masses in Ethiopia, artists including Bob Geldof of the Boomtown Rats, members of Duran Duran, George Michael of Wham! Boy George of Culture Club, Bono of U2, Sting of the Police, and many others recorded "Do They Know It's Christmas?" a charity single aimed at raising funds for the drought-stricken region. It got heavy airplay on MTV and would inspire some Cleveland-area musicians and media people to help in their own way as well.

Morning Zoo funnyman Kenny Clean was getting television time as its representative to push for local donations on that year's *Lou Rawls Telethon*. Plus, the questions about WMMS's apparent shift toward more contemporary "pop" music was proving to be a wise move, as Prince's "Erotic City" and Madonna's "Like a Virgin" became two of the most requested songs on the station. After a turbulent year with some very important victories, the powers at WMMS were feeling very confident about the station and its future.

Live Wire

It was a show that usually had little to do with music, but it became one of the most compelling segments on WMMS and Cleveland radio. Steve Church may have been the station's engineer, but he also proved to be a skilled talk-show host, and his Sunday night show *Live Wire* drew a large and enthusiastic following.

Steve says, "I'd been doing the show in Buffalo and Indianapolis prior to Cleveland. It was a huge success in Indianapolis, with something like a 50 percent share—

admittedly, on Sunday nights, where the competition is not that serious. But still, it did open some eyes, I think." It certainly did draw attention. One Halloween the Indianapolis program centered on a spoof promotion called the "screaming bridge," so named because every year in late October one could hear a woman's screams. The show told about the legend of the bridge, with a woman, her baby, and a train meeting in a tragic end. On the appointed night, a crowd of ten thousand people showed up at the bridge, bringing the local police in that small town to near panic and adding to Church's growing reputation.

Engineering was Church's day job and his primary concern at WMMS, but the Sunday-night program offered him the chance to show off his creative side. Looking back, he says, "It was strange at that time [1985] to have a talk show on an FM rock station, and WMMS was a pioneer. This was the more-music, less-talk era. But WMMS was not a normal FM rock station. Denny, who was pretty heavily involved in programming decisions, once said in a trade newspaper article that his goal was to have WMMS sound so good with the elements between the records that people would want to listen even if there was no music."

The show ran from 9:00 to midnight on Sundays, though it could vary from time to time, and it often featured an eclectic array of guests. For example, Church says,

> One of my favorite guests was Linus Pauling, the two-time Nobel Prize winner [for] peace and chemistry. He was promoting a book called something like *How to Live Longer and Feel Better*, and he was a living example. In his mid-eighties, he stayed the full three hours, and his handler was falling asleep, but Pauling was feisty to the end. One of his prescriptions is high doses of Vitamin C, and he sure convinced me. To this day, I take a couple of grams a day.
>
> Another was the x-rated movie star Marilyn Chambers, who asked me if she could do the show in the nude, saying she would feel more comfortable that way. She did, and it was certainly distracting.
>
> Yet another was Timothy Leary, best known for his promotion of LSD as a useful mind-expander. He was actually a very bright and complex guy, with a lot of interesting ideas, and not the "major baddie" the media usually painted him. But this night was not easy for him. Leary was apparently embroiled in a major spat with his wife, which was carried on with much gusto and verve during the commercial breaks and audible only to me and the show producer. When he came back on the air, he was cool and composed, as if nothing had happened.

Church says the best shows had no guests other than the callers.

> For some reason, the topic often seemed to drift to ghost stories, strange events, and the like. People still care about this stuff—witness the *X-Files*—and I used

to do the show with the lights off to get the right mood. I was never much of a believer in the supernatural, but one night almost convinced me.

A mother and daughter fan team from my days in Florida called one night and said they had "prayed that they could receive the WMMS show in Cocoa Beach," and it was working. FM simply does not travel that far in any normal condition, and there are other stations more close on the channel. So I didn't think they had much chance, but they said they were devout Christians, and the power of prayer had not failed them. After a few weeks of trying, to no avail, they called one Sunday to tell me that God had delivered the program to them, as requested. They held the phone to their radio, and, sure enough, there was WMMS. Being skeptical, I insisted to call them back at the Florida area code number, and they were surely there in the Sunshine State. I still think about that one sometimes.

While the Sunday night shift is often viewed as the public-service show graveyard, the station let Church experiment and get away with a little more than other disc jockeys at different times might be allowed. Also, people can be open to a lot more novelty, because Sunday nights are in between the final hours of the weekend and the start of the new workweek. Church stresses,

> Only one time did anyone ever tell me what I should do or say, and I never had any problem with management trying to control anything. The one time was when I named an upcoming show "How to Make a Madonna Song," and started to promote it. The idea was that you could do it without any real musicians. It was all programmed synthesizers. I was asked to change the name to "How to Make a Pop Song." I imagine that the station didn't want to alienate a major star or her record company. That show was actually a lot of fun, and one of the most memorable for me. Kirk Yano, who has gone on to be a major New York recording engineer, and Dennis Lewin of the band Beau Coup, programmed all the gear and set it up in the studio. When the time came, Kirk touched off the motley assembly, and a near exact reproduction of the "Like a Virgin" music track was heard. We added live vocals from Dennis's sister, and it sure did sound like a Madonna song.

When Church left WMMS, he took *Live Wire* with him, although the station wasn't about to let it go without a fight. As he puts it, "WMMS decided to keep the show going with the same name and format. They stopped it after a couple of months, but for a while there were two *Live Wires*, both on at the same time. That must have been a little confusing for the listeners." Church eventually won that test of wills, and *Live Wire* lived on for years to come.

The Legend Grows

On January 1, 1985, Akron found itself with a new rock station. That's when longtime radio personality Fred Anthony announced that WAEZ-FM would now be called WONE-FM, playing a mix of rock standards and some harder-edged current releases. The station's Denver-based owners made the change after researching the Northeast Ohio market and how targeting specific demographics was reaping big benefits for other stations. WLTF-FM's so-called "lite-rock" format was doing extremely well with young female listeners, while WZAK-FM owned the urban market and WMMS's "contemporary" or "current hits" format was a ratings magnet. WONE program director Ward Holmes set his station's sights on the album oriented rock (AOR) audience, putting in toll-free request lines and state-of-the-art technology to make WONE as listener friendly as possible.

On that first day of 1985, following the final newscast from WAEZ, there was a slight pause followed by the announcement, "This is WONE-FM. Cleveland. Akron. Canton." Anthony cracked the mike, wished his audience good morning, and said,

> Over the past eleven years, WAEZ has served the Akron, Canton, Cleveland listening audience with an easy-listening format. All things change with time, including radio stations, and the time has come to change ours. Beginning this morning, WAEZ-FM will become WONE-FM. The goals for WONE-FM will be consistent—to fulfill the needs of the listener, not only with music, but with meaningful public service and community involvement.
>
> To those faithful listeners who spent countless hours with WAEZ, we say "thank you" and invite you to join the excitement that will be generated by this totally new concept.

At a minute before 6:00 A.M., Deeya McKay played Bob Seger's "Old Time Rock and Roll" and welcomed listeners to the new "ONE," promising more variety and less repetition.

The first WONE staff included McKay, Jim Chenot, Mike Michelli, Tim Daugherty, Jeff Daniel Kunes, Brian Fowler, and Brett Summers, with Rich Barnett from WAKR-AM handling news duties. Fowler, Daugherty, Kunes, Chenot, and Barnett all had deep local roots. Fowler, coming from a news position at WLAN in Lancaster, Pennsylvania, had come from Parma and studied at Ohio University. Kunes was already familiar in Cleveland radio, as he did internships at both M-105 and WMMS while attending Brush High School in the late '70s.

Although it wasn't likely this upstart station, that could barely be picked up on Cleveland's West Side, could make a sizable dent in the WMMS audience, it was still deeply committed to its format. Kunes says that the general attitude was, "You weren't going to out-'MMS the Buzzard. We were just going to be an eclectic radio station, like WXRT in Chicago, and just target twenty-five-to-fifty-four [aged] adults. No one at that time was playing classic rock. WMMS was CHR, and WGCL was Top-40. The thought process was to get the older demo."

Daugherty adds that WONE capitalized on the anomaly of WMMS adding Michael Jackson and Madonna to their format. He recalls looking through WONE's initial playlist—with songs from Elton John's *Tumbleweed Connection*, Traffic's *Low Spark of High Heeled Boys*, Sea Level, and the Allman Brothers—thinking there was a whole "classic rock" audience that had been underserved up to that time. As Daugherty remembers it, right from the start, "The audience was small, but they were extremely loyal, rabid fans." Word about the station started to spread quickly.

The influence of WMMS could not be disregarded or ignored by advertisers, competing stations, or the entire radio industry. Arbitron's fall book, which came out in January 1985, gave WMMS a 10.9 rating for listeners twelve and older, with the morning show posting an eye-opening 13.5 share. But the real story was in the demographic breakdowns. The station had more than 44 percent of men in the big-money age group of eighteen to twenty-four, more than 35 percent of women in that same category, and more than 38 percent of men ages twenty-five to thirty-four. The push by WMMS to win the Jacksons show may have been a big help. Even though WZAK owned the urban audience, both that station and competitor WDMT-FM lost listeners in the fall book, most likely to WMMS.

Meanwhile MTV's growing popularity also helped solidify the format and philosophy of WMMS with the average Northeast Ohio radio listener. MTV aimed its programming predominately at the eighteen-to-thirty-four-year-old market, a very important target to advertisers looking for consumers with expendable cash. MTV's video deejays (veejays), including Northeast Ohio native Nina Blackwood, J. J. Jackson, Alan Hunter, Mark Goodman, and Martha Quinn, were viewer friendly and pioneers in a brand new medium.

The Michael Stanley Band was very aware of MTV's influence. Film and tape from their Blossom Music Center shows were fashioned into music videos promoting "My Town," "Take the Time," and "He Can't Love You," in addition to a long-form video release for Sony. They debuted their efforts at a special WMMS-sponsored, invitation-only event at the Flats' nightclub Peabody's Down Under. Some of those videos, notably "My Town," did get airplay on MTV, with Quinn in particular giving the band and its music an enthusiastic thumbs-up. Still it was frustrating for Stanley, the band, the record companies, and local radio—particularly WMMS—that the group wasn't breaking out to a bigger national or international audience.

In an effort to expand its audience while still holding true to its roots, WMMS expanded its *Classic Rock Saturday Night* programming to the entire weekend, with great success. Then in mid-January, in what had become something of an annual tradition, *Rolling Stone* once again named WMMS its Best Radio Station of the Year. There were stations around the country that wanted to wrest that honor away from WMMS, particularly in Dallas and Pittsburgh, but could never manage to pull it off. The station took top honors year after year. But the real winners were the Cleveland radio audience, because the annual honor had come to mean another series of free, high profile concerts in appreciation for their continued loyalty supporting WMMS in the annual readers' poll.

It was clearly time for WGCL-FM to regroup. The ratings and press going to WMMS made it painfully evident that WGCL needed to generate some excitement. The ratings surge of the previous summer may have been a fluke, despite WGCL's continuing popularity, but to an advertiser, it still comes down to consistency. But the big gun in the station's arsenal against WMMS was soon to return, as talks resumed to bring Danny Wright back to Cleveland from Kansas City's KBEQ-FM, where he headed after leaving Washington the year before. The station heralded his return at the Welcome Back, Danny Wright bash at the Rascal House bar on the Saturday before his first show.

On February 11 Wright cracked open the mike to begin what would soon become WGCL's last stand against WMMS. Wright had more than just a love of Cleveland compelling his return. He had married a Cleveland woman just four months before and decided to settle down in Northeast Ohio. Looking back, Wright now says, "I realized two things. I made a mistake turning my back on the popularity I found in C-town, and although it would be tough to prove, it appeared that [the competition] paid someone off to get me out of town! When I went to Washington D.C., I thought I was the p.d.'s darling. Within a short time, he was on me every single day—meetings, air-check sessions, showing me negative letters from listeners. Their whole attitude changed overnight! When Kim Colebrook called to offer me a chance to return with a wonderful three-year deal, I jumped at the chance."

Even though the Jacksons' *Victory* Tour might have given WMMS a good amount of exposure the year before, the station's listeners were far more loyal to the Buzzard than to Michael Jackson and his brothers. The annual WMMS–*Plain Dealer* Rock Poll showed that acts like Wham! Cyndi Lauper, and Madonna were winning accolades, while Michael's solo efforts as well as the tour with his brothers were panned in categories including Worst R&B–Soul Band or Performer, Worst Large Hall Concert, Worst Album, and Biggest Letdown. Just a year earlier, Michael had won honors in three categories, but now it became evident that the hype surrounding the *Victory* Tour had worked against him and his family. It was becoming obvious that while MTV and WMMS were promoting Michael as a huge crossover artist, the musical tastes of Greater Cleveland listeners didn't see him or his brothers quite that way.

WMMS continued to make strategic moves to assure its growing dominance, including the return of the popular *Coffeebreak Concerts* from a new midday concert site, Peabody's Down Under, on February 15. The concerts had proven to be a popular Wednesday getaway, and the return of that series was welcomed with open arms.

Television and radio had drawn talent from each other for decades, but a meeting on March 13 saw plans emerge for a special project that would join some of the biggest names from those fields, in addition to the local music scene, for a common cause. Present at the meeting were Denny Sanders, John Gorman, and Jim Marchysyn from WMMS; Dale Solly from WKYC-TV; and Marc Benesch, the local representative for CBS Records. It was only a preliminary discussion, but tentative plans were drawn up to release a special charity single similar to Band Aid's "Do They Know It's Christmas?" Jackson, Bruce Springsteen, Bob Dylan, Ray Charles, and a large group of American artists had just produced "We Are the World" to help the cause called U.S.A. for Africa. At the same time Canadian artists Bryan Adams, Joni Mitchell, Geddy Lee, Neil Young, and others recorded "Tears Are Not Enough" for the Northern Lights effort to feed the world's hungry. Wheels started to turn for the Cleveland project, and although it might not have the international media push of an MTV—which was crucial to the worldwide success of "We Are The World" and "Do They Know It's Christmas?"—it was still sure to draw a good amount of local publicity and maybe even raise money for Northeast Ohio food banks and centers around the globe.

A week after the initial meeting, a formal announcement came about the project, with Gorman calling for local talent, ranging from members of the Michael Stanley Band to Rocco Scotti (who was a Cleveland Browns institution with his charming rendition of "The Star Spangled Banner"). The working title for the project was Cleveland Artists Recording for Ethiopia (CARE). Ten percent of the proposed record's sales would benefit local hunger banks, while the rest of the profits would go toward the Ethiopian relief efforts. Cleveland was the first city to do its own recording of this type, and the artists were being contacted about a suitable recording date.

But the WMMS-backed project would soon have some company. A Rock and Bowl event sponsored by rock photographer Janet Macoska and local record promoter John Awarski, benefiting U.S.A. for Africa, gathered representatives of all the local radio and television stations. It brought together deejays from a variety of stations, including Wright, JoAnne "Mother Love" Hart, and Bob Brown from WGCL; Kenny Clean, "Spaceman Scott" Hughes, and Ruby Cheeks from WMMS; Dean Rufus, Matt Morgan, and "Kenny C" from WDMT; and representatives from WZAK and WMJI, along with television personalities, musicians, and sports figures. The Cleveland Rocks Against Hunger project benefited Eastlake soup kitchens and drew a formidable crowd, raising more than $3,000 and a small mountain of canned goods.

Back on the radio dial, WGCL continued to hammer out a full-service morning show to go one on one with WMMS, bringing in Dave Baron reporting traffic from the "WGCL-icopter." Baron had distinguished himself as one of the finest traffic reporters in the Midwest, and WMMS would regularly take down his information for use on its own program. Even with the new commitment of a morning team, WGCL stood little chance of making a dent against WMMS.

In early April the WMMS-backed Ethiopian relief project had set recording dates at Beachwood Recording for the 15 through the 26. Donnie Iris signed on to help, as did members of the Cleveland Orchestra, the Dazz Band, and Beau Coup. Benjamin Orr of the Cars, a native Clevelander, had agreed to participate as well. The song, titled "We Can Make it Happen," was drawing media attention from print, television, and even other radio stations. By April 10 WGCL's Wright had put together his own famine relief song "A Cry from Africa," recorded with his band Project 98. "It was a deal put together by the WGCL promotion staff and Bruce Seifert, the owner of Great Tracks Recording Studio," Wright explains. "It wasn't bad considering how quickly we put it together, and [the song] ended up raising a few bucks, but nowhere near what we thought it would." Despite Wright's song being recorded with the best of intentions, not even the announcement of a single release could draw much interest away from the WMMS project.

On April 16 the Michael Stanley Band laid down the WMMS song's backing tracks at Beachwood Studios, while soloists were scheduled to come in over the next few days to record their parts. Attending the sessions were Joe Vitale, Alex Bevan, Rickie Medlocke of Blackfoot, former Raspberries guitarist Wally Bryson, Frank and Tom Amato of Beau Coup, and even more coming in every day. The original song had been scrapped in favor of a new tune, "The Eyes of the Children," penned by Donnie Iris; the Cruisers' Mark Avsec; and Stanley, Kevin Raleigh, Danny Powers, and Bob Pelander of the Michael Stanley Band. Local hunger centers got a bigger cut of the proceeds than originally planned, as local efforts and U.S.A. for Africa each received 50 percent.

As the charity song was receiving lots of local attention, the ever-changing WMMS was getting more national press as well. In the April 5 issue of the industry trade magazine *Radio and Records* an article pointed out, "Over the past 18 months, WMMS has evolved into a hyper hybrid, playing about 80% currents, including flat out CHR tracks by Sheena Easton, the Commodores, and Isley Jasper Isley." The uncredited author of the article spoke with operations manager Gorman, who said that the term AOR had become too narrow for his station. Gorman added, "It's not that we have abandoned the format; it's more that the format has abandoned us."

The station may have evolved over the years, but no one could deny that it had broadened its listener base in the process. Some thought that WMMS had forgotten its true mission, but in the radio industry the mission is to get ratings, generate revenue, and stay in business. Despite the criticism from some corners, WMMS was still drawing huge audience numbers while continuing to grow at a staggering rate. The WMMS staff had every reason to feel like stars. They showed their appreciation to their fans by showcasing 'Til Tuesday, John Cafferty and the Beaver Brown Band in a free Public Hall show to commemorate the station's latest win in the *Rolling Stone* readers' poll. Around this time word had spread that WMMS's Jim Marchysyn was relocating as well, leaving the station after three years to become director of advertising and marketing at KSHE-FM in St. Louis. For Marchyshyn it was simply time to move, especially when WMMS became so successful. As he recalls, "There were a ton of chiefs, and not enough Indians. I could see that there were five consultants in the building at any given time, and that was not going to work."

Veteran newsman John Webster started getting attention at high-profile appearances for WMJI, where he had been heard doing news (and would soon be an important fixture in its future plans). Mike McVay was one of twelve finalists for Radio Consultant of the Year in the fourteenth annual *Radio and Records* Pop Music Survey, nominated for his consulting work across the country. McVay's star continued to steadily rise, and his influence on Cleveland radio would soon have an even more profound effect.

Hunger was still a major concern as Northeast Ohio entered the summer of 1985. On June 26 at exactly 4:00 P.M., the CARE project's "The Eyes of the Children" recording made its radio debut on WMMS and other stations across the Cleveland area. A video for the song also debuted the same day, premiering in the second half of WKYC's 6:00 newscast. The television station's Solly had been a prime mover behind the song, and though he was an accomplished musician, he didn't play on the recording. Instead Solly and WKYC produced the video, and he promoted the song on programs and in the press.

As much as local and national causes could help, hunger relief was still a major concern in Africa. A massive benefit concert was to be held simultaneously in London and Philadelphia on Saturday, July 13. Television and radio promised blanket coverage of Live Aid, featuring the biggest lineup of stars ever assembled. Even on this grand of a scale, local stations were trying to do their part to help. WCLQ-TV and WEWS-TV aired the event on television, as did MTV, and WGCL broadcasted the show in its entirety. The radio station's news anchor Sharon Taylor accompanied the *Plain Dealer*'s Jane Scott to the Philadelphia concert to host the WGCL broadcast. The Voice of America network chose only two personalities to anchor segments of its broadcast, Kid Leo and *Rockline* host Bob Coburn of KLOS-FM in Los Angeles. Leo carried plenty of weight in the music industry, so he was able to score exclusive interviews with stars such as Paul Stanley of Kiss and actor Jack Nicholson, who in fact hosted part of the Philadelphia show. Not to be outdone, WMMS's Sanders was able to catch Mick Jagger and Tina Turner rehearsing their songs that Friday night, giving WMMS listeners a sneak preview of what they could expect the following day. The massive fundraising concerts raised tens of millions of dollars for Africa's starving population, with the overwhelming majority of that money locally being generated by MTV's non-stop broadcast.

On July 18, not even a week after the global charity event, a group of local officials and businesspeople flew to New York. Headed by Mayor George Voinovich, the group included representatives of the New Cleveland Campaign, the Greater Cleveland Growth Association, Belkin Productions, and others. Their mission was simple: to pitch Cleveland's Lake Erie shores as a possible home for a proposed Rock and Roll Hall of Fame. Their story was heard, so they flew back to Cleveland with high hopes and fingers crossed.

It was an exciting time to be in Cleveland that July of 1985. Major concerts and events were scheduled around Northeast Ohio, and the Sohio Riverfest was drawing tens of thousands to the Flats for a celebration of summer. The Sohio event in particular grew to be so popular that it quickly became an annual tradition, becoming bigger and bigger every year. WMMS remained on top during the spring's ratings race. It made big gains in the over-thirty-year-old market, and it ranked as the most listened-to station in town with an 11 overall rating. On the morning show front WMMS's 14.2 rating far out-distanced the station from its closest competitor WDOK-FM, which had an 8.7.

WMMS was also looking ahead, expecting to reap the benefits of a Bruce Springsteen show that was announced for Cleveland Municipal Stadium that August. Springsteen was already an incredibly huge artist whose music had a set Cleveland home at WMMS, especially with Leo continuing a long tradition of kicking off the weekend every Friday afternoon with "Born to Run." More than 70,000 tickets had sold only hours after they went on sale, but there was a major roadblock developing that could have prevented the concert from taking place. The Service Employees local, representing stadium ushers, ticket takers, and security personnel, had voted to strike the venue. A large part of Springsteen's fan base was made up of blue collar working stiffs, so "The Boss" made it clear that he would not cross a picket line. The union was so taken by this gesture that its membership voted to negotiate a one-event agreement with Belkin Productions, allowing the concert to go on as scheduled. The show was a three-and-a-half hour tour de force, with Springsteen even getting in a plug for the Greater Cleveland Community Food Bank.

As cool as the idea was of Bruce Springsteen working out labor disputes, the next day's *Plain Dealer* would put rock and roll into the forefront of Cleveland's collective consciousness in a way that it had never been done before. An article, practically buried in the newspaper, mentioned that Cleveland was being considered as one of five finalists for the Rock and Roll Hall of Fame. A major campaign was set in motion to ensure that the city would get this distinction.

In late August rumors had reemerged that John Lanigan was considering a return to Cleveland. A format change at Tampa's WMGG-FM forced Lanigan to

Bruce Springsteen at Cleveland Stadium (1985). The Boss makes his triumphant return to Cleveland with two sold-out shows that even put a work stoppage by stadium employees on hold. Photo by Janet Macoska.

consider his options, so he decided to leave the station. Lanigan was reportedly talking to stations in Chicago, Los Angeles, and of course Cleveland, where he had always earned strong ratings on WGAR-AM. Radio stations put their "big guns" on in the mornings, and his reentry into the market would further complicate the choice for drive-time listeners. The rumors were soon confirmed to be true, as it was announced that Lanigan would replace Dan Deeley, teaming up with Deeley's partner (and wife) Kim Scott for WMJI's morning show. WMMS was about to face one of the biggest, and most enduring, challenges in its history with Lanigan's return to town.

By September WMJI's owner Larry Robinson had sold both it and sister station WBBG-AM to Cincinnati-based Jacor Communications. The sale price was over $13 million, more than double what he and his partners had paid just three years before. But a move on the West Coast that October would involve yet another former Malrite employee who decided to try his hand at station ownership.

In the middle of October 1985 Carl Hirsch left his position as president and chief operating officer of Malrite Communications. Just a week later he paid $44 million for Los Angeles station KJOI-FM, at that time the single highest price ever paid for a station. At such a high price, there were skeptics who thought that Hirsch's company Regency Communications had vastly overpaid for the property. Commenting on the purchase, Malrite chairman and owner Milton Maltz said that his company had turned down an opportunity to buy the station for a far lower price. Still Maltz wished Hirsch well, though it was said that Hirsch's departure from Malrite was not well received. Malrite also owned stations in Los Angeles, KLAC-FM and KZLA-FM, so they were poised to compete with Hirsch's KJOI. Hirsch packed up and moved to Los Angeles, amid great speculation as to what he planned to do with KJOI's "beautiful music" format.

There were no doubts of audience loyalty when the summer ratings came out, as WMMS completed its first two-year run at the top. The *Buzzard Morning Zoo* was in double digits again, while WGCL posted very strong overall numbers as it tied for the number two spot with WLTF.

After nearly four months of being held up in legal wranglings over publishing rights and other legal issues, the hunger relief song "The Eyes of the Children," by the WMMS-sponsored CARE group, finally saw a national release on November 18. The cover of the twelve-inch single featured more than forty musicians. Half of the proceeds went to the Greater Cleveland Interchurch Task Force.

WMMS continued to sponsor listener appreciation shows, offering its fifth of the year with Jefferson Starship at Public Hall on November 26. To obtain tickets, a self-addressed stamped envelope had to be mailed to the station's studio at 12th and Euclid. This resulted in a mountain of mail, as more than 30,000 requests were sent in to get into the 10,000-seat venue.

It was in that envelope-flooded studio of WMMS that a new voice was emerging, that of Ruby Cheeks. With her quick wit and equally quick tongue she became an instant favorite on the *Buzzard Morning Zoo*. WMMS wasn't the first station Cheeks had worked at, but still she was petrified before her first shift on a Saturday morning following Len "Boom" Goldberg. Cheeks recalls, "Shaking, and convulsing, and being dry mouthed! Boom held my hand the whole damned show, and God love him for it!"

That month Gorman was invited to Bob Dylan's birthday party at New York's Whitney Art Museum. Gorman later said it included a twenty-second "meet and greet," but even that short amount of time became memorable. He met up with rock and roll luminaries such as Billy Joel, Pete Townshend, David Bowie, Iggy Pop, Roy Orbison, Allen Ginsberg, members of the Band, and dozens more.

In the final days of that eventful year the influence of WMMS was being felt on the West Coast in a most unlikely way. *Howard the Duck*, focusing on the adventures of an "average" alien fowl who lands in Cleveland, was a runaway hit for the Marvel Comics Group. The book drew such an underground interest that producer George Lucas and director Willard Huyck started filming a live action feature about the bird. Only a few of the movie's exterior shots were actually done in Cleveland, while the majority of the filming took place near Lucasfilm's production complex Skywalker Ranch in Marin County, California. Sets and outdoor locations now had to be made to look like Cleveland. That didn't sit well with segments of the radio community in nearby San Francisco. *San Francisco Chronicle* columnist Herb Caen wrote that the city's intersection of 10th and Irving looked enough like the North Coast already that several cars with Ohio plates were quickly brought over so the crew could start outdoor filming. Caen had added that "some very San Francisco-looking people" were asked to stay out of the shots. Huyck also put up WMMS billboards near Marin County to add to the authenticity. That didn't sit well with San Francisco's KFOG-FM, which had been bested year after year by WMMS in the *Rolling Stone* readers' poll. Rumors even started to circulate that WMMS had put up the billboards to rub KFOG's nose in Cleveland's past victories. The out-of-town billboards didn't help that balloting for the upcoming readers' poll was still under way.

As Cleveland was drawing hatred from a city on the other side of the country over a talking duck, an incident was about to take place back at home that would send a horrifying chill throughout the local radio community.

On Christmas evening WZAK's Lynn Tolliver ended his afternoon drive shift at the studio on East 17th and Superior. He decided to stop by his office to catch up on some paperwork on a new format being developed for a station his parent company had just purchased in Atlanta. The neighborhood where the station was located had seen its problems in recent years. Employees were getting robbed, there was vandalism, and cars were being stolen. As Tolliver walked down the fourth floor hallway he came upon a man who somehow had gained access into the locked building. The trespasser was wearing sunglasses and a wide-brimmed hat. Staring straight at Tolliver, he held out a handgun fitted with a silencer. Tolliver told the gunman that he would give him anything he needed and that he didn't want any violence. Suddenly three shots tore through Tolliver's left arm, right leg, and chest. Tolliver managed to get to an elevator and hit the button for the first floor, but then returned to the fourth in an attempt to warn the deejay on duty and to get medical help. The attacker escaped, while Tolliver told fellow disc jockey Jeffrey Charles that he had been shot. Charles kept playing records as he called for an ambulance and Cleveland police. Tolliver was in surgery just after midnight.

Police on the scene couldn't determine a motive, saying that robbery was an unlikely cause. The next day Tolliver met with reporters. Though he was in obvious pain, he still wanted to get back on the air. He had even proposed a remote from his hospital bed. Radio stations across the city tightened security. Although no further violence was recorded, it provided an unsettling end to the year 1985.

The New Face of the Enemy

By the mid-1980s WMMS had established itself as the long time dominant force in Cleveland radio. The station was able to stay on top by softening a good part of its playlist, yet it was still maintaining its rebellious image. The last time WMMS had failed to dominate a ratings book was in 1983, when program director John Gorman started weeding out some of the harder-edged music and filtered in groups such as Wham!, Madonna, the Pointer Sisters, and other acts that might appeal to women and older demographics. The gamble of maintaining the contrast between playlist and image paid off, as indicated in the Arbitron ratings book released in January 1986. It showed WMMS posting a stunning victory over its competitors, with more than 14 percent of listeners ages twelve and over. The station's morning drive numbers were no less than astounding, as host Jeff Kinzbach drew an amazing 17.4 percent of the audience, more than twice the numbers of second place's Larry Morrow at WQAL-FM with 8 percent.

One-time rival WGCL-FM dropped like a stone to a poor 4.9 percent, losing almost a quarter of its listeners from the previous book. The Arbitron report was so disturbing to the folks at WGCL that program director Tom Jeffries resigned that same day. The fall book shows numbers for the most important ratings period of the year, and it often sets the pace and advertising rates for the next twelve months. It became obvious that WGCL's sales department would have a very tough time ahead selling spots. Though Jeffries resigned as a result of the ratings, he was hardly the sole culprit. It was painfully evident to Jeffries and others inside the station that his hands were tied and other management was second-guessing him and often pushing him into directions that were obviously not going to increase the audience. The problems were only amplified with the station's limited promotional budget and the fact that it was up against the monster WMMS. Jeffries was dealt a hand that he just couldn't play.

The overwhelming success of WMMS and the stunning failure of WGCL weren't the only surprises in the fall ratings book. John Lanigan's return to WMJI-FM helped push the station's morning show to an impressive third place rating of 7.1. Lanigan was off to a great start. Thanks to a WMJI promotion that January, many people were phoning in their votes to *Cleveland Magazine* to make him the city's Most Watchable Man. However that honor eventually went to Michael Stanley, who was getting lots of television time on *PM Magazine*.

As WMMS continued its efforts to make its upcoming winter ratings as good as its fall ratings, the station was informed that it had topped the *Rolling Stone* readers' poll for the seventh straight year. It easily beat out KFOG-FM in San Francisco and KSHE-FM in St. Louis. The winning station immediately started work on yet another series of Appreciation Day concerts, which was kicked off with a show by Night Ranger and Beau Coup at Public Hall in February. The station wasn't only receiving accolades as a whole, as WMMS's Kid Leo won acclaim with a salute on *The Rolling Stone Readers and Critics Award Show*, broadcasted locally on television newcomer WOIO-TV. It was not a good time to be a competing rock station in Cleveland.

Gorman recalls a tradition that Leo and he would share at the beginning of every year. They would review all the previous year's events, assessing both the station's many accomplishments and its few losses. The conversation would usually end with the question, "How long can we keep this thing going?" Every year they would usually agree that it would continue for as long as they liked and that they would work hard to accomplish that. For whatever reason 1986 became the first time in many years that the annual review did not take place.

The WMMS lineup had remained fairly intact since 1974, and its core group of disc jockeys rode the wave of success for nearly twelve years. But in early 1986 one of the deejays decided it was time for a change. From 1976 to 1983 Betty Korvan had done evenings at WMMS. When Tom Rezny eventually replaced Korvan, she was moved over to weekends. She was also able to work in a noon to 3:00 shift on WHK-AM. But at midnight on Saturday, March 15 Korvan said her goodbyes to Cleveland. She soon headed to the Black Hills for a job in Rapid City, South Dakota. Looking back on the move, Korvan recalls, "I always wanted to work some place besides Cleveland. I felt I had gone as far as I could, and I didn't want to make my home there without seeing the rest of the country. I had visited the South Dakota station—"K-Sky"—and I was fascinated by it, and that part of the country. They called me and asked, 'Do you want to do afternoon drive?,' and I packed my Subaru with books and records, and off I went!" Despite her yearning to venture elsewhere, she does admit that it was still a difficult and emotional move to make.

"Spaceman Scott" Hughes took over late night duties from midnight to 2:00 that March, while the popular *BLF Bash* continued its lock on overnights. The spring ratings period was set to start just a few days after Korvan's last day, so stations had plenty to prove in the next three months. At WZAK-FM Lynn Tolliver continued to recover from his mugging nicely and was able to join the station's morning drive.

A former disc jockey and now programmer named Jim Harper from Detroit had been calling Gorman over the past few months with a proposal to start programming a new station in Cleveland. Gorman said that he simply refused to listen the first dozen or so times Harper called. He started to change his mind after relations strained with some members of Malrite management, including president Gil Rosenwald. "Everything started falling apart at once," recalls Gorman, "and it was no longer an enjoyable experience going in there." Harper's persistence would soon wear Gorman down.

Cleveland's slow but sure renaissance continued to draw interest from around the nation, particularly from its rock and roll audience. Early that May local production began on *Light of Day*, a film starring Michael J. Fox, Michael McKean, and Joan Jett. Scenes were shot at a west-side factory, on Coventry Road, and other locations around town. Radio stations leaked when and where filming was scheduled to start, causing huge traffic backups in Cleveland Heights. The movie also featured local musicians and bands such as the Motion, Eddie and the Edsels, the Fabulous Flashbacks, and Norm Cotone and Friends. Cotone was already an established part of the Cleveland music scene, creating a regional following with the band Rainbow Canyon on the Capitol label. Collinwood High School became the setting of a scene featuring Peabody's Down Under owner Dewey Forward playing a sex education teacher. A key sequence of the film was shot at the Euclid Tavern, where Fox and Jett performed with their group. As filming wrapped Fox and the crew met the WMMS deejays—calling themselves the All-Stars—for a private softball game at Center Junior High in Strongsville.

Jett wasn't the only the big rocker in town that spring. WMMS continued its series of appreciation shows by presenting a free concert with Blue Oyster Cult and Platinum Blonde.

By the time the winter Arbitron numbers were released that April there were dramatic changes under way in Cleveland radio, on and off the air. WMMS remained at number one in double digits. The morning show of Kinzbach, Ed "Flash" Ferenc, Goldberg, and Cheeks posted an amazing 18.6 share, increasing its fall lead by more than a point. In an effort to explain the appeal of WMMS, Cheeks points out, "The station broke every rule of radio. At critical times it was an innovator, and even a revolutionary radio station, as opposed to reactionary to the market. It set the pace, set the place, and was more a lifestyle than something you listened to in your car or on the clock radio." WMJI's Lanigan jumped to the number two spot in his second ratings book since returning to town, but he was still trailing behind the *Buzzard Morning Zoo* by a little more than ten points.

As WGCL continued to lose ground, something strange was happening that May at its sister station WERE-AM. Management at the station tried to persuade its air staff to dump the American Federation of Television and Radio Artists (AFTRA) as its union. Some staff members were taken to dinner at fancy restaurants, while some of the more vocal union loyalists were only taken to a lunch counter down the street. Though they couldn't make any promises to the staff, the management kept stressing how much better things would be without AFTRA. On May 30 a vote was taken, and the staff voted overwhelmingly to keep the union intact, giving the management a stinging slap in the process. The result was seen as a stunning defeat for the station's upper and middle management, especially since much of the union's support came from longtime staffers. It would soon become obvious why the front office was so keen on getting the union out of the building.

The Fight for the Rock and Roll Hall of Fame

There are a lot of good arguments as to where rock and roll originated. It had its roots in Memphis and Delta blues, Southwest country, East Coast and New Orleans jazz, and on and on. But some things are certain. The first popular use of the term "rock and roll" can be linked to Leo Mintz, the owner of Record Rendezvous on Prospect Avenue, who backed Alan Freed's *Moondog Rock and Roll Party* on WJW-AM, and Mintz often handed him the records to play.

There had been a call for a Rock and Roll Hall of Fame in Cleveland as early as 1981. Members of the band Marionette suggested it could be located at the site of the Cleveland Agora, and start regionally by honoring groups like the Choir, the Outsiders, the Michael Stanley Band, and others. They even outlined thoughts about displays and suggested local bands could play benefits to raise money for the project.

One of the cornerstones of the campaign was the legacy of Freed, host of the first real rock and roll concert at the Cleveland Arena, and the riot that followed. The city had a long-standing reputation as a major market for music. But no matter what, no one wanted the Rock and Roll Hall of Fame more than Cleveland, and WMMS in particular led the long charge to bring it here. But many don't realize just how close the city came to losing the project in the nine-year battle to finally open its doors.

The term "rock and roll hall of fame" had been used by a lot of different disc jockeys over the years, including Cleveland's Norm N. Nite. In fact Nite played a critical role in bringing the project to Cleveland, along with Hank Lo Conti, who toyed with the idea for a hall as far back as 1974. He recalls,

> When we drew up plans for our new club that we were going to build across the street from the Agora, it was supposed to be in there. I really got serious about it in 1978, '79. I went to New York and gave a presentation, figuring that if we got two major record companies to back us, that would be enough. I picked Columbia and Warner Bros. [Former Cleveland finance director] Joe Tegreene was working with me at the time, and I brought him too. After our presentation to Columbia, their answer to me was, "There is going to be a Rock and Roll Hall of Fame someday, but when it's built, it will definitely be located in New York." When I came back, there was an article in the *Plain Dealer* at that time where Dick Clark was in Cleveland, and one of his comments was, "Boy, there should be a Rock and Roll Hall of Fame, and it should be built in Cleveland." I wrote him a letter asking him to support our efforts, and Clark was kind enough to write back, saying, "Yes, there will be a Rock and Roll Hall of Fame, but it definitely will be in Los Angeles!" Later on, Bill Graham said it would be in San Francisco.

Lo Conti got busier with his other clubs around the country and didn't really pursue the idea until it started becoming a hot topic again in the mid-1980s.

By December 1984 WMJI had started promoting its vision for a "Rock 'n Roll Hall of Fame" by naming Elvis Presley its first inductee the following month and displaying a special plaque in the station's lobby. There had been hopes for a museum similar to the Pro Football Hall of Fame in Canton, but the first serious push to establish a permanent museum and repository started in May 1985.

A former Clevelander, Eddie Spiezel, had moved his agency to San Francisco during the recession of the 1970s. It was during a visit to Cleveland that he disclosed plans to friends at WMMS that legendary concert promoter Bill Graham was hoping to put up a rock hall near Pier 39 at San Francisco's Fisherman's Wharf. The idea sounded intriguing, and the folks at WMMS wanted to hear more. So they contacted Tunc Erim at Atlantic Records in New York, who let them in on plans by company CEO Ahmet Ertegun for an annual induction ceremony honoring the pioneers of rock. The basic groundwork had already been mapped out for the project, though most thought it would be located in Manhattan rather than Graham's home base in California. WMMS, Lo Conti, and many others knew it belonged in Cleveland, but it would be a hard, hard sell.

Calls were made on a number of different fronts to people and companies that might make the hall a reality in Cleveland. Mike Benz of the Greater Cleveland

Growth Association was among them, and he called Lo Conti to ask him to be part of a meeting. Benz reportedly told Lo Conti that he had heard about the hall, and that there was a chance that Cleveland would be the site. Lo Conti told him, "You have one shot in a million, but if anyone could help it would be Norm N. Nite." Soon after, Nite agreed to meet with Lo Conti and Benz, though he made it clear that no other city was likely to sway the New York board.

Cleveland in 1985 was a shadow of the tourist spot that it has since become. Sports teams were a disappointment even to hardcore fans, the job market was stagnant at best, and just about everything that could be said about a rustbelt city could be linked to Cleveland. But the city also had some elements that could match just about anything New York and San Francisco could muster: some great public relations firms, politicians and civic leaders who wanted to resurrect Northeast Ohio, and a very powerful radio station in WMMS. It would be a good compromise location outside the rival East and West Coast factions of the recording industry. If Cleveland were to be reborn, it would need a cornerstone, something to focus worldwide attention on the area and what it had to offer. That centerpiece for the twenty-first century was the Rock and Roll Hall of Fame.

Among those attending that initial planning meeting were Benz, Lo Conti, Nite, Jules Belkin of Belkin Productions, Bill Smith of WMMS, and Kim Colebrook of WGCL-FM, following Lo Conti's suggestion that all of Cleveland radio would have to be united in this campaign. Nite broke the news that New York Mayor Ed Koch had given the Rock Hall board a building on East 42nd Street, but agreed to ask for a meeting. A week later he reported back that chairman Ertegun heard the request, and was told the location would not even be discussed. The hall would be in New York City, period. Nite was able to persuade Ertegun to, at the very least, hear Cleveland's presentation, and that was enough of a spark to ignite the huge cooperative campaign to bring the Rock Hall to Northeast Ohio. Radio stations across the region agreed to cooperate, though WMMS had the clout to really spearhead that effort.

After Benz's comprehensive presentation, word got to the press that the New York Hall of Fame board was open to other cities bidding for the project, and Chicago, Philadelphia, and New Orleans quickly lined up for the chance. The board planned visits to check out their proposals, and plans were soon under way in Cleveland for a petition drive to show the public's interest. Buzzard creator David Helton designed the logo, and agencies and businesses ranging from the Northeastern Ohio League of Savings Institutions to the Growth Association agreed to help with the drive. Board member Tim La Rose pledged to get the petitions to all of the House of La Rose Budweiser distributorships in a seven-county area, and that totaled more than 6,000 locations.

The deadline was September 10, and more than 660,000 names were collected, including just about every rock star that came through town. More meetings between the city and the New York board soon followed, with a secret trip in December 1985 for another major presentation to Ertegun and the rest of the committee.

The campaign ended 1985 with cautious optimism when the *Atlanta Journal,* in a story about the New Orleans bid, ranked, "Alan Freed's Cleveland, Ohio," as a strong frontrunner. By mid-January 1986, there were already rumors where the hall would be located if Cleveland got the nod. Speculation centered on existing locations like the old Allen Theater on Playhouse Square, the old U.S. Post Office headquarters closer to Public Square, the former Society Bank headquarters, the Central Market site, and even the Greyhound Bus terminal on Chester Avenue. The official designation was still months away, but people in Cleveland were anxious to hit the ground running if Cleveland were to be named the city.

Lo Conti strongly favored redesigning the old Masonic Temple, as the empty lot down the street was the site of the former Cleveland Arena, where the first rock concert was held. Lo Conti got together with an architect friend to map out what a possible Rock Hall complex might look like. Lo Conti explains, "That building is absolutely beautiful! You got a 2,200-seat theater there, which is acoustically perfect. Below it, you can put a thousand people for dinner. They have four floors, and three of the floors have theaters in them! The building is in excellent condition. You could have bought the building for a million and a half, and you could have bought the Arena property from Bowling Green [State University] for about the same, and [the Rock Hall] would have owned their own property." He adds, "What they would have done for that area would have been phenomenal! They talk about developing the Euclid Avenue Corridor. They could have developed that corridor ten years ago if they would have done that."

But when Lo Conti presented his plan to one of the key politicians in the Cleveland campaign, he recalls being told, "Before they take it away from Tower City, [that person] would rather lose it!" Everyone had their favorite location, and some had bigger voices than others. But the idea had a few critics as well, including WCLV-FM's general manager Bob Conrad, who wrote a letter to *Billboard* magazine that read, in part, "A museum implies that whatever is housed in it is worth saving... Just what would such a museum contain? A collection of boom boxes? Twisted Sister T-Shirts? The definitive assortment of WMMS bumper stickers?" Conrad found all of the attention amusing, telling the *Plain Dealer*'s Jane Scott, "Quite frankly, I was just having some fun with the museum. I talk about them on my Saturday night show with comedy and folk music and other tunes."

The New York board was seeing extensive publicity about the project, due in great part to stories about Cleveland's efforts. It was a good situation to be in because it was helping raise awareness of the project as it headed toward the very first induction ceremony at New York's Waldorf-Astoria hotel. *USA Today* did a survey of its own about where the Rock Hall should be located, asking readers to call a 900 number to vote for their favorite city. On Monday, January 20, the newspaper published the numbers for each city, and the voting times, and Cleveland radio jumped at the opportunity.

WOIO-TV put Starship's "We Built This City on Rock and Roll" to scenes of the Cleveland skyline in on-air promos for the telephone survey, and the results were overwhelming. *USA Today* had so many calls for Cleveland that it extended the voting time by fifteen hours to accommodate as many callers as possible, which did not sit well with some in the local media. Denny Sanders told the *Plain Dealer* that it was "like a football game with a minute to go in the last quarter, 32–6. Then someone runs out on the field and says, 'Let's expand the game to give the losing team more of a chance.'" Benz agreed, asking Clevelanders to keep voting, and predicting, "We'll bury 'em again today!" Late that Monday, Cleveland had a sizable lead with 48,728 votes, followed far behind by number two Memphis, with just 2,282, and six other distant runners-up. The final tally left no doubt which the favorite was: Cleveland came in at a big number one with 103,047 votes! Memphis never came close at number two with only 7,000 votes.

Cleveland's enthusiasm, and the local Rock Hall committee's willingness to cooperate with the New York board, drew praise from Ertegun, and some other key players echoed his applause. There was heavy local media coverage of the first induction ceremony that January, where many of the pioneers of early rock were honored. There was no word on the permanent site, but Cleveland was praised for its civic pride in the *USA Today* survey. WRQC-FM's Jan McKay offered an impressive two-hour special, *Cleveland . . . Where Rock Began to Roll,* featuring rare interview footage with Freed and other key players.

The FM assault continued to grind away. Songwriter Eric Carmen joined forces with his brother, Fred, to pen "The Rock Stops Here," which WMMS debuted before the official premiere date. It was followed by television exposure on WJW-TV's *PM Magazine.* Carmen also presented the first pressing of the song to Mayor George Voinovich at a City Hall news conference, as the mayor officially declared March 7 "The Rock Stops Here Day" in Cleveland. (Later that summer the song would eventually hit number seven on the Japanese charts for two weeks.)

Stations got together later that month for yet another demonstration of Cleveland's passion for the project, with local and state proclamations designating March 21 as "Rock and Roll Day in Cleveland." More than thirty sites around the city hosted events on the thirty-fourth anniversary of the very first Moondog Coronation Ball. Cleveland rocked well past 2:00 A.M. with "twist king" Chubby Checker on West St. Clair, Carmen and Chuck Berry at Tower City Center, Nite spinning records with celebrities at the Palace Theater, dozens of local bands, James Brown look-alike contests, a showing of Freed's *Go Johnny Go* at the Hanna Theater, a "Hound Dog Howl" on Public Square, and to top it all off, fireworks in the Flats. Even if the board didn't notice, which it in fact did, it was a great way to get a crowd downtown to welcome the first official day of spring. It also resulted in a bizarre stage pairing a few weeks later when State Senator Lee Fisher took the stage at Blossom Music Center to sing "Strike Up the Band" with the Michael Stanley Band.

Stanley has fond memories of that appearance, saying, "Lee is a rock and roller! I didn't know what he did at the time, but I knew he was a politician. This guy was so into rock and roll, and looked so unrock and roll! He said, 'It would make my life if I could sing on stage with you,' and who am I to keep him from his dreams? That was the first time I ever met Lee, and we've been friends ever since. I swear to you, right now, if I were to say, 'Lee. Come up to the Odeon or someplace on a certain date and we'll do "Wooly Bully" or something,' he'd do it!"

Another indication of Cleveland's chances came on April 25 when a number of Rock Hall trustees flew into town for Congresswoman Mary Rose Oakar's "celebration" at the Palace Theater, where she was roasted by a list of celebrities that included Kid Leo, Danny Thomas, and Phil Donahue, among others.

The long wait dragged on for weeks as Cleveland waited to hear if it won the Rock Hall project. The local committee had presented a comprehensive plan to build and sustain the Hall of Fame, though the New Yorkers remained tightlipped. Finally word came down that an announcement was expected on Monday, May 6, and Cleveland media was invited to fly up on two private jets to witness the event. It looked very, very good, and sent local radio and television scrambling to get the scoop before the next week's announcement.

That confirmation came on Thursday, May 2. WERE-AM aired the *ABC Talkradio* program hosted by the British-born Michael Jackson, and as luck would have it, one of his guests was Arista Records' Clive Davis, who was also a member of the New York committee. Davis said, "Cleveland really put on an ambitious and aggressive campaign. Cleveland was the source of great moments in rock and roll over the years in various categories, so that their enthusiasm and aggressive coveting of this role really paid off for them." Those comments broke the seal of secrecy that was agreed to the day before when the city got the official designation. Sensing the cover had been blown, and the national impact of the following Monday's announcement would be diminished, Rock Hall committees in Cleveland and New York offered little, if any, comment about Davis's remarks.

The Associated Press and UPI picked up the story, and it made headlines around the world. The following Monday in New York, the media gathered at the appointed time in Ertegun's office at Rockefeller Center to hear him say with a chuckle, "To absolutely no one's surprise, and contingent on certain requirements, the foundation board has chosen Cleveland to be the site of the Rock and Roll Hall of Fame and Museum."

All assembled broke out in applause. Two private planes flew the Cleveland media crews and politicians back to Burke Lakefront Airport, where a hastily scheduled rally drew a huge response. Even CNN ran headline stories about the city landing the honor. It was a tremendous boost for Northeast Ohio, but other towns were not happy about it.

Chicago backers tersely shot back that they would continue efforts to build their own hall, with Albert J. Copland of the Chicago Committee for the Rock and Roll Hall of Fame saying it was located in Northeast Ohio because Cleveland had, "a long headstart on a big and complex project... a strong radio station [WMMS]... and a stronger need to have it, a more single-minded focus." He still predicted the Windy City would have its shrine to popular music up and running in about three years. Paul Grushkin headed the committee for the Rock Museum of San Francisco, and said his facility could share national honors with the Cleveland hall. Legendary promoter Graham also sat on the New York board, and when asked if the annual induction ceremony would move to the shores of Lake Erie, he said, stressing no offense to the city, that all he knew about Cleveland was "Mike and Jules Belkin, and of course, WMMS." Grushkin also suggested that a good part of Graham's extensive collection stretching back to the days of the Fillmore might end up in a Bay Area museum. Philadelphia backers had talked about a museum there even before a Rock Hall was suggested, and said their project could be operational in two or three years.

Initial estimates put local costs for the project at about $20 million, with an annual return of $25 million as more than 325,000 visitors were expected to stop by the museum in its fifth year. It was also suggested that door admission would be $4, with an average $2 per person being spent on souvenirs, with a windfall for hotels and restaurants from tourists visiting the hall. Those numbers would change drastically over the next few years.

The Rock Hall's New York Trustees traveled to Cleveland in the last week of July 1986 to inspect sites in the Flats, the Lakefront, University Circle, and other spots around the city. Representatives of the Hard Rock Cafe flew in to check out possible locations for one of their theme restaurants. The hard work had clearly paid off.

The Cleveland delegation flew to New York on January 14, 1987, for a meeting on the Rock Hall's location, with a plot of land near Tower City being favored initially, and two lakefront sites also in the running. The New York board traveled to Cleveland twelve days later to inspect them, with a final decision due any day. Meanwhile, the Rock Hall's design by I. M. Pei was due to be unveiled to the press at the second annual induction at the Waldorf-Astoria that same month.

Like with many of his designs, Pei favored lots of glass, and similar to his controversial addition to the Louvre in Paris, his plan for the Cleveland Rock Hall featured a pyramid. The press didn't know how to react when the model was unveiled. Some looked on with amusement, others seemed confused, while it drew raves from the rest. Pei's work is often greeted with skepticism, only later to win widespread acceptance once the project is actually built. The cost of the project, at that time, was now expected to go as high as $35 million.

Unofficial Rock Hall groundbreaking (1987). Frustration over stalled plans to start construction of the Rock and Roll Hall of Fame resulted in WMMS jocks and listeners taking matters into their own hands. Photo by Bob Ferrell

The wait for an official groundbreaking date kept getting longer and longer, and even the most loyal fans of the project were showing their frustration. It was a topic on the *Buzzard Morning Zoo* for some time. One day Jeff Kinzbach took matters into his own hands, telling the audience, "We're going to dig the damned hole ourselves!" Kinzbach, Ed Ferenc, and some engineers traveled down to the site near Tower City, and dug their shovels in under the sign proclaiming Cleveland, "The Home of the Rock and Roll Hall of Fame." Within minutes, a listener pulled up with a flatbed, a tractor, and a backhoe. As Kinzbach recalls, "He started seriously digging a hole! I'm thinking, 'Hmm. We didn't call the utilities about the gas lines! It's gonna really be 'Blow Something Up' this morning!' It was a lot of fun, and it worked out really well. A lot of people stopped by with their shovels, cause they just wanted to put their shovels in the dirt and throw some. People really were frustrated with all the politics that went on with New York and Cleveland."

Some hall officials were none too pleased by the stunt. Kinzbach says it could have stemmed from their own frustrations, calling them, "A bunch of babies! Limp-wristed crybabies that couldn't put up with it. Every once in a while you have to go down there, pull the rug out from under them, and shake them up a little bit. We were famous for that."

There were drawn-out discussions about financing, the ever-rising cost of the project, and even relocating it to a new site on the shores of Lake Erie. Finally a secret meeting was called for Saturday morning, May 20, 1989 at the offices of the local board's legal advisers on Lakeside Avenue.

The Cleveland board had been as frustrated as the rest of Northeast Ohio about the long delays in breaking ground for the project. The local folks wanted some kind of assurance that the project would not be pulled, but the New York board had its own frustrations, and most of them centered on funding. The state, city, and county would be picking up the vast majority of the cost, and New York wanted some verification that all the funding would be set in stone. The two sides agreed on a six-month deadline to get the funding set, or there was a very real possibility that the hall could go elsewhere. Tens of millions of dollars had to be raised by November, and it promised to be a very difficult job.

There was a very real threat that Cleveland could lose the project. Dia Stein was still recovering from her skiing vacation accident that badly damaged her leg, and she was due to go into the Cleveland Clinic for surgery. Her father flew in to look after her, but the night before that operation, she found herself writing press releases in case the hall project was pulled. Looking back, Stein recalls, "There were two drafts of that release. If it stayed in Cleveland, Yeah! A vote of confidence from the New York board. If we lost it, the other release pointed out all the work the station and people of Cleveland did, only to have the rug pulled out from under us. They wanted it understood that WMMS had done its part to try to get it built." Stein's father delivered the drafts to the station before dropping off his daughter at the Clinic.

That summer and fall were very busy for the Cleveland board as it scrambled to line up every penny for financing, which was in the tens of millions of dollars. This was going to be big news at WMMS, because it had covered the story so closely right from the start. It was decided that on the appointed day, the station would send its broadcast truck to the news conference to air it live as it happened.

When the day arrived that November, and rumors were flying that the board had come just under the wire to meet its goal, a press conference was scheduled for Stouffer's Inn on the Square.

Fortunately Cleveland was able to meet the funding goal, the Rock Hall was built, and that project may have been the major catalyst toward bringing Cleveland back to prominence.

All Aboard the North Coast Express

Though the WERE-AM staff was successful in keeping its ties with AFTRA, serious changes were still going to come to the station. A week after the vote the *Plain Dealer*'s front page announced some major moves in two fields of the local media. GCC Communications of Cleveland had sold both WERE and WGCL-FM to the Detroit-based Metropolis Broadcasting for $10.7 million, pending approval from the FCC. At the same time Malrite Communications announced plans to acquire controlling interest in WOIO-TV.

The purchase of WOIO wasn't the only television news in the area. NBC sent Willard Scott to Cleveland in early June to broadcast his *Today Show* weather segments, during which he praised Cleveland's renaissance. ABC followed just a few weeks later to scout locations for its *Good Morning America*.

Despite the television exposure, it was still WMMS that was continuing to win both local and national accolades. At the fifteenth annual Bobby Poe Pop Music Survey Awards in Atlanta, the station's Kid Leo was named both Best Music Director and Air Personality of the Year. John Gorman also took home honors for his work at WMMS, specifically for Program Director of the Year and Consultant of the Year. WMMS itself took the prize for Radio Station of the Year. With the waves of positive publicity locally for the station's ratings wins, innovative promotional campaigns, and high-profile contests, plus this national award, it seemed as though WMMS was unstoppable.

As the awards and honors affected WMMS's prestige, changes and additions were also affecting the staff. Denny Sanders was named assistant program director. Leo took on the additional responsibilities of creative services director, helping oversee the station's promotions and publicity. Maria Farina also joined the staff as a fill in disc jockey.

In mid-July, the public was promised a look at the mysterious Mr. Leonard. John Rio's character had been part of the *Morning Zoo* for a while, but never made public appearances, nor was seen in any promotional photos. Mr. Leonard kept up a very funny running gag as to why he could never make it to work. The character took Cleveland by storm, becoming so popular that he was scheduled to be one of the pitchers in an upcoming softball game between the Michael Stanley Band and the WMMS Media All Stars.

The current WMMS spirit and personalities quickly became local institutions. *Morning Zoo* co-host Jeff Kinzbach made a return visit, albeit unidentified, to the *Funky Winkerbean* comic strip in July. Even more unique was WMMS's entry into the food business that same month. Local stores such as the Convenient Food Mart chain started selling a snack called Buzzard Bites, a type of extra-soft corn puff packaged in a special WMMS Buzzard bag designed by Buzzard creator David Helton. Even with all the awards and promotional tie-ins, WMMS didn't forget about its listeners or the music that made the station so popular. Yet another free Appreciation Day concert was hosted at Blossom Music Center, this time with the Del Fuegos and Australian band INXS. WMMS also hosted a free show by the Fabulous Thunderbirds at that year's Riverfest.

Across town WGCL was trying to get on the special-event bandwagon, hoping to get its call letters out in front of the public without buying expensive advertising as the station's funding had been dwindling due to the recent sale. This resulted in WGCL hosting unsuccessful events such as a kite-flying contest that drew only two contestants, both of who quickly left when they saw the lack of interest.

National media interest in Cleveland was starting to grow. On Friday, August 1 MTV flew veejay Alan Hunter into town so he could challenge WMMS's Kinzbach to a pie-eating contest. The winner was to be named Toughest Jock in America. The contest took place in the parking lot of Chester's restaurant in Independence, with Kinzbach's WMMS co-worker Leo acting as referee. Hunter had joked with a crowd the night before at the Cleveland Beach Club bar that he was planning to "cheat like hell." Perhaps in playful retaliation, on the day of the contest WMMS's "Spaceman Scott" Hughes snuck under the table to help eat Kinzbach's pie.

Leo declared Kinzbach the winner, and the entire city was caught up in the glory. While in town Hunter also threw out the ceremonial first pitch at the Indians and Detroit Tigers game, although he was more so hoping that he would be asked to sing the national anthem. As their veejay was eating pastries and throwing baseballs, MTV's crew was busy scouting possible Rock Hall sites. They had also interviewed folks around town for a feature they called *Amuck in America*.

Kinzbach wasn't the only WMMS talent receiving national media attention. Leo was informed he would be spotlighted in the January 1987 edition of *Playboy* in an article titled "The Best of . . ." He was also selected to join a group of five other disc jockeys to review new music every month in *USA Today*.

With all of the press and honors that WMMS and its crew were receiving, it makes sense that other Cleveland stations were attempting to make headlines as well. WGCL's Danny Wright was able to get some favorable press coverage when the station campaigned to raise funds to replace equipment stolen from the Great Peace Marchers during a Cleveland stop. He took in donations totaling more than $3,100, including a $500 check from legendary deejay Casey Kasem, whose syndicated show *American Top 40* ran on WGCL. Still WMMS overshadowed its local competitors in print, yet some of its future headlines were going to be based on more than just publicity.

Word had started to spread outside of Cleveland that WMMS's Gorman, and possibly Denny Sanders and other key personnel, might be contemplating a move. The station had already been keeping tabs on Metropolis Broadcasting's plans for

Cleveland since the WGCL-WERE purchase. While speculation centered on whether Gorman would be program director or even higher up in the management chain, there was no rush to get him out of the Statler Office Tower. The concern had become so great that at least one member of station management came right out and asked Gorman if the rumors were true. Gorman said that there was no basis for any of it, at best passing them off as "exaggerations." Despite his candor there was said to be overwhelming evidence from local record company officials and radio industry insiders that the opposite would soon prove true, and preparations were made for the inevitable.

Many couldn't figure out why people like Gorman, Sanders, and others would even think about abandoning the station and format they helped create, nurturing WMMS into one of the most successful radio outlets in history. According to Sanders, he had no choice. He says,

> Malrite had grown to owning several stations and hired a national program director, the first time they ever had one. His name was Jim Wood, and he couldn't figure out what the hell we were doing! We were playing Psychedelic Furs and OMD, as well as the little pop smidgens that we put in. Couldn't figure it out. He used to give Gorman a hard time, saying, "You're an AOR station! What are you doing playing this crap?! You're supposed to sit in this little slot!" We explained to him that, "A) We could get away with it, B) We did get away with it, and C) We've got the highest ratings we ever had! Where do you see a problem?"

Wood responded, "I've got research that shows that this shouldn't work!"

Sanders looked him straight in the eye and said, "Tell you what. You and I will jump into a time machine and go back to Decca Records in 1963, and we'll sit on that board and tell them they're 100 percent right! Don't sign the Beatles, because nobody's going to buy a British act with long hair! Get me some surfer boys! That's what the research says we should do! Don't be stupid! This is show business! If everything could be figured out mathematically we'd all be rich and geniuses."

The bottom line was that the folks at WMMS had a gut feeling what the Cleveland audience wanted, and their format worked. Wood simply didn't agree.

According to Sanders, the difference of opinions were amplified during a meeting with Malrite president Gil Rosenwald. Rosenwald asked Sanders, Gorman, and other key members of the staff, "What do you think of this television campaign?" He popped in a video tape of a spot produced by the Nashville based Filmhouse. The promo spot featured two wrestlers clashing in a ring, an exploding radio, and what Sanders describes as "all these AOR-hard rock, stupid, adolescent, drop out images." It was reportedly more akin to the WWF than WMMS. Rosenwald asked for their opinions. Gorman and Sanders both shrugged it off, explaining that the station had evolved past that. Rosenwald said, "Well, that's your new spot! Z100 in New York wants that spot, and we got a discount if we bought it for two markets instead of one." Gorman would actually later grow to accept the spot because *Saturday Night Live* alum Don Novello, a friend of the station, had worked on it. Still Sanders says that at the time they saw it as a major blunder, causing them to question what the future might hold at the station. The friction also opened the door for Metropolis Broadcasting out of Detroit to persuade some key WMMS personnel to jump ship.

A staff meeting was scheduled early that August at the Statler Office Tower, and it was stressed that attendance was mandatory. Tension grew as the employees wondered what the focus of the meeting would be, with some thinking that WMMS had been sold. As the staff assembled station owner Milton Maltz began going over the station's past challenges and victories, bringing special attention to the spirit that made WMMS overcome nearly insurmountable odds to become one of the premier broadcast outlets in the country. Maltz reportedly knew exactly what Gorman was going to do, and who would likely be going along with him. As to how Maltz found out, Gorman believes,

> I had told a couple of people at the station in confidence that I would be resigning on a certain date. I was also trying to get Leo to delay signing a contract. I couldn't tell him why, but I was saying, "Just don't sign anything yet!" They gave him everything he wanted, and I'd wanted him to be part of this new thing.
>
> What happened was one of the people that I told, and I believe I know who it is, probably told some other members of the staff, and it was taken to Gil and Milt. That's when we had the "big meeting."

As Maltz continued his speech at the meeting, he brought up loyalty and betrayal, making a point to make direct eye contact with each person in question. The underlying theme he tried to convey was that no one was bigger than WMMS, and the station would meet future challenges in the same aggressive manner as it had in the past. Some of the people he visually singled out started clearing out the very next day.

Others stayed at WMMS longer but were still planning to leave. In the weeks before the rest departed, questions had come up about Metropolis Broadcasting. Sanders recalls, "John was very high on them, and I asked him repeatedly if he checked them out. He said up and down, 'Yes, I checked them out,' but Steve Church, who was leaving with us, was said to have had some concerns. Steve was pretty tight in the engineering community. He warned me that, 'I put some feelers out in Detroit, and they're not paying their bills!' I went to John and told him, and John's reaction was, 'Oh, no. That's bullshit! They're good people.'"

Chief engineer Steve Church remembers the situation a bit differently. "They were reasonably well regarded as up and comers—a spunky, clever outfit that had potential. I visited Jim Harper at the station in Detroit, and he impressed me as a creative and personable guy. The station wasn't much to look at though!"

On Friday, August 15 Gorman made a move that would not only change his life, but also alter the course of Cleveland radio. He resigned his position as WMMS operations director, saying that his passion for the job was gone and that he wanted new challenges. Gorman told the *Plain Dealer* that it was a move he had been mulling over for some time, saying he planned to devote all of his energies to his consulting firm Gorman Media. He also told the paper that Sanders, who first brought Gorman to Cleveland, would join him and outline plans to aggressively pursue other clients for his firm. Gorman left the station immediately, though Sanders would stay on for a few weeks on his evening shift. Rosenwald said the company hoped to promote from within. Soon after, longtime staff member Lonnie Gronek took over the general manager's position. Maureen Duffy, another native Clevelander, filled the promotions job. As for the program director's job, the company was planning to hold off the selection in order to bring in someone from outside the city.

Gronek issued a memo to the staff announcing the two departures, with very little mention of Gorman. It read,

> John Gorman has resigned his position as Operations Manager of WHK/WMMS, effective immediately. John will devote full attention to his consulting business. Denny Sanders will be joining John Gorman and has resigned his position as Air Personality/Assistant Program Director of WMMS, effective August 29th. As a bright, creative contributor, Denny will leave WMMS and the Malrite Family with our highest regard and respect. His sincere passion for WMMS and great radio, over the years, has always been clear. His integrity with others around him has never been questioned. I'm sure you'll want to wish him well . . . and good luck as he leaves the station at the end of August.

Malrite's John Chaffee was never known to pull punches, and even today calls Gorman a traitor. He says,

> I have great respect for John and the work he did regardless of all that was underneath it. I have great respect for what he did in leading that radio station, and I will not take that away from him. But the moment he felt he was bigger than WMMS, which nobody was, and that he could go over to another station, take people with him and go against us, the guns came out. We just said, "No way." Leo took over programming at that time, and we put a lot of money and muscle behind him. Gorman also made a major mistake by trying to take the WMMS format with him and put it against us. You can't do that against a 10-share station, because then you're just seen as a copy. He tried album rock with

a mix of Top-40 and extended mixes, and we were already the station that established itself doing just that.

Gorman recalls how difficult the decision to leave really was. He says,

> You have to understand how emotional we all were about the station. This was our creation, and a big part of our life. I was thirty-five years old, and my entire adult life is rolled into this radio station. To leave WMMS, and do something across town—although the original format was not up against 'MMS—was very difficult. Plus, I was at the station thirteen years. I didn't want to just walk away with two weeks severance, which is all I got. It just wasn't worth fighting. I had signed all the deals to join Metropolis, except one. There was that one thing I didn't sign that I could have gotten out of it. The day I resigned from WMMS, I gave them a chance to say, "No, we don't want you to leave." I told them, "I really don't want to do this, but we're destroying the very thing that made us a radio station." The reaction was like, "The elevator's at the end of the hall. Nice knowing you." I walked out the front door and never turned around."

But Sanders was treated a bit differently. He remembers,

> Milt Maltz wanted to see me. He said, "We'd like you to stay as program director. We'll let you do whatever you want." I said, "I didn't sign a contract with the new operation. Just a handshake, but I honor my handshakes, Mr. Maltz." He said, "You don't have to answer now. Just think about it." Offered me a lot of money to stay, too. I wasn't making a hell of a lot even in those later days. I told John, "The money that we agreed to have me work for at WNCX is only very slightly more than I'm making here at 'MMS, and I'm going to be program director and do all this air work and everything. I know you said this was an investment in the new company, and that you're doing the same thing. If you're doing it, I'm doing it. All for one! I appreciate your guts, and I'm willing to take the risk if you are. But they're offering a lot of money, and a tremendous amount of freedom. This could be an insurance policy. You go do 'NCX, and I'll do this, and whichever wins we can join each other and really have some power in this market." John got pissed. I said, "Okay, John, but I'm nervous about these people."

The decision was made to fly to Detroit for a face-to-face meeting with Metropolis. Sanders adds, "Later, when John left WNCX, it was reported that his Metropolis money deal was quite a lot larger than he led me to believe."

The owners of Metropolis Broadcasting showed their books and confidently said, "Don't believe what you hear. We're solvent!" Gorman was reportedly convinced. Sanders reaffirmed his commitment, but felt a bit apprehensive about his initial thoughts of staying with WMMS. Church had also pledged his support. He explains that he did feel uneasy about the move, but "Denny was a friend, and I really respected his radio chops. When he went, I wanted to be on his team—but I had friends who stayed at WMMS, too, and that was hard." This was the end of an era at WMMS.

The news hit the local media like a bombshell. Kinzbach recalls,

> The station had been going for a long time, and divisions had started. I look back and I realize why it happened. A lot of it was internal. We had tried to keep the management out of the operation of WMMS for a long time—Milt, Carl, and all those guys—and they made the best effort they could to run the station right, but in a lot of instances micro-managed when they shouldn't have. That could be argued both ways, but the truth of the matter was the product was a great one and I don't think they quite understood it. I think what led John Gorman, Denny Sanders, and that group to leave was the fact that they were just tired of fighting it. They were tired of constantly having to fight with John Chaffee, Milt Maltz, Carl Hirsch, and the rest about what the station should be, and what direction it should be going in.
>
> It took a lot of the heart out of WMMS. I remember talking to Denny Sanders, and I know a couple of other people talked to him before he left, saying that he ought to stay, that it was the place to be. He left, and it was unfortunate. I have the utmost respect for Denny, because he was the visionary. He was also the diplomat, and the guy that put it together. It's very rare to find somebody like that in broadcasting. I didn't want to compete against them, but at the same time, you have to survive. There wasn't anything in the morning they threw at us, but nobody could touch us there.
>
> Kid Leo wasn't about to let them take an inch of our turf! Things did get super charged after that!

Gorman definitely had a vision, and it resulted in WMMS becoming one of the most successful radio stations in the country. It also helped make Malrite a media powerhouse. Still Gorman could be rude to his co-workers and flat out ruthless to his competitors, so there were a good number of people who were happy to see him leave. Even the WMMS sales department opened bottles of champagne when news of his departure got back to their office. Reminiscent of the dirty tricks played on and off the air against WMJI, M-105, and WGCL, rolls of toilet paper were printed with Gorman's photo and placed into the staff toilets. Just about everyone knew that Gorman wasn't content to be a consultant. When it came to radio he hungered for so much more.

Gorman says it became evident that things had changed drastically at the station, so it was time for a personal change as well. He explains,

It had outlived its usefulness! It had gone as far as it could go. It had become the establishment, and it was time to shake things up. It was also at a time when the station was faced with a problem. After years of not paying people what they should have been paid, people went out and got agents like Ed Keating, who was a top sports agent. I don't think anyone was paid what they really deserved until the agents came into play. But what happened was the agents weren't handled well, and I don't think the company did a good job of negotiating and the agents ended up dictating a lot of things they didn't have to.

It probably wouldn't have happened under Carl Hirsch, who would have come up with equitable solutions for everybody. You don't try to make up for the past sins and the agents try to dictate. There were a couple of times that I critiqued some of the air staff, and I'd get called into Gil Rosenwald's office and I hear, "I just got a call from so-and-so's agent saying you were too tough on him." Too tough! I'm just telling them they shouldn't be doing something in the studio that went on too long!

Gorman wanted something new, so he took the step that would alter his life. He adds that his planning to shake things up with a new station was not a new idea. He recalls,

At one time in 1985, I'd proposed to Gil that WDBN was for sale, and I was going to put something together myself. I wanted to buy it! It was a prototype of what would eventually be called a duopoly. We wanted to take the younger end of the rock audience, because you couldn't take WMMS back to where it was. There would have been an understanding between 'MMS and the new station that 'DBN would be the rock station, and 'MMS would still have its huge audience. I was told it was insubordination.

I was looking at it from the greedy end, too, because WMMS was completely sold out for spots. This station could have lived off the spillage!

Unfortunately the management had rejected that idea.

New ownership at a station usually means widespread changes, so staff members at both WGCL and WERE were getting ready for them. Rumors were already circulating that Gorman would be the new man in charge. However the first order of official business was adopting a new look for WGCL. The new call letters WNCX were proposed, short for "North Coast Express," though that took an unexpected turn.

Sanders describes the process,

Somewhere in September, Jim Harper called the little office which we were renting before we put the station together. He said, "Well, fellas! What are you going to do?" We said we'd decided on the call letters WNCX, which was my selection. We also said it would stand for "North Coast Express," and he thought it was a great idea. Harper said he would apply for the call letters, since we already knew that they were available. We also said, "By the way, Jim. Make sure you service mark 'North Coast Express' for the company." Harper assured us he would.

A couple of weeks go by, and we open a copy of *Radio and Records*. There's a quote from Jim Harper in the "Street Talk" section saying Metropolis was putting on a new station in Cleveland called the "North Coast Express." I turned to John and said, "Did he service mark that?" We got on the phone to Harper, and he told us, "Well, um, I don't know if we got around to it yet." I went ballistic! After we hung up, I remember saying, "Fifty bucks says that's on 'MMS by six o'clock!" Sure enough. The ID later that day on 'MMS was, "The North Coast Express—WMMS, Cleveland." They were not going to use that, but went out and got the first use service mark. They put it up on the air a few times so they could roll tape on it, document that they had first use, and apply for the service mark. They completely blew us out of the water on that one!

Now I'm really having some second thoughts about whether these people have any broadcast savvy. John still thinks they're terrific.

The service mark snafu was soon joined by another unsettling dilemma. Church had been allowed to order equipment for the new station, but Metropolis management later reportedly cancelled some of those orders without informing him.

Despite the problems both stations continued operating as usual. Though the employees were assured that there wouldn't be any expected major staff changes, plenty of tapes and resumes were being sent out to prepare for the inevitable.

When the FCC finally approved the transfer of ownership of WGCL and WERE, the new owners flew in from Detroit to meet with personnel from both stations. On October 15 Metropolis's Steve Joos came in to assume the general manager's position. The company's owners, Lorraine Golden, Jim Harper, and Harvey Deutsch, accompanied him. They gathered the combined staffs of WERE and WGCL in the office's second floor conference room. The mood was predictably dark as Deutsch took questions from the staff. Though he assured that there wouldn't be any wholesale changes at WERE, he did admit that the current staff of WGCL would be replaced. Perhaps trying to lighten the mood after the bombshell he just dropped, Deutsch grinningly predicted that in exactly one year, "The Buzzard will be dead!" Everyone knew that would be a tall order, one that an out-of-town group might not fully understand. With mixed emotions and unsure expectations, half the room accepted their fate.

Gorman and Sanders met with some of the WGCL staff before saying their final goodbyes. When the time came to let go of Danny Wright, his farewell from the station was anything but simple. Sanders said the company wanted to buy out Wright's contract. He remembers,

> The agreement that we had with Metropolis as we went into 'GCL and turn it into 'NCX was the only person under contract was Danny, and he had a big, nice contract. Deservedly so. He's very good! But he wasn't right for us. At that time, Danny was doing a teen act, and we wanted to get away from that. The stereotype of G98 was that it was a teen station, and it wouldn't help to have Danny on 'NCX. It also wouldn't have been fair to Danny to say we were going to make him over into something he's not. He had a thing that he did, and he was very good at it.
>
> Metropolis was supposed to buy Danny out. He had six or eight months left on his contract, and they were going to write him a great big check and say, "Thank you, and good night!" Right before we went on the air, Harper or Golden or Deutsch—one of them—said, "We can't write Danny the check. You're going to have to use him!" We said we couldn't use him because he wasn't right for us. I don't recall who thought it up, but we said, "If we have to technically keep him on the payroll from two to six, then we'll put him on overnight, and do not publicize him at all. He will not open the microphone. He'll be the world's highest paid board op! We won't tell the press, and we won't even tell people that he even works here."
>
> Keep in mind that Danny is a very talented writer, and I told him, "Bring in your typewriter. Make a pot of coffee, and write all night in the control room. All you have to do is segue music. You're not going to go on, but we're not going to shame you. We're not going to say Danny was demoted to the overnights. We're not going to do any of that stuff. Please, hide in there for six months, make a tremendous amount of money, write, and please be happy."

Wright doesn't see it that way. Without going into specifics, he remembers, "It wasn't pleasant—real boring. Real waste of time. Every communication between myself and them was mostly done through attorneys, and when my six-month 'sentence' was over, I was free! I used that contract money for a nice down payment on a house and never much thought about them again."

According to Sanders, "He went to the press, and told everybody that he was demoted and so on, and there was a problem. I don't think he quite understood the intentions. This was done for the exact opposite reason, so he wouldn't be humiliated in any way. We could have put him on and said, 'Knock off the Dancin' Danny act. We want you to do this and that.' That's not what we did. We wanted him to just hide in there, and not tell anyone he was still with us. Just write! He was the one who made it public."

The Original WNCX Staff (1986). (Clockwise from lower left) "Spaceman Scott" Hughes, Rhonda Kiefer, Kenny Campbell, Nancy Alden, John Gorman, Bernie Kimble, Denny Sanders, Shane Hollet, Charlie Seitz, Paul Tapie, and Gina Iorillo. Photo by Janet Macoska

A new staff had been assembled. Word had leaked to the press that it would include local comic Paul Tapie, who resigned his morning slot at WGAR-AM and FM. The new staff also included Nancy Alden from WKDD-FM and Bernie Kimble, who was part of the group that recently defected from Malrite.

As had been rumored, key promotions person Gina Iorillo, music coordinator Rhonda Kiefer, and Church were also part of the new lineup, as were Gorman and Sanders. Almost immediately fax machines at the WNCX studios at East 15th and Chester started pumping out sketches and photos depicting Gorman as a pig, or "Porkman" as they were occasionally titled. The propaganda machine was turning on one of its most effective practitioners, and there wasn't anyone who didn't expect some kind of retaliation.

The final hours of WGCL were scheduled for just before midnight on October 19. A group of station disc jockeys met at the studio to gather up their belongings and reminisce. It wasn't long before the somber event took on a party atmosphere as the group passed around beers and jokes. During the farewell get-together some of the deejays decided to take a few souvenirs for themselves. It started off as just photos from the hallways, and then soon expanded to Gold Records, the record library, and even the station's CD player. When the board operator who was scheduled to change over the station arrived, he found only a couple of worthless albums and a departing deejay waving goodbye. A few minutes before midnight the new call letters made their debut, and shortly after the station went off the air.

The new WNCX-FM put on an all-Beatles format for a few days while equipment was repaired and preparations were made to launch the new staff and format. Gorman explains, "It was strictly to accommodate the repairs and updating needed for all the equipment there. You had turntables and other equipment that were way out of phase, and the easiest thing to do in the transition—and to keep the competition guessing—was to put on the Beatles catalog. If we would have found a stack of Stones albums, we would have done that instead. We just found the Beatles albums, and went with it." The station ran imaging statements proclaiming it "The North Coast's New FM, 98.5, 'NCX," but the legal ID on the air was still "WGCL, Cleveland." A number of commercials that aired also continued to mention the support of WGCL. The new format debuted the following Wednesday, October 22. At 5:50 A.M., the impromptu Beatles format ended and the "North Coast's new FM" officially began with Neil Young's "Comes a Time" and Rod Stewart's "Passion," both of which had been chosen because of their underlying themes.

Kinzbach remembers, "The amazing thing was that when they went on the air, it wasn't exactly what we thought the station would be. It was an unusual mix of music. It wasn't progressive rock. It wasn't really classic rock. It was a mixture of urban and progressive and classic. It was really unique! To make a long story short, I remember that when I heard the station I thought, 'I don't know if this is going to work or not!'"

Shortly after the mass departure to WNCX, WMMS hinted at a world exclusive they would premiere that Friday. Gorman and Sanders were able to find out that WMMS had used its pull to get a copy of Bruce Springsteen's upcoming live boxed set. "Spaceman Scott" Hughes was at WNCX at that time and he recalls, "We tried to tape it off the air, but Leo came in over the record now and then to whisper 'a WMMS world premiere exclusive,' which killed any plans we had to air it that night on our station. But it was funny, here we were trying to steal from a station that we had just left." Though the record was flown in to WNCX the next day, WMMS still had the scoop. WMMS also continued to sponsor its free appreciation concerts into the fall, hosting one with Southside Johnny at Public Hall that October. WNCX counteracted by hooking up with MTV for a taping of the network's *Closet Classic* series at Peabody's Down Under.

The WMMS staff was able to handle the defections to WNCX very efficiently, but a small error did manage to slip through the cracks. Before the station debuted its 1987 Buzzard Rock 'n Roll Calendar that October, the product was redesigned in order to eliminate former staff members who followed Gorman to WNCX. The cover-up was only partially successful, as Church's photo still appeared in the "Family Album" section on the December page. Despite these little errors WMMS managed to keep its publicity machine rolling along nicely. Kinzbach and Ed Ferenc were featured on a two-page spread in the *Plain Dealer* when a group of investors opened the club Noisemakers at the old U.S. customs house in the Flats. Across town the wide range of music being played on WNCX not only allowed artists such as Stacey Q to be interviewed on Sanders's show, but also still featured local music such as Richard and the Heartbeats. WNCX expanded its prominence by publishing a weekly "hit list" to distribute to local record stores. It featured music, news, and other information about the station.

As the two radio stations were locking horns over listeners, Cleveland's music scene was about to lose one of its most popular acts. An announcement came in early November that the Michael Stanley Band was calling it quits. Stanley's role on the entertainment television series *PM Magazine* had been getting increased attention, and the grind of producing fresh music and touring was taking its toll. The group had been together for more than a decade, producing eleven albums during that time. Everyone in the group wanted to go out on top, so they scheduled nine year-end shows at the Front Row Theater. The decision to break up came at a meeting after the band's recording contract had expired, and prospects for a new label seemed dim. Cleveland radio helped make the Michael Stanley Band stars, and the group did its best to live up to that support. Still they knew it was time to simply step aside.

Throughout the fall WMMS tried a variety of new advertising stunts. To promote Boston's new *Third Stage* LP the station held a contest that offered one lucky listener a trip around the world with stops in San Francisco, Vancouver, Hong Kong, Bombay, London, and Amsterdam. Though hardly with the same glamour as a free world trip, WMMS was also handing out hundreds of free t-shirts, usually as giveaways for public-service efforts such as cleaning up outdated political-campaign signs. They also continued a long-standing campaign with a Knock the Stuffing Out of Hunger drive to benefit local soup kitchens, in which giveaway prizes were offered including sold out concert tickets. Len "Boom" Goldberg hosted Friday night festivities at the Hilarities comedy club, deejays were showing up at nightspots around the city, and Dia Stein (who seemed to be the "go to" person for just about everything at the station) started co-hosting the "Cleveland Breakout" segment with Steve Talbert on certain WEWS-TV newscasts.

An animated discussion backstage at Blossom (1981). (Left to right) Michael Stanley; Howard Tesnick, East Coast promotion director for EMI Records; John Gorman; Kid Leo; and Don Grierson, vice president for EMI Records. Photo by Janet Macoska

The Michael Stanley Band's series of farewell shows at the Front Row Theater continued to dominate the local music headlines. The concert dates were scheduled from December 16 through January 3. WMMS continued supporting the band right to the end with an Opening Night Rhythm Section call-in contest. Ten listeners were chosen to join the Michael Stanley Band on stage to back them up with instruments ranging from tambourines to congas. Winners also received specially produced Budweiser "bow tie" logo guitars commemorating the event. The guitars were courtesy of the House of La Rose distributors, who had been partners with WMMS for many years and were big supporters of the local rock scene. The contest provided some excellent publicity for the sold out shows. The Cleveland Browns were red hot that year, so nose tackle Bob Golic made a surprise appearance at the first show by jumping on stage to back the group on "My Town." In an attempt at local cross-promotion, the Stanley-featured *PM Magazine* series focused on the Michael Stanley Band during the run of the concerts, airing memorable moments from the band's history.

As WMMS helped close a chapter in local music, the newly christened WNCX continued to shape its on-air identity. Church continued his Sunday night interview series, while Sanders featured in studio visits from a diverse selection of artists as new as Regina and as established as Graham Nash.

Despite these strides, the future of WNCX was still in question. The *Plain Dealer*'s Jane Scott tipped off Gorman and Sanders about a flight she'd recently taken in which she found herself sitting next to Metropolis Broadcasting's Jim Harper. He expressed to her his disappointment with WNCX, not knowing that Scott and the station's crew were friends. Harper told her, "I don't know what these guys in Cleveland are doing! It doesn't sound right. They're mixing dance music with rock, and playing new stuff with old stuff." Sanders, Gorman, and the rest felt that they were preparing the station for the future of the rock format, which would eventually be called rock CHR. Nonetheless the management in Detroit didn't have a clue what WNCX was all about. The station's group took Scott's story very seriously.

As the Greater Cleveland area enjoyed the holiday season, WMMS continued its streak of gimmicks and promotions following the contests for the Michael Stanley Band concerts. Leo brought 1986 to a close by awarding a new fully loaded 1987 Silver Bullet Corvette. Contestants could enter by simply sending in a postcard or registering at the sponsoring dealership. The station's promotions continued to garner high interest, which often translated to big ratings. In its first face-off with former co-workers who defected to WNCX, a car giveaway promotion could be a critical point in the battle to win the next Arbitron book. In a timely move to fan the "esprit de corps" and bolster morale at WMMS, Leo issued a staff memo. In it he commented on an old quote by Gorman that was published in that week's *Radio and Records* which said, "I believe in monopolies and Cleveland should be ours ... WMMS will never get caught with its pants down ... The only way we'll ever be beaten is if we beat ourselves." Leo added in the memo, "We will never be 'caught' in any compromising situation, nor, shall we ever compromise. We are the best and I expect the very best out of everyone of you. We know no other way. That fact is why all the others get out of our way. As far as Porkman is concerned, I doubt if that fat ass can bend down far enough to put his pants on one leg at a time. In fact, he's not man enough to do it."

It had been one of the most eventful years in Cleveland broadcast history. Both local and national writers were watching the Cleveland radio scene closely, trying to figure out which station would emerge as the victor. They wouldn't have long to wait for an answer.

Bar Blasters, Browns, and the Revenge of the Buzzard

Northeast Ohio's radio and concert market had nurtured a very healthy music environment throughout the area, but there were so many types of popular music that one station couldn't possibly cover them all. Both WMMS and WNCX stretched the boundaries of their formats to include Top-40, Urban, New Wave, and even dance music. Even with all the additions of genres, there still wasn't an outlet for serious hard rock and metal. That changed on January 1, 1987.

Groups such as Motley Crue, Iron Maiden, and Quiet Riot drew big crowds to their shows but got surprisingly little air play outside of college radio. Sensing there might be an untapped audience among the "headbangers," Lorain's WBEA-FM changed its format at midnight on New Year's Eve. The new station was fed by the Satellite Music Network out of Dallas, and it took the call letters WCZR-FM. Referring to itself as "Z-Rock," the station promised "music to give you a headache, and you'll keep coming back for more." The format was the brainchild of veteran broadcaster "Wild Bill" Scott, who spent most of his career working at stations around the country playing music that could shake down walls. In late 1986 Satellite Music Network approached him about a format he helped pioneer at KNAC-FM in Long Beach. Within a month "Z-Rock" was on the air.

The Greater Cleveland area was the third and largest market to date to adopt that format. The station said right from the start that the programming wasn't based here and was being piped in from out of town, but the music it played was universal. Scott teamed up on the air with his producer, who was known only as "Boobie Bondage." If the music was hard enough it found air time no matter who the artist might be. Classic rockers such as Cream, Grand Funk Railroad, and even the Beatles could be heard along with newer acts, possibly in a move to open its format to older listeners. A lingering problem remained of Lorain's signal reaching into Cleveland and the eastern suburbs. Long Beach's station was faced with a similar situation, yet ratings skyrocketed nonetheless. Both time and listener loyalty would determine WCZR's eventual fate in Cleveland. Almost overnight spray painted graffiti saying "Z-Rock" started appearing on buildings, bridges, and even parked cars. The station's listeners saw themselves as "promotional guerillas," getting the word out about the format.

WMMS welcomed both a new face and a familiar face that January. Brian Phillips from Charleston was brought in as the new program director. Station darling Michael Stanley joined the air staff to host *Michael Stanley's New Review* on Sunday nights. Stanley also offered his "My Town" commentaries every Tuesday and Thursday on Kid Leo's show. Even though the station was moving ahead with confidence, morning man Jeff Kinzbach could see a trend developing. He remembers,

> Key people would leave, and they would not replace them, or they started replacing them with people from out of town. I remember a lot of conversations with the higher ups, saying, "We've got to promote someone from within, or, somebody from Cleveland has got to do this job. We can't just hire somebody from Florida, or out of town, to do this. This is a special radio station. A special product. A special breed." That's when I knew. That's when the paint started peeling.
>
> The guys who left the station were really the ones who put up the "big fight" to keep management's hands out of it. When they had gone, there wasn't too much left to keep the powers that be from coming down to dictate how the station should be run. We started seeing people hired that had no experience, or were hired because they work cheap, or people who were brought in from other parts of the country.

Though it's the nature of the radio business to see people move from job to job, at WMMS new faces were few and far between.

Off the air, there was still a lot of enthusiasm to get the Rock and Roll Hall of Fame located in Cleveland. A group of forty local personalities and industry folk headed to New York on January 21 to see the second annual induction at the Waldorf Astoria. The warm mood of the evening was set when Bruce Springsteen rushed up to Leo before the ceremony to introduce his new bride Julianne Phillips. Many industry notables singled out Cleveland's efforts to win the Hall of Fame project. Inductee Carl Perkins supported Cleveland by saying, "Any city that united like Cleveland did should have it. It's in great hands!"

Conversely there were those who expressed some misgivings about locating the project in Cleveland. Sun Records founder Sam Phillips thought it should be in Memphis. Springsteen joked that he didn't care what Phillips thought, teasingly adding that it might as well be in Springsteen's hometown Asbury Park, New Jersey. When a slide of the prototype of I. M. Pei's Hall of Fame design was shown on a giant screen, reactions were mixed. Pei and other Rock Hall officials didn't seem to worry, as they were convinced that it would eventually be accepted. A $35 million funding drive for the Hall was also announced.

Back on the North Coast a different type of funding drive was under way for the Hall of Fame. WMMS had been playing a customized local version of the Joan Jett and the Blackhearts song "Roadrunner," featuring mentions of the Shoreway, Murray Hill, and the station itself. Mr. Hero restaurants started selling a single of the new version for a limited time. A portion of the single's profits went

Joan Jett and Michael J. Fox in town for the *Light of Day* premiere (1987). Fans at the Appreciation Day concert got an unexpected treat when the *Family Ties* star walked onstage during the Joan Jett show. Photo by Janet Macoska

to the Hall of Fame drive. Jett was due in town at the end of January for a couple of important reasons. WMMS had won top honors in *Rolling Stone*'s readers' poll for the eighth consecutive year, so the station hosted an appreciation concert with Jett at the Phantasy Theater. She was also in Cleveland to attend the world premiere of her film *Light of Day* the following night at the Ohio Theater. The premiere attracted 125 movie critics from around the world, including Roger Ebert and Rex Reed, and Jett's co-stars Michael J. Fox and Michael McKean. Fox spent some time on stage before the premiere, joining local rockers Nation of One at Peabody's Down Under and making a surprise appearance at Jett's WMMS show. Fox and Jett treated the audience to an encore duet of *Light of Day*'s title song, which was penned by none other than Bruce Springsteen. For the many who were unable to spend $225 to attend the movie premiere, this was their only chance to see the *Family Ties* star while he was in town.

The premiere itself also brought together most of the local people who helped get the production off the ground eight months before, including the manager of the Euclid Tavern, local extras, and the Motion, who played at the formal dinner party following the movie that same night. Reviews of the film were mixed, but still Cleveland briefly felt the touch of Hollywood glamour.

Keeping Cleveland in the national eye, WNCX sponsored a contest in conjunction with cable music channel VH-1. The winner would receive a spot as a guest veejay. Former WGCL-FM deejay Tim Byrd helped judge the contest, and he was also the VH-1 personality the winner would sit in for.

Even with the attention the VH-1 contest brought in, the real news at WNCX was going on behind the scenes. The station was in for yet another major upheaval. On February 9 Metropolis Broadcasting put secret plans into motion to dump the format debuted just three and a half months before by John Gorman and company, replacing it with a classic hits sound.

Denny Sanders remembers telling Metropolis that the station wouldn't see any real movement until the spring Arbitron book. Though the November book showed a ratings dip in the transition from WGCL to WNCX, it also started to show increases in certain demographics. Sanders likens it to CBS canceling *Captain Kangaroo* in order to concentrate on more adult programming. He recalls, "By the time we had done a face lift at WMMS in '86, 'GCL had deteriorated into a twelve-to-seventeen pop screamer. We wanted to blow that audience off."

As heavy snow fell on Cleveland that day, Gorman and Sanders made their way into the station. They were immediately called into a meeting with Harvey Deutsch, Jim Harper, and Lorraine Golden, the Metropolis hierarchy who had traveled back to Cleveland from their Detroit base. Sanders remembers, "Deutsch did all of the talking, saying 'We don't think you guys know what you're doing. We've had a bunch of people listen to this radio station, and they don't think you know what you're doing. We've decided to change format. Right now, we're pulling all your music, and we have all these new carts from Detroit.'"

The staff heard about the change at 9:00 A.M., and the new taped IDs and music were in place and on the air about an hour later. The station now billed itself as "The North Coast's Classic Hits," which might best be described as a 1970s pop format that heavily featured Cat Stevens, Paul McCartney, and Elton John, but nothing that might be considered too hard. When Sanders heard that Mike McVay would be consulting the station, he quickly pointed out that McVay might have a conflict of interest. Sanders asked, "How can you have him consult this station? Doesn't he have deals cooking with other stations in town?" But in retrospect, some people close to the situation say it was a good move, and that McVay proved to be a lot smarter than the folks running the station at that time. One person notes, "He saw that Metropolis was inexperienced, and compulsive. McVay swooped down on WNCX, a radio station that, fully developed, could get some female numbers away from his other potential or existing Cleveland clients." McVay and Gorman had also been rivals, so it seemed as though it was a personal victory for him as well.

McVay recalls speaking to the Metropolis hierarchy a couple of months before they pulled the plug. He says, "Steve Joos, the general manager, asked if I was available in the market. I had ended my relationship with WMJI, and agreed to do some research on WNCX. From that research, we determined that WMMS was not doing a great job of playing classic rock. They were reluctant to go full ahead classic rock, and looked instead at classic hits." It wasn't long before station management made its move. McVay believes Metropolis might have been more prepared for what was in store for Cleveland if they would have done the proper research before Gorman launched his vision for WNCX. He also believes that Gorman had something valuable in his format, but Metropolis didn't give it nearly enough time and it could have also used some fine-tuning before it went on the air. Gorman also stresses that Metropolis assured him the format would have time to establish itself. It obviously never got that chance.

Gorman explains,

I don't know what they were thinking in Michigan. You've heard of people with a bi-polar condition? This was a company that was bi-polar! First they hire us saying, "You're free to do the format you think is best," and then within weeks they're saying, "Well, here in Detroit, we're getting more response doing something else!" Their station in Detroit debuted very well, but then Metropolis started tinkering with it, and they ended up blowing it up! I told them you have to stay a course. You can't be jumping all over the place.

They also felt that because we had all this former 'MMS staff, that we should go up against WMMS. That was not our intent! I told them they weren't going to beat 'MMS. Believe me, nobody knew that better than I did. It was a time when alternative was starting to come into its own, and dance was getting very big, and there was a big hole in the Cleveland market at the time. That was the format we wanted. It was a very current, modern, dance-leaning, female oriented station, because the most vulnerable part of the 'MMS audience were women. WMMS had huge numbers, but that station was stretching itself to the extreme, because you can't be all things to all people. It was the same format we would have done if I would have bought that station in Medina.

The powers at Metropolis Broadcasting didn't see that line of reasoning.

"Spaceman Scott" Hughes woke up that morning and heard oldies coming out of his clock radio. He remembers, "I called Bernie on the warm line, and asked 'What's going on?!' and he just said, 'I think you'd better talk to John!' John told me to come on down, I was on for a couple more days, and that was pretty much it. They gave me some severance, and said 'See ya!" Gorman was also on the way out. His replacement was Harry Lyles, who had worked in Louisville, Indianapolis, St. Louis, Columbus, Cincinnati, and Miami.

WNCX didn't invent the so-called "classic rock" format, in which proven past hits were aired with little or no new product. Detroit already had a station like that, so it was almost inevitable that one would spring up in the Cleveland market. Sanders elaborates,

> When I was there, WMMS did not want to be a classic rock station! We could have very easily said we're not going to play anything new ever again! We're just going to start playing our library forever. We knew we didn't want to do that, but we also knew that if we continued to be 85 percent library, somebody was going to do 100 percent library and we stood the chance of getting killed! Some of those stations had 15 or so percent newer acts, most of which sounded old or were old acts with new product. Just like WNEW in New York, WBCN in Boston, KMET in Los Angeles—those stations were blown away the second that a full time classic rock station went on the air, because they let their defenses down.

Prior to the mass departure in 1986 there was a concerted effort to feature new music on WMMS, even if it was something as controversial as Michael Jackson. Still it was too early to tell if WNCX's new format would affect WMMS the same way as Sanders describes in other cities.

But the overwhelming power of WMMS clearly made Metropolis Broadcasting nervous about introducing and nurturing an untested format, despite its track record elsewhere. A look at the Arbitron numbers around that time was enough to make sales people at competing stations consider new lines of work. In the fall of 1986, when WNCX started up, WMMS had a 15.7 share of listeners in the twelve-and-older category. The number dropped to a 12.1 the next winter, and then went back up to a 12.9 in the spring. The *Buzzard Morning Zoo* posted a spectacular 20.4 ratings share, with WLTF-FM as its closest competitor at 6.6. Leo and Matt the Cat's shows were also in double digits. The fall 1986 book posted that WMMS's morning show pulled a stunning 40 share in the eighteen-to-thirty-four-year-old age range. Kinzbach calls the report one of his most memorable moments in broadcasting.

The *Buzzard Morning Zoo*'s great success was a result of long term strategic planning by the company, an energetic and creative promotions and marketing staff, and most of all chemistry. Kinzbach used his production talents to his greatest advantage, knowing how to keep the show upbeat and moving. Ed Ferenc could act as a laugh track when needed. Goldberg and Ruby Cheeks both had strong personalities that added immensely to the show's spirit. Mr. Leonard's huge fan base also continued to grow as he added a fresh injection of unpredictable humor to the *Zoo*. WMMS also owned the afternoon due to the strength of Leo, who had popular features such as "Bookie Joint," in which a lucky listener could win a very expensive limited edition jacket. Anyone hoping to battle WMMS would have to be a programming genius or out of their mind.

News about the changes at WNCX, particularly Gorman's departure, spread like wildfire. Minutes after the abrupt transition was announced, the WNCX fax machines started humming again. This time they started spitting out sketches of a buzzard circling a rotting pig carcass. In the background was a gravestone that read, "RIP . . . J. Porkman." It also contained the caption, "WNC-Xtink!" It wasn't likely that Gorman was around to see this final insult, but it stood as an unsettling warning to those he left behind.

It was quickly becoming apparent to the WNCX staff that a bitter lesson was being learned. Sanders remembers, "Certain people at Malrite were drilling holes in the boat, and it was obvious it was going to sink! It was not wrong to leave WMMS. It was wrong to leave WMMS for this station, what would become WNCX."

So WNCX didn't live up to Harvey Deutsch's original prediction that the "Buzzard would be dead in a year." Eerily enough, Deutsch himself had passed on during that time, the victim of a brain tumor.

As Gorman's tenure at WNCX ended, Michael Stanley's non-musical star continued to rise. He was selected to co-host *PM Magazine* with Cathy Brugett. He had started as a freelance reporter for the show just ten months before, but the publicity that accompanied the Michael Stanley Band's final concerts had made him a hot property. The producers hoped that his fun-loving charm would draw younger viewers.

Emerging unscathed from his experience at WNCX, Danny Wright took over morning show duties at WRQC-FM. He started on March 9, giving new hope to a station that had been fighting for ratings respectability for years. Wright expressed relief that his arrival did not result in anyone losing a job, though it did result in some shuffling of the air schedule. Wright soon brought in his old WGCL pal JoAnne Hart to help out on the show. Wright remembers, "Jammin' 92 was okay, and there were some good people there. But the parent company was pretty much clueless, and extremely tight with a buck. I was doing mornings, but I wasn't enjoying it. First, I'm more of a night person, and was having trouble adjusting to the new schedule. Secondly, the money was horrible, and I started resenting that." It's not surprising that Wright didn't stay at the new station long.

WMMS continued to have a strong visual presence on the Cleveland streets. Billboard advertising was everywhere, the forty foot long Buzzard Bus made an appearance at Cleveland's St. Patrick's Day parade, and personalities such as Maria Farina hosted parties at local clubs.

All was not fun for everyone at the station. Dia Stein was injured in a serious skiing accident in Steamboat Springs, breaking her leg and wrist. She was sent to the

Cleveland Clinic to recuperate. Tom Rezny took over Stein's evening shift while she recovered. Problems arose in the studio as well when WMMS was contemplating whether to host the Beastie Boys, causing serious debate among some staff members. Leo ultimately rejected the idea, instead welcoming Jon Bon Jovi and his band. On the air Bon Jovi announced the winner of an autographed Kramer guitar that was to be given backstage at the group's Richfield Coliseum concert that night.

WLTF continued its successful efforts to carve a niche with lite-rock listeners, going so far as awarding hundreds of thousands of dollars with its ongoing Free Money Contest. The station had a solid personality staff that included morning man Doug Sutherland, Ted Lux doing middays, Bruce Ryan on afternoon drive, and Jay Hudson overnight, with Ken Dardis, Pam Godfrey, and Jim Crocker filling various roles and shifts as needed, as well as Ryan's wife Tracy St. John doing news. Program director Dave Popovich guided the station into promotions such as Coats for Kids, which collected hundreds of new and used outerwear and tens of thousands of dollars for Cleveland's needy children. The need for children's winter coats was so great that the station would distribute the clothing at Public Hall every year a few weeks before the Christmas holiday. It resulted in some prime television and print publicity, spreading the word about WLTF.

On the air WMJI continued to show strength with John Lanigan. The adult contemporary station encouraged its deejays to improvise and chat on the air, unlike other outlets that were making a point to promise more music. WMJI's philosophy seemed to work, resulting in solid ratings. The morning team of Lanigan and John Webster continued to sneak up on the leading *Buzzard Morning Zoo*.

After what was beginning to seem like an endless amount of time of WMMS holding a lock on the Cleveland market, the local press was starting to tire of "the Buzzard." The *Plain Dealer*'s report about the winter Arbitron ratings was headlined, "Buzzard Clipped in Ratings." The article noted that WMMS dropped almost three points since the fall. The station's overall share had also dropped from the fall's 15.7 to a 12.1, still topping the market but just a reachable four point gap ahead of its nearest competitor, easy listening WQAL-FM. Individual WMMS shifts had a little more breathing room. The *Morning Zoo* had nearly twice the numbers of the number two spot's Lanigan at WMJI. WMMS in fact led all time shifts except 7:00 to midnight, which belonged to WZAK-FM and Jeffrey Charles's *For Lovers Only* show. The *Plain Dealer* article suggested that targeting certain demographics could have cut WMMS's lead from the fall. So while WMMS's light wasn't shining as bright as it once had, the Buzzard was hardly "clipped," at least not to the point that anyone was panicking at the Statler Office Tower.

As Northeast Ohio headed into the summer months, WMMS announced another of its appreciation shows with Lone Justice at the Agora's Metropolitan Theater. The band had been opening for U2 and was able to fit in the show during the tour's American and European schedules.

Ratings were starting to slide at WNCX since February's big change, and the station was constantly readjusting through its long transition. Promotions director Gina Iorillo left for New York's National Artists Management. Staff members such as music director Rhonda Kiefer and Catrina Severson (formerly Brandy Kellogg), who had returned to Cleveland from Osaka, Japan, where she had attained superstar status with an audience topping several million listeners, were furloughed. This was obviously not the same 98.5 where Severson began her career years earlier. Church's Sunday night *Live Wire* show was cut, but he stayed on as chief engineer. Midday host Bernie Kimble turned in his resignation to pursue a teaching career. Disc jockey Nancy Alden was also dismissed and resurfaced a few weeks later at WDOK-FM, another changing station that was starting to lean toward more contemporary music.

WNCX wasn't the only station experiencing changes. WLTF's Sutherland announced he was leaving to pursue his true passion, poetry. He was truly an artist at heart, hoping to bicycle around the country to read his work. Program director Popovich gathered the staff in the conference room to meet Sutherland's replacement, introducing a longtime deejay from a long-established, or "heritage," station. He was "Trapper Jack" Elliott out of Pittsburgh. A former gag writer for comedian Joan Rivers, Elliott was scheduled to co-host the morning show with Sutherland for a week before going solo. However those plans were about to hit a temporary legal snag, and that was a contract he was apparently still bound to at his old station.

Malrite's recently acquired WOIO-TV used familiar voices for its commercials. Though personalities from sister station WMMS received the lion's share of WOIO work, the young television station pulled talent from all of local radio. The voice of the *Morning Zoo*'s Kinzbach popped up the most in garden variety station identification bumpers, saying "You're watching channel 19, Shaker Heights, Cleveland" and the like. Kinzbach's morning competitor Webster from WMJI also did voice work for WOIO. The fact that they had employed a personality from another station didn't bother Malrite, as they treated their television and radio stations as two separate entities as opposed to one conglomerate. The company in fact opened the television station to promotional affiliations with a number of local radio outlets.

The WMMS promotional machine kept running. All of the station's attention wasn't due to giveaways and gimmicks though. WMMS's Skyway Patrol pilot Pat Brady made headlines when she returned from a fifteen-day tour of China, where she was one of fifty delegates asked to be part of an information exchange on aviation. Brady went at the invitation of the Ninety-Nines, Inc., the international organization of female pilots. It seemed that not a day would pass without some mention of WMMS or its personalities in a media other than radio. It was a very good time to be with the station.

The station tried to put its local power into action when a longtime WMMS favorite was bypassing Cleveland on his current tour. Even though David Bowie's

Glass Spider Tour was only planned to stop in a handful of U.S. cities anyway, some felt that the artist owed a concert to the city that broke him in. Leo sprung into action. He announced yet another petition drive on his afternoon drive show, asking for signatures, cards, and letters demanding that Bowie's management schedule a Cleveland date. Four sisters from Parma delivered 500 names, while hundreds of more responses poured in over the next couple of weeks. The campaign served to promote Bowie and get people talking about yet another WMMS campaign, though regrettably it failed to lure the artist back to the shores of Lake Erie that year.

Changes continued rapidly at WNCX. Church turned in his notice as chief engineer, deciding to fully devote his efforts to his own company Telos, which supplied high quality equipment to the broadcast industry. Church had made a reputation for himself as an engineering trailblazer, designing new equipment and techniques on the spot to solve problems. He no doubt felt that his drive and innovation would better suit a company of his own. In time Telos would set the industry standard.

As great of a loss as Church's resignation was, rumor started to spread that a bigger change would involve the station itself. Word was going around that both WNCX and its AM sister station WERE were on the market. The stations' Detroit-based owner Metropolis Broadcasting was said to be eager to get rid of the Cleveland properties to better concentrate on the problems at its flagship. WDTX-FM in Detroit had slipped to an embarrassing seventeenth in the ratings. The problems were so severe that disc jockey Jim Harper, one of the three co-founders of Metropolis, was rumored to be talking about moving to another station in the Motor City. It had become painfully obvious that the Cleveland market was too much for the management in Michigan, and the dream of crushing WMMS was quickly turning into a nightmare.

While WMJI's Lanigan and Webster sometimes skirted the borders of good taste on their morning show, the station was still able to advertise itself as family oriented, a sentiment echoed in many of its promotions. WMJI continued what would become a long tradition that July by sponsoring Cleveland's giant Fourth of July fireworks display at Edgewater Park. The city provided permits and helped with security, and WMJI offered on air promotion and picked up the then-$40,000 tab for the show. The Festival of Freedom Majic Fireworks to Music often drew more than a quarter million people, all of whom were asked to bring radios so they could experience the aerial display synchronized to the station's programming.

Summer was also a big season for WMMS and its personalities. Kinzbach and Ferenc traveled to local beaches, getting footage for the "Beach Week" insert bumpers that ran between shows on WOIO. Leo drew on his passion for golf by launching a contest seeking a third to go a round with the deejay and Starship vocalist Mickey Thomas.

Cleveland's rock and roll reputation caught the attention of Peter Gabriel, who decided to shoot a video during his appearance at Blossom Music Center at the end of July. The song, "Biko," from Gabriel's album, *So*, focused on black South African activist Steven Biko, who died in prison in 1977. It was a five-camera shoot, and the finished product ended up on MTV.

WMMS continued its series of listener appreciation shows. Eddie Money was featured at the National Rib Cook-Off to benefit the Rock and Roll Hall of Fame project, which received $20,000. The station also hosted Glass Tiger at the Nautica Stage, and a concert with Bruce Hornsby and the Range was announced for Blossom Music Center. WMMS remained a powerful force in the music industry, and the record companies would make it a point to offer acts for free shows in an effort to premiere them in Cleveland. Not that WMMS always received newcomers, as it could use its clout to help spotlight promising new artists while also getting established acts. The prestige of WMMS benefited everybody—the label, artist, station, and listener.

Perhaps the biggest surprise of the summer came from WRQC's Danny Wright. Just five months after taking over the station's morning show, he announced that he was retiring from radio to become a real estate agent. Wright started studying for his license during his final days at WNCX. When he passed the state board, Wright signed on with one of the bigger Northeast Ohio real estate companies. He promised to continue his writing, but said his radio career was over. A lot of fans hoped he would reconsider.

Wright's retirement announcement wasn't the only unique change to Cleveland radio. WLTF's Bruce Ryan aired a very ambitious project on his Sunday night *Golden Age of Rock and Roll* program. It was the three-hour retrospective "1967: Sgt. Pepper and the Summer of Love," featuring not only music from that era but also comments from Beatles publicist Derek Taylor, Apple Records' Peter Brown, Mark Volman of the Turtles, Donovan, and the *Plain Dealer*'s Jane Scott. Ryan put the show together with others including production head Joe Gunderman and staff researcher Susan Locke. The "Sgt. Pepper" episode received terrific reviews and won the prestigious Twyla Conway Award from the Radio and Television Council of Greater Cleveland, designating it as the year's best entertainment program. Locke was a tremendously talented and well-liked member of the WLTF staff. Tragically, her name was thrust into the headlines in 1999 when an intruder took her life at the downtown office building where she worked.

The summer ended with a major coup for WMMS. At noon on August 27 Matt premiered one of the most anticipated releases in years. The station became the first in the nation to play Michael Jackson's new album *Bad*, the follow-up to his history-making *Thriller*. WMMS had the title single days before any other station, obtained through the good relationship the station developed with Jackson's camp after it sponsored the *Victory* Tour. Though urban and CHR stations played just as

Ruby Cheeks (1987). Always outspoken and quick with a line, Cheeks was an instant hit when she settled in with the *Buzzard Morning Zoo*. It wasn't long before Cheeks was hosting her own show. Photo by Janet Macoska

much, if not more, of Jackson's work, record labels were always impressed with WMMS's huge lead in the ratings. Even the companies that had complaints about the station knew that sending new material there first could only help their artists.

In August 1987 consultant Mike McVay severed his ties with Metropolis Broadcasting. Looking back McVay explains that he felt taken advantage of by the company, a feeling echoed by Gorman after he left WNCX. After his relationship with Metropolis ended, McVay turned his attention to WLTF, reuniting with his old friend Popovich.

On September 25 a native Clevelander opened his mike for the first time at WNCX, as Bill Louis came on board the classic hits station. He remembers, "As far as I was concerned, I was just stopping in for the Browns season. I started out part time, and if things didn't work out at 'NCX, I was going to just head back on the highway. But the Browns in '87 were a team that I couldn't miss."

Back on WMMS Leo used his pull to debut Bruce Springsteen's "Brilliant Disguise" a full day ahead of the competition. In the cutthroat world of radio twenty-four hours could be an eternity if a rival station has a hot release, and a lot of angry programmers were already desperately trying to make up lost ground against WMMS before this premiere came along.

The Springsteen premiere wasn't the only news at WMMS in the fall. The very popular Cheeks was moved to the 6:00 to 10:00 P.M. slot on October 5, replacing Stein. Cheeks had been a member of the *Buzzard Morning Zoo* for two years, but she wanted a show of her own after having great success with an evening shift in Florida. Cheeks's wit and personality had established her as one of the most recognized female personalities in Cleveland, and Leo told *Scene* magazine that he expected her to dominate the new time slot. Cheeks was able to take the new shift after Stein moved into management, allowing her to work on special projects but also pushing her on-air time to weekends.

Just a few days later the *Morning Zoo* announced Cheeks's replacement with the addition of Roberta Gale, a stand-up comic out of Tuscon and a broadcast veteran of such programs as National Public Radio's *All Things Considered* and the syndicated *Dr. Demento Show*. The station also extended the contract of the voice of the morning show's Buzzard Skyway Patrol, Pat Brady. Brady was the only female helicopter pilot and reporter east of the Rockies, had an easygoing delivery and natural sense of humor, and fit in well with the *Zoo* crew.

Though Gale was the only new face joining the station, WMMS also welcomed back a very familiar face. Back at the studio was Malrite's corporate art director David Helton, who returned to Cleveland after spending some time at the company's WHTZ-FM in New York. Helton was now able to supervise Malrite's graphic arts department from his studio at the Statler Office Tower. In a memo announcing Helton's return to Cleveland, general manager Lonnie Gronek invited the staff to "Stop by David's new ceement pond, sit a spell, do some wit'lin and say howdy!"

While the *Morning Zoo* was undoubtedly a success, a good segment of the audience wondered why there were so many voices on the show instead of just focusing on Kinzbach and Ferenc. Gorman offers this opinion, "[Kinzbach] has a great talent and an excellent voice. But Bob Hope doesn't write his own material. Hope knows how to deliver it, and it's the same for Jay Leno, Letterman. You reach a point where you're doing material day after day, and you can't write your own. Jeff was very good at doing others' material. That's why, at one point, we had Len, and Ruby, and Kenny Clean in there, because you could bounce things back and forth and keep it moving. It was like a multi-level *Pong* game, where the balls are bouncing all over the place all morning."

WMMS's morning show wasn't the only one enjoying great success. Over at WMJI Lanigan, Webster, and newcomer Jimmy Malone celebrated the one-year anniversary of their popular "Knuckleheads in the News" segment with an hour-long special. Lanigan had seen Malone at the Cleveland Comedy Club doing a bit

Alice Cooper (with Kid Leo) hosts *Classic Rock Saturday Night* (1987). Photo by Brian Chalmers

about bizarre stories taken from actual newspaper headlines. He invited Malone to do the same on the show, noticed the response, and soon added him to the WMJI morning team.

The U2 concert at Cleveland Municipal Stadium on October 6 drew a tremendous response. Prior to the sold out show the band's guitarist David "The Edge" Evans put in a call to Leo's show. He wanted to compliment the city and especially WMMS for the great support the group received back when they were just playing clubs.

The Edge wasn't the only musician to show appreciation for WMMS. As the station's *BLF Bash* maintained its late night superiority, host Bill Freeman was given a rare Saturday night off. WMMS wanted a high-profile personality to fill in that night, so they asked none other than Alice Cooper. Cooper was about to go on tour, and scheduled a Cleveland stop for the following month. He was flown in to promote the tour and new album, and he took to the WMMS airwaves as if he'd been there for years. He played some of his favorite music, took dozens of calls, and did an admirable job filling in for Freeman, though some fans actually called in to ask where the regular host was!

The crew of the WMMS *Zoo* expanded its off-the-air notoriety on November 12 with the opening of the club Cadillac Beach in North Ridgeville. Describing it as the "sister club" to Noisemakers in the Flats, Kinzbach and Ferenc promised to make regular appearances inside the pink Cadillac that served as the disc jockey's booth.

Hard times were in store for hard rock fans to the west of Cadillac Beach, as Lorain's WCZR dropped its "Z-Rock" format on November 15. Less than a year after its debut, the station decided to switch to a light jazz satellite feed called "The Wave." The new format had already been getting a good response across the country, proving to be far more profitable than hard rock. The Lorain station made the switch official by changing its call letters to WNWV. A grass roots group called Z-Rock Rules quickly started a petition drive in an effort to save the old format. They collected more than 1,500 signatures at various heavy metal venues and staged a march from Public Square to Public Hall, where Cooper was appearing that night. Though it was doubtful that WNWV would go back to its old format, hope was rising that another station might take a chance on hard rock.

In an article in *Scene* magazine, the movers behind "Z-Rock's" initial entry in Cleveland said that while Northeast Ohio fans were among its most rabid followers, that devotion just wasn't reflected in the ratings. Boobie Bondage told *Scene* about a trip she made with program director Bill Scott to a rained out drag strip promotion, but still saw the raw energy of the station's fans as they relocated to Shadows Bar for an ear numbing evening of hard rock and roll. The station returned that enthusiasm by playing records by local metal artists that stations in Cleveland wouldn't touch, and the response back to the Satellite Music Network was very encouraging. There was still hope for "Z-Rock's" return, but until that happened, metal fans went back to the college dial where ratings did not determine programming.

WMMS's *Classic Rock Saturday Night* series welcomed a guest host on November 21. Joe Walsh and his friends stopped by the studio to celebrate the Eagles guitarist's fortieth birthday. Walsh and longtime musical partner "Rick the Bass Player" Rosas took over the airwaves. They started telling jokes about friends like Eric Clapton but then promised to hold their tongues when they spoke later in the show. Walsh also played an acoustic set for a group of fans in Dayton who held a birthday party in his honor every year. Whether or not they were able to hear the broadcast is another question.

While on the air Walsh had some fun taking playful jabs at a few Cleveland friends in what can only be described as an improvised comedy dialogue. It started as Walsh announced, "It's 7:56 [sound of crashing cars] and that's our time chime at WMMS. You know Eric Carmen? You ever meet him?"

"No, thank you," Rosas answered.

"Personal friend of mine!"

Joe Walsh (1980). The Clown Prince of Rock and Roll, Walsh embraced Cleveland as his adopted home and was a favorite guest on local radio. Photo by Janet Macoska

" ... okay."

"We used to play with those guys ..."

"Those guys. . . . ?"

" ... those guys Wally Bryson, and Jim Bonfanti, the drummer. Hey Jim! How you doooing? I was in a band with Jim and Danny Klawon. Where's Danny lately? Eric went ahead to become, uh ... Barry Manilow's protege, I guess."

Walsh and Rosas started up again when Walsh started mentioning a WMMS personality. Loudly chomping his gum into the microphone, he proclaimed, "It's

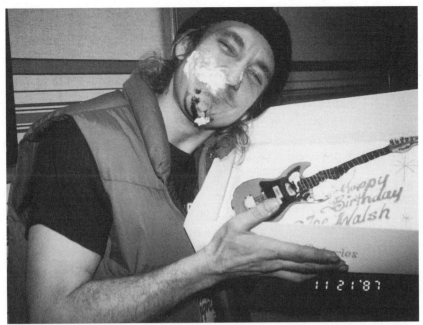

Walsh at his fortieth birthday party at WMMS's *Classic Rock Saturday Night* (1987). Photo by Brian Chalmers

my birthday weekend extravaganza . . . no thanks to Kid Leo! That bum didn't even show up."

"That's a fine attitude," Rosas responded.

"Where do you think he is anyway?"

"I don't know. He's a bum if you ask me."

"Maybe he went down to Houston to sing the national anthem for the game, because Cleveland's in . . . I doubt it, but maybe."

"Could be at the Jim Nabors concert, too."

"I'm not touching that one . . . or him."

Walsh continued to goof around after the show had ended. Just before the photo of Walsh and his birthday cake was snapped outside the studio, he whispered to Leo, "You can push my face in the cake, if you want. It might make a better picture!"

When WMMS didn't have big time celebrities in the studio, the station enjoyed the local recognizability its own personalities were getting. Tom Batiuk's *Funky Winkerbean* spin-off *Crankshaft*, also set in Northeast Ohio, was due to premiere in December. Word had leaked to the press that the *Morning Zoo*'s Kinzbach and Skyway Patrol pilot Brady would get mentions in the strip.

This kind of coverage didn't stop other stations from testing the marketability of their staffs. WMJI released *Knuckleheads in the News: The Book*, which benefited the Boys and Girls Clubs of Greater Cleveland. Lanigan, Webster, and Malone headed out on the road to promote it with book signings.

As WMMS and WMJI continued to make individual names for themselves, they joined other Cleveland stations in airing a song by six-year-old Sharon Batts. Titled "Dear Mr. Jesus," the song was produced to fight child abuse. While it did get some heart-tugging response from listeners, it was still an odd choice to air on the morning drives, especially on WMMS. Kinzbach called Batts's mother Jan to discuss the song on the air. The mother said, "Everybody tells me it's a miracle, and I think it is. I think God did it!" Kinzbach called it, "A powerful song, maybe because it was done so innocently." Gale, the newest member of the *Morning Zoo*, decided to show her news sense by asking Sharon Batts if she fully understood what she was singing about, as lyrics included "Please don't tell my daddy, but my mommy hits me, too!" The young singer admitted that she basically just sang the words, stressing that she herself was not abused. She then favored the audience with a few off-key choruses over the phone, much to the enthusiastic response of the *Zoo*.

It was right around that time that a full-sized heavy punching bag was hung outside the WMMS studio. It was adorned with stickers from WNCX, WPHR-FM, and others, and featured dead-on caricatures of Lanigan and Webster by Helton. Employees and visitors were invited to take a swing at the competition before entering the studio. The jovial game symbolized a year in which the station had thrown plenty of punches, knocking out a lot of key competitors.

Most Clevelanders however were concentrating on another sport. Football was the main focus that December as the Cleveland Browns continued to battle their way through the NFL. The showdown with their dreaded division rivals the Pittsburgh Steelers was just about the only thing on peoples' minds the week leading up to the game. The *Buzzard Morning Zoo* took advantage of that interest with an on-air bet that raised plenty of eyebrows. Gale told Kinzbach that the Steelers were too strong a team and that the Browns would go down in flames. She was so sure Cleveland would lose that she pledged to go out in the streets without clothing if the Browns won. She was sure surprised when they did just that. A big crowd, including television camera crews and the Browns' Hanford Dixon, gathered the following day at East 9th and Euclid to see Gale wear nothing but team pennants and a smile.

As Northeast Ohio approached 1988, WMMS was in store for some changes. Personality John Rio (Mr. Leonard) moved to Richmond Heights from Houston, while still paying frequent visits to New York City's WHTZ so he could be closer to Malrite's corporate center. This allowed him to start personally popping up in the studio a lot more frequently. In addition to phoning in Mr. Leonard bits to other Malrite stations, Rio's lengthy stays in the WMMS studio often stole the morning show from Kinzbach and company.

As WMMS regained a star it lost one of its key personnel. Brian Phillips resigned as program director, claming his Southern blood couldn't get used to the

Gene Simmons of Kiss and Kid Leo belt the competition (1988). The station's aggressive nature could even be seen in the hallways, with caricatures of radio rivals adorning the heavy bag. Guests and employees were all encouraged to take a symbolic crack at the competition. Photo by Brian Chalmers

Cleveland winter and that he was yearning for a warmer climate. He told *Radio and Records* that his new position as operations manager and program director at Orlando's WRJW-FM would give him a "shot at creating a legendary station, not just helping to maintain one." WMMS was once again forced to look for a new program director, a position Gorman had held for an unheard of length of time before he defected to start WNCX.

Another key player headed to Cleveland that month as Brad Hanson moved in to take over the music director's post at WMMS. Remembering the time, Hanson admits he was slightly intimidated,

> It was a station I'd heard about for years and years, especially when I was at Bowling Green from 1979 to '83. It wasn't just a powerhouse, but had assumed

this legendary status at the same time. At that time, Cleveland had had its share of problems, and WMMS was one of the few bright spots. I first visited Cleveland when I was still in college, and it wasn't exactly a tourist mecca. But when I came back in 1988, and Leo interviewed and hired me, I admit I was intimidated. It was also a pretty strange time, because I was listening to the station as I drove in for the interview, and heard the station go from Yes's "Starship Trooper" into a Tiffany record! Then I heard Whitney Houston. When I came in for the interview I had a lot of questions, but the one I didn't ask was "Why?" This was supposed to be this great rock station, and they were playing stuff like Gloria Estefan! I was pretty confused, along with intimidated, and the size of the station and number of people who were working there—and some who weren't working—made for an interesting mix. It was also a very exciting time, especially for my career.

The Browns' march to the Super Bowl stopped suddenly on Sunday, January 17 when the Denver Broncos slipped past the team once again. As one Cleveland institution failed, another continued its reign. The fall Arbitron ratings book put WMMS at a 12.5, which was more than two points ahead of its nearest competitor, second-place WQAL at 10.1. Overall WMMS dominated its seventeenth straight book, while it topped the eighteen-to-thirty-four-year-old market for forty-one books in a row. The *Morning Zoo* and Leo ruled their respective morning and afternoon drives, and Cheeks pulled up a close second behind WZAK's Jeffrey Charles. Stations around town were hard pressed to predict when WMMS would finally start to show any sign of weakness. The ratings book distinction was soon followed by another honor, as WMMS had been named Radio Station of the Year for the ninth consecutive time in the annual *Rolling Stone* readers' poll.

Just months after her nigh-nude run following the Browns' victory over the Steelers, the *Zoo*'s Gale was again called into public service on February 2, being declared Cleveland's official groundhog in a proclamation from City Hall. Gale announced she would enter a manhole at Public Square wearing a groundhog outfit at 8:00 A.M. She would then emerge to look for her shadow. As simple as the stunt was planned out, Gale was able to work through a couple of problems. A groundhog suit couldn't be located, so the station had to settle for a beaver costume. By the time Gale arrived to crawl into the manhole at 8:00, the sun had been out for some time, already casting shadows. However the show must go on. Gale acted as planned, though she admitted that she didn't want to stay in a sewer hole any longer than she had to.

South of Gale's little excursion down a manhole, Akron's WONE-FM was going through some changes. Brian Taylor was settling into the spot of program director that he took over for Ward Holmes a couple of months before. Taylor started shuffling personnel around to strengthen the on-air sound. Although WONE depended heavily on classic rock and a slight offering of new music, it looked to its personalities to sell the station. He switched Brian Fowler from midday to morning drive, giving the 10:00 to 2:00 slot to Deeya McKay. McKay was now followed by Jim Chenot until 6:00 P.M., where Mike Michelli picked up until 10:00. Overnight belonged to Jeff Daniel Kunes until 2:00 A.M. and then finally Debbie Vincent until the morning. Taylor had questions about how effective veteran Bill Hall's on-air chemistry was with Fowler, and he decided another voice should be added soon.

WMMS was also making changes here and there. Jeff McCartney of St. Louis's KSD-FM was brought in as the new program director. Leo announced the addition in the February 26 issue of *Radio and Records*, "Like Newman and Redford, Ali and Frazier, and Sean and Madonna, look out for the knock out team of the Kid and McCartney!" McCartney was said to be the unanimous choice of the search committee, but he would have no idea what he was about to jump into.

The Buzzard Droppings Hit the Fan

There are certain benchmarks in time that defined WMMS's history, and many of them are very positive. However, stuffing the ballot box in *Rolling Stone* magazine's annual readers' poll is not one of them.

Few people know exactly what went on behind the scenes before the *Plain Dealer* expose that haunted WMMS for many years to come. The story focused on closed-door meetings and a level of cutthroat competitiveness that few industries would ever approach, let alone face on an almost daily basis. The old saying at WMMS was "We don't go to work, we go to war!" The *Rolling Stone* debacle could very well have been likened to D-Day.

WMMS and particularly the *Buzzard Morning Zoo* were riding the crest of a ratings tidal wave. They weren't simply beating the upstart WNCX in the ratings, they were humiliating them. The power of *Morning Zoo* co-host Jeff Kinzbach and his troupe, helped in large part by a well financed and savvy promotions staff, was so awesome that they had become a ratings juggernaut, flattening the steady stream of morning deejays that came and went through WNCX. WMMS prided itself on winning the *Rolling Stone* readers' poll for Best Radio Station (Large Market) nine years in a row. How the station kept winning year after year was one of the worst secrets in Cleveland radio. "Spaceman Scott" Hughes recalls, "That was one of the first things I did when I got there. I was handed a check for $3,000, and took an unmarked WMMS van to all the B. Dalton bookstores, newsstands and Dairy Marts. I'd get as many copies of *Rolling Stone* as I could, and bring them in the back way. All the jocks and promo people got together in the conference room, and we filled out all the forms."

Former WMMS program director John Gorman remembers getting copies from the band American Noise's road manager, who also worked for local newspaper and magazine distributor Klein News. Gorman explains, "The unsold *Rolling Stones* would have the title banners torn off and returned to the magazine for credit, and they would toss the rest of the issues away. We'd get the discarded copies, and advertise a show by saying you should fill out a ballot when you arrive. I don't think that was illegal." The station's promotions staff even opened the magazine to the exact page for concertgoers and handed them pencils. It was a full-fledged annual campaign to win that title. It became a very important distinction to win that poll, as the clout was a valuable sales pitch for national sales accounts.

When the job of getting the ballots filled out became too big for the promotions staff, it was decided that every single employee at the station would each receive envelopes with ballots inside. In addition to the big prize the WMMS staff was also given suggestions on what to write in the survey's other categories. They were then instructed to write the names of past WMMS contest winners in the envelopes' return address fields.

The ballot stuffing for WMMS wasn't just going on in Cleveland. Frank Foti was part of the Cleveland contingent working at Malrite's WHTZ-FM New York, where the promotions department was filling out ballots for WMMS as well. In 1985 WHTZ came just a handful of votes short of legitimately beating WMMS, leading some of the New Yorkers to jab at their Cleveland colleagues with "You guys are stuffing the ballot box, but the real winners are here!"

News of this process was leaked when a disgruntled employee at WHK-AM had been enjoying a few cocktails prior to his shift. Chatting with another bar patron, he mentioned his affiliation with Malrite Broadcasting. The alcohol had loosened his tongue a bit, so he casually mentioned the annual *Rolling Stone* ballot campaign and explained exactly how WMMS obtained the prize. This piqued the interest of the guy sitting next to him, a WNCX employee.

It was no secret in Cleveland radio that WMMS was more than likely helping its own cause, but the competition never had the proof to act on it. This changed when an innocent mistake proved to be the station's undoing, though the ethics of how the evidence was obtained has been questioned as well. Maureen Duffy was the promotions director at WMMS, and she was very attentive to details. She often spent long hours and many weekends on the job. When it came time to get more *Rolling Stone* magazines for the campaign, she jotted down a note for the station's business manager Judy Goldberg and left it in Goldberg's slot in the mailroom. The handwritten note, dated November 14, 1987, read: "It is important that *Rolling Stone* magazines from Rosie's (800) be available for pick up by Friday. When they are at Rosie's please let Steve or John know (I'll be out Wed.–Fri.) and they'll get John to go do them."

The memo was left on a Saturday afternoon, but when Goldberg arrived at work that Monday it was not in her mailbox. It has been assumed that a low-end staff member, probably loyal to friends at WNCX, swiped it. As open as the campaign used to be, Gorman warns, "If you're going to do something like that, you don't put it in writing."

The truth came out on February 25, 1988. The *Plain Dealer* headlines left nothing to the imagination. They simply read, "Rolling Stone Poll is Rigged by WMMS." The paper had tried to take swings at the station in the past. This however was a swift punch to the gut.

The story, written by Michael Heaton and David Sowd, reported WMMS general manager Lonnie Gronek admitting to the scheme after he was presented a copy of the memo. He openly stated, "We buy the magazines and distribute them to staff, to our family and friends and relatives. We take copies of the magazine to outside events and distribute them. I've filled out a number of them myself."

The paper also quoted *Rolling Stone* managing editor and former Clevelander Jim Henke, who said WMMS received about a thousand votes and won by a margin of about 200. Gronek told the *Plain Dealer* that if WMMS didn't buy the magazines, "Some other city would. We just do it bigger and better." Runner up stations WNEW-FM in New York and WMMR-FM in Philadelphia quickly disagreed, both saying they would never stoop to that level of competition. Critics lined up quickly.

Gorman contradicted management at his former station, telling the paper, "Winning [the poll] was everything . . . an incredibly hot sales positioning statement . . . an extra sales point to tack onto the ratings. It can help close a deal." General managers at WMJI-FM and WGAR-FM agreed, though Gronek said that in the past nine years the honor did not result in ratings and additional revenue. WZAK-FM's general manager Lee Zapis called it a "lack of judgment." WLTF-FM said it was "not good business practice." WONE-FM's programming head Fred Anthony stated, "It would be unfair to say we all do it."

The negative reaction didn't stop there. The Associated Press picked up the story, and stations in Pittsburgh, Baton Rouge, and other cities across the country ran it. Back in Cleveland, it was the talk of radio and television.

Kid Leo appeared on television news reports to defend the station, joking about writers cramp. WMJI's morning co-host John Lanigan interviewed Hughes, who openly admitted to everything. Hughes reportedly said the staff members would take the ballots back to their home areas for different postmarks. Following the interview Hughes received a call from his girlfriend, Lori Lyall, who worked in Malrite's corporate office. She told him that the top management there was furious. Hughes soon received interview requests from the *Wall Street Journal*, *USA Today*, and other major papers, but he decided to take Lyall's advice and not offer any further comments.

The *Morning Zoo* crew called Gronek before they went on the air, asking him to come down to the station and answer listener calls. Looking back, Gronek says, "Overall, and when all the dust settled by day's end, it did not leave the big black mark on WMMS that our competitors hoped it would. The calls were extremely positive. The *PD* had spoken to us, but we didn't know it was going to run the next day. You have to understand that the competition between WMMS and WNCX was mean, and vicious, and loaded with venom. We more than held our own."

Despite Gronek's positive spin on the situation, *Zoo* co-host Kinzbach says it's one of the few mornings he wishes he could have back. He remembers, "I had Milt Maltz calling me, and Lonnie Gronek calling me, and the program director—all the cooks were stirring the stew that morning. If I could do it over again, I would never have answered the hot line. I would have just put listeners on the air that morning, and let them say what they wanted and told the real story of how it really happened."

As to the motives why the story was leaked, Kinzbach speculates, "Our past 'comrades' were at 'NCX, and they wanted a particular concert. We were sponsoring the show, and they basically said, 'Hey, we have a note from one of your promotion people' that implicated us. They came over to the station for a big meeting, and someone from our management said, 'Do what you want! We're not giving you the concert.'"

Other parties have also mentioned the concert angle, but no one seems to recall the particular show. Kinzbach continues, "I was in the middle of this damned thing, and I should have just said, 'Screw all you idiots! This is the way we're going to handle it,' and gone on the air and told what really happened. They tried to put a spin on it. The station had already been playing Michael Jackson and Madonna, and now this! The credibility was already shaken. We had an audience that trusted us, and believed in us. Unfortunately, the radio station basically took that for granted, and it hurt."

WWWE-AM asked Kinzbach to appear on the station's afternoon drive show for an interview, but he declined by telling the executive producer the situation was "between the newspaper and WMMS." But WMJI's Lanigan did appear, saying it was "pathetic." He went on to say WMMS "didn't do it for the good of Cleveland. [They] did it for the bottom line," referring to money. Lanigan did admit WMMS was a tough competitor, adding that Malrite Communications was one of the finest broadcast companies in the country. He pointed to a recent survey by the industry newsletter *Inside Radio* to pick the ten most respected radio stations in America. Lanigan said, "Malrite had two of them in the top ten. Stations like KISS in Los Angeles and WHTZ in New York, which is a Malrite station, those were all considered the top stations, and WMMS was in there also. I think that's obtained by winning consistently in the ratings, and by a ten year run as far as *Rolling Stone* goes." He wondered how the broadcasting company would feel if it did the survey after the news of rigging the contest. Lanigan concluded, "It is very important to that company to be one of the most respected, and I think that's why they did it. I don't believe it was for Cleveland. I believe it was for the corporation and for the image."

WMMS went into immediate damage control, starting with a full-page ad in the *Plain Dealer* the next morning. Headlined "Secret?" in large bold type, it read:

> For the past nine years, WMMS has campaigned . . . and campaigned hard throughout the Northeast Ohio area to win the *Rolling Stone* magazine award for Best Radio Station in America.
>
> We have openly solicited the populace at local clubs, various shopping malls, concert stages throughout the area, and other events such as Riverfest and the National Rib Cook-Off. We have passed out official *Rolling Stone* ballots at many of these locales. Many of you have participated in the campaign over the years. Secret? Hardly.
>
> WMMS has always urged its own employees to participate in the balloting—all at the station's expense, keeping in mind that the rules allow multiple entries. It is not a referendum. Jim Henke from *Rolling Stone* stated, ". . . best really means most popular."
>
> Earlier this year, WMMS aired a tribute to the Rock 'n' Roll Hall of Fame inductees and openly talked about our marketing campaign for the *Rolling Stone* Annual Music Award. Secretive? No way. Why have we gone through all of this year after year . . . just to win a popularity poll? No. WMMS has seen the impact of this award on the city and realized long ago that the positive recognition of the station translated as a source of pride for Cleveland. Contrary to a *Plain Dealer* article on Thursday, February 25th, WMMS has never "rigged" this poll. It is common knowledge that major radio stations across America have conducted similar marketing campaigns both off and on the air to try to win this award.
>
> It seems odd that after all these years as we approach the benchmark of the 10th successive award, that suddenly our marketing tactics are questioned. As a market leader, we enjoy the challenge to remain on top. Jealous competitors often take desperate steps to "knock off" #1. WMMS will not be intimidated.
>
> Here's the bottom line: We serve one master . . . You the audience. The overwhelming positive response to Thursday's *Buzzard Morning Zoo* on WMMS and subsequent calls to our office (in excess of 95% positive) leads us to believe that you want us to continue in this effort for the city and the station.
>
> As usual, WMMS will act in partnership with you to maintain Cleveland's image as the rock 'n' roll capital of the world.

That same day in the *Plain Dealer*'s *Friday* supplement Heaton reinforced his stand, citing the station's "long and glorious history" and suggesting that, "Without it, we might not have gotten the nod for the Rock and Roll Hall of Fame." He recalled his own history listening to the early WMMS, as well as the disc jockeys who established it as one of the premier radio outlets in the country. Heaton talked about the great pride he had coming from a city that produced WMMS, and how he would defend its honor when other stations tried to dump on the station or Cleveland. While living in San Francisco he challenged a station that did just that. He wrote,

> I was pretty honked off. I called the KFOG station manager and ripped into him, saying that WMMS was a great radio station and had been for a long time. The station manager, who had worked at 'MMS, chuckled and said that the key words were "had been." He asked me when was the last time I had listened to 'MMS. I told him about five years.
>
> He said things at 'MMS had changed. He informed me that 'MMS had gone to a Contemporary Hit format. They had playlists. They had a "wacky" morning zoo crew. They played (gasp!) Lionel Ritchie. I didn't believe him. I called 'MMS myself. It was all true.
>
> What I couldn't figure out was how they continued to win the *Rolling Stone* Reader's Poll every year. At first I blamed the city. I thought the people of Cleveland were so spineless, so wishy-washy and indiscriminate about their music that they'd listen to anything that goes by the name WMMS. That I found very depressing.
>
> So now it's all about aggressive marketing and merchandising. 'MMS buys the right to be called "the best radio station" in the nation. So big deal. It doesn't mean anything to the listeners. Just to the people who sell advertising time. Now instead of new and interesting rock music we get hit singles from the *Billboard* chart. We get Jeff and Flash telling their dirty little "token" jokes, yuk-yukking it up. We get the kind of stuff that was embarrassing in junior high school.
>
> I don't want a radio station that wins awards. I don't want a radio station that wins anything. I want a radio station that plays good music. Plain and simple. I think a lot of other people do, too. How about it?

In many ways Heaton's article was more damaging than the front-page story the day before. Deep down a lot of people felt the same way, but they weren't comfortable admitting it.

Meanwhile WMMS found an unlikely ally in *Scene* magazine, which had faced its own battles with the station over the years. In a full-page editorial in its March 3 issue titled "Much Ado About Nothing," associate publisher Keith Rathbun wrote,

> Next, maybe, the *Plain Dealer* can do an expose that Cory Snyder's family and friends filled out Gillette All-Star ballots at the Stadium last year. . . . In fact, I hope the *PeeDee*'s Heaton doesn't find out that . . . oh, OK, I confess: I filled out

multiple entries to win a free skiing trip at last November's Ski Show. I know the sign said only one entry, but I thought it was OK because for every contest *Scene* does and requests just one entry per person, everybody sends in at least a couple. And anyway, I wanted to win the trip. I even filled out a couple in my wife's name. Yeah, I forged her name. I'm coming clean because I know it's just a matter of time until the *Pee Dee* uncovers the truth.

Nobody in the media has watched the WMMS machine more carefully—nor been more critical of the station—over the years than *Scene* has. We—*Scene* and WMMS—are not the best of friends; we both have the battle scars to prove it. But *Scene* saw absolutely nothing of substance in this story to report. The *PeeDee* uses terms like "cheated," "rigged," and "fixed." Those are words used to describe "Watergate," college basketball point-shaving schemes, and Panamanian presidential elections. But not a radio promotion that is focused on winning an unofficial "Best Radio Station" award in an unscientific poll by a subscription-driven magazine.

There was nothing illegal about what WMMS did. The fact is that there is nothing more to this story than reporting on a radio promotion. *Scene* recognized that fact nine years ago, the first time that WMMS approached us about doing a feature on the award. We didn't write the story nine years ago, and we turned a deaf ear on the station's same request every year since . . . Primary among *Scene*'s editorial priorities is to separate promotion from news.

One publication that has even a more difficult time with that than even the *PeeDee* is the *Rolling Stone*. How else could they include such a meaningless category in nothing more than a self-serving poll in the first place, and for each of nine years since? *Rolling Stone* Managing Editor Jim Henke even admitted to the *PeeDee* that the primary reason the magazine includes the category in its poll is to increase promotion for the magazine in general, and for the Readers Poll, specifically. Henke also admitted that the Readers Poll is highly vulnerable to "ballot stuffing." This year the category was expanded to include sub-categories for "Large," "Medium," and "Small" markets, thus allowing not only for greater promotion of the magazine but increased probability of "ballot stuffing," as well.

Maybe the promotional benefits to the *PeeDee* justified its editorial placement. Let's see, the story resulted in getting advertising revenues of about $4,500 from WMMS for a full-page rebuttal on Friday, and probably in the neighborhood of a free $30,000 radio promotion blitz as WMMS and each of the station's competitors wasted their valuable airtime discussing the *PeeDee*'s story. But, nah, that couldn't have been a consideration. That would be unethical.

But *Scene* had its critics, too. In subsequent issues readers sent in letters suggesting, "WMMS ceased being a 'Rock Station' five years ago, yet has continued to pass itself off as an innovative, trend-setting radio programmer . . . no amount of zany morning jocks or $10,000 birthday giveaways is going to alter that status." Others said, "We all know that if it had been any of the 'competitors' who had stuffed ballot boxes to win, WMMS would have spoken out very strongly about it, saying 'cheat, unfair, etc.'" WMMS had actually done just that years earlier when the station found out that crosstown rival WGCL-FM had been speeding up music to get in more songs per hour, freeing up additional ad time. WMMS exposed WGCL on the air.

Another letter suggested that *Scene* stood up for WMMS because it was a source of advertising revenue. Yet another said it was simply because *Scene* didn't break the story first. Other letters targeted the station personnel, with one saying, "One has to feel kind of sorry for Jeff . . . Flash, Kid Leo, et al., aging has-beens reduced to playing Michael Jackson and Whitney Houston for housewives and teeny-boppers, trying desperately to buy back their once-glorious past." Another reader noted an illustration on the station's latest t-shirt, pointing out, "It shows the Buzzard character leaning against a nailed up crate labeled 'Rock and Roll.' It's too bad they don't open the crate and allow people to hear it . . . What about those of us who are fed up with stale Stosh jokes, Mr. Leonard, and two guys who sit and play disco dance music in a pink Cadillac every weekend as a morning (or is it mourning) show." Not all of the letters were anti-WMMS, as some suggested that the station get a second chance. Ultimately the ratings would determine which station was in fact the "best."

WMMS wasn't the only party affected by the controversy. *Rolling Stone* saw the annual WMMS win as a springboard for promotional opportunities with the station, having representatives call WMMS to propose various tie-ins to the poll. Plans for a regional issue with a special WMMS cover were quickly scrapped. *Rolling Stone* responded to the uproar by announcing plans to remove the Best Radio Station category from future readers' polls, at least for the foreseeable future.

And of course winning the poll instilled a great sense of pride among the WMMS staff. Calling back to his days as assistant program director, Denny Sanders recalls that WMMS won the contest fair and square in the first years. Then-business manager Judy Goldberg corroborates this, "As the years went on, we said, 'Okay. Let's make sure we win that contest,' and we sat around and filled out a bunch of ballots and sent them in. It might have been just the last couple of years that we even filled them out. We just got a big kick out of it, and we never felt we were doing anything illegal. WMMS never got anything from it. We didn't even get a plaque. What's the difference with filling out all star ballots every time you go to a baseball game?"

Goldberg also points out that Malrite was a public company at the time, and it didn't look good to stockholders when the story hit the *Wall Street Journal*. She explains, "What they didn't understand is that these were a bunch of kids at a rock and roll station having fun, and all we ever wanted was the notoriety for Cleveland and WMMS."

It was common knowledge among many WMMS staff members that the station was openly going to accept the designation for ten years, and then pass it on to another station. But it was in that tenth year that the story broke, ending the hope of that plan succeeding.

WMMS management wasn't oblivious to what was going on. The station was carefully run, and nothing happened in a vacuum. There were many people in on most of the decisions, right or wrong, however only a few were held responsible. Gronek recalls, "I knew there was always some manipulation of [the *Rolling Stone* poll], but it wasn't until I got to the general manager's position that I understood the scope of that effort. I remember sending a memo out to the six or eight people who were influential in deciding issues about the station. To a person, I got all those memos back . . . and from people in high positions . . . saying we unquestionably had to continue doing it."

Even with the full knowledge throughout the station and its management, a problem such as this wasn't going to go unpunished. Promotions director Maureen Duffy and contest coordinator Steve Legerski soon found themselves out of jobs. The official reason for Duffy's dismissal was because of reported "attitudinal problems." However she was told her firing was part of a "staff realignment." Legerski was dumbfounded, saying his department was always efficiently run. Duffy was replaced by Rich Piombino, who would later rise to the position of program director and later, operations manager.

The station itself didn't get out of this mess unscathed. WMMS weathered the storm and though ratings were still very strong, its image took a battering that would last for years to come.

A mandatory staff meeting was called at the Statler's Competitors Club on an evening in early March. Its intention was to introduce the new staffers, to discuss some new promotions, and mostly to address the controversy around the negative headlines. There was a sense of urgency and electricity in the room, with many angry staff members calling for some sort of retaliation. After Kid Leo introduced the new members of the crew, he gave the floor to station owner Milton Maltz. Maltz was a very engaging speaker with an occasional flair for the dramatic. Most of his employees, especially the veterans, had a deep and unwavering loyalty to him. Everyone felt an unreal tension just before he spoke, as if the room would explode at any second.

Maltz started by describing the backhanded threats by the competition that led to the *Plain Dealer* expose. He then calmly but very firmly vowed that WMMS would prevail as never before. The staff erupted into cheers of approval, with everyone applauding, yelling, and even stamping their feet. The noise level could be likened to a crowd at a rock concert. The buzz of the evening's speech continued as the staff lined up at the bar, ready to toast the future of WMMS.

The *Plain Dealer* still had its say in the days after the initial fallout. In an article headlined "Throwing Good Money After Bad," David Sowd wrote in his column that WMMS ran a half-page ad in the new issue of *Rolling Stone* that stated, "Thanks for Your Votes!" The ad was said to have cost about $17,000. Sowd noted, "An interoffice memo would have been cheaper!" Talk of the controversy wasn't the only mention the station received in the newspaper that day. A week after WMMS printed its response in a full-page advertisement, the *Plain Dealer* printed another full sized ad. Bold lettering stated, "A new Buzzard is about to land in Cleveland." It was an Ameritrust Bank ad, featuring a David Helton drawing of the Buzzard holding the newly offered WMMS MasterCard. Advertised as "the most rock-solid credit card in Cleveland," WMMS was the first station in the country to offer that kind of promotion, something the station's founders would never have dreamed twenty years before.

A credit card wasn't the only product the station affiliated itself with. In a bizarre yet ingenious product tie-in, WMMS teamed up with the Pick-n-Pay supermarket chain to premiere Buzzard Morning Zoo Corn Flakes in the spring of 1988. The grocery chain simply packaged its generic brand of cereal in a Helton-designed *Morning Zoo*-themed box. Pick-n-Pay picked up the costs of production, packaging, and shipping. The chain got all of the profits, as well as increased consumer traffic from the frequent mentions on Cleveland's dominant morning radio show. The novelty cereal didn't cost the station a dime, yet it put its call letters and flagship program right into people's homes. The box featured caricatures of Kinzbach, Ferenc, Roberta Gale, and Len Goldberg against a backdrop of the Cleveland skyline. It also offered a recipe from Mr. Leonard. And in true breakfast cereal fashion, there was even a send-away offer for a *Buzzard Morning Zoo* cup and bowl set on the back. Orders for the cups and bowls, as well as for *Morning Zoo* t-shirts designed by Helton, poured into the Statler Office Tower.

Promotion for the cereal didn't stop there. Since it was still essentially the store's generic brand, the eighteen-ounce box sold for just $1.29, well below commercial brands. Only 25,000 boxes were printed in the first run, as the promotion was only scheduled to run for six weeks. This created an immediacy to snatch them up before they were all sold out. The *Morning Zoo* crew continued to help promote the cereal by doing three, two-hour, in-store appearances, autographing boxes and giving away prizes with games similar to *The Price Is Right*. An invitation-only get-together with the morning team was held and broadcasted live on the air. Winners were served a special breakfast of Buzzard Flakes, receiving boxes to take home as well. The cereal promoted goodwill with a major advertiser, put the call letters in front of the shoppers, and increased store traffic. The expected flak came once

Morning Zoo Cornflakes (1988). The WMMS influence was so great at its peak that it even made its way to the aisles of local grocery stores. Photo by Brian Chalmers

again from Sowd, who griped about the promotion in his column. In a piece titled "Gag Me With A Spoon" he wrote, "The big bird is back in your gullet with its own brand of breakfast cereal . . . a food stuff that gives new meaning to the term empty calories."

When the promotion finally ended, there were still plenty of unsold boxes. Pick-n-Pay marked them down to move them out. Years later those same cartons were going for as much as $200 among cereal box and nostalgia collectors.

The Buzzard Morning Zoo Corn Flakes cereal was the last promotional hurrah the *Morning Zoo* crew enjoyed with the company of Gale. Sitting just a couple of blocks away at a company-leased apartment on 13th and Chester was the woman who had been hired to replace her. Lisa Dillon was getting up very early to listen to every minute of the *Buzzard Morning Zoo*.

Dillon was working in Detroit when she heard that WMMS had liked her tape and that the station wanted her to come in for an audition. She recalls, "I had three interviews before I got the job, with Jeff and Flash first, then Jeff McCartney and Leo, followed by John Chaffee and Gil Rosenwald. They hired me in April, but I didn't go on the air until the end of May because I sat in a suite at the Reserve

Square apartments listening to Roberta for an entire month trying to figure out what they didn't like about her. It sounded like she was trying to be Joan Rivers, but I decided the best way to approach the job was to just be me. It worked out very well."

Dillon wasn't the only new voice on the local airwaves. In Akron, WONE's Brian Fowler was paired up with an old friend, Joe Cronauer. A native of Cleveland's Southeast side, Cronauer had quite an interesting connection to the history of radio. A distant relative of his was controversial Armed Forces Radio Network deejay Adrian Cronauer, who was portrayed—albeit inaccurately—by Robin Williams in *Good Morning Vietnam*. Joe was called up from a position at a Christian station in Nashville. While it took some time to perfect a formula and chemistry, the team of Fowler and Cronauer were poised to become a major force in Northeast Ohio radio.

Scene magazine decided to have some fun with the local stations in their April Fool's issue, with articles such as one jokingly suggesting that Leo was leaving WMMS to program an all-Springsteen station in Asbury Park, New Jersey, called WBOS. Another piece featured in the regular "Makin' the Scene" column suggested that local writer Peanuts, aka "The Host of the North Coast," was actually the voice of WMMS's mysterious Mr. Leonard. It drew an immediate response from both parties. Mr. Leonard wrote in, "How can you dare confuse two such diverse people! . . . I'm often referred to as a lean, mean, loving machine, a master power flash connector, a yearning churning, burning hunk of funk and an all around ultra chic and ready to peak kind of guy of the '80s." Peanuts wrote in that same issue, "You can tell us apart in person because I'm much better looking . . . Many of us were under the impression the March 3, 1988 issue of *Scene* was the April Fool's Day one. You mean Keith Rathbun's editorial on page two wasn't a joke?" It wasn't smart to write about a station that had that big a voice in the community, and could slam its enemies in front of hundreds of thousands of listeners at once if it chose to. WMMS also had a long memory when it came to whom its friends and allies were.

What wasn't a joke was WHK-AM's plans that month. The station, which had adopted the 14K "Top-40 Oldies" music format that had brought it fame in the heyday of AM radio, was trying a bold experiment in the wake of hard rock station WCZR-FM's demise. It was a collaboration between WHK and Belkin Productions called Powercord. It premiered on April 16, taking the 9:00-to-midnight block. In a *Scene* article announcing its debut, Leo called it a "highly interactive program." Belkin's Barry Gabel said that heavy metal fans "are the most loyal listeners. Maybe they don't get [Arbitron] diaries, but they're out there. And the reason we know, is that without having a place to advertise these metal shows, they sell out." Leo likened AM radio to the emergence of cable TV. He explained, "Where they would superserve a small segment of the populace. You're going to see more and more AM being the specialized formats, and FM will be more mass appeal format." He continued, "It's not a question that we didn't feel it was right for 'MMS, but rather, to look at the challenge of doing something for the AM side. We're definitely committed to the show. It's something we find thrilling in radio again."

The Powercord shift featured three special programs. The first was the *Big Band Theory*, tracing the career or genesis of a metal act. It was followed by *Strap Yourself Into the Electric Chair*, which asked listeners to judge two different acts. The last show was *Dr. Destructo's Screamer of the Night*. A take-off on Goldberg's WMMS character "The Boom," the host looked at a new artist or song that promised to be huge. The original host was not a Dr. Destructo but in fact Iron Mike, portrayed by WMMS's Brad Hanson. Ric "Rocco the Rock Dog" Bennett, the part-time disc jockey who was also part of the crack engineering crew at the station, later joined Iron Mike. Harkening back to the days of radio dramas, it was up to the listener to determine what the hosts might look like. As the station's normal Top-40 programming ended for the day, a humorous disclaimer preceded the transition to hard rock, warning the listener what was about to transpire and instructing "Get out now!" in a heavily modulated voice. The Powercord format wasn't guaranteed to be a permanent fixture, as WHK management was looking at number of different formats to revive the station. Heavy metal was only one option. Ratings would be the deciding factor.

As WHK was trying to reinvent itself, a former Cleveland radio personality was hoping that the city could forgive. Former M-105 morning man Marty Sobol had left Cleveland radio six years before for Cincinnati's WSKS-FM. When he was interviewed by industry magazine *FMQB* he was asked, "Did you leave your Michael Stanley albums in Cleveland?" Sobol replied, "Absolutely!" This jovial comment would come back to haunt Sobol in 1988 when he came in to interview for the job left by the departure of program director Brian Phillips. As he approached the desk of receptionist Verdell Warren, Sobol was greeted with an autographed photo of Stanley. He didn't get the job.

Seeing the success WMMS had with its Buzzard character, WNCX attempted to promote its "Classic Hits" format by introducing a cartoon mascot Classic Cat. The character showed up on some t-shirts and print ads, but was retired in a fairly quick fashion.

Meanwhile WMMS continued its successful promotion techniques. The station presented more Appreciation Day Shows, which was odd considering they were usually done to thank listeners for the annual *Rolling Stone* honor. Nonetheless WMMS hosted a concert with Michael Bolton at the Front Row Theater. Bolton was already a WMMS show veteran, appearing years before with the hard-edged group Blackjack. Though he was now making a name for himself on VH-1 and other outlets as a soft rocker, WMMS still embraced Bolton as a chart topper. It was still free music for the public, and to some that's all that mattered.

The fresh voice of Mr. Leonard had proven to be a hit on WMMS, breathing new life into the *Morning Zoo*. But no one knew what he looked like. Listeners were teased that May when WOIO-TV's horror hosts Allen "Frank" Christopher and Robert "Drac" Tokai promised an "appearance" by the station's "assistant public service director." Of course, Mr. Leonard ended up avoiding the camera in his usual creative way. Many listeners didn't know that Mr. Leonard would sometimes do his radio bits live on the street, visible to the public, but no photos or video ever made it to the media. Adding to his mystique and "position" at WMMS, listeners would call the station and occasionally hear Mr. Leonard utter off-color comments.

Mr. Leonard wasn't the only WMMS personality that was making a name outside of radio. Bruce Springsteen honored Leo with a special invitation to co-host a worldwide satellite concert from Stockholm on July 3. It would be the largest audience for a single artist ever in broadcast history. Leo shared hosting duties with WNEW's Scott Muni, formerly of Akron's WAKR-AM. This wasn't the first time Leo was asked to help out on Springsteen's Tunnel of Love Tour. He had already acted as the carnival barker at Springsteen's two sold-out concerts at the Richfield Coliseum earlier that year. After a quick tour of Stockholm, Leo returned home to his afternoon show at WMMS.

WMMS itself was doing quite well, hammering away at the competition through the summer of 1988. The station presented Lita Ford in a special show at its Beach Party '88. This was also where WMMS introduced the latest in its marketing line, a full-sized beach towel featuring the Buzzard and a date dancing on the shore.

Just weeks later the giant BP Sohio Riverfest was set to get under way. Radio stations across Northeast Ohio scrambled to make their presences known. This event presented an opportunity to be seen by hundreds of thousands of people celebrating the season on the banks of the Cuyahoga River. WMMS put the *Buzzard Morning Zoo* on location for a live broadcast to open the event, which in itself usually drew an impressive crowd, and Ruby Cheeks was in the middle of it later that Friday. Knowing it would draw an overflow crowd the station announced yet another listener appreciation concert, hosting Henry Lee Summer and Timbuk 3 at Nautica Stage. In just three short years the Riverfest had become one of the premier summer events of the Midwest, dwarfing longer-established festivals like those seen in Buffalo, Baltimore, and other towns. Along with a heavy dose of rock it offered a free performance by the world class Cleveland Orchestra, as well as a huge parade of lighted boats, hundreds of displays, and even a "Lobsterfest." A great deal of its success and growth could be linked to the support shown by local media, which used it as a showcase for its own products. The last weekend of July 1988 drew more than 700,000 partygoers to downtown Cleveland.

August followed with a surprise move when longtime WMMS personality Matt the Cat was inexplicably pulled from his midday shift and put on nights, switch-

Matt the Cat and Lou Reed (1988). Reed was in town to perform when he stopped by the WMMS studios for an intense chat on the air with Matt the Cat and to talk about his book, *Magic and Loss.* Photo by Brian Chalmers

ing places with Tom Rezny. While Rezny had a similar easygoing and relaxed style, it didn't seem to make sense as to why Matt was pulled from a shift that he held for more than a decade. He was one of the most knowledgeable disc jockeys in Cleveland radio, had established himself as the very capable host of the *Coffeebreak Concert* series, and always delivered a thoughtful and entertaining show. Matt took the move in stride and did an equally professional job in his new time slot. But others wondered about the move nonetheless. There were rumors that his move was linked to a personal dispute with management. Not long after his shift change Matt got a call from the general manager at WNCX, offering him a midday show at the station. Matt was still under contract, so he politely declined.

Over at WHK, Jim Butler had proven himself to be a first-rate newsman, but by the late summer of 1988 he felt he needed a change of venue. He accepted an offer to be WERE-AM's afternoon news anchor. Butler seemed to be uncomfortable with the switch, and it wasn't long before he was looking for something on the FM dial.

Butler wasn't the only one in radio testing waters elsewhere. Murray Saul, the man who became famous by telling WMMS listeners to "Get Down" every Friday, signed on as *Scene* magazine's director of sales, helping to market the weekly publication throughout the entertainment community and the public in general. The magazine was eager to welcome him aboard, as his reputation from his days at WMMS preceded him. Saul's long experience in sales, broadcasting, and the record business made it the perfect match for both him and *Scene*.

But it was another potential move that became the biggest focus of conversation throughout the entire radio industry. Trade magazines such as *Radio and Records* had been trying to contact Leo all summer about rumors indicating that he was ready to leave WMMS. *Radio and Records*'s Joel Denver wrote in his August 19 "Street Talk" column that Leo wouldn't comment. Station program director Jeff McCartney responded to the *Plain Dealer*'s own inquiries, saying the paper should have simply asked Leo. McCartney said, "Have you ever known Kid Leo not to have a comment about anything?" Leo had been a pillar of strength through WMMS's periods of turmoil, a signature star in the station's finest hours, and had a solid presence on the national scene. Rumors no doubt grew from the fact that Leo had been working without a contract since the previous December, though his agent had continued talks with the station. A lot of people were crossing their fingers that those talks would be fruitful, as Leo's departure under any circumstance could prove damaging to WMMS's image.

In the middle of the speculation of Leo's future at the station, another WMMS personality decided to leave in early September 1989. On his final day a huge farewell party was held at Noisemakers for Bob Wright. His quick sense of humor and streetwise delivery made him a natural on-air foil to Mr. Leonard, and he was able to create characters such as the superhero "Blackman" during the craze surrounding the recent release of Tim Burton's *Batman*. In his time at WMMS Wright had worked his way up from the mailroom to promotions, even doing some weekend work in the air studio. But the time had come to move on to a better paying position outside the company. Wright's leaving may not have had as big an impact on-air as it did on the staff, as the sense of family was very strong at the station and the loss of yet another popular co-worker was very unsettling.

Wright wasn't the only WMMS veteran making a change, as a familiar name was set to debut on WMJI. Denny Sanders cracked the mike on September 19, ending a long hiatus that started the previous October when he was relieved of his duties at WNCX. He took over the 7:00 to midnight shift. WMJI management promised that Sanders would have greater leeway to expand his playlist to include artists outside of the station's regular programming. WMJI management took a slap at WMMS while announcing Sanders's arrival, telling *Scene*, "It's just another sign of the extinction of the once-great animal, the Buzzard."

WMMS was hit again on October 7, but it was more worrisome than a competitor simply taking cheap shots. The *Lake County News Herald* ran a piece featuring comments from the *Morning Zoo*'s Kinzbach. Headlined "WMMS's Kinzbach Admits He's Unhappy with the Format," columnist David Glazier quoted the morning co-host as being "dissatisfied with the direction the station has taken," further speculating that he "might be weighing his employment options elsewhere."

The comments were reportedly made to Glazier in an interview following the *Buzzard Morning Zoo* show the previous day. The article ultimately suggested that Kinzbach was not happy with the musical content favored by program director McCartney. Glazier quoted Kinzbach as saying, "Now we're playing a harder brand of rock and roll. Let's face it. Listeners in the 25-to-54 age group don't like getting their heads bashed in at that time of day, and a lot of them are tuning out . . . If we're going to continue to narrow down the music, then we might not be the top-rated morning show anymore." The article continued to say that Kinzbach couldn't reach a compromise over the *Morning Zoo*'s musical content, and it was suggested that it could influence his decision whether to re-sign with WMMS when his contract was up the following year. Kinzbach was quoted as saying, "This is still a great station, but there's no doubt about the fact it's down trending . . . Not everything great lasts forever. If the down trending continues, if they still ignore warnings by people who know this market, then I'd be crazy not to consider another offer if it was there . . . I'm a competitive guy. I like to win. Sometimes, when you hit a snag, it can get a little disheartening."

Kinzbach's comments at face value raised some eyebrows throughout Cleveland, particularly at the Statler Office Tower. Hoping to debunk any speculation of his leaving, Kinzbach explained in a memo to the staff,

> The conversation took place about a month ago. As usual, I denied the rumors that have floated around and answered some questions. There were some comments I made off the record to the reporter who I thought I could trust. Unfortunately, they printed those remarks and focused on them rather than the other more positive aspects of our conversation. It's just more MMS bashing, however, I am responsible for what I say to these people and hope that none of you will think it's true. The article basically dealt with me being disenchanted and

perhaps leaving. Sorry to say . . . I intend to stay. And, as you might have seen last week, the ratings seem to have turned around. Let's put it this way: I wouldn't want to be at NCX or MJI right now.

I would not use the press for anything but positive results for this station. Unfortunately, the papers tend to dwell on the negative. I am sorry it was printed and hope none of you will take it seriously. Nothing is easy. It's been a tough year for the Buzzard. But believe me, I can't wait to see their faces after this station rises and sticks it up their noses. Now if I can just keep the lid on this Robin Swoboda stuff. . . .

The last line was making light of published reports in the press that Kinzbach was seen dating the WJW-TV news anchor.

The WMMS staff rallied around Kinzbach. Still the article could prove damaging in the hands of rival stations, where sales departments could use the information to drum up potential business.

Malrite's television acquisition WOIO used a lot of sister radio station WMMS's talent. WMMS personalities would host WOIO marathons of *The Adventures of Superman* and *The Twilight Zone*, among others. That October, WOIO filmed a spot that even surprised and delighted folks at WMMS. It featured Rocco made up as a hunchbacked Igor-type character, ambling around an eerie-looking castle setting. As Rocco sauntered past a coffin, its lid slowly opened with a loud creak. A shadowy vampire started to emerge. It suddenly became apparent who was portraying the vampire, none other than Leo. He always seemed to be very image conscious, so this was a great departure for him. Reports from the television station came back saying Leo had a great time doing it, and he really seemed to be getting into the role.

There were yet again some new faces behind the scenes at WNCX. The Detroit-based Metropolis Broadcasting learned a bitter lesson coming into Northeast Ohio, apparently grossly unprepared to battle the ratings power and capital held by WMMS. The company sold both WNCX and sister station WERE to a trio of Cleveland broadcasting veterans. Former WIXY-AM owners Norman Wain and Robert Weiss, partnering up as Metroplex Communications, joined former WMJI owner Larry Robinson to assume control of the station on November 3. Considering the past victories Wain and Weiss had with WIXY, and Robinson's successful use of airtime for his jewelry business and influence on WMJI, it could be assumed that the three had high hopes for the station.

The headache of the *Rolling Stone* balloting disaster recurred at WMMS that November when the magazine's latest readers' poll hit the newsstands. *Rolling Stone* originally claimed that it would eliminate the Best Radio Station category after the *Plain Dealer* expose the previous February, but plans had changed. Managing editor Jim Henke told the *Plain Dealer*, "There was an outcry from other stations complaining that we were blaming all of radio for what WMMS had done. The feeling here is that we get a lot of promotion from radio stations, and there's no reason at this point to alienate them. So what we decided was to re-institute the category and define the rules more clearly, and enforce them strictly." Those new rules included the statement "Multiple entries would be discarded" and explained that all ballots would be inspected by an outside agency to ensure they were legitimate.

Malrite president Gil Rosenwald issued a memo to the WMMS staff, stressing that any official statements made to the media could only come from the general manager's office. It also offered a statement that could be used as a sort of guideline for any questions that might arise from friends, clients or listeners. The memo read,

> At no time did WMMS break any rules. The fact is that multiple ballotting was indirectly encouraged by the magazine much like multiple voting is heavily encouraged in major league All-Star voting in several sports. Unfortunately, there were some who saw this as an opportunity to try to taint or belittle WMMS for their own personal gain or find news where there was none. We feel the whole situation was much ado about nothing. We also feel there are those who will continue to attack us whether we participate in this year's survey or not. Thus, we've chosen not to participate. WMMS is known around the country as a great radio station without the benefits of a magazine survey.

It was true. WMMS was well known outside of Cleveland, though it couldn't be denied that the *Rolling Stone* ballot mess would give the station a black eye that people would snicker about for years. Despite the problems WMMS faced when the story first broke, things were still going pretty well. Ratings were solid, the station was still receiving good press, and the deejays were getting lots of exposure. Lisa Dillon even appeared on stage with Rod Stewart at his Richfield Coliseum concert when the rocker presented a $5,000 check to Cleveland's Health Care for the Homeless project, one of several such groups to which he pledged a portion of the royalties from his single, "Forever Young."

The station continued its own promotions through the end of the year. The *Buzzard Morning Zoo* was handing out thousands of dollars in prizes every Thursday in its Easy Money Birthday Contest. A sealed envelope was opened on the air to reveal a birth date. Those who shared that birthday could call in to win. There were also big prizes offered weekly for the best Token Joke of the Morning, as well as other call-in opportunities for records, CDs, shirts, and a seemingly endless flow of giveaways. WMMS also continued its long-standing free concert series, presenting a promising new Dallas band at the Phantasy Theater called Edie Brickell and the New Bohemians.

A special guest stopped by the WMMS studio in early December when *CBS This Morning* weatherman and entertainment reporter Mark McEwen did his reports from Cleveland. Though he had family to visit in the Cleveland area, the former disc jockey made it clear that the highlight of his stay was being on Leo's afternoon program. While having fun chatting and goofing around with Leo, McEwen even stayed to do a special weather report for the show. McEwen had been a WMMS fan for years, and when he would drive from Detroit to his Maryland home he made a point to slow down through Northeast Ohio to catch as much of the station as he could.

If there had been any sign that things were about to change, then Leo hid it well in jovial segments such as that with McEwen. But still no amount of radio fun could soften the blow felt by the Cleveland radio community when the announcement came on December 15 that Leo was leaving WMMS. The next day's show would be his last.

PART FOUR
THE CHANGING OF THE GUARD
1988–1999

Baby, I Was Born to Run

It was no longer a rumor. Kid Leo was in fact on his way out. The bottom line was that Leo had a richly deserved reputation, not only as a disc jockey but also for spotting and breaking new talent. This reputation gave him a lot of close friends in the record industry. CBS Records president Tommy Mottola had approached him in August about taking a position with the company's division of promotions and artist management. The job would mean that Leo would have to relocate to New York. After weighing the benefits for his wife and kids, as well as himself, Leo decided to take Mottola up on his offer. It was simply an offer he couldn't refuse.

As he put his family's best interests forward, there was still another family Leo was loyal to, the WMMS staff and the Cleveland radio audience that embraced him for fifteen years. The station tried to offer its listeners a silver lining by promising to eventually feature Leo in a weekend show called *Cleveland Classics*, taped at Malrite's New York station WHTZ-FM. In the meantime this still meant that someone else would obviously be taking over his afternoon drive show.

Leo's departure was officially announced to the public on the *Buzzard Morning Zoo* and in a *Plain Dealer* article. Reaction among the WMMS staff members could best be described as quiet disbelief, and the prospect of losing one of their key players hung like a black cloud over everyone's heads. Theories were traded in the halls about why he was leaving. Was there a disagreement with management? Was there some cataclysmic change in the works that Leo didn't want to be part of? Or did he just want something new? It was anyone's guess, but there was one thing for certain. It would be a very emotional afternoon at WMMS.

Leo came into the station as he did everyday. Everyone on the staff from sales to programming eventually made their way to the studio to say goodbye and a few words of congratulations for a job well done. It promised to be a very intense final show.

The calls started coming in right from the start, with Neil Giraldo, Billy Joel, Southside Johnny, Eddie Money, and so many others phoning to wish him their best. The Cleveland Browns' Bob Golic came by the studio as well. John Mellencamp called to salute Leo, saying, "When I first started making records, there was only one place in the country playing them, and this guy here was one of the guys that took a shot, and we've been friends ever since. I hate to see him leave Cleveland, cause nobody will play my records anymore!"

As star-studded as this final show already was, there was still some question whether Bruce Springsteen might call to wish Leo well. Springsteen was said to be very distraught over the passing of his good friend Roy Orbison, so there was the chance that he might instead call after the show. Still Leo asked if the afternoon news reports could be postponed on the off chance that Springsteen called during one.

A few minutes later the phone rang, and sure enough a familiar Jersey accent was heard asking, "Hey, Kid! How you doing? You comin' to New York?" Springsteen always made a point to spend a few quality moments with the deejay who helped make him a star in Cleveland, and their mutual fondness could be heard in their voices. Leo and Springsteen spoke and joked around on the air like two guys from the neighborhood talking about the good old days. Springsteen asked, "If you're this close to New Jersey, does this mean I gotta see you all the time now?" Leo responded by inquiring about the guest room at the musician's house. Turning serious for a second Springsteen pointed out, "Cleveland was a good town for us right from the start, and I want to take this opportunity to thank you for your support, and for playing all my records all the time." Leo's final show would still have been special, but not nearly as memorable without the tribute from Springsteen.

There was obviously no set playlist on Leo's last day, as he played everything from Willy De Ville, Roxy Music, and even Nat King Cole's "Christmas Song." In the final forty minutes Leo admitted that the emotions were all starting to catch up on him, a feeling very likely shared by a lot of the audience. He became far more subdued as he went into that last segment of his show.

Leo kept stressing that his farewell bow shouldn't be considered a funeral but more like an Irish wake. Nonetheless, when he went to the commercial break to "punch out, wash up, and come back to wrap it up," there were no doubt listeners glued to their radios to hear what would likely be one of the saddest moments in local broadcasting. When the break ended Leo announced,

> For the final time, I'm saying "Later, people." Before I do, it's been sixteen years, and the decision to go to Columbia Records was a tough one for a couple of reasons. One, the family I have here at WMMS. It's an exciting opportunity for me, but leaving these people is not easy. It was a decision I tossed and turned over for a few months. The other thing that's so hard is leaving Cleveland. The buildings, yes, but the people, that's the major thing. We've been in this together. As I've said a few times on this show, I have an opportunity for a great career challenge, but only because you took me into your homes. You liked the music I played, and accepted me as someone who knew about rock and roll, and loved it. That's for sure. Therefore, you gave me a claim in the industry, and you allowed me to have challenges like this.
>
> It's exciting to have gotten to know all these people, like the ones who called in today, and they're all very special to me, and it meant a lot.
>
> The final thing is, there are a lot of ways that people measure wealth in life. But if you measure it by friends, then I consider myself one of the richest men

Kid Leo exits the studio after signing off for his last WMMS show (1988). Photo by Brian Chalmers

in the world. I thank you for sixteen great years. They've been the greatest years of my life. I can't think of any better way to end it, and you wouldn't have it any other way on a Friday, than "hup, two, three, four." You're off, and running, and I will see you. Later people. I love you. Thank you.

And with the opening chords of Springsteen's "Born to Run," the Cleveland audience heard Leo draw the curtain on his long and successful run at WMMS.

Leo's departure left an immeasurable impact on WMMS. The exit just two years before of Denny Sanders, John Gorman, Steve Church, and others to WNCX had already struck a terrible blow, but the station still managed to hold it together and prosper. Leo's announcement came as a roundhouse punch. His swagger, confidence, and panache made him a true Cleveland original, and no one could fill his chair. Leo attended a party in his honor later that night, and when he returned to his car he found his office chair in the back seat, courtesy of some very clever undercover moving from engineer Rocco the Rock Dog and promotions man Jim Oktavec.

In a memo to the staff, Leo poured out his heart, stating:

There's no good way to say goodbye, so I won't even attempt it. What I want to say and what must be said is that I will always consider WMMS my home and all of you as my family. It happens that people do leave home but they never, ever leave their family. We've shared a lot including our rise in achieving the #1 slot in Cleveland (and the nation as far as I'm concerned); the days we did other stations' promotions better than they did; the time I almost broke Verdell's leg by slamming a car door on her ankle; the Buzzard Appreciation Days; convincing Milt and Gil that buying out the Who's Coliseum show was a guaranteed return on our money (I had a flight booked that night); rationalizing to Milt and Gil the fact that even though we lost money on buying out the Michael Jackson show we'd make it back tenfold in publicity (I still owe 'em on that one); being called off of the golf course because Jeff and the Zoo aired a tape of expletives from a local record rep and gave out the record company's local number on the air in retaliation (while I was about to shoot par for the first time in my life)—I almost shot Kinzbach; hearing the announcement of my children's births upon the air (my proudest moment as a proud papa.).

We've shared, and I care. Sure I've screamed, yelled, swore and threatened your lives with my baseball bat, but only because I cared and I've felt that caring returned one hundred point seven times over and over again.

As I've stated: "There's no good way to say goodbye . . . so I'll say 'Later People.'"

Ruby Cheeks was selected to replace him, at least for the time being. Cheeks recalls Leo's high-spirited nature. He was sometimes prone to express himself in tense situations by angrily throwing things in the studio, what Cheeks called "Leo Fits." She would tease him by saying, "You know, Leo? One of these days you're going to have one too many Leo Fits, pop a blood vessel and have a cerebral hemorrhage. You're going to stroke out, and I'm going to get your show!" Fortunately she got his slot on happier terms. Cheeks was terrified at the prospect of taking over for a local radio legend. She remembers station owner Milton Maltz coming up to her in the hallway just before airtime and saying, "Rube. If anyone can do this, you can. Just remember one thing: Onomatopoeia!" She apparently did remember it, quickly becoming very comfortable before she started her new show.

Leo stayed in town for a few more weeks, tying up loose ends before heading to New York. He sat down for a long interview with *Scene* magazine to reflect on his career. He explained the new job was designed especially for him, saying that he might use his name recognition to go out as a sort of "rock and roll ambassador" for some of the acts he would represent. Leo also commented on the 1986 staff defection to WNCX, likening the ensuing battle to "a Civil War." He also took a swipe at corporate radio, "I think broadcasters should run broadcast outlets, not bankers."

WMMS had continued and prospered after a major departure before, and it would no doubt do so again. But no one could deny that the station had changed forever at 6:00 P.M. on December 16, 1988.

The Numbers Game

WMMS and WMJI-FM had been locked in a long and bloody battle for ratings and especially ad revenue, so no one really expected the strange alliance that took place on January 1, 1989. It was a joint operating venture between Malrite and WMJI's parent company, Legacy Broadcasting Partners. Called Radio One Marketing and based at the Statler Office Tower, it offered one stop shopping to advertisers by combining the sales and client service staffs of both stations into one unit. The stations would each have separate ownership, but the combined sales staff could give sponsors the opportunity to advertise on both. It also reunited Malrite founder Milton Maltz with the man who started Legacy, Carl Hirsch, who had left Malrite years before to start his own broadcasting firm. The danger in that agreement soon became evident, as the sales staff would now hear rumors, staff movements, and promotions from both stations, and that information had a way of getting back to the opposing side. One rumor had been that WMMS's *Buzzard Morning Zoo* would relocate to an unnamed New York station. The departure of Kid Leo left an uncertainty about the future of WMMS and its key players, so anything was possible at that point. Actually it was program director Jeff McCartney who left the station on January 20, replaced by promotions director Rich Piombino.

The results of the fall Arbitron ratings book were the cause for jubilation at some stations and concern at others. The *Buzzard Morning Zoo* and Tom Rezny's midday show, both on WMMS, far outdistanced their nearest competition, especially in the big-money age group of twenty-five to fifty-four. But afternoons and evenings on WMMS took a hit.

The *Cleveland Edition* took a few editorial swipes at the continuing relevancy of WMMS and rumors about its future in the paper's January 26 issue, prompting an angry letter from *Morning Zoo* co-host Jeff Kinzbach. Kinzbach stated that he was staying in Cleveland, applauded the recent staff changes, and stressed that no other station in town cared more for its listeners.

Other local publications took shots at the station. The *Plain Dealer*'s David Sowd called WMMS's afternoon show "particularly forgettable" and lamented the departure of personalities such as "Spaceman Scott" Hughes. The search for a new afternoon personality was well under way, with names such as Rocco the Rock Dog and Michael Stanley coming up for consideration. Ads were even placed in the trade magazines.

There was news concerning other shifts as well. A familiar voice from the station going all the way back to its earliest days became a full fledged staff member on February 11, as Alex Bevan brought his wit and music to host the 6:00-to-9:00-A.M. Saturday show. Unfortunately Bevan was eventually forced to give it up because of his recording and concert commitments. Saturdays also saw *BLF Bash* host Bill Freeman take on additional duties hosting *Cleveland Breakout* from 8:00 to midnight, a program showcasing local talent and re-establishing the station's reputation for breaking new acts. On the weekday front WMMS signed Kinzbach to a new five-year contract, ensuring he would continue with the station during a very critical time to come.

In the middle of all this shuffling, re-signing, and adding of personalities, speculation started to center on the future of another WMMS staff member. John Rio's Mr. Leonard character had become one of the most popular segments on the *Buzzard Morning Zoo*, and Rio's move to Richmond Heights at the end of 1987 was seen as a confident sign that he intended to stay with WMMS for a long time. But that was before his good friend Scott Shannon, who helped create the *Morning Zoo* format while at New York's WHTZ-FM, decided to head to Los Angeles to become vice-president of programming for the Westwood One Radio Group. Shannon was also to take over morning duties at KIQQ-FM, which the company had recently acquired. He was expected to make the same impact in Los Angeles that he did in the Big Apple, for a reported $2 million a year. The industry magazine *Inside Radio* suggested that Shannon might lure Rio to the West Coast with him, bringing his tenure at WMMS to an immediate end. Rio had already been heard on Shannon's syndicated *Rockin' America Top 30 Countdown* series, so if he left he would prove to be a key player in the Los Angeles market. The only problem was Rio was still under contract to Malrite, and the company wasn't likely to just let him up and go to a competing firm. With all of the major departures the station had seen in recent years, rumors such as this planted a seed of doubt among everybody keeping tabs on WMMS to see who might jump ship next.

The potential loss of Mr. Leonard was the least of problems for WMMS management, sales, and air staff in mid-February. The *Plain Dealer* ran an article by Sowd and Michael Heaton, the reporters who broke the *Rolling Stone* poll scandal the year before. Headlined "Analysts Circling as Buzzards Look for Answers" and "Discord at WMMS Leads to Massive Staff Shakeups, Departures," it suggested that the station's troubles really began with the staff defection back in 1986 to start WNCX, and that WMMS may not have fully recovered from that migration yet. They spoke with WMMS's former operations manager John Gorman and former assistant program director Denny Sanders, both of who were at the front of the major move to WNCX, and former general manager Walt Tiburski. Tiburski hinted that the success of Malrite's WHTZ might have been a deciding factor in WMMS's freefall. Tiburski said, "With the unbelievably quick success of Z100 in New York, Cleveland and WMMS became second class citizens. And that was Malrite's mistake: to continually shove Z100 and Scott Shannon in the face of kids who had knocked themselves out and got paid slave wages to make WMMS. It caused so much resentment and hostility you could almost cut it with a knife." He pointed out that listeners didn't see WMMS as "anti-establishment underdogs" anymore, saying they might have outgrown the format. The article didn't paint WMMS in a particularly favorable light. Though some at WMMS claimed it was a tad biased, the majority of the staff whispered that much of it was right on the money.

The station took another hit in the press on February 24 when the *Plain Dealer* headlined a story "MMS Plays fair, Drops to 4th." It concerned the latest *Rolling Stone* readers' poll featured in the magazine's March 9 issue. The poll showed WMMS placing fourth in the Best Radio Station (Large Market) category. The station had made it clear that it would do no campaigning for a tenth straight win, yet it still placed fairly high in that survey. Surprisingly the stations that placed first and second were Philadelphia's WMMR-FM and New York's WNEW-FM, which were both owned by former Malrite president Carl Hirsch.

A fairly familiar face had settled into Northeast Ohio in late February, as former Humble Pie drummer Jerry Shirley relocated to be near his girlfriend. Shirley had a new version of the group lined up, but he was able to make time to take over the Saturday morning shift at WNCX. He called the two-hour show *The British Invasion*, in which he played music and spun yarns about his life in the music industry. Station management announced they would fly Shirley in from any part of the country he might be touring, so he could do the show live in the air studio.

Shirley wasn't the only person in Cleveland to get a brand new show. Maria Farina had been doing weekend and fill-in shifts at WMMS since 1986, in addition to a lot of personal appearances and club work, but she always wanted a chance to host her own show. In early 1988 she got a call from rival WPHR-FM. Farina remembers, "I wasn't a big radio head, so I didn't really listen to any other station in town. All I knew was that I wasn't happy at WMMS, because it was very tense there with all the changes. I was still a novice to radio, and didn't feel like I was betraying anybody. I didn't know station politics, but I knew I would be able to stay in Cleveland."

Maria Farina (1987). She came to Cleveland to be a Buzzard, but when the opportunity arose to make a name on her own, Farina made the tough decision to do mornings at WPHR. Photo by Brian Chalmers

She turned in her resignation to WMMS on February 28 and was on the air at WPHR two days later. But Farina felt a backlash from some of her former co-workers, who felt she'd stabbed them in the back by looking for work at another station.

Things went a little better for another female personality at WMMS. Ruby Cheeks officially took over the afternoon drive slot vacated by Kid Leo a few months before. Cheeks had a quick sense of humor, but most of all she had attitude. That combination promised to make her a very formidable player in the ratings wars. Elsewhere on the schedule the station had resumed its successful *Coffeebreak Concert* series with Melissa Etheridge at Peabody's Down Under.

As stations around town were adding new shows to their schedules, WMJI took one of its established flagship programs and presented it in a way never before done in Cleveland. The station announced that it would simulcast John Lanigan and John Webster's morning show on WQHS-TV. The recently reformatted television station was the local non-cable outlet for the Home Shopping Club. Adding the four-and-a-half-hour WMJI show was beneficial to both stations. For WQHS, fans of Lanigan and Webster might keep their television sets tuned to sample the on-air shopping channel. And for WMJI, since it didn't matter to Arbitron if a station was heard on the radio or seen on television, it would still receive ratings points for the broadcast. The new television show would air a video feed of the live in-studio segments with Lanigan, Webster, and their guests, while showing scenes of the Cleveland area during the songs. The premiere broadcast was set for May 4, though a snag did develop with the local office of the American Federation of Television and Radio Artists (AFTRA). The union complained about the voices used in radio spots, demanding additional payments if they were also aired on television. Since neither the television nor radio station needed negative publicity from an AFTRA contract dispute, the situation was soon resolved. Viewers received the extra advantage of seeing what went on behind the scenes at a radio station, catching visual bits such as Webster holding up a stuffed buzzard doll with its head in a noose.

March ended with WMMS taking broadsides from both Webster's on-air antics and a new television campaign for WNCX. It announced, "When you outgrow the Buzzard," leading into a shot of a flaming guitar poking a bird in diapers. Time would tell if the ad would prove to be effective.

April 17 was a day to celebrate at WMMS, as it marked fifteen years since the Buzzard character first appeared in print. The occasion was marked with a special birthday cake for the staff and an impromptu party in the lunchroom. Buzzard illustrator David Helton was heartily congratulated by his co-workers for creating such an enduring local icon.

Other stations around town weren't feeling as festive. Sensing that WMMS was still vulnerable after the events of the past year, and knowing that just about any station that was clever enough could benefit, some strategic realignment began for a number of different competing shifts. Jimmy Malone was added to the Saturday morning show at WMJI. The station's weekday morning co-host Lanigan took to heart the spirit of Cleveland's Brotherhood Day on April 26. He joined WZAK-FM's Lynn Tolliver and Ralph Poole from the steps of City Hall for a canned goods drive that was simulcast on those stations and WQHS.

Summer has always been a critical time for rock stations. WMMS added to its roster, bringing in Craig "Killer" Kilpatrick from WQFM-FM in Milwaukee as the permanent evening host. The station also had the good news that Mr. Leonard would be staying on the air for a long time to come, thanks to a five-year contract signed by his alter ego, Rio. That ended speculation in *Radio and Records* that he would soon be joining Shannon in Los Angeles.

With the departure of Paul Tapie in April, WNCX faced the warm weather months without a morning show to anchor the day. Station program director Paul Ingles joined Mike Trivisonno for a time, but a permanent replacement was still needed. A very expensive full-page ad had been placed in trade papers. It read, "The Ultimate

Those Guys in the Morning (1989). Todd Brandt and Rick Rydell had their hands full when they came to Cleveland to do battle with the Buzzards, not to mention the local press. Photo by Bob Ferrell

Rock 'n' Roll Challenge . . . Mornings in the Rock 'n' Roll Capital of the World! Morning Drive Hero Needed. No Beginners. Only Experienced Morning Show Winners!" General manager Dave Urbach said he was willing to pay serious money to the right candidates, and tapes and resumes started to pour into the station.

WNCX continued to search for a show that could make any kind of dent against the long-established *Buzzard Morning Zoo* and WMJI's Lanigan and Webster. The station's latest entry in the morning war debuted on Friday, June 16. Todd Brandt and Rick Rydell out of Portland, Oregon, calling themselves "Those Guys in the Morning," were introduced in a marathon twelve-hour show starting at 6:00 A.M. that did get some press coverage. Unfortunately the gimmicky stunt wasn't enough to get the kind of numbers needed to be any real threat to the ratings giants. It also didn't compel listeners to tune in when the duo told *Scene* magazine that they couldn't pinpoint a specific goal for their show, they don't really prepare for it, and they pretty much just planned to ad lib their way through the morning.

WMMS continued the *Coffeebreak Concert* series from Peabody's Down Under with a special broadcast featuring the Indigo Girls. However the big show of the summer promised to be the Who's twenty-fifth anniversary tour, set to stop at Cleveland Municipal Stadium that July. The station promoted the tour hard, air-

Dia Stein (1987). A polished and seasoned performer, Stein also used her ace production skills and vast musical knowledge to add greatly to the Buzzard's overall air sound and credibility. Photo by Janet Macoska

ing the band's Radio City Music Hall performance of *Tommy* live and presenting "Who Moments" segments between Cheeks's "Hooligans" twin-spins of Who songs every afternoon. WMMS's heavy Who enthusiasm caught the attention of promoters, record companies, and the band's management. Usually this would give a promoting station extra opportunities at the concert in question. In this case WMMS was poised to make out in spades.

Just a few days before the Who's Municipal Stadium show, the Rolling Stones announced a concert date for that same venue scheduled for September 27, with tickets going on sale on July 15. This announcement promised a very exciting concert season.

The Who performance was set for Wednesday, July 19. The WMMS promotions staff was there early in the day to assure that everyone walking into the stadium would see the call letters just about everywhere. The station's logo was seen on light-up signs, the scoreboards, special t-shirts, and huge banners. Hanging from the upper decks, the banners proclaimed, "WMMS Welcomes The Who." Cheeks came on before the band to thank the crowd, the only disc jockey in Cleveland allowed that distinction. WMMS deejay Dia Stein filed special reports from the concert site, and WMMS even picked up the concertgoers' parking tabs with a special pass. All in all it was a huge victory for the station.

The Who show also proved to be very embarrassing for WNCX. Deejay Bill Louis, incensed both about the access given to WMMS and about the poor seats his station was issued to give away to listeners, took to the air the next day. He angrily lashed out, "Belkin and the Buzzards are in bed with each other!" He ran down a litany of things he took issue with from both parties, but a little reserved judgment might have saved an embarrassing situation. Word traveled quickly about the outburst. Belkin spent a lot of money promoting its concerts, and now that money wasn't likely to be heading to WNCX. A produced spot was hurried onto the air with the sheepish voice of Louis apologizing to Barry Gabel and Belkin Productions, though not to WMMS.

A new voice joined the *Buzzard Morning Zoo* that July with the addition of Cleveland native Tom Bush, who had recently parted ways with WWWE-AM. Bush had a lot of experience in television and radio, was a gifted impressionist, and could be fall-down funny at times. There had already been talk that WMJI's Lanigan might be considering him for a recurring role on his show, so WMMS acted quickly and brought him into the fold. Bush fit in nicely with the *Zoo*, working well with Mr. Leonard and the rest of the gang. He also had a somewhat unnerving habit of making people laugh hysterically just before the mike came back on. Bush joined up just in time for the BP Sohio Riverfest in the Flats, offering a prime opportunity to introduce himself at a high profile and heavily attended event.

The spring Arbitron ratings book showed WMMS still in the lead, but with strong challenges from WNCX, WMJI, and WLTF-FM. Even with other stations gaining WMMS could still muster up a good deal of excitement, such as with Cyndi Lauper's visit on July 26. Lauper got a major push from the station and especially Leo when her debut album *She's So Unusual* shot up the charts. She sat in with Rocco, took calls on and off the air, and generally hung around the station chatting before joining the staff for lunch at the unofficial WMMS hangout, Apple Annie's. Lauper always seemed to enjoy her visits to Cleveland, and she made a point to thank the station, recalling her *Coffeebreak Concert* appearance of a few years earlier.

Lauper wasn't the only star to casually show up. On July 28 Alice Cooper arrived at the station on a garbage truck. He was accompanied by a few of the dancers from the Tiffany's Cabaret strip club, visiting the *Morning Zoo* to deliver a copy of his new album, *Trash*. After some oddball news stories were read, such as one involving former mayor Carl Stokes's shoplifting trial, Cooper took each story and reread them silently. During the commercial break he said, "Cleveland really has the coolest news!" The following Tuesday, one of the pioneers of Cleveland's FM rock scene appeared on network television. Rick D'Amico, now at KTSP-TV in Phoenix, filled in for weatherman Mark McEwen for a week on the new segment of *CBS This Morning*.

WMMS had prided itself on an air staff that had remained in place for more than a decade, but that ended in 1986 with the mass defection to WNCX and later with Leo's departure. Changes came a lot more frequently by 1989, and the roller coaster ride with the ratings didn't help keep personalities in one place. In September of that year, Cheeks decided to leave after a four-and-a-half-year run to look at stations in other markets, but she really didn't expect the change. Her agent had marched into general manager Chuck Bortnick's office, announcing that she had offers in the Boston market and that she would walk unless certain contract demands were met. The general manager didn't blink, so Cheeks soon found herself leaving WMMS. It was an extremely emotional time for Cheeks, who truly loved her job in a town she considered home. Lee Zapis, the station's Cleveland-based owner, recalls, "We hired Ruby during the time John Gorman worked as our consultant at WAAF. John was one of the team responsible for putting us on the active rock path. Prior to that, WAAF was a very unfocused rock radio station." Cheeks would be one of the cornerstones of the new WAAF-FM.

Cheeks's replacement was Lisa Dillon, who had left the *Morning Zoo* and was basically doing fill-in work. This wasn't the only change going on at WMMS, all evidently resulting from the station's decline in the ratings. Tom Rezny replaced departing production director Tom O'Brien, with Rezny's 10:00 A.M. to 2:00 P.M. shift going to Kilpatrick, and Rocco taking 6:00 to 10:00 P.M. Len Goldberg was reassigned to the promotions director spot, but could still be heard on various shifts and in voiceovers and commercials.

Major changes continued at the Statler Office Tower with the exiting of another longtime staff member, one who helped define the station's visual identity. Helton's Buzzard character came to symbolize WMMS like no other mascot in the medium. In the final days of August it was announced that the artist would be heading south to Atlanta, reportedly to do special projects for the Malrite stations and to pursue a freelance career. His assistant Brian Chalmers would take over his immediate duties at WMMS.

Yet another change occurred when Stein had caught the attention of Westwood One Radio. She had been in New York to host Paul McCartney's press conference to announce his U.S. tour. The hosting offer seemed too good to pass up. It was a huge decision for Stein, though the choice seemed like a logical one. She recalls,

Things had really come to an end, and when I look back at my time at 'MMS, it was so wonderful. I remember the times we would just be sitting around John Gorman's office. John would be at his desk, Denny would be stretched out in a chair talking to Matt, and Betty would be hanging upside down on the couch. Pretty soon we'd all fall into the same conversation, get an idea for producing something, and all run off to the production room to do it. We were all so different, but it worked! We loved each other, and we hated each other, and in some cases we dated each other. It was just sort of one of those great convergences of people. We were so young, and so impressionable, and we didn't even know how big of a thing we were making. Everyone had a very distinctive style that they brought into this mix, and it became part of this greater thing that was 'MMS.

The time had come for her to try to re-create that magic elsewhere.

The Rolling Stones brought their much-anticipated Steel Wheels Tour to Cleveland Municipal Stadium on September 27. As always, every serious rock station in town was on hand to show its stuff to the crowds. The concert also marked the saddest day in broadcasting for the departing Cheeks. After considering a number of jobs, she officially took the afternoon drive position at WAAF. Not that long before, Cheeks had introduced the Who at that same venue to a great response. But now she found herself sitting as a spectator with Zapis instead of doing the live broadcast she'd been scheduled to do from the show when she was still with WMMS. The end came without warning for Cheeks, and she never got a chance to say goodbye to her audience. As Mick Jagger launched into "Ruby Tuesday," the stadium crowd joined along, and tears started to stream down her face. To this day Cheeks still considers it one of the most profound experiences of her radio career.

At WNCX Brandt and Rydell were scrambling for any kind of publicity and ratings they could muster. So they climbed to the roof of the station's building at 15th and Chester, vowing not to come down until they'd raised $10,000 for a local charity. It was actually a carbon copy of a promotion WLTF's "Trapper Jack" Elliott did just a couple of weeks before in a Cleveland Browns-styled "Dawg House" in the Flats. It was becoming painfully obvious that WNCX needed a new morning show, and Brandt and Rydell weren't in the station's long-term plans.

At the end of September WMMS ran full-page newspaper ads announcing its "new Fall line up." But as some columnists pointed out, in television the line up changes every fall, and change was not something the station wanted listeners to think about. WMMS needed to do something, as it fell again in the summer Arbitron book. WMJI, WLTF, WZAK, and even WNCX were picking up some of the station's listeners.

WPHR was able to pick up something else from WMMS, a $1,000 check. In one of the oddest promotional pushes ever in Cleveland radio, WMMS offered that amount to anyone who was able to get its call letters shown on a banner during

ABC's coverage of the Browns-Bengals game in Cincinnati. WPHR's promotional staff took up the challenge. Sure enough, ABC caught sight of their banner and put it on the air. The WPHR staff's banner used the network's call letters to spell out "Another Bengals Championship," followed by WMMS's call letters. It obviously implied that WMMS was rooting against the Browns. Nonetheless the station made good on its bet, and WPHR gave the money to a local charity.

Norm N. Nite was instrumental in the campaign to bring the Rock and Roll Hall of Fame to Cleveland, and he was equally persuasive in getting the New York board to announce its list of 1990 inductees at a special presentation at the Palace Theater on Playhouse Square. It happened on October 24, with WMMS breaking in with special live programming from the site to announce the honorees. The Grand Hall of the theater was packed shoulder to shoulder with local, national, and even international media, and the WMMS news department staffed it heavily. Paul Shaffer, music director for the annual induction ceremony, traveled to Cleveland with other Rock Hall dignitaries, acting as master of ceremonies for the event. While WMMS had taken its lumps over the years, it was still recognized as the station that did more than any other to bring the Rock Hall project to Northeast Ohio, and it kept up its reputation as the definitive source for Hall updates by bringing events such as this to the listening public as it happened. The broadcast went without a hitch and was a well-polished example of live radio at its best.

The *Cleveland Breakout* show went through a transformation that same month, being re-christened as the *Scene Breakout*. The weekly entertainment magazine's Mark Holan became one of the new hosts, as did WMMS music director Brad Hanson. Former station personality Murray Saul, now with *Scene*, also gave entertainment updates. The updated program centered mostly on alternative music, with some local acts getting air time as well. But it was stuck on Sunday evenings from 6:00 to 9:00, when that kind of audience was relatively low. Part of the attraction for Holan was the lure of free form radio. He explains, "Anything. Local stuff, or even national. As long as it was new, British, American, I could play whatever I wanted!" He says that drew a lot of attention from local representatives of the national record labels, too.

The very first broadcast coincided with the release of Nine Inch Nails' *Pretty Hate Machine*, with the twelve-inch single for "Head Like a Hole" coming out that week. Group frontman Trent Reznor stopped by the studio to debut it. It would be the only time Reznor would come by the station, even after it started playing a heavy alternative format in later years.

An addition to the WMMS staff in late fall would prove to have a quiet, but profound, impact during his time at the station. John Filby came on board as a production assistant on November 5 and quickly grew into the role of producer for Dillon's afternoon show. The Taylor University graduate spent most of the second half of the 1980s studying for a career at WMMS. He was able to land a brief internship with Leo, who had a deep influence on him. Filby's organizational talents and ability in the studio soon became very evident, and he proved to be so valuable that he was brought on full-time after concluding his internship.

The future of Cleveland radio was the focus of a market profile in the industry magazine *Radio and Records* that November. Writer Joel Denver profiled local stations, their formats, and program directors. WMMS's Rich Piombino talked about "having something to prove" with the station's staff. He also discussed the tremendous loss of Leo, whom he likened to a "spiritual glue," and the possibility of maybe playing artists like Paula Abdul and New Kids on the Block.

WMMS was already undergoing a transformation. The station's programming was heading into a distinctly different direction. Rumor had it that WMMS was trying to battle the increasing popularity of WNCX's classic rock. So WMMS changed its format back from the contemporary hits genre (CHR) to album-oriented rock (AOR), and that change was evident just by tuning in for a half- hour or more. The station was playing more testosterone-laced guitar rock and was featuring Boston, the Rolling Stones, .38 Special, Led Zeppelin, Bob Seger, and the Beatles. Not long after, in December, Dave Urbach left WNCX to explore other opportunities, including the possibility of station ownership.

The Christmas season proved to be a wealth of innovative programming. Most stations usually go to a heavy music rotation during the holidays so as many staff members as possible can take the day off. WMMS was no exception, but the station did break the music for a news special at 5:00 P.M. on Christmas Eve. Titled *We Built This City on Rock and Roll*, the Starship tune was a convenient theme for that one-hour show. It was a comprehensive look at Cleveland's uneven history and how the Rock and Roll Hall of Fame would help guide the city's fortunes into the twenty-first century.

Taylor Hall, Beatle Paul, and Paint Falls

The 1990s kicked off when a familiar face, Kim Colebrook, returned as general manager at WNCX. He had left a little over three years before following the station's changeover from WGCL. Colebrook was rehired by Metroplex-Robinson Broadcasting, which took over the station from Metropolis Broadcasting. It was a vindication of sorts for Colebrook, who was at the helm when WGCL beat WMMS in the Arbitron book a few years before.

Colebrook recalls the day he left the station in 1986,

> When [Metropolis] came in to take over the station, the first thing they did was hold the door open for me to leave. That came as no surprise, and every-

thing was in order for them. It was very upsetting for me, because I'd been there for fourteen years, starting from an entry level sales job to president and g.m. I had a lot of emotion tied up in that station, and gave it everything I had all the time. But before I'd left, I put a G98 bumper sticker in the knee pit of my desk, underneath a desk drawer. When I moved back into my office, I looked under there and, sure enough, it was still there. I peeled it off three and a half years later!

A year after the departure of Kid Leo, the *Plain Dealer Sunday Magazine* ran a lengthy update about his new career at CBS Records. Leo told the paper he was enjoying life since WMMS, but he still missed the station. There was also a surprising revelation. The paper claimed Leo and former WMMS operations manager John Gorman were betting to see how low the station's ratings would go before it bottomed out. Leo reflected on the old days of radio, when it was an "art form" and before Wall Street types started using stations as quick money investments. Though his comments rang true to anyone in the business, they didn't exactly instill confidence in the everyday listener who wasn't sure what WMMS was about anymore, nor where it was heading.

Although there was a lot of movement on the FM dial in just the first few weeks of 1990, one thing that didn't change was the loyalty some artists held for WMMS. On January 23, David Bowie held a satellite press conference from London's Rainbow Theater to announce a major U.S. tour following his band's European dates. This was going to be the final time Bowie would perform many longtime fan favorites, which had become tiresome in light of the tremendous amount of work he had done since their initial releases. Bowie ended the announcement by performing an acoustic version of "Space Oddity." WMMS was the only station to carry Westwood One's coverage of the press conference, with additional reports coming through the next week.

Not even an hour after the press conference, Bowie's management called WMMS so that the artist could make additional comments just for the station. Bowie said he was looking forward to stopping in Cleveland, as he was sure a concert date would be announced soon. Looking ahead to the possible Rock and Roll Hall of Fame groundbreaking in the very near future, Bowie was asked what he thought of it. He mentioned that he wasn't exactly fond of the Hall of Fame, but for Cleveland he would be happy to comply. He congratulated the city for its successful campaign with a hearty "well done," proving once again why he is one of the special artists that will always have an audience in Northeast Ohio.

Bowie's personal call would be a welcomed highlight considering what else was going on that month. The results of the latest *Rolling Stone* readers' poll hit the press, showing that WMMS didn't even make it into the top five. The station kept its commitment to not participate in the poll in the wake of the ballot-stuffing scandal from two years before. The best course of action was to continue concentrating on the local audience, and WMMS was in a relatively good position there.

However, it wasn't exactly a love fest at the Statler Office Building according to the trade newsletter *Inside Radio*. A few weeks before, operations manager Rich Piombino was suspended. The details covered in *Inside Radio* said that he had gotten into a very loud argument with a female employee, which ended with Piombino calling her a very graphic name, and which resulted in him getting a stern warning in the process. Soon after, it was reported that Piombino got into another exchange in general manager Chuck Bortnick's office, one that could be heard down the hall at Malrite president Gil Rosenwald's office. Rosenwald was said to have arrived just as Piombino was storming down the hall. Rosenwald then told Bortnick to suspend him. When Bortnick reportedly refused, Rosenwald said, "If you don't suspend him, I'll find a GM who will!" Piombino found himself with time off, saying Bortnick told him to take a week's vacation.

WNCX's *British Invasion* host Jerry Shirley was able to pull some strings with one of Humble Pie's former roadies, now a personal assistant to Paul McCartney, to set up a one-on-one interview with the former Beatle. McCartney was in the middle of traveling through America with a stadium tour. The talk was a big plus for WNCX, but it was a long tour and WMMS could still take advantage of it. Word soon leaked about a McCartney date that July at Cleveland Municipal Stadium, with tickets soon to be available. WNCX added a new voice to its lineup in mid-March, as the *Plain Dealer*'s Michael Heaton signed on to do an hour-long show on Sunday evenings. This did not come as good news to WMMS, which had been targeted in Heaton's columns in the past.

Kid Leo returned to Cleveland and decided to make a round of radio appearances while in town. He was interviewed on WWWE-FM's *Becker and Heggs*, on WPHR-FM, and of course on WMMS. Leo spoke frankly about the state of radio, his reason for leaving the city, and his relationship with other deejays in town.

Former WMMS deejay Dia Stein contacted the station in April, asking if it could put together a half-hour show for Westwood One's *Source Report* about the twentieth anniversary of the Kent State University shootings that May. Most of the principal players that survived that infamous day were still around Northeast Ohio. After a flurry of interviews, a program was put together that later went on to win the Silver Medal in the International Radio Festivals of New York competition. The higher ups at the *Source* were so impressed that they wanted a show every month about various topics of interest, but that also led to a very memorable broadcast the morning of Friday, May 4.

WMMS staff members such as operations manager Rich Piombino, *Buzzard Morning Zoo* co-host Jeff Kinzbach, producer Murray Oden, and others discussed how the station should commemorate the anniversary. WMMS was around at that time, but playing a far different style of music in 1970 than its current format. It was decided to turn back the clock and use *Billboard*'s Top-40 lists as a guideline. That worked because so many of those artists had made an impact on listeners at that time. Many of those artists such as former Clevelander Edwin Starr, the Beatles, Jimi Hendrix, and Janis Joplin shared audience on both the AM and FM dials. WMMS was expecting a lot of calls that day from folks with where-they-were-and-what-they-were-doing stories when they first heard about the shootings. WJW-TV's Carl Monday, a longtime friend of the station, was a young disc jockey in Kent at the time of the shootings, and he had amazing stories of his own concerning the events leading up to that confrontation. Still the real action was taking place at the Kent State campus, so it was decided to do live cut-ins throughout the morning from the remote truck and mini-studio.

A lot of preparation went into that broadcast. WMMS prepared a special report to break into programming at the exact moment as twenty years before when the National Guardsmen opened fire on the students, as well as produced lead-ins and liners telling the audience about that morning's programming.

The day before the broadcast, engineers Mike "Mondo" Knisely and Brian "Sparky" Emery took the remote truck to Kent to test the signal. The microwave link had to make it back cleanly to the antenna on top of the Terminal Tower. After a couple of hours they realized there was only one place on campus where the mast could be extended high enough to relay that signal. It was next to Prentice Hall, just steps away from where the students had been gunned down in cold blood. Knisely knocked on a dorm window, told a female student what the station was going to do, and asked if they could plug the power line into her wall outlet. He explained that the crew had a generator, but it would probably make a lot of noise and they didn't want to disturb anyone. The student was more than happy to comply and asked how many people would be working that morning, promising to bring coffee and donuts for everyone. Knisely said that was a very gracious gesture, but really wasn't necessary because they would be onsite by 2:00 A.M. She smiled and said it wouldn't be a problem. As she explained it, "We're just getting back from the bars at that time, and we want to watch it anyway!"

True to her word, her group was there right at 2:00 sharp. They received WMMS t-shirts for coffee and pastry, and then the crew got down to work. While Emery and Knisely plugged in and set up the microwave link, a crowd gathered around the corner at an all-night candlelight vigil. Students too young to remember Kent State, if they had even been born, were standing motionless and in eerie silence, holding candles at the exact spots where the students lay dying twenty years before. Eyes were swollen from steady flows of tears.

The *Morning Zoo* opened with Crosby, Stills, Nash, and Young's "Ohio." It was wisely determined beforehand that the song would only air twice that morning, as it stood to be overplayed through the day. Calls soon came in from former students,

Vietnam vets, and even former Raspberries guitarist Wally Bryson all expressing some deeply felt sentiment about the anniversary and what it meant to them.

Listeners continued to recount where they were when the news broke and how it affected them afterwards. Some were in tears, and many thanked WMMS for devoting the morning show to the anniversary. A number of key people at the campus for that day's commemoration agreed to on-air interviews with reporters at the scene. Alan Canfora, who was wounded by guard fire twenty years before, recalled how WMMS had been such a big part of campus life. He remembered hearing the station coming from open dorm windows in the days leading up to shootings, and how many kept the dial locked there for the latest information after they occurred.

Two more guests who were directly involved joined the station at the remote site. Charles Fassinger was a Lieutenant Colonel in the Ohio Guard that fateful day in 1970, and Dean Kahler was one of the wounded, now confined to a wheelchair because of his injuries. The two had become friends after numerous discussions and public appearances in the years following the Kent shootings. Now Fassinger was guiding Kahler's wheelchair down the path to the remote truck. When they arrived it was obvious from even their casual conversation that both held a deep respect and admiration for the other.

Former Senator George McGovern and friends and relatives of the victims also made it to the microphones at Kent State. At 10:00, the station capped a very emotional morning for both the listeners and crew. At exactly 12:24 P.M., twenty years to the minute since the Kent shootings, Len Goldberg broke into programming to air a special report. WMMS had gotten hold of a recording made by a student who was in the middle of the crowd, rolling tape just as the shootings began. Twenty years later, the sound was heard of the protestors in pandemonium, screaming and scrambling for cover as the rifle shots rang out as clear as a bell. It was followed by thirty seconds of silence in honor of the students, and their cause. Every element, including the voices of students then and now, the comments from that day's ceremonies, and the clanging of the Kent State bell for the victims, made the day's special broadcast a memorable and moving tribute to a local event with international ramifications.

Despite all the negative press over the last few years, WMMS was still a major contender and one of the most respected stations in the Cleveland market. Loyalty came through once again when Bowie announced the Cleveland date for his Sound and Vision Tour 1990, again calling the station to ask to do an interview. But the upcoming McCartney date was the main topic of concern at WMMS. Just about every station with any kind of connection to the former Beatle's music was trying to link up to the tour. WMMS was determined to own his Cleveland show, but the station had a lot of competition. Plans were forming all over town at every station as soon as that concert was announced.

WNCX was still tinkering with its morning show, hosted by Todd Brandt and Rick Rydell. The station expanded the lineup to include sports from Mike Trivisonno, who had first been heard on Pete Franklin's WWWE show as "Mr. Know It All" (who would soon add a Sunday night talk show to his resume).

The morning program was already a favorite target for criticism by the *Plain Dealer*'s David Sowd. So Brandt and Rydell took out ad space on the page of Sowd's column. It stated in bold letters, "Dear David Sowd, We're Taking Monday Off," referring to the Memorial Day holiday. The duo tried to play on the lack of praise by running a second ad that stated, "We Don't Suck." Their act just wasn't catching fire, and stories started to circulate that WNCX was ready for a serious change in morning drive. WNCX continued to search for identity with a concept from its creative director Mark Gullett. The station's television spot centered on its morning team playing baseball, only with a live puppy instead of a ball. Sowd quoted Gullett, "I don't think anyone's going to take exception to seeing the puppy. Not a male, anyway. And we're not a female based radio station."

WMMS also continued to get television time, only with its promotions. A group of women traveled to the station agreeing to shave their heads live on the air for Sinead O'Connor tickets. How anyone could tell if they were really bald by listening to the radio is another question. WMMS's *Scene Breakout* program was starting to promote shows of its own, hosting a simulcast of Hot Tin Roof and In Fear of Roses from the Phantasy Nite Club in Lakewood.

The club scene expanded that July when brothers John, Tony, and Mark Ciulla, the former promoters at Peabody's Down Under, took over the Ritz Niteclub on Sumner Court and renamed it the Empire. They planned to showcase national as well as local acts, with John acting as principle booking agent. He quickly started putting together a schedule for the expected opening on July 12. In the volatile club scene of downtown Cleveland, the Empire was destined to make a big impact in a relatively short time.

Over at WNCX time was running out for Brandt and Rydell, and the negative press following their Flag Day stunt didn't help. There had been a lot of recent coverage in the press about the legality of burning an American flag, so Brandt and Rydell announced that they would burn a flag themselves. A small group of onlookers saw the two drag a remote mike outside the station's Huron Road entrance, where they torched a Japanese flag. They followed it up with what Sowd described as "blatantly racist remarks," reaffirming their faith as "real Americans." They even played Bruce Springsteen's "Born in the U.S.A." It clearly seemed that this show was in a freefall, and privately there was heavy speculation—both in the local radio industry and the press—about the exact date the station would excuse Brandt and Rydell.

Bowie hit town for his two sold-out Richfield Coliseum concerts, and there was a huge sensation on the first night. The rest of the band was hidden from

Murray Oden, John Rio ("Mr. Leonard"), and Adrian Belew (1990). David Bowie's guitarist clowns with the *Morning Zoo* during his visit to the show. He confidentially tipped off the folks at WMMS that members of U2 would be taking the stage with Bowie that evening. Photo by Brian Chalmers

view, and the only other person to be seen on stage with Bowie was guitar whiz Adrian Belew. Producer Murray Oden was able to arrange a visit by Belew on the *Morning Zoo* the morning of the second concert date. Belew had fun with the *Zoo* crew, did a few runs on the guitar, and promised to come back. But after the mikes were off, Belew let the staff know about some expected guests. He said the members of U2 had planned to show up at the Coliseum the night before, but decided instead to pop in on that night's show. Sure enough that night Bono and The Edge traveled to Richfield. When Bowie's encore segued from his "Jean Genie" to the Doors' "Gloria," he pulled Bono on stage to close out the show. It was an amazing performance from two of the most exciting figures in rock, and it was a complete surprise to everyone, except the folks at the *Morning Zoo*.

Piombino's embattled tenure at WMMS came to an end that June. The station was forced to look for its fourth program director in just two years. As music director Brad Hanson describes it, "With every leadership change, there was a slight change in format. There had been a real lack of direction for some time. Strangely enough, this turned into one of our strongest periods, because there was no one really in charge and we all grouped together in the programming department and made it work."

A number of candidates had come under consideration prior to Piombino's departure, such as Doug Podell out of Detroit. Finally it was decided to give the position to Maple Heights native Michael Luczak, who had done ten years at Cincinnati's top-rated WEBN. For Luczak it was a shot at making it big in his old hometown. But the WMMS he remembered was changing rapidly, and he would be a major part of the transformation that aimed at re-establishing WMMS as the undisputed king of the ratings. He would start the latter part of July.

It was a critical transition point for WMMS. The station had weathered changing musical trends, challenges from upstart competitors, and the departure of some of its key personnel. As Luczak saw it,

> My primary job was to focus the station as a rock station, which was relatively easy to do because it had been a hybrid CHR. Plus, the disc jockeys had been there for years, so there was enough of the "old school" to legitimately say we'd been rocking for more than twenty years. I came in with the belief that WMMS was built on being the current music station, and always on top of the latest, hottest thing. That was a major component. Even though we were a rock station, there was a very wide audience, with an extremely strong morning show. The marketing machine was up for the job, and it would be relatively easy to focus on eighteen-to-thirty-four numbers. Let's face it: You couldn't be the same station to a fifty-four-year-old that you were to an eighteen-year-old, but we were able to set our sights on that main group.

Paul McCartney's *Flowers in the Dirt* Tour was quickly heading to Cleveland. WMMS couldn't just have a big presence at the concert, it had to be the only station that mattered. Piombino sent members of the WMMS staff to Washington, D.C., where McCartney was to perform a Fourth of July concert. They had a mission: to come back with an interview, and to nail down a WMMS exclusive when the tour hit Cleveland. The philosophy around the station has always been, "No matter how difficult the task, just get it done, and do it better than anyone else."

Armed with tape recorders and promotional goodie bags full of WMMS and Buzzard t-shirts and merchandise, they headed to RFK Stadium to catch McCartney as he arrived to perform. There they met up with a familiar Cleveland face, Dale Solly. Solly was now working with WJLA-TV in Washington, and he agreed to help bombard McCartney with questions hoping to monopolize his time with WMMS. When McCartney strolled in to meet the press, he was immediately pounced on with questions about Cleveland, his family, the Rock Hall, and just about anything to limit his time with other reporters. McCartney was very accommodating and charming. He promised, "See you in Cleveland."

McCartney wasn't the only one cornered by Cleveland-based reporters. His publicist was approached and was told about WMMS's plans to use a segment of the interview every night up to the Cleveland date to promote the show, which still had a number of seats available. The publicist was given all of the promotional "Buzzard Bags," each one containing WMMS contact information. He was very grateful, took down the station's call letters, and promised to put a bag in each of the band's trailers.

The week McCartney hit Cleveland, a satellite studio had been set up in Malrite's loge at Cleveland Municipal Stadium. The signal from the makeshift studio could be relayed to a remote truck that was parked outside. WMMS had already been heavily promoting the concert even before the journey to Washington, and calls were being made at the station to nail down final arrangements for an exclusive interview with McCartney. Every station in town was trying to get at least a sound byte from the former Beatle when he finally arrived in Cleveland, but the publicist on McCartney's private jet sent word out that there would be no press availability. It was announced that McCartney only had time for one interview that was already arranged weeks before. McCartney would only speak to WMMS.

WNCX attempted to counter WMMS's exclusive by rerunning the interview Jerry Shirley did with McCartney a few months before, but it had been aired so many times already that no one was really fooled into believing that it was a new piece. Malrite was having a private party at Alvie's stadium restaurant for a number of clients, and WMMS was looking very good after scoring the exclusive.

WMMS was still willing to spend money when Luczak came on board, but it did so a lot more conservatively. At one meeting with general manager Bortnick, Luczak was told in no uncertain terms that these were not "the free spending eighties." There were a lot of overhead expenses that could be trimmed at that time. The station owned a lime green, custom-painted Ford Pinto with the logo and Mr. Leonard's name painted on the sides. The annual cost to maintain the vehicle was as much as $30,000, even though it was only exhibited a few times a year. There was also an elaborate Buzzard Bus, among other vehicles. The "need versus cost" of each was evaluated very carefully, with most falling to the budget axe.

By late July a couple of important names had announced their new direction, with Walt Tiburski and Tony Ocepek leaving the RP Companies, Inc. radio group for other opportunities. These opportunities included new radio acquisitions.

Another radio duo getting lots of attention were Tiburski's former co-workers at Malrite Communications, who were now attempting a big move back into Northeast Ohio. Former WMMS executives Carl Hirsch and Dean Thacker, with the Los Angeles-based Legacy Communications, were expecting to close on the deal to take over WMJI-FM from Jacor Communications. There had also been much speculation that Kid Leo might be persuaded to come back to Cleveland's airwaves at WMJI once that deal was closed. That rumor was laid to rest with a mention in the industry magazine *Radio and Records*, stating that Leo had just bought a house in the New York area and would not be joining his former bosses on the shores of Lake Erie. Similar rumors were surfacing about Piombino, who was also in New York City programming WQCD-FM's New Age format. The speculation around Piombino was that he might be coming back to Cleveland as the new program director at WNCX.

There was a continuing call for local music on the air, but most programmers didn't think it would make that much of an impact. Many of the recordings submitted by local acts just didn't have the production values suitable for airplay, and in far too many cases the music just wasn't good. WMMS did have its *Scene Breakout* show, and its *New Music Showcase* featured local acts such as the Floyd Band, Starvation Army, the Pimps, and several others performing at the Phantasy Nite Club. The future of the *Breakout* show had been the topic of much concern, as rumors started to circulate about its impending demise at the hand of new program director Luczak.

The *Scene Breakout* underwent a transformation in early August, becoming a one-hour showcase on Sundays at 10:00 P.M. Host Mark Holan invited other *Scene* magazine members to share their opinions on important new releases. Luczak announced that Lisa Dillon and Rocco the Rock Dog would be featuring *Scene Breakout* tracks on their respective shows, which gave the publication additional nods during the week. The station had new plans for *Breakout* co-host Matt the Cat, moving him out of the Sunday night and evening shifts and putting him back on middays after a two-year absence. Craig Kilpatrick went over to weekends. Dillon followed Matt, remaining on afternoon drive until 7:00. Rocco took it from there, handing the airwaves over to Bill Freeman at midnight.

A bold experiment was about to get under way at WMMS, with the very funny Rocco launching *Gonzo Radio* on August 20. It was designed to be a freewheeling, no holds barred, spontaneous rock and roll radio show, showcasing Rocco's sometimes-bizarre tastes in music and comedy. Rocco says it was, "A highly produced, free form radio show, and recalled early underground progressive formats. A lot of writing went into the show, but however I felt when I walked through the door became the theme of that night's show." *Gonzo Radio* soon became one of the most talked about shows on the dial. The program featured live music and a nightly visit from the syndicated "Mr. Manly," a *National Lampoon*-type character doling out advice on life to young men and women. Big name guests visited the *Gonzo* studio, including Gene Simmons and Paul Stanley of Kiss, who drew a crowd of women in lingerie within minutes of their taking to the air; guitar aces Steve Vai and Yngwie Malmsteen; and comedian Steven Wright, among many others. The show's underlying theme was, "Gonzo is; therefore, we are."

Another radio voice expanded his duties at WNCX, as Shirley took over morning drive duties from Todd Brandt and Rick Rydell on September 17. Trivisonno continued his role on sports. It was a bit unusual to put a relatively inexperienced disc jockey on morning drive in one of the most competitive markets in America, but it seemed to make sense to put a classic rock musician on a classic hits station. Shirley brought in his old Humble Pie bandmate Peter Frampton for a couple of days that first week to help him out. Shirley and Frampton held autograph sessions at the Halle Building food court and joined the band Armstrong-Bearcat for an evening show at the Sahara Night Club. Frampton didn't help just WNCX while in town. He stopped by the WMMS studios the night before Shirley's debut, appearing on the *Scene Breakout*. On the show he reminisced about his 1982 *Coffeebreak Concert* performance, commenting on the great support the station had shown him through the years.

WMMS had made it a long-standing practice to get station identification bumpers from visiting stars, and WNCX disc jockeys Shirley and Michael Stanley were well represented in the archives. As tempting as it might have been to use Shirley's "Hello Cleveland! I listen to the Buzzard!" ID, the station decided to instead go after the morning show and beat it in the ratings.

On his first day with the new *Mad Dogs and Englishmen* morning show, Shirley gave heartfelt thanks to the station management and listeners for the opportunity. But his departure from the evening *British Invasion* show left a programming gap. Program director Doug Podell filled it with Midwest rocker Michael Stanley, whose *In the Heartland* premiered on September 17. The station even started a twenty-four-hour Heartland Hotline to solicit requests from listeners. Stanley also gave a weekly commentary on Shirley's morning show. Stanley admits that it took a lot of guts for WNCX to give an untested disc jockey such a prominent role, but he had plenty of experience on television, the concert stage, and at WMMS. He was confident he could pull it off. A couple of years before, around the time that Leo left, WMMS had talked to Stanley about doing his own show, and Dia Stein taught him how to run the board. Those plans however never came to fruition. Now Stanley's chance had finally arrived.

Unfortunately, Michael Heaton's Sunday night *Radio One* program became a victim of station shuffling. Podell was quoted in *Scene* as viewing Heaton's show as a "bit self-indulgent." In an effort to keep him at the station, WNCX offered Heaton a one-hour Friday spot on Shirley's new morning show, which he reportedly turned down. Heaton later commented to *Scene* that his stay at WNCX was a great opportunity, "I got to learn how to do radio, and part of learning how to do radio is getting cancelled, so it was a complete full circle experience."

WMMS art director Brian Chalmers found himself swamped with work since the departure of David Helton. The station hired on a temporary assistant, Brigid Krane, to help ease that burden. But Chalmers never anticipated the massive project awaiting him at a special promotions meeting on September 18. Plans were unveiled for a billboard, the biggest in Ohio, to be painted on the side of a cold storage building to greet motorists traveling into downtown Cleveland on I-71. There would be a similar billboard on I-77, and a series of smaller signs with the same theme around the city, all aimed at showcasing the wide range of artists on the WMMS playlist. Some outside ad agencies had been asked to submit possible designs, but the station decided to keep the work in house. WMMS approved a very ambitious preliminary design submitted by Chalmers with forty-three different musicians, the Buzzard character, and Bart Simpson as a reference to Malrite's Fox station WOIO-TV. Each of the musicians' images would be hand painted by Chalmers in fine detail. When asked how long it would take to come up with the finished design, Chalmers took into account his current workload and the intricacy of the work needed for that sign. He responded, "The way it should be done, and if I budget my time out just right, it can be done in six months." He was told the billboard had to be in place within three months, by the middle of December. Chalmers was then informed that his design work had to be done and submitted to the billboard company in just six weeks! It would take a team of highly skilled people to do the final design with that kind of turnaround, but one of the unspoken rules at WMMS was "Just get it done, no matter what." Chalmers set to work immediately on one of the biggest challenges of his professional life.

His first order of business was to get his department on a strict production schedule, handing much of the peripheral work to Crane. He started work on the billboard project by assembling the source material, gathering classic photos and poses of stars ranging from Chuck Berry to Axl Rose. Chalmers sometimes spent thirty-six straight hours on the job, taking quick naps on his studio floor and only returning home to get toiletries, mail, and fresh clothing.

Nonetheless Chalmers was still falling behind, so he had to push up his work schedule even further to make the deadline. He spent so much time at the drawing board that he started losing weight, relegating his sleep to just half-hour naps, wearing the same clothes for days on end, chugging coffee, and wandering around the WMMS offices in a disoriented stupor. When it came time to finally present his work, Chalmers walked into the promotions meeting looking like he had aged twenty years. His skin was a sickly shade, his clothes just hung off his body, and his eyes were barely open. He spoke only in whispers, propping himself up against the wall, but he carried the finished designs under his arm. The final renderings were spread across the conference room table and drew unanimous praise. Within days the

Patrick Media company was hard at work painting his designs on the side of that building overlooking the freeway. Chalmers was told to go home and get some rest after a job well done. He wasn't sure if he was able to make the fifteen-minute drive home without falling asleep at the wheel, seriously considering checking into a hospital for extreme exhaustion. He did fortunately make it home, where he took his phone off the hook and spent the entire next week in bed. Chalmers returned to work nine or ten days later still a bit shaken, but anxious to start on his next project.

As one station was embarking on a remarkable ad campaign, another soon found itself without a popular duo. After refining what had become a very entertaining act in Akron, WONE-FM's morning hosts Brian Fowler and Joe Cronauer headed west to Denver's KAZY-FM. It wouldn't be long before they returned to their hometown.

By mid-September the *Morning Zoo* was going through another change, as impressionist Tom Bush left his full-time gig to become an "independent contractor," contributing his talents on an interim basis. His loss was noticed almost from the start. Solon native Mark "Munch" Bishop was hired on as WMMS promotions director after a four-year stint in Columbus radio. Bishop was a high energy sports and music fanatic who brought a new enthusiasm for the station, not only to the staff but also to the people he dealt with on the Cleveland scene.

The Persian Gulf conflict brought on a new sense of patriotism across the country. The *Buzzard Morning Zoo* started each broadcast day with the "Star Spangled Banner," alternating between a traditional recording and Jimi Hendrix's incendiary version from Woodstock. Stations began sending tapes of their programming to servicemen overseas, and the troops responded by sending photos back to their hometowns. Most of the photos had station call letters emblazoned on a bomb or gun turret. WMMS fans went a couple of extra steps in their pictures, with renderings of the Buzzard going to war or a WMMS t-shirt showing up in exotic ports of call. The station also sponsored appearances of a limo that was supposedly designed for Iraqi leader Saddam Hussein. For a charitable donation, listeners were offered a chance to take a swing at the car with a sledgehammer to show their patriotic spirit. It was a far cry from the anti-war fervor seen when the station first went on the air in 1968.

Hussein-bashing wasn't the only source for charity. Rocco called attention to the plight of Northeast Ohio's homeless the first week of October when he abandoned the Statler Office Tower studios for a tent on the roof of the Empire Concert Club. He vowed not to come down until a substantial amount of cash and food was pledged. Facing the impending arrival of winter weather, Rocco planned to stay in the tent on his new perch until a semi-truck trailer was filled to benefit Cleveland's Food Bank. That goodwill gesture would soon explode with publicity thanks to politics and local media rivalries. Rocco was able to get a truck filled in just five days. *Scene* magazine was quick to point out that while it was all for a good cause, the idea wasn't exactly original. WNCX's Podell countered the WMMS promotion by saying he had done a similar annual event in Detroit for the past seven years. The station's morning host Shirley would soon follow WMMS's example by moving to a parking lot house trailer for his Thirty Days in the Hole promotion. That didn't sit well with WMMS management.

In a staff memo dated October 15, WMMS general manager Bortnick wrote,

> As you may know, WMMS management is very concerned about the current position of the *Scene* in regards to a recent story that appeared about Rocco's food drive effort. I can assure you that Michael [Luczak] and I are not taking the situation lightly.
>
> It is one thing for the *PD* to take a shot at the station. We are not in a promotional partnership with the *PD*. However, the *Scene* taking a cheap shot while we are helping them promote their publication will not be tolerated.
>
> I will be having a meeting with the *Scene* next week to address their agenda vs. our agenda, and whether the station and the publication can get on the same page.
>
> Until we reach a disposition, I would expect that everyone will handle the situation with professionalism. We will continue to run their ads and their [*sic*] will not be any attacks on them over the air. I will report back to you as soon as I have had the meeting and we reach a final decision on their future role with WMMS. Your patience in this situation is greatly appreciated. Remember, considering everything that this station has been through in the press, we are still having the last laugh.

The station went into that meeting with some ammunition. Despite the broadside hits it had taken in the last couple of years, WMMS was still a major player. Its strength was backed up by the summer Arbitron ratings book that showed WMMS at a strong number one, followed by WZAK-FM, WLTF-FM, and WMJI. The *Morning Zoo* was a winner by three percentage points over second place Lanigan and Webster at WMJI. WMMS program director Luczak credited the win to the station redefining itself as an AOR station. Podell at WNCX suggested any gains his station made were in part because WMMS was drawing more interest to the rock stations.

Paul Ingles departed WNCX soon after. Replacing him as co-host of *Mad Dogs and Englishmen* was Skip Herman, the former morning voice at WGTR-FM in Miami. Rumors started circulating that former Cleveland radio executive Carl Hirsch was negotiating to get back into the market, possibly by taking over WMJI. The on-again off-again talks with the Cincinnati-based Jacor Communications resumed in earnest that fall, and word was that Hirsch was anxious to wage war on his old battle grounds.

In any event momentum was on the side of WMMS. Tensions were running high with *Scene*, and the magazine could be a crucial ally in the upcoming radio wars. There was an ugly incident that would divide the two sides even further, concerning the loyalty of certain *Scene* personnel who were seen mingling with folks from WNCX at a promotional party for the newly released Led Zeppelin boxed set. It was only natural that members of the WNCX staff were there since the station sponsored the party, but the aggressive nature of WMMS didn't take to *Scene*'s staff even acknowledging the hosts.

The results of the high-level meeting WMMS and *Scene* magazine were soon evident, as the entertainment weekly decided to end its affiliation with the *Breakout* show. In early November the paper issued this statement,

> Not wanting to compromise *Scene*'s editorial integrity by aligning itself too closely with any one station, *Scene* has decided to cancel the *Scene Breakout* show on WMMS. If you were listening to the station last Sunday night in eager anticipation of the latest releases and up-to-the-minute information on your favorite *Breakout* artist, you may have been a bit surprised. Instead of the voice of *Scene* editor Mark Holan, Big Mike (Program Director Michael Luczak) and Scooter (Music Director Brad Hanson) were at the microphone playing breakout releases, which included a track from John Mayall. Noticeably absent was music from local artists. Will the show now take a "Classic Breakout" lean, with new songs from established artists? WMMS was unavailable at press time.
>
> *Scene* has decided to end what has, in essence, been a year of radio education for both the magazine and the listener. The show literally brought the pages of the paper to life, with in-studio interviews with such artists as Michael Penn, REM's Peter Buck (with Kevin Kinney), Steve Wynn, the Innocence Mission, and the Blue Nile. The *Scene Breakout* began last October as a three-hour counterpart to the pages of *Scene*, and featured emerging local, national and international bands and solo artists. Almost immediately, another hour was added, and longtime WMMS personality "Matt the Cat" became the show's co-host. It remained a four-hour show until two months ago when, citing the show's poor ratings, program director Michael Luczak cut it down to one hour (10–11 P.M.). In addition to this hour on Sunday nights, Luczak added one song a week as the "*Scene Breakout* Track" into the station's "prime time" programming. In that time the show kept to the paper's promise to cover the local music scene. The *Scene Breakout* not only played many releases from Cleveland's talented bands, but it also took that a step farther by broadcasting "live" performances from the Phantasy Nite Club and Peabody's Down Under, which

included Cleveland's own Hot Tin Roof and Detroit's metal funksters Second Self. WMMS will continue their version of the *Breakout* show on Sunday nights.

Stories began to circulate that this was far from a happy split, which would turn even more ugly in the months to come. The affiliation with *Scene* wasn't the only loss at WMMS, as after a year and a half Craig Kilpatrick turned in his resignation to head to Indianapolis. The station was banking on its Fast Car Fantasy promotion to help it win the important fall ratings book, but it also had some big help from old friends and visiting artists. Leo stopped by to introduce the band Love-Hate, with additional stops from Damn Yankees, Social Distortion, and Toy Matinee, among others.

WMMS wasn't the only station with some tricks up its sleeve. Over at WNCX Shirley's Thirty Days in the Hole promotion began on November 30, with Ricky Medlocke of Blackfoot joining Stanley and various sports celebrities to help kick it off. Shirley also went on a modified hunger strike for charity. Within just two weeks he reached his goal, earning another two trailer loads holding more than twenty tons of food. Pranksters from the WMMS staff invited local media to stop by as it rolled up in the station's van to offer the drive cases of the ill-conceived Buzzard Breath Barbecue Sauce, but they were quickly turned away by WNCX personnel and pretty much ignored by the press. WNCX had established something many radio stations only dream about, a high profile, publicity-friendly promotion that could also benefit public service. It may have been done before, but WNCX made the best use of the idea.

By December 11 the giant WMMS billboard was about to be officially unveiled, but due to its size it was impossible to keep under wraps. So painters did their work in the open, while the station mulled over ways to inaugurate the new campaign. Patrick Media's press materials about the huge sign showed the magnitude of the project, which covered nearly 11,000 square feet of space. Eight painters and ten operations staff on the Patrick crew worked 1,900 hours to complete it on time, using thirty-four colors and 132 gallons of paint. It quoted numbers from the Traffic Audit Bureau that estimated that 121,700 people every day, more than 3,650,000 a month, would drive by the imposing signage on their way into downtown Cleveland. It was decided that the *Buzzard Morning Zoo* would broadcast live from the site and apply the final brushstrokes to officially complete that project.

What occurred at the official unveiling that morning of Friday, December 20 is still being debated, especially what happened afterward. All that is known for sure is what transpired at the broadcast site, live on the air. *Zoo* co-hosts Jeff Kinzbach and Ed Ferenc had been on location, giving information about the new sign and doing interviews. At the appointed time they got into the basket of a cherry picker to be lifted to the place where the final strokes would be applied.

"We got Mr. Leonard on the roof," Kinzbach said on the air. "Can you see us down the side of the wall?"

"Yes, I can Jeff," Mr. Leonard answered. "I'm looking down right now, and I can see all the cars, east and west on the—the seventy—Alright! Honk if you're horny! Ah-hahahaha!"

The request was greeted by a wave of car horns from the West Shoreway.

"There seems to be quite a few horny people, Mr. Leonard," Kinzbach concluded.

"Would you quit foolin' around!" Ferenc said.

"Look, here's what Flash and I are going to do," Kinzbach explained. "We are officially going to finish the wall here, with the paintbrushes and everything, and finish the 'WMMS-FM.'"

"Wait, Jeff," Mr. Leonard said. "Before you did that, I was wondering if you would allow me just a minute on the radio to remember this moment by singing a little Christmas song with the fellas, and the people who are involved with the building and the Patrick billboard company."

"Now what!?" Kinzbach asked.

"It's a little song that my Grandma used to sing around the Christmas tree, with the Leonard family warming our hands by the fire, and sipping eggnog and cim-mom-mon apple juice," Mr. Leonard explained.

"I guess it's pretty good the . . ." Kinzbach started.

Mr. Leonard cut him off to clear his throat and count off, "One, two, three.

I found the brains of Santa Claus underneath my bed.

I found them in a pickle jar. Wonder if he's dead?

He smelled like rotten tuna, and looked more green than red.

I found the brains of Santa Claus underneath my bed.

One more time!"

After a short break for traffic and spots, the live broadcast resumed. They were using remote mikes, as was Mr. Leonard, who was poised on top of the building. When the bucket of paint was being lowered to the co-hosts so they could dip their brushes, mayhem ensued. Promotions people were located just above the image of Chuck Berry, where the final strokes were to be applied. As they were letting the paint can down foot by foot, Mr. Leonard and *Morning Zoo* producer Murray Oden could be heard struggling.

"Wait a minute!" Kinzbach said. "What's going on?! Something's going wrong!"

"Oh, my god," Ferenc gasped.

"What are you doing!" Kinzbach called out. "You're getting paint all over the sign! It's going all over the side of the billboard!"

The Wall of Fame billboard (1990). A comedy bit for the *Morning Zoo* goes horribly awry as the wind spatters paint spilled by Mr. Leonard across a parking lot full of cars. Photo by Brian Chalmers

Mr. Leonard was recklessly letting the paint can swing in front of the two when it hit the wall, spattering paint down the sign and across some very high-end automobiles and equipment that were parked down below. Patrick Media had guarded against such an incident when the wall was being painted by hiring someone to keep any stray drops from damaging cars, and there weren't many because most drivers were redirected to other parts of the lot during the project. But no one expected this kind of mess. Almost immediately dozens of calls started pouring in to WMMS management, threatening one lawsuit after another. It originally started as just a comedy bit, but nobody expected it would go as horribly awry as it did. John Rio, the man behind the Mr. Leonard character, says no one expected the wind to be as strong as it was that day, "The wind blew the paint, and if you look at the sign, it's like a big swash. But the worst part was the paint that fell on the cars on the street below."

By day's end the vast majority of those complaints had been successfully resolved, but there was still the matter of how to deal with Mr. Leonard. Many listeners, such as the staff listening back at the station, were sure it was just a bit the *Zoo* had worked up for the broadcast. But extensive damage to the sign was evident, not to mention very expensive to repair. A curious memo appeared in staff mailboxes at the end of the day, addressed to "All WMMS Personnel." Concerning the "Status of Mr. Leonard," it read,

> Due to the unfortunate events this morning at the WMMS Wall of Fame (paint spill on the billboard, paint damage to several vehicles, loading dock and clothing of onlookers) caused by Mr. Leonard's gross negligence that has resulted in public embarrassment and potential litigation to WMMS, I have no other recourse but to immediately suspend Mr. Leonard, without pay, for one week. In addition, Mr. Leonard must pay all damages out of his pocket. A final decision on Mr. Leonard's employment status is pending.
>
> Until that time, Mr. Leonard is not allowed in the offices or studios of WMMS. Building security has been notified.

There was question as to how legitimate the memo was. Station management wouldn't just issue random memos to the staff, and there was a good amount of damage reported. But at the same time, the memo never referred to Mr. Leonard by his real name. WMMS wasn't about to let its biggest star go. Predictably it turned into a running topic of conversation the next week, and as could be expected Mr. Leonard soon returned to the *Buzzard Morning Zoo*. If it was just a publicity gimmick, it turned into a very costly one.

The Buzzard Flies West

The early days of 1991 saw a number of major players return to the Cleveland radio scene. The long-rumored sale of WMJI-FM to Carl Hirsch became a reality on January 16, and he wasted no time in taking the station to a new competitive level. Another former Malrite executive, Dean Thacker, was named the new general manager. J. R. Nelson became production director. And in a move aimed at striking fear at its competitors, John Gorman took over as program director. Hirsch, Thacker, and Gorman were reunited with Denny Sanders and other WMMS refugees, and competing stations braced for what was expected to be a free-for-all going into the winter ratings race. Upon his arrival Gorman found that a lot of the station's music library was from source material, vinyl and analog tapes, which had deteriorated over the years. Some of the music had even originated from tapes found in the old WIXY-AM archives. It had to be replaced to improve WMJI's overall sound, and Gorman set about replacing as much as he possibly could. He added a lot of regional hits that might not have made the same impact elsewhere. Songs such as the Outsiders' "Respectable," Richard and the Young Lions' "Open Up Your Door," and many others soon found their way into the rotation along with some of the biggest hits of the rock and roll era. Gorman took over a very healthy radio station with a popular format, and his re-entry into Cleveland radio promised to bring a new level of excitement into the market.

WMMS still had plenty of ammunition. *Gonzo Radio* was getting more popular, with an in-your-face format and music meant to be played at a level aimed at giving most parents a pounding headache. But the real strength of *Gonzo* was host Rocco the Rock Dog, whose demented wit and love for music with an edge gave a new personality to the station's evening fare. He also received some well-deserved press for coming to the aid of a family of five who were left homeless by a fire, as he solicited clothes, cash, and other donations to help them out. He even did some broadcasting from their West Side home. The station showcased *Gonzo* in the morning when Jeff Kinzbach and Ed Ferenc took a winter vacation, putting Rocco and Mr. Leonard in the A.M. slot. Among the highlights of that memorable week were Mr. Leonard's live cut-ins from outside the Statler Office Tower's East 12th Street studios. They featured Don Wright, a local Elvis Presley impersonator, and the two announced they would be handing out jelly doughnuts and giving away a "big ol' Cadillac" in memory of Presley. It caused an immediate bottleneck down East 12th as motorists came by as much to see what Mr. Leonard looked like as for a chance at the car. Cleveland traffic police raced to the street, confronted the two, and were not amused when Mr. Leonard tried to bribe them with jelly doughnuts. They were told in no uncertain terms to move on or face arrest, so they grudgingly complied and took the show a few blocks down to East 9th Street. Fearing another face-off with the cops, the two quickly gave out the rest of the doughnuts and awarded the Cadillac, in actuality a Matchbox toy car, before heading back to the safety of the Statler's 12th-floor studios.

Despite the demise of *PM Magazine*, Michael Stanley was getting television time with WJW-TV's new *Cleveland Tonight* program, which debuted on January 21. Stanley was already a recognizable face on television, and this latest venture also helped raise awareness for his radio work on WNCX.

WMMS announced plans that January to produce a charity CD of local music, soliciting tapes from unsigned bands. The *Northcoast Buzzard Tracks* album, to be issued later that spring, was produced to help the Tri-C Foundation and National Music Center Scholarship. The station's request for tapes resulted in a huge rush of entries, and program director Michael Luczak gathered a bunch of staff members to trim down the list of finalists. The project got almost immediate flak from the local music community, as WMMS wouldn't commit to putting any of the songs from that proposed CD on its playlist. But there was another change at the station drawing widespread attention, and this one was very visible.

For years David Helton's design for the Buzzard character stood as the continuing symbol of Cleveland rock and roll, the embodiment of what WMMS was all about, and a link to the days when the station was a ratings monolith. But Luczak decided the bird's image needed some freshening up, ordering art director Brian Chalmers to do a makeover that included a new hairstyle. That order came despite the fact that a classic Buzzard measuring seventeen feet tall and fourteen feet wide was unveiled among the stars on the Wall of Fame just the month before. Chalmers argued his case that the Buzzard should be left as it was, but his impassioned pleas fell on deaf ears and he was told to move ahead with the new design.

Gonzo Radio's Rocco was a busy man, joining the band Trouble Tribe onstage at the Cleveland Agora to sing Queen's "Tie Your Mother Down." It was becoming evident that Rocco was doing something right in his timeslot, and he was getting some well-deserved press for his efforts. WMMS had also partnered with the Empire Concert Club for a series of broadcasts, including one of local talent Neil Zaza's appearance. But not all was well at the station. *Buzzard Morning Zoo* producer Murray Oden was relieved of his duties to work exclusively on bits for Mr. Leonard to use in all his affiliate appearances. Luczak was quoted as saying Oden was replaced on the *Zoo* because, "I just felt that the chemistry between Murray and Jeff (Kinzbach) wasn't there."

Columbia Pictures decided to pull its film distribution office, Triumph Releasing Corporation, out of Independence that March to relocate to Chicago. That left David Spero without a position outside of his WNCX Saturday show, but Joe Walsh made sure that he wouldn't be idle for very long.

Spero remembers,

He'd called me up and said, "Listen. I got a new record, I want to do a couple of dates, and I wonder if you could help me put a few things together." He was looking for a manager, and asked, "Do you know anybody who's interested?" I sat there and thought, "Wow, maybe I'm interested! I'll come out and talk to you about it."

Joe left me a first class plane ticket, and a limousine picks me up at the airport. It took me to his house, and the driver said, "Just go right in. Door's unlocked." It was like 11 A.M. West Coast time, and he's sound asleep in a futon in the corner, wrapped up in a comforter, with the phone about two inches from his hand. There's no furniture at all in this house. Joe had just moved in, there's no place to sit, and the phone is ringing incessantly! He never answers it. People like Ringo calling, saying "Oh, Joe! I found a place to get a great pair of boots!" and Pete Townshend, "I'm coming out in a couple of weeks. Gimme a call!" There's nothing to eat or drink in the entire house, I've got a pounding headache, and I'm afraid to wake him up. Like at three o'clock his time, he finally wakes up, and says, "What are you doing here?" I said, "You told me to come out!" and Joe recalls, "Oh yeah! We got a meeting at four with my business manager and my lawyer."

We get in his car, and Joe says, "We've got a lot of work to do!" All of a sudden, I'm sitting down with the lawyer and the business manager, and they're like, "Okay! Here's how it's going to work!" What's going to work? I didn't realize I already had the job. I thought we were just going to talk about it! I started that day, and we went from there to the record company to talk about a release plan for the album.

The album was *Ordinary Average Guy*, and it wasn't long before Walsh had a hit single.

Saint Patrick's Day was always a huge day for WMMS. The station set up live broadcasts for morning and afternoon drive, usually with a club in the Flats, and they always drew big interest. The station simply sent out the remote truck, set up some bar games with big prizes, did the Token Jokes, and brought in special guests to entertain the crowd. The *Morning Zoo*'s remotes packed them in as soon as the doors opened at 6:00 A.M.

The 1991 remote may still be in a class of its own, and not for what happened on the air. Rocco had lined up some great entertainment for a live broadcast in the afternoon drive at Mooney's on Cedar and Lee in Cleveland Heights. A few of the Browns came by to have some fun with the crowd, and Alex Bevan came out to entertain as well. There was always a big turnout from the WMMS staff, and this was no exception. Crewmembers such as John Filby, Mike "Mondo" Knisely, and Bill Alford attended, along with a packed house of listeners clinking glasses and eating corned beef.

Filby recalls,

> The show was just about over, and the weather started getting real lousy. There was freezing rain, and biting cold temperatures, and the wind was starting to kick up, too. We all pitched in to get the equipment packed up so everyone could get home, but we forgot one major thing. The "Rover" truck had its mast extended twenty feet so we could get a signal back to the station, and it stayed that way for nearly six hours while we did the show. The mast was frozen solid, and we tried everything we could to bring it down. Hydraulic oil. Hot water. You name it, we tried it! One of the engineers even climbed to the top and tried hanging on it. Nothing worked.
>
> Finally, it's 2:30 in the morning, and the decision was made to drive it to [the staff member's] house, and worry about it tomorrow. The driver didn't think there were any trees or wires he might hit on the way home, so off he went in a truck with WMMS painted on the side, and a mast sticking twenty feet in the air! Talk about a red flag!
>
> The next day, I walked into work and heard the driver came upon a bridge that he didn't expect to see. It was a fourteen-footer. The mast had been sheared clean off! Thousands of dollars damage, the driver was lucky he wasn't hurt, and it didn't look like we'd be doing the next remote later that day.

A feature story in the *Plain Dealer* at the end of March caused some serious ripples in the radio landscape. Headlined "Radio's Women of Rock Fight 2nd-Banana Image," it quoted former *Morning Zoo* co-host Roberta Gale as saying she was let go from WMMS for being too funny. She made the charge that the radio industry was still an "old boys club." The article featured comments from former WMMS program director Gorman, who speculated that Ruby Cheeks's tenure on the *Zoo* might have ended because she was getting more fan mail than the men on that show. The station's afternoon deejay Lisa Dillon said that the entire radio industry was essentially "anti-female," charging Kinzbach and his crew with putting women down and locker room humor. Former WMMS disc jockey Dia Stein worked numerous jobs from behind the mike to special projects to management, and she concluded, "If you're only able to perform as a bimbo, maybe that's all you'll get!" Regardless of how each individual WMMS employee truly felt, the article would cause plenty of talk in the halls of the Statler Office Tower in the coming days.

As WMMS was getting hit in the press, another local station was getting saluted. On the steps of Cleveland City Hall, WMJI was honored for its live, on-air food drives, particularly the efforts of John Lanigan, John Webster, and the rest of the air staff to raise more than $22,000 and to collect hundreds of cases of food for local hunger centers. Mayor Michael White declared March 22 of that year to be WMJI Day in Cleveland. Gorman celebrated the prestige by hiring former WMMS art director Helton, bringing in his old friend to do graphics for WMJI. This led many to think Gorman was getting ready to battle WMMS with the best of its former employees.

The winter ratings race propelled WMMS back to the top, which was no small feat considering the station had placed a dismal fifth in the fall Arbitron ratings book. The *Buzzard Morning Zoo* was number one as well in double digits, and the Persian Gulf war may have played a big hand. A breaking news story of national interest usually drives listeners to news and talk outlets, but WMMS heavily covered Desert Storm, suspending regular programming the night the first missiles were launched against Iraq. There were additional special reports on the hour, and the station even cut to live network coverage of President George Bush's comments about the war. There was simply no need to turn anywhere else. But the big winner that ratings period was Rocco's *Gonzo Radio*. The program earned a staggering 22-plus share of men from 7:00 to midnight, which was an increase of 13 percentage points from the previous ratings battle. Rocco had been long overdue for a shift of his own, and these numbers were a vindication against those who said he couldn't pull it off.

WMMS hoped to capitalize on Rocco's *Gonzo* notoriety by naming him host of *Buzzard B-Movies*, which ran Saturday afternoons on Malrite's WOIO-TV. Along with the bottom-of-the-barrel film fare, the series featured a different guest WMMS personality every week. It usually worked within a general theme, such as when the Who film *The Kids Are Alright* was introduced with clips of Len Goldberg babysitting Rocco's brood.

Over on WNCX it was a virtual potpourri of voices on Jerry Shirley's *Mad Dogs and Englishmen* morning show. Along with Shirley, Mike Trivisonno, and Skip Herman, WNCX added Paul Tapie and the quick-witted Sally Ride (aka Sally Woznicki) on traffic. But the station was about to spring a huge surprise on the Cleveland listening audience, and especially its competition, when it announced the addition of the irascible Ruby Cheeks to its evening lineup. Cheeks had returned from a gig in Boston, and her re-entry into the Cleveland market promised to drum up a good amount of excitement on the dial. She kicked off her return show on June 3 with Elton John's "The Bitch is Back" and Ian Hunter's "Cleveland Rocks," the latter of which WMMS had pretty much claimed as its own over the past few years. WNCX also put program director Doug Podell in the afternoon drive seat just before Michael Stanley's *In the Heartland* at 6:00, with Cheeks taking over afterward until midnight.

Cheeks left the Boston radio market because she didn't feel she could express herself as creatively in a very rigid format, and she flirted with a return to the record industry. But while heading out to Los Angeles to look into that possibility, she briefly stopped in Cleveland and spoke with Podell, who put her on the air for an in-studio visit with deejay "Tony C" Colter. It was during that layover that Cheeks decided that Cleveland radio was where she really belonged.

Cleveland radio stations didn't have many opportunities to face off against each other in public, but the annual BP Riverfest had become a perennial battleground for just about every outlet in town. The Flats event had grown beyond Woodstock-like proportions by drawing more than a million people downtown, and stations scrambled to grab any of the highly visible attractions during that July weekend. Among them was WMJI's Majic Parade of Lights, a long procession of highly decorated boats and yachts making their way down the Cuyahoga River. WMMS countered with the WMMS-*Scene* Doo Dah Gonzo Parade. Similar to Philadelphia's Mummer's Parade, a long procession of oddly dressed marchers making their way from one of the Flats' better known bars to Settler's Landing. Some of the entries included housewife drill teams, synchronized land swimmers, and many more colorful participants. It was the place to be seen and heard that wild weekend in the Flats, and Cleveland radio did all it could to make its presence known to anyone who was a part of it.

The Premiere Radio Networks approached WMMS about a promotion sending two of the station's disc jockeys to Los Angeles to broadcast live from the NBC studios, just a few steps from Johnny Carson's *Tonight Show* set, and interview a large group of stars assembled by the promoters. A promotion like this usually involves a cattle call for stations across the country to head to Los Angeles, with as many as thirty shows broadcasting at the same time. The sponsors were very generous to all the radio crews, setting them up at the Hotel Mondrian on the Sunset Strip. Food and alcohol kept coming at all hours, and at one point the promoters even brought in Wolfgang Puck to make lunch for everyone.

True to their word, Premiere sent an impressive lineup of celebrities to WMMS's crew for interviews. John Entwistle of the Who, Jeff "Skunk" Baxter of the Doobie Brothers, and Mickey Thomas of Starship handed the station a scoop on their upcoming unnamed "super group," although Entwistle did confide that it didn't have much of a chance. Bob Hope gave WMMS more time than any other station once he found out it was from his old hometown.

Seinfeld cast members Jason Alexander and Julia Louis-Dreyfuss stopped individually to talk about the upcoming season, which was just starting to show strength in those early years. The real surprise came when Jerry Seinfeld also sat down with WMMS, but he didn't seem as if he was in the mood to be doing a satellite tour. Before the station's crew had a chance to start the interview, Seinfeld quickly asked, "How's my good friend John Lanigan doing back in Cleveland?" Seinfeld held Lanigan in high regard, not only for his own comedic skills but also for giving him exposure when the New York comic was first doing standup in clubs. WMMS didn't give the comedian the same kind of respect until he was getting hot, but by then loyalties had already been established.

There might have also been some resentment from a trick played on Lanigan and Seinfeld some time before. A few years before Seinfeld hit it big with his show, he had established himself as a very popular comedian on the club circuit. WMMS had become famous for its St. Patrick's Day broadcasts with the *Morning Zoo* from various bars around the city. One particular St. Patrick's Day had the *Zoo* at Tony Roma's in North Olmsted. Since Seinfeld was in town for an appearance, efforts were made to get him down for a few words with the crowd. But Seinfeld's loyalties were with Lanigan, who had been with him from the start, and he promised an appearance on WMJI. Engineer Frank Foti recalls,

> Jeff was upset that Seinfeld was going over to Majic, and I simply told him, "No problem. If you know where he's staying, I can make sure he doesn't go on the air. Leave it to me." We found out he was staying at the Holiday Inn-Lakeside, and I walked over to a pay phone, and asked for his room. At the time, he could still register under his own name, and sure enough, Seinfeld answers the phone.
>
> I said, "Jerry, this is Bob Johnson. I'm one of the engineers at WMJI radio, and I just wanted to call and say we're having some technical problems this morning. We know you were supposed to come over here for an interview, but we thought it might be best if you didn't waste your time. Just go back to bed, and we're sorry to get you up this early cause we know you had a show last night. We're really very sorry."

Seinfeld seemed a bit confused, and told Foti, "Well, your limo driver is here in my room!"

"Hey listen. Since we inconvenienced you by getting you up this early, just have him take you out to a nice breakfast on us!"

"Gosh. Thanks! See you later."

And that was supposed to be the end of it. But later that day, Foti came back to the station, and there was a note from Lonnie Gronek saying "Please see me." He walked into the office, closed the door, and Gronek said with a slight smile, "Frank, I need a little bit of help. You wouldn't happen to know why Phil Levine [the WMJI general manager] called me all upset that Jerry Seinfeld didn't show up this morning." Foti smiled back and said, "Lonnie, I was out all morning working the remote. How the hell would I know why he didn't show?" Gronek casually mentioned, "Well, Foe-tie. It just sounded like something you would do." They left it at that, but perhaps Seinfeld was still bitter by the experience when he was speaking at WMMS's Los Angeles remote.

That same September WMJI's Lanigan tried his hand at a new medium in a cameo role in the Playhouse Square production of *Buddy*, a musical about the career of Rock Hall inductee Buddy Holly.

It was no secret in the industry that WMMS was going through some major changes, and an article in the trade paper *Radio and Records* explained them in detail. In a piece titled "Still No. 1 . . . WMMS in the 90's," the magazine called the station "musically confused" until it "officially declared its hybrid days over and returned to mainstream AOR." It quoted program director Michael Luczak,

> The previous management made some mistakes that I've had to fight; one was that they left the door wide open for Classic Rock. While they were still doing their P1A CHR thing, they had a program called *Classic Rock Saturday Night* that got 40-shares. By doing that, they basically seeded the market.
>
> When I was at WEBN-Cincinnati and a Classic Rock threatened to come on, [WEBN operations manager] Tom Owens immediately put on positioning statements like "We play the most classics" and "You never have to wait for a classic." Everybody on the street was going, "Hey, you guys are playing more classic rock." But we weren't. We didn't really change anything. We just positioned it differently."

Luczak also commented on the success of Rocco's *Gonzo Radio*,

> When I got here, our night numbers were dismal. So I said, "Let's rock. Let's play some current music and have an attitude." So we put on *Gonzo Radio* which I'll admit we stole from [KLOL/Houston's] *Outlaw Radio*.
>
> When we first started *Gonzo* we played a lot of "hair bands," but about six months into it we realized we needed to play more alternative rock bands. That's when we [discovered] that bands like Jesus Jones, School of Fish, and Alice in Chains actually researched better than Great White or White Lion, or—for that matter—the second Allman Brothers track.

He went on to applaud the drawing power of acts such as Elvis Costello, the Indigo Girls, and the Replacements, saying they tested "just fine" for the twenty-five-and-older age group. Not everyone at the station shared Luczak's opinions of the format and the direction it was taking.

Kinzbach and Ed Ferenc marked fifteen years together on December 28. Despite the serious challenges from other stations, it appeared as if they might see twenty or even twenty-five years together. But there was a dark challenge looming on the horizon, and its impact would change Cleveland radio history.

The Heart of Rock and Roll

At the beginning of 1992, there was a very unsettling rumor that made its way back from Las Vegas. It would prove to be chillingly true. It happened a couple of weeks before in a Vegas hotel room on the day after Christmas. Michael Stanley, fresh from two well-attended and highly praised Front Row Theater dates, decided to take up former bandmate Jonah Koslen's invitation to visit him out in Los Angeles. Stanley and his fiancée, Mary McCrone, headed west, making a quick stopover in Las Vegas for a visit with McCrone's sister and her husband, entertainment legend Wayne Newton.

The forty-two-year-old Stanley was an admitted junk food junkie. When he lived in Shaker Heights Stanley would sometimes drive several miles to a Taco Bell in Warrensville Heights just for a little variety in his obsession for fast food. He was also a three-pack-a-day smoker and was used to battling occasional discomfort from those habits. But when he awoke at 3:00 A.M. in the Las Vegas Hilton, he automatically knew the pain that shot into his left arm wasn't heartburn. After some hurried calls, Stanley found himself lying on a hospital gurney, with doctors working to save him from the effects of a serious heart attack.

Stanley recalls the frightening ordeal,

> I had just met my soon to be brother-in-law, Wayne Newton, so it was a very psychedelic time just at that level! Wayne had graciously arranged for us to stay at the Elvis suite at the Hilton, because he knew what a big Elvis fan I was. I had my heart attack in the Elvis Presley suite!
>
> I still remember when I realized what was going on. I knew what it was, and the first thing that went through my mind was, "This is supposed to be happening to Keith Richards, or someone like him. This isn't supposed to be happening to me!" I can remember bouncing back and forth between being completely freaked out and hysterical, and then being very calm and very outside the thing. I remember thinking, "You know, Michael, this could be it! This could be the way it all goes for you! I'll miss seeing my wife, and my kids and granddaughter, my family and friends." But, on the other hand, I wasn't going with a lot of regrets. I'd seen the world, and pretty much did what I wanted to do as far as making a living. Hopefully, I brought some pleasure to people, and I wasn't any kind of negative force in the universe!

He remembers the emergency room in chilling detail,

> I was on the gurney, and the doctors were trying to relieve the elephant standing

on my chest. There was a guy next to me, a Hispanic guy covered in blood, who was in a knife attack. His girlfriend or wife was there, and she was in hysterics! I could hear people talking, and they were saying, "I don't think he's going to make it," and frankly, I was hoping they were talking about the other guy! He sure looked worse than I did!

I was going in and out of consciousness, and all of a sudden, I open my eyes and there's a priest standing over me! He said, "Son, I'm Father So-and-so. May I administer the last rites?" Okay, so this is kind of serious! I told him, "Father, I want you to know I really appreciate this, but you're going to start saying this stuff, and I'm pretty sure I'm supposed to say something back. I'm not Catholic, you know." The priest seemed a bit surprised, and said, "Hmm. Maybe it's this guy over here!"

Doctors raced to stabilize his condition and prevent any further damage. They administered a clot-dissolving drug in an effort to open the main artery of his heart, which was completely blocked. It didn't work. They hurried to insert a tiny balloon for an angioplasty aimed at pushing aside that blockage. That move proved successful. But Stanley remained under close watch in the hospital's cardiac unit for the next week, and he flew home for further treatment at the Cleveland Clinic in January. The effects of heavy smoking and high cholesterol levels from his diet were the main culprits, and doctors set about assessing the damage and working to strengthen Stanley's chances for long-term survival.

It came at yet another pivotal point in Stanley's musical journey. The recently formed label Razor and Tie Records was about to release a Michael Stanley Band greatest hits compilation, and Stanley's new group, The Avalons, was shopping songs around for a record deal. Stanley was ordered to take it easy for a while, get some exercise, improve his diet, and throw away the cigarettes. So often after an eye-opening experience such as this the survivor is anxious to resume life, and Stanley was no exception. He was itching to get back on the radio and told friends he looked forward to heading out for some basketball. But doctors told him if he took it slow and followed their advice he would be fine. Stanley was never known for inactivity, but he did comply by sitting on the sidelines instead of playing in a charity basketball game between Euclid Mayor David Lynch's team and a celebrity lineup that included Cleveland comic and television personality Steve Skrovan, Michael Belkin, and others.

There was a continuing call for local music on the airwaves, though many programmers still had to be convinced that listeners would support it. WMMS had issued its *Northcoast Buzzard Tracks* and aired cuts from the compilation, even featuring some of the artists live on the air. But despite the quality of the artists on the CD, it just wasn't taking off the way people at the station had hoped.

Organizations such as the Cleveland Music Group led the charge for additional local music, but it was making little headway with programmers.

Over the years Scott Hanson had drawn a huge regional following in bands such as Boy Wonder, Hanson: The Movie, Slam Bam Boo, and Son of Slam. He saw the fight local musicians went through first hand, and he knows a number of key reasons why Cleveland radio didn't play home grown bands. It wasn't because of a lack of local talent. He says,

> Part of it came from national promoters, and there are only so many minutes to fill per hour for every station. Local musicians cut into promotional considerations that would come from national industry people who want their music played. But there was a responsibility on the bands' side as well.
>
> The odds are stacked against you in Cleveland because we're so isolated in the Midwest. There's no real scene here like in New York or L.A. Labels fight over you there because they want the newest and hottest bands, but it doesn't happen here. You have to go where the labels are.

Hanson had worked with people such as Michael Cartellone from Damn Yankees, former Exotic Bird Andy Kubiszewski, and Nine Inch Nails' Trent Reznor, and they all left Cleveland for that very reason. There are also a lot of acts in bigger cities. But Hanson points out that bands playing original music aren't making the money that cover bands do. It comes down to support from the promoters and public alike. As he points out,

> People want to remember the good times, but even when my bands were touted as being the next Duran Duran, we'd play shows and have twenty people show up! Cover bands were making a thousand a show, and we're making a hundred. When local bands got the opening spot for a national act, nobody ever told you they'd make you dress in the bathroom, you didn't get any backstage conveniences, and as soon as you were done, you had to get your shit out of there. Look at Nine Inch Nails. Trent had a plan, and he knew Cleveland didn't have what it takes to break out here. The scene just isn't here. Part of it was a lack of radio support, but just a part of it.

By January 13 Stanley had enough inactivity, so he began taping his *In The Heartland* shows again for WNCX. Touched by the outpouring of cards and letters he received during his downtime, he was anxious to show he was ready to go back to work, though he was still cautioned to take it slow.

While their anniversary date was officially three weeks before, Jeff Kinzbach and Ed Ferenc finally got around to celebrating fifteen years together on January

21. Kinzbach and Ferenc joked with their radio audience and a cast of co-workers such as Patty "Janet from Another Planet" Harken, who had come up through the ranks starting in the newsroom and whose sharp wit paved the way for more air work. The two reflected on their lives together and the similarities they shared, including second marriages, their ages, and even where they lived. Kinzbach was quoted in a feature article in the *Plain Dealer* saying, "Oh, I might want to hang up my headphones when I'm 45 or 50. There are lots of other jobs in radio." But there were very few like his position at the *Buzzard Morning Zoo*, and wheels were turning at other stations to show him just how few jobs there were.

Former WMMS art director David Helton continued his work for WMJI. WMMS had been blessed with two gifted artists during its heyday. Helton and Brian Chalmers, who took over Helton's WMMS responsibilities when he headed south to start his freelance career, remained close friends. The combined work of both artists improved their respective stations' print exposure in local newspapers and magazines. Each artist had a distinctive art and lettering style, though both knew the parameters in working with the Buzzard character. Whether it was evident at that time or not, it was a wise move to give work to Helton at "Majic," and keep Chalmers as busy as possible at the "Buzzard."

Chalmers was already losing hair that February by the time conversations started concerning the design of the upcoming WMMS spring t-shirt. He had a great amount of creative input and control in the past, but Michael Luczak had his own concept how those shirts should look. When those ideas were illustrated by Chalmers and subsequently rejected by the rest of the staff, Luczak took matters into his own hands.

In a memo dated February 17 and labeled "Top Secret," Luczak's wife made a request to an outside artist to come up with concept designs based on some crude outlines, using the image from that year's WMMS calendar as a reference point. The memo instructed the artist to design a shirt that is "so appealing and 'cool' that when people see one, they want to buy one.... Remember... be an artist, be outrageous... this is rock and roll... let your imagination go wild and have fun!" Chalmers continued to work on his own design, unaware that another group of artists was coming up with its own ideas independent of the station's art department.

Finally, out of desperation Chalmers did a tongue-in-cheek takeoff featuring an angular depiction of the Buzzard with sharp corners and bright colors reminiscent of some of Picasso's work. Oddly enough Luczak liked it, and it soon went into production as "the only shirt suitable for framing." Chalmers could be very passionate about his work, but Luczak could be equally committed, and the two often went back and forth in promotion meetings with Chalmers sometimes emerging worse for the wear.

Popular music was going through yet another change at the time, as the alternative sound continued to pick up steam. WMMS capitalized on it that April by showcasing its *Modern Music Show* from 11:00 to midnight, hosted by Brad Hanson. The promise was made that if any of the new music caught on with listeners, it would be put into the daily rotation. But Luczak downplayed any possible shift to an alternative format by saying, "Personally I don't think anyone in town has the balls to do it!" A surprise came that month with the emergence of WGAR-FM as the number one station in the winter Arbitron ratings. The Cleveland market was now wide open, with oldies, urban, lite-rock, AOR, and country all having risen to the number one spot over the past couple of years.

Fans of WPHR-FM were stunned on May 11 when they tuned in to hear R.E.M.'s "It's the End of the World as We Know it (and I Feel Fine)" playing over and over. Station deejay Maria Farina was among them. "There had been rumors everywhere, but I remember driving in that day because the new p.d., Lyn Abell, had asked me to do a midday shift. I heard 'It's the End of the World' and thought, 'Wow! That's a great tune. I haven't heard that one in awhile!' Then I heard it again, and thought, 'Uh-oh! Somebody made a mistake.' When I heard it a third time, I immediately thought, 'Oh no! I'm out of a job! I'm fired!'" When she arrived at the station, Farina was taken immediately to the program director's office, prepared to cry and walk out. Abell simply asked, "How would you like to do middays?" Farina practically jumped over the desk to hug him. There had been several offers to return her to Pittsburgh, but Northeast Ohio was the place she now called home and this was an opportunity to stay in a job she loved as well.

At 6:00 A.M. the following morning, a voice announced the birth of WENZ, nicknamed "The End," offering the "best of yesterday, today and tomorrow" and leaning toward an alternative mix. The station was still playing some classic artists, such as Genesis and Queen, but a lot more new bands along with some of the more artsy and techno work from the 1980s. Abell, who had been at the station for about a month, gave a vague description of the new format as "contemporary hit music for Cleveland," downplaying talk of full-blown alternative programming. Abell later revised those comments to the *Plain Dealer*, calling the format "less hit-oriented contemporary music." In explaining the reason for that abrupt change, he told the paper that WPHR played "Top-40 very badly," and in a listener-driven industry, ratings sometimes force drastic moves.

While some WPHR fans were unhappy, a number of advertisers and parties linked to promotions were absolutely livid. Roger Price and RAP Promotions had partnered with WPHR to sponsor the North Coast Con, celebrating comic art and popular culture, with some major guests scheduled to fly in for that event. But when WPHR switched to WENZ, many of those commitments went out the window, and playing the R.E.M. song didn't keep people listening in the final weeks before the event. Despite that glitch, Price's event was a huge success based mostly on word of mouth advertising from his past conventions. The station also hosted a free show featuring Toad the Wet Sprocket at the Empire Concert Club a few weeks later. Listeners wearing shirts reading "I Was There For The Beginning of The End" took home CDs and cassettes from the group. The change seemed to make sense. Like heavy metal, alternative wasn't getting a lot of airplay outside of college stations, while groups of that genre like Nine Inch Nails, Live, and Henry Rollins, among others, were selling out show after show.

There was a programming change of a different sort on WNCX. Stanley was named host of the afternoon drive show, expanding his *In the Heartland* duties from 3:00 to 7:00 P.M.. He replaced Doug Podell, who stepped away from the mike so he could spend more time as program director. Stanley was a marquee name for Cleveland's classic rock audience. He started his new duties on June 1, and even though he was still recovering from his heart problems more than five months before, he was confident the show would fly. As he told the *Plain Dealer*, "Hey, it's not like I'm lifting bricks for a living. I was anxious to do more radio." Ruby Cheeks got an earlier shift, starting at 10:00 A.M. The station was still struggling for ratings against the giants, and it hoped the changes could spark more prominent numbers.

The big rumor in early summer was the impending arrival of Howard Stern, whose syndicated show out of New York had already drawn huge numbers in Philadelphia and Los Angeles. Stern was looking to expand into other markets. Rumors circulated that WENZ was talking to Stern's agents, but unwilling to meet his million-dollar price tag. Stern took no prisoners, as he had a reputation as a devastatingly funny albeit downright vicious competitor. But it wasn't certain if a syndicated morning drive show could work in Cleveland, especially against heavy hitters such as the *Buzzard Morning Zoo* and John Lanigan and John Webster. Other cities were watching closely what Cleveland might do, and how Stern might be accepted once a deal was struck.

It was turning into another busy summer for WMMS. Work had been under way to prepare new studio and office space at Tower City overlooking the Avenue shopping complex. Rocco the Rock Dog, Mike "Mondo" Knisely, and Bill Alford, along with some engineers brought on as hired hands, worked around contractors putting up walls and offices to build working studios in just six-weeks time. It would usually take a minimum of six months work, but Rocco and Knisely spent up to eighteen hours a day, seven days a week, stringing miles of wire and hooking up computers, and coming back a few hours later to start all over again. The two moved out of their homes into a downtown hotel for the majority of that time, but they had a deadline of June 28, and they were determined to meet that goal.

Meanwhile everyone back at the Statler Office Tower was busily packing up equipment, paperwork, and other belongings for the day Commercial Movers would come in to haul it all away. The Statler was an old building, and newer office locations were making serious bids for Malrite to relocate its radio operations at very competitive prices. WMMS had been settled in at the Statler for a good seventeen years. Relocating everything was a massive job, and it all had to be there in the course of one weekend. Some staff members thought the change of scenery would be good, but for many others the Statler was home and leaving would not be easy. The offices had history. This was where Kid Leo, Denny Sanders, and John Gorman reigned for years, bringing the fledgling rock station to unmatched prominence, and where lifelong friendships had been made not only among the staff but also with the listeners. Those relationships would continue, but when the doors on the twelfth floor closed for a final time, a special time in Cleveland radio would pass into history.

Rocco, Knisely, and Alford worked to the brink of exhaustion, even supervising the movers that first weekend. They held their breath at midnight Sunday, and when Bill Freeman threw open the mike switch, the station went on the air at the new location without a problem. It was a major effort, one that no one wanted to ever repeat, but a job well done and a great tribute to an exceptional engineering crew.

The spring Arbitron ratings book held good news for WMMS, with the *Buzzard Morning Zoo* taking the number one spot. WNCX took serious hits, only showing strength in middays. The morning drive show there was circling the drain, and it became clear that changes had to be made in that time slot.

By early July it became evident that cuts at BP America, and the city of Cleveland's refusal to pay for police overtime, had spelled the end of the Riverfest. The festival drew half a million people to downtown Cleveland, and it was impossible to monitor every part of the city during that weekend. But it was still a surprise that no one would pick up the cost to ensure that the event would continue, as it was the premier summer event in the city. On July 11 a press conference drew the final curtain on that summer spectacle, but there were immediate attempts to revive it. Mayor Michael White, developer Jeff Jacobs, and others promised a smaller, but no less prestigious, event to take place the weekend of July 24 to 26. A number of stations had pledged their support. Cleveland had salvaged itself in the past, though it remained to be seen whether Riverfest, or whatever it was destined to be called, could be resurrected as well.

WENZ wanted to be a major radio outlet. The format switch was expected to draw new listeners, but the failure to win Howard Stern left the station wondering who could fill the critical morning drive slot. The answer came on July 13 with the return of Northeast Ohio natives Brian Fowler and Joe Cronauer. The former WONE-FM morning voices left Denver to return home, but they were in an uphill battle against the old flagships of Cleveland radio. They had a lot to prove, and their resources were slim against the big-money stations such as WMMS, WMJI, and others. Fowler and Cronauer did have a big advantage, their own talent and their drive to make it work.

They would need it, as rival WNCX had an ace up its sleeve. In a last-ditch effort that would either catapult the station to new heights or force a drastic change in its programming philosophy, the station decided to introduce Northeast Ohio to a man known as the "King of All Media."

The New Kid in Town

By the summer of 1992 the powers at WNCX had pretty much thrown up their hands in despair at trying to play on the same field as John Lanigan or the *Buzzard Morning Zoo*. Both WMJI and WMMS had a firm lock on their respective audiences, though numbers at WMMS had proven to be a bit soft in recent years. Rumors built that WNCX was preparing for a major shakeup, maybe even a format change, if they couldn't make a dent in the morning wars. There was an option, but it was a major risk because the show was so controversial and it wasn't local.

Mountain guitarist Leslie West claims he brought up Howard Stern's name during a visit to Michael Stanley's afternoon show. He said Stern had branched out into Philadelphia, Washington, and other major markets, and he would be a strong contender in Cleveland. Stanley downplays that story, though program director Doug Podell was well aware of Stern, having worked with him in Detroit. Station co-owner Robert Weiss says the bottom line was, "We wanted to win, and we weren't winning! We obviously had to do something."

Whoever it was that finally decided to take the risk, it was obvious that they were willing to do just about anything to make WNCX a contender. General manager Kim Colebrook obviously had a hand in that decision, saying,

> We had tried and tried to put a morning show together to make headway against WMMS. Having done that for a while, I knew that was no easy chore. Jeff and Flash were well established, and we had our team. Trivisonno was with them, and a guy named T.C. doing a traffic bit. I added up the payroll and the benefits, and I realized we could probably afford Howard Stern, and that would be like dropping an atomic bomb on the market! Podell and I got pumped up about it, and went to Norm [Wain] and Bob [Weiss] and started pitching the idea to them. When they got behind it, we eventually made it happen.

Finally, after months of speculation, it was announced in early August that Stern would be taking over the morning drive slot by the end of the month. Metroplex Communications served notice to Trivisonno, Paul Tapie, and Skip Herman that they would be replaced. Although very little was said about the transition prior to their final days on the air, there were enough veiled goodbyes to let people know the big change was coming. Jerry Shirley was reassigned to an 11:00-to-midnight slot, in addition to a Saturday morning show, and he was named the station's community affairs director. Soon after he was given full-time overnight duties.

Colebrook said talks had been under way with the Stern camp as early as June. The "King of All Media" promised to deliver big ratings, but he didn't come cheap. After lengthy negotiations over price and length of contract, an agreement was finally reached by both personality and station. Colebrook would later tell the *Plain Dealer* that Stern had been signed to "a pricey, multi-year deal that will cost the station zillions of dollars." Stern's show also promised to make back that money if it proved to be the hit everyone expected. He appealed mostly to men twenty-five to fifty-four, the core group for a classic rock station, and Stern didn't have any set hours. He took to the air at 6:00 A.M. and might stay into the afternoon if the mood so hit him.

There were those in the local press and the industry nationwide who thought Cleveland was far too conservative a market to handle Stern's outrageous act, and the sales department had to argue against that to its clients. He was going head to head with long time favorites Jeff Kinzbach at WMMS and Lanigan at WMJI, and both would be expected to rise up against the out-of-town threat. Metroplex was thought to be financially strapped at the time they hammered out the deal with Stern's agent Don Buchwald. The company was betting the farm on him.

On August 12 WNCX announced Stern's arrival in a press release, with Colebrook saying: "Not since the pre-television 'Golden Days' has radio seen a personality as popular and entertaining as Howard Stern, and we're delighted to be the station that brings this incredibly successful talent to the Cleveland airwaves.... With his noticeably fresh dimension of entertainment, and in combination with Ruby Cheeks (10 AM–3 PM), Michael Stanley (3 PM–7 PM) and Bill Louis (7 PM–11 PM), we are expecting he will do the same in Cleveland."

The projected August 31 start date gave other stations time to plot strategy. WMJI's Lanigan didn't seem concerned, and he appeared ready to deal with Stern the same way he did with every other challenger. Malrite Communications hoped to divert some of the publicity surrounding Stern's arrival by ignoring him on its airwaves, instead mounting a campaign to bring the proposed Navy payroll center to Cleveland to create hundreds of jobs in the process. They hoped to mobilize listeners, politicians, and dignitaries to rally forth and demand the U.S. government locate that center in Northeast Ohio. The campaign was originally called Operation: Payroll, later changed to Operation: Paycheck after fears that it might imply that the staff was on someone's payroll and after realizing that "payroll" sounds a lot like "payola." It was set to kick off the very day Stern debuted on WNCX. That Monday one of the first calls Kinzbach took on the air was a voice that simply said "Stern!"

Broadcasting from his New York studio to WNCX, Stern opened his first broadcast in the usual way, greeting the audience and saying hello to his new affiliate in

Cleveland. He promised a live news conference in the final hour of the show, taking jabs at his new competition along the way. But when he did speak to the local media, it was no secret whom Stern wanted out of the way first.

"To the people of Cleveland," Stern announced. "To the people of the press. I am the 'King of All Media.' I congratulate Cleveland on my arrival."

"You do?" asked his newsperson and on-air sidekick Robin Quivers.

"Yes! And now you can bask in the presence that is me. Now I will take my first question."

After a few minor technical glitches were worked out, Stern assured the media that he knew very little about the city but promised that in a few short months he would be number one in Cleveland. Stern suggested he would consider coming to Cleveland when his show hit number one, but wanted to see how well a similar Los Angeles "funeral" went before committing to any future appearances. Then he took aim at the competition.

"The guys on WMMS," Stern commented, "Flash and Frizzbone? Those are who I'm going to beat in Cleveland. They're going to be the quickest to go down. They have absolutely no talent! Doug [Podell] sent me a tape of theirs. I don't know how people can even listen to that garbage. We will win in Cleveland! I don't need to suck ass! You'll see what will happen."

A memo had been passed to Malrite employees prior to Stern's first show stating in no uncertain terms that no one was to acknowledge him or speak to anyone with the press about his arrival. Kinzbach says he was extremely frustrated, "We were given orders not to acknowledge Howard Stern, and the morning he debuted we were talking about this DFAS center, and it was the most boring thing to your audience. Once again, shot in the foot! But I did something right. I knew Stern was a power to be reckoned with. I also knew that, at that particular point, WMMS was in trouble. There were rumors that we were going to be sold, so I could see the handwriting on the wall. I said to myself, 'I don't know what could possibly happen here, but I'm going to plan for the worst.'"

He continues, "I had been listening to Stern one morning, a tape of him, and I said, 'You know, this guy is really good! When the smoke clears, he is funny.' I went out and bought stock in his company. Two thousand shares, and when it was all said and done, Howard Stern made me a hell of a lot of money! It split three or four times! Worked out pretty good."

Stern did have supporters in the press, including the *Akron Beacon Journal*'s Bob Dyer, who wrote, "You have to admit he's flat out funny.... Stern has rejuvenated Cleveland's morning drive time.... How much longer can Jeff and Flash do the same tired bits at WMMS? How much longer will John Lanigan & Co. wear us down every Thursday on WMJI with the numbingly redundant Knuckleheads in the News? Stern has shaken up the status quo, and listeners are benefiting." Dyer also pointed out that while WMMS wasn't going to mention Stern's name, "Lanigan is, as usual, more than willing to square off with an attacker." Lanigan was quoted in Dyer's article as calling Stern a "heavy-metal wannabe," and then dropped a bomb on the audience and his competition when he addressed Stern, "Just remember you were their second choice, Howard. They [WNCX] asked me first. I turned them down!" As Dyer pointed out, "Lanigan may come on too strong for some listeners tastes, especially so early in the morning, but the man has a long history of honesty, and there is no reason to believe he fabricated that tale."

Plain Dealer radio and television columnist Roberto Santiago had listened to Stern in his native New York since the mid-1980s. He pointed out about Stern's arrival,

> [It] pumped WMJI morning man John Lanigan and sidekicks John Webster and Jimmy Malone into giving their most exciting performances. . . . While unable to beat Stern at the insult game, Lanigan scored points over WMMS-FM's Jeff and Flash by becoming a hot print and television news item. . . . Jeff and Flash are ignoring Stern and have toughened their screening of crank calls from Stern fans. The duo is positioning itself as community activists. This approach has diluted Jeff and Flash's humor and, by default, has given advantage to Lanigan and Stern. If the morning radio wars ride on laughs, Jeff and Flash are losing.

Michael Heaton was an early supporter as well, writing in the *Plain Dealer Friday* magazine, "Howard Stern is the primeval everyman. He's got 57 channels in his head and all of them are on. At any one given time, some are funnier than others, but he's constantly turning the dial.

"So will Howard Stern become No. 1 in the ratings in Cleveland? Who cares? He doesn't. That's what makes him so funny. But what's wrong with Cleveland radio is something that's beyond Howard, beyond Lanigan, beyond Jeff and Flash. You know what's wrong with Cleveland radio? The music stinks. Baba-booey."

Heaton didn't share those feelings with some others in the *Plain Dealer* staff. Santiago took his lumps for his critique of Stern's first show. An editorial cartoon that ran the day after his debut showed two fisherman looking across the lake and commenting, "What's shocking is not that radio station WNCX 98.5 hauled that garbage barge all the way in from New York City! . . . It's that they're paying $500,000 a year to keep it here!!!" A caption above a tugboat with a "98.5" on its stack read, "Think of it as our @#*! gift to the @#*! city!" Stern would either have a huge impact, or WNCX could be in for a major change, maybe even starting from scratch.

But Stern also had some fans in unlikely places. The WMMS engineers wired the WNCX signal into the production booths, newsroom workstations, and their own office. Stern could be heard with the punch of a button, which could be quickly punched out if someone from the *Morning Zoo* or management was nearby.

It soon became clear that WMMS, and all of Cleveland radio, was facing what might have been its biggest challenge ever. In an attempt to counter all the publicity surrounding Stern's arrival, WMMS management decided to push its twenty-fifth anniversary celebration. The twentieth anniversary a few years before got some great press, despite the *Rolling Stone* ballot-stuffing controversy. Everyone got to work on plans to celebrate the Silver Buzzard.

It started on September 28, the date the FCC granted WHK-FM the new call letters reflecting Metromedia Stereo. Kinzbach and company kicked it off with a flashback show featuring music, news, and personalities from that era. The station contacted Billy Bass, who declined to be interviewed but was represented in archival tape, along with Bruce Springsteen's Agora show and other highlights. A long history of WMMS and its importance to Northeast Ohio was being assembled. An on-air request was made to anyone who had anything audio-related to contribute.

A listener stopped by the station with a large box full of seven-inch reels, explaining that he picked up the tapes at an estate sale for a buck. But when he looked them over, he noticed that because of age and poor storage, the tapes were brittle and starting to curl up on the sides. He also noticed they had radio stations listed on each box, and dropped them off on the chance the station could use them. The box contained air checks from Bass, Doc Nemo, and David Spero, along with tape from WHK, WIXY-AM, and KYW-AM. It turned out that the person who taped all these stations from 1964 to 1974 was a radio fan from the Near West Side of Cleveland, who hoped to get on the air himself but fell seriously ill and passed away before he realized that dream. Most of the tapes were recorded at 1⅞ speed, which is very slow but allows up to eight hours on a reel. As many tapes as possible were dubbed over, but a few of them were recorded at the 15/16 speed and machines with that capability are almost impossible to find. One of those very slow tapes had the inscription "Jimmie [sic] Hendrix interview, March 1968, Cleveland." That was put into a special place to be preserved for history.

The anniversary programming continued that day, as Matt the Cat showcased the music of 1969 to 1972. In afternoon drive, Lisa Dillon relinquished the mike to the man the station made a legend with the return of Kid Leo. He came through the door like a conquering hero, with old friends running to shake his hand and embrace him, and newer folk looking on with admiration. Leo had been back to the city and WMMS quite a few times since his departure in late 1988, but this was the first time he would take over his old shift, and it promised to be a very special afternoon. It was like Leo never left.

A call came from "Southside Johnny" Lyon of Asbury Jukes fame, who jokingly said, "I thought you already had a job!" After some heavy reminiscing about some past shows at Blossom Music Center and their adventures with Springsteen, it was like a trip back in time to when Leo reigned supreme in the afternoons. Many in the station staff just sat back and listened, riveted to every word. He wouldn't have been able to get out of the studio without his signature weekend kick off, and Leo promised to make good on that as well.

The show progressed with a call from Eddie Money, who talked about his long history with Leo and his upcoming unplugged shows. It was obviously a kick for Leo as well, as he played the songs he loved and the station made famous, such as Meat Loaf's "Paradise by the Dashboard Light," the Clash's "Stand by Me," and Lou Reed's "Walk on the Wild Side." Leo was in his element, telling the audience to call Michael Luczak and "demand ten free concerts this year!" As he put it, "I can get away with that because I'm going back to New York tomorrow, and I don't have to worry about that stuff!" After thanking his audience, and admitting he enjoyed getting back behind the mike, Leo played a few more favorites before closing the show with a song that came to identify him in Cleveland as much as its author, Springsteen's "Born to Run."

It was a great day for local radio, but there was one obvious problem that came to light early that morning. The station went on the air with its new format and call letters in 1968, which was twenty-four years before, not twenty-five. WMMS was obviously rushing its silver anniversary. One confused listener called the *Buzzard Morning Zoo*, and was told on the air that it signaled a two-year observance. The actual anniversary date would be the centerpiece and midway point of the celebration. It may have been a stretch, but WMMS faced big competition with the continuing threat of Lanigan and Webster and the rest, not to mention Stern.

Stern was drastically different from anything else on the Cleveland radio dial. He had plenty of detractors on the street as well, and like the Detroit woman who took on Fox for airing *Married . . . with Children*, it backfired into even more publicity for the intended target. A Lakewood grandmother began a public campaign to boycott Stern's sponsors and drive him from the local airwaves, scheduling a meeting and rally at the local Public Library in early October. Stern had been kept apprised of the campaign via WNCX, and even requested an on-air interview with the woman, who quickly declined. Stern sent two of his stalwart sidekicks, "Stuttering John" Melendez and "Fred the Elephant Boy" Schreiber, to cover the meeting. About 150 Stern fans showed up to get a glimpse of his on-air cronies and to voice their support for WNCX. Melendez tried to gain entry to the meeting, but was turned away at the door by Lakewood police. Instead Melendez and Schreiber joined Podell outside the library for an impromptu news conference. That very vocal crowd drowned out the proceedings inside the library, forcing the meeting to be cancelled. It drew heavy television and print coverage. Segments of the rally were broadcast on Stern's show the next day, including a Melendez interview with one of the anti-Stern loyalists. He vowed that a boycott of WNCX was still a very viable option.

The summer Arbitron ratings book showed *Lanigan, Webster, and Malone* a strong number one in double digits, followed by the *Buzzard Morning Zoo*. Even

The Changing of the Guard   389

though Stern had only been on the air a few weeks, his strength was already starting to show.

WNCX received a bill from AT&T for its 800 number to call the New York show. It showed more than 41,000 people in the listening area tried to call Stern in his first three weeks on the air, averaging out to 610 tries per hour. In that time fifty-seven made it on the air, and there was speculation as to how long fans would listen if their chances were slim of getting through to the show.

In the two months since his debut on WNCX, Stern was fighting a huge fine from the FCC, got additional sanctions leveled against his Los Angeles affiliate, tore up the *MTV Video Music Awards* as his Fartman character, got an interview show on the E! cable network, and humiliated his Philadelphia competition by interviewing and fondling John DeBella's estranged wife live on the air.

But there was still a glimmer of hope for WMMS. It continued to entertain a large and loyal fan base. Perhaps they weren't generating the numbers the station had seen in the past, but it was still a very impressive share of the audience. The Chicago-based company Strategic Radio Research/Accuratings said its mid-October survey of about 1,000 Northeast Ohio listeners ranked the *Buzzard Morning Zoo* as the number-one-rated show with a 15.5 share. It was rumored that the survey was contracted by WMMS, and Stern went on the air to say the results actually showed him beating Kinzbach and Ed Ferenc within weeks of his arrival. Accuratings president Kurt Hanson quickly denied Stern's claims, though some industry columnists claimed Stern's show was meeting its greatest competition from WMJI, with the *Zoo* falling to fourth place. Arbitron would have the final say in the fall ratings, and that wasn't due for a couple of months.

WNCX was getting a lot bolder in staging its public events, with a prime example occurring on October 27. Joe Walsh was due to play a show that weekend at Playhouse Square. WNCX pulled some strings to have him appear at Tower City's Avenue for a special presentation by the Rock and Roll Hall of Fame to talk and play a few songs live on the air with his old friend Michael Stanley. A pretty good-sized crowd gathered around the fountain for that performance, right under the WMMS studios. WMMS bannered its windows, but the real show was on the stage with Walsh and Stanley. The WNCX promotions crew was also very cautious of any last-minute kamikaze attack by WMMS, so they kept in touch via two-way radio to keep an eye on the opposition. Nothing happened, but what the WNCX crew might not have known was that every word they said on the two-way was being monitored by WMMS.

Malrite Communications was also making headlines, but in a very different way. There had been rumors through the previous summer that WMMS's parent company would merge with either Emmis or Infinity Broadcasting, the latter of which owned Stern's show. Rumors spread of Malrite having to account for a $130 million debt load by the end of December 1992. Malrite chairman Milton Maltz downplayed those rumors, but any speculation of that type tends to filter down, and it caused a good amount of uncertainty among some members of the WMMS staff. Malrite was seen as the Cadillac of local broadcasters. For many longtime WMMS staff members, it had been the only broadcast firm they'd ever worked for. Any rumored change in Malrite made the rounds quickly, and combined with Stern's arrival, caused a strain in station morale.

Maltz got some ink in the industry paper *Crain's Electronic Media* when he was quoted as suggesting Stern's show was pornographic. According to the article, Maltz said, "I think the FCC needs to exercise the authority it has been granted by Congress to expunge this kind of programming from the airwaves." His comments drew an immediate, and very surprising, response from Stern's agent Don Buchwald, who told an industry paper that they had serious conversations with Malrite about simulcasting Stern in Cleveland before they ultimately sided with Metroplex.

The news would soon be a lot worse, and add even more stress to the WMMS staff. It happened on December 11. No one wanted to believe it, but word traveled fast, and it proved to be devastatingly true.

When assistant music director Doug Kubinski got to work that day he was told that he would do Matt's midday shift. Kubinski sensed that something was going down. The only instructions he had were to be prepared for an interview with Joe Satriani. It would turn into the worst day of his broadcasting career.

Matt had been told that the station no longer employed him.

Matt was well loved by just about every person on the staff. He was an encyclopedia of information about radio and popular music, and had been loyal to WMMS for nearly twenty years. Most saw him as one of the cornerstones of the station, and when the rumors were confirmed, it stunned his co-workers as if they'd been served notice themselves. Matt was asked into a meeting after he signed off, and Luczak and Chuck Bortnick delivered the verdict. Matt left the station to reflect on his own rather than stand around to chat with shocked co-workers. Kubinski, like a lot of others, came close to tears. He recalls, "T. R. walked in, and he felt it too, but he said, 'Dude. You just have to keep going. Be a professional, and get through it.'" Tom Rezny, like so many others, was putting on a brave front, although he felt just as devastated as the rest of the staff.

Bortnick issued a memo to the staff that same day. Simply titled "Changes," it read:

As I'm sure you would agree, both in business and our personal lives, we are constantly being pushed to the limit and challenged at every turn. No company has been challenged more than Malrite and WMMS/WHK. We continue to achieve many great successes, yet, the times we live in require the stations to do whatever is necessary in order for all of us to grow in the future.

Thus, I am announcing that Matt "The Cat" Matt is leaving WMMS. Matt has been a dedicated Buzzard staff member since 1974. Matt's wit and dry humor will remain in our hearts. He understands the concept of "economy of words." Matt acted as the station historian and keeper of the archives. We'll miss his consistent attitude and commitment. His name will always be synonymous with the legend of the Buzzards.

As a result of this change, Lisa Dillon will be moving into the 10am–3pm period, with Ric "Rock Dog" Bennet taking over the 3pm–7pm.

Please join me wishing Matt the very best.

If Bortnick honestly thought this memo would sufficiently explain Matt's dismissal, or ease the pain of seeing a longtime friend and co-worker exit the building, he was sadly mistaken. There was talk that another heritage staff member, Len "The Boom" Goldberg, might also be asked to leave. There were tears shed, angry words, and most of all questions, how two of the most recognizable and respected voices at WMMS, and Cleveland radio, could be treated so shabbily. The most ironic aspect of it all came in a letter to *Scene* magazine, with WMMS soliciting the paper's help in celebrating the station's silver anniversary with a special edition issue. It noted the station's long and illustrious history and the staff of seasoned veterans who had grown with WMMS and helped make it a legend. It listed people such as Matt, Kinzbach, Ferenc, and Bill Freeman. It was dated December 11, Matt's final day at WMMS. *Scene* rejected the bid to do a "Silver Buzzard" supplement, but *Northern Ohio Live* picked up the proposal. *Scene*'s "Mouth of the Cuyahoga" column promised that the paper would do its own "official" *Unauthorized WMMS 25th Anniversary Supplement* a full month ahead of *Northern Ohio Live*, saying it would show the real story behind WMMS. Needless to say the often-strained relations between the station and *Scene* did not lead WMMS management to believe it would be a favorable review of the station's history.

As 1992 drew to a close there were reports in the *New York Post* suggesting Malrite was about to merge with a company out of Burbank, California, Shamrock Broadcasting. The employees at WMMS had gotten a pretty thick skin after hearing the endless stream of stories from the rumor mills. The general attitude was "If it happens, it happens," but still many had concerns about their futures if Malrite would no longer control the stations.

Transition

WMMS management had decided the *Buzzard Morning Zoo* needed some fresh blood. Program director Michael Luczak had originally hoped to shuffle the station's existing news anchors around, but ultimately it was WMMS newcomer Laura Farrell who started on January 4.

When asked why he added another person to the *Morning Zoo*, Luczak told the *Plain Dealer* that the show hadn't had a female voice since Ruby Cheeks, although he seemed to have forgotten about Roberta Gale and Lisa Dillon's tenures during that time frame. He predicted, "She will bring more female listeners to the show along with the male listeners." The station faced its toughest competition ever, as Howard Stern on WNCX and John Lanigan on WMJI grabbed up the male audience and WMMS promotions such as the well-intentioned albeit lackluster Operation: Paycheck campaign failed to catch fire.

Some of the biggest news that winter focused on FCC action, when WMMS and WHK-AM were granted license renewals. Those outlets had been fighting discrimination charges concerning their hiring practices, even gathering their collective staff for sensitivity training sessions. Still the FCC served notice that it would keep an eye on those stations' minority recruitment records, the same as it did for WENZ-FM when it renewed its license a few months before. That would clear the way for a possible sale of WMMS in the future.

The fall Arbitron ratings book showed what many had expected. *Lanigan, Webster, and Malone* topped the morning drive race, the *Buzzard Morning Zoo* ranked second, and Stern came in strong at number four.

WNCX mainstay Jerry Shirley was getting recognition for his continuing efforts to feed the poor, winning the U.S. Mayor's End Hunger Award for his Thirty Days in the Hole campaigns. Shirley's yearly food drive had earned him a reputation as one of the "nice guys" of Cleveland radio, and despite being put on overnights in favor of the syndicated Stern show his star continued to rise. WNCX could also be heard updating its sound by playing newer music from some of its artist roster, claiming "it doesn't have to be old to be a classic." Groups such as Living Colour, the Black Crowes, and Queensryche hardly had the kind of history to be classified as "classic rock," but they were heard on WNCX along with dozens of other contemporary artists. There was a lot of speculation about that programming move, including rumors of a format change, though most of the stories concerned freshening up the playlist to keep Stern fans listening or to get more support from the record companies.

An expected takeover was going smoothly, as the California-based Shamrock Broadcasting expected to announce its merger with Malrite Communications in a deal in which Shamrock would pay anywhere from $180 million to $200 million for the Cleveland-based company's stations. Coincidentally WMMS general manager Chuck Bortnick announced his resignation, effective February 28, moving on to become regional director of marketing for Metro Traffic Control's Chicago office. An announcement naming his replacement was expected soon.

As promised, *Scene* magazine produced its own WMMS twenty-fifth anniversary supplement that February. It included, among other articles, a critical evaluation of the station's history, an interview with Kid Leo, and overall a pretty fair assessment of what WMMS meant to Northeast Ohio. It was an honest, well-balanced portrayal, and although it wasn't anything close to the hatchet job some thought it would be, it also didn't hold back in its criticism of the station. A lot of the staff let out a collective sigh of relief. With the spotty history WMMS had with *Scene*, it could have been a lot worse. Sure enough, the very next week the paper made yet another reference to the ballot-stuffing scandal that tainted the station's reputation for the past five years.

It didn't take long to find out who would be manning the helm at WMMS. On February 24 the staff gathered in the station's conference room for the big announcement, and the new general manager walked in to be greeted by enthusiastic applause. It was Walt Tiburski, returning to the position he had left in 1984 to make his mark as a successful station owner. Tiburski came back to WMMS, at the invitation of Malrite owner Milton Maltz and John Chaffee, to oversee the station's transition in its merger with Shamrock Broadcasting. His arrival instilled immediate confidence among the staff in the station's future, and a comment he later made to the *Plain Dealer* would send shivers down the spine of its competitors. Tiburski had worked as a consultant with Ameritrust Bank in its Broadcasting and Trust departments, so he knew what kind of revenue stations needed to survive and prosper. As he told the paper, "I know the financial limitations of WNCX and WENZ, so I know what they can and cannot afford to do."

The Silver Buzzard campaign wasn't exactly reaping the promotional bonanza the station had hoped by pushing up the celebration almost a year, but there were flashes of excitement. One of them happened on March 2 at the Agora Ballroom. Queen guitarist Brian May had been promoting a solo release and touring as the opening act for Guns N' Roses. During a brief layover in the tour it was decided that May would do a free concert in Cleveland, offered to WMMS as a Silver Anniversary show, although all of the rock stations were promoting the event anyway. May stopped by the Skylight Office Tower early that day to talk about his new CD and the untimely death of Freddie Mercury. He proved to be an absolutely charming guest, happy to accommodate anyone who approached him for an autograph or photo. May also had an uncanny memory of every show he played in Cleveland, along with the date and venue, and he truly seemed to enjoy his visit to the station.

On March 11 the inevitable happened with the announced stock-for-stock merger of Malrite Communications Group with the Burbank-based Shamrock Broadcasting. The combined value of that transaction was substantial, more than $300 million. It was a difficult time for a staff that had been stable for so long, but now faced an uncertain future. As music director Brad Hanson remembers it, "Nobody knew if they were going to have a job the next day! How could you expect to be successful if you don't know if you're going to be there tomorrow? How could people put their hearts into it when the situation is changing all the time? With Malrite, there weren't really any surprises, but when a new company comes in, people start to worry about themselves and not about the station. That's just human nature." In the end the employees just decided there was nothing they could do to ensure their future employment, but still expected to rally around the new owners.

By the middle of March, a familiar voice had returned to the airwaves when Matt the Cat started a weekend show on WNCX. It was hoped by his friends and the average radio listener alike that he would see a lot more time on the air soon.

The weather proved to be a problem in the final days of winter, and several Northeast Ohio concerts had to be cancelled because of snow. The Wallflowers couldn't get out of Pittsburgh to make their Peabody's Down Under date, and the second of the Grateful Dead's two nights at the Richfield Coliseum fell victim as well. Knowing the band's penchant of providing for its fans, the Grateful Dead partnered with WMMS for a hastily arranged live broadcast of the first show. Because of the weather Rocco the Rock Dog headed out early to the Coliseum in the station's "Rover" remote truck to do call-ins leading up to the show. He drove headlong into a blinding snowstorm, and it was getting uglier all the time. But he was still able to get a signal back to WMMS. As he recalls,

> They decided after I'd gotten there that the first show would go on as planned, but the next night would not, and there would be a lot of angry ticket holders for that sold out show. Someone with Belkin Productions suggested to Walt [Tiburski] and me that we might be able to broadcast the show live, and give the Dead fans a chance to hear, if not see, the band while it was battling the Northeast Ohio winter. I knew we were only set up for a mono broadcast. It was just supposed to be me and a microphone doing some breaks from the show, and there was no left and right channel.
>
> The next thing I know, I'm pow-wowing with Jules Belkin and Walt, and the longtime sound engineer for the Dead. This guy was the professor of rock and roll sound! He invented equipment that would come into common use around the world! Yeah, I was a little in awe, but I've also got a problem on my hands. The Dead wouldn't allow the show to be broadcast if it wasn't in stereo. How do we pass up this opportunity? I did what any good radio person would do in a situation like this.

The Changing of the Guard   **395**

I lied, to save the show!

He asked me directly if I could put out a stereo signal, left and right channels, the whole bit. They were giving us the board mix, a stereo feed, and I just kept nodding my head. I could only play stupid for so long, when this guy pulls out an audio analyzer, and checks the Marti frequency and the output of the Rover! I was busted! He knew we only had a single carrier, not the twin carrier needed for a stereo broadcast. He looked me right in the eye, and I knew he caught me. But I'm standing there with Walt, Jules Belkin and, at this point, we're at the edge of the cliff. We've come this far, and can't turn back. It was twenty minutes to seven, and the Dead were going to start playing at 7:30.

This "professor of rock" turns around and yells for a guy named Howard, and they have a few private words between themselves. He comes back to me and says, "I know you're bullshitting me, but we're going to do it anyway. I'm going to give you a separate mono feed that Howard will mix down for you." Cool! That was also the first time I'd had the opportunity to work alongside with Walt, and failure was not an option.

For years afterward both Rocco and Tiburski would still fondly recall the events surrounding the tense moments leading up that broadcast. To mark the occasion, the Grateful Dead kicked off that live broadcast with the song "Cold Rain and Snow."

Stern's morning show on WNCX continued to gather strength, but the station found itself in a very precarious position following the tragic deaths of Cleveland Indians players Steve Olin and Tim Crews in a boating accident on Florida's Little Lake Nellie during a break from spring training. Stern's "Shipwrecked Bodies" song, set to the tune of the *Gilligan's Island* theme, had sound effects and lyrics poking fun at the circumstances surrounding the horrible accident, including references to alcohol use on the boat. Program director Doug Podell was able to cut the song before it made it over the airwaves, as he was contractually allowed to do, but the publicity surrounding that song drew a tidal wave of angry calls and letters.

Dave Jockers was running the board for the Stern show at WNCX, and he shrugged it off as just a tasteless radio parody. But the phones at WNCX started lighting up with complaints. Later during Robin Quivers's news segment there were more comments, and the complaints started flooding in. As Jockers recalls, "The show had only been on the air for six or seven months at the time, and advertisers actually complained to Norman Wain and Bob Weiss, who came down hard on the general manager, and that went to the program director, Doug Podell. It got all the way to me, asking 'Why didn't you cut it off?!'" Jockers had not been given any set guidelines for dealing with situations like this, which might have been considered censorship, but nonetheless he was suspended for three days. Still no one else at the station really knew how to run the board smoothly to put it on locally, so the day after the suspension Podell asked Jockers to come back.

Podell phoned the Stern show to say advertisers were "up in arms" and the reaction was "very disturbing," though he was later quoted as saying there had been no loss in advertising.

WMMS's *Modern Music Show* saw its final broadcast on March 26, just one month short of its one-year anniversary. Despite good press it just wasn't pulling any numbers in the 10:00-to-midnight slot. General manager Tiburski called the cancellation "an unfortunate decision that had to be made." It was decided to focus the programming on tried and true elements of mainstream rock such as the Who, the Rolling Stones, Jimi Hendrix, and even Pearl Jam, which had served the station well in the past. Much of the music could be heard on college stations, though alternative sounds weren't likely to post any significant numbers on the commercial side.

WENZ's Maria Farina took a step to the other side of the music business when she flew to Los Angeles to sing back up on Thomas Dolby's song "Cruel," and she appeared in the video as well. Farina had gotten to know Dolby from her station's support of the artist, who was also due to deliver the keynote address for the Undercurrents '93 music conference in Cleveland that May.

When change is imminent, rumors start to spread almost immediately. That was the case in mid-April when stories started to circulate that *Buzzard Morning Zoo* co-host Jeff Kinzbach would most likely be leaving WMMS after the Malrite-Shamrock merger. Kinzbach's contract was due to expire that December, and speculation centered on him possibly leaving WMMS to be replaced by Don Imus within the next four months. The *Plain Dealer*'s Roberto Santiago wrote on April 15, "Kinzbach is fighting for his career. Last month, the WMMS host met privately with a WMJI-FM/105.7 executive to get ideas on how to improve his morning show. WMJI is the station that carries ratings champ John Lanigan." There had also been speculation that Kinzbach might be heading to Philadelphia. Adding fuel to those rumors was the apparent name change of the morning show from the *Buzzard Morning Zoo* to the *Morning Show*, with staple segments such as the Token Jokes being sidelined for the time being. There was a reported rumor in *Scene* quoting Kinzbach as telling a friend in New York he'd "be gone by the end of the year." Time would tell if Santiago or *Scene* would be proven right.

Kinzbach was not happy about the reports. He wrote to *Scene*, "Before you printed this tabloid trash, why didn't you just pick up the phone and call me? There are human beings and families involved here. Some of us are trying to refinance our homes to get lower rates. What do the banks think when they read this stuff? Hey, in case you haven't noticed, it isn't easy making a living nowadays. I guess I just expected a little more from the *Scene*. One more time . . . I am staying in my hometown, staying at WMMS, and we are planning some pretty spectacular morning radio!"

In the middle of rumors and promises, WMMS was plotting strategy to dominate the upcoming Rock and Roll Hall of Fame groundbreaking. Along with a month long series of *Hall of Fame* one-minute bumpers featuring comments on the Cleveland project from some of the biggest names in music, the station set about assembling a backlog of greetings from names such as David Bowie, Mick Jagger, Robert Plant, and many others to give the on-air perception that the artists had aligned themselves with the station as the definitive source of information about the Rock Hall. There was also a lot of preparation for a day-long broadcast of the groundbreaking, which included airing the ceremony live, commentary before and after the presentation from station personalities acting as on-scene reporters, and wrapping it up in the afternoon drive. Prior to the event, Rock Hall officials suffered a major publicity snag when they were forced to cancel the all-star concert to be held the evening of that long awaited milestone.

A front-page article in the *Plain Dealer* announcing that show speculated that Paul McCartney would be headlining a concert at the Cleveland State Convocation Center, with acts such as Little Richard, Carole King, and other artists spanning the history of rock and roll. It made sense because the closest that McCartney's latest tour would bring him to Cleveland was a date at Cincinnati's Riverfront Stadium, and the former Beatle never made it to the band's induction to the Rock Hall back in 1988. A show honoring the Hall of Fame might show his support for the museum, despite the apparent snub a few years before. But the article touched off a firestorm of controversy, with Rock Hall officials refusing to comment on the lineup until all of the artists had been officially signed. McCartney's publicist Joe Dera called reports that McCartney would headline the show "irresponsible." On top of that there reportedly was a problem of securing "minority acts" for the bill, with Little Richard said to be refusing to take part in the concert because the New York officials failed to invite him to the Los Angeles induction ceremony.

It was later revealed that plans called for McCartney to perform with Little Richard in the show's finale, though McCartney supposedly never agreed to headline the concert. It was soon announced he wouldn't be part of the lineup. This was supposed to be a major fundraiser for the Rock Hall, with ticket prices expected to be in the hundreds of dollars for the Convocation Center and less expensive seats for the simulcast to nearby Public Hall. Without McCartney at the top of the bill, it was extremely unlikely the Rock Hall could demand that kind of revenue, and the show was cancelled.

The initial confusion arose when *Plain Dealer* writer Mike Norman was said to have asked his sources at the Rock Hall if McCartney was the biggest name on the bill. The Hall of Fame reportedly confirmed it, but did not state McCartney would headline the show.

A press conference was scheduled for May 13, and a remote WMMS broadcast was quickly set up. There was a lot of tension in the air, and the press was seated in front of the dais to hear the status of the groundbreaking concert. As expected, the announcement was made that the all-star show was not to happen, but the good news centered on bonds to finance the Rock Hall selling out the first day they were offered. Some of the other plans to celebrate the long-awaited groundbreaking included a silly citywide "dress up like your favorite rock star" event, but those were overshadowed by the importance of the upcoming festivities. Cleveland was finally going to build the project that would help take it into the twenty-first century, and it would be another cornerstone in its renaissance as a world-class city.

WENZ continued to struggle in the ratings, but still found time to celebrate its first anniversary with a free concert at the Nautica Stage in the Flats, with acts such as the Posies, Lemonheads, and Judybats. It drew an impressive audience of 4,500, showing that while young listeners usually didn't fill out ratings books they still supported their station and its music. The challenge for WENZ was getting that message across to its advertisers, and getting the kind of revenue needed to keep the station healthy.

The morning ratings race was turning into a very heated battle between WGAR-FM, WMJI's *Lanigan, Webster, and Malone*, and WNCX's *The Howard Stern Show*. Stern was scoring impressive numbers despite a very strong lead by Lanigan and crew, while the renamed *Buzzard Morning Show* was starting to show signs of ratings fatigue. There was a good bit of controversy at the WMMS program when quick-witted traffic reporter Patty Harken was dismissed. There was a good amount of speculation why, with both the *Plain Dealer* and *Scene* suggesting it stemmed from an on-air remark that poked fun at Kinzbach's sex life. Whatever the real reason was, Harken was reassigned to another station and was soon replaced by the golden tones of local singer Maria Jacobs.

A lot of attention was centered on the June 7 groundbreaking for the Rock Hall. After years of waiting, the city of Cleveland basked in the international spotlight as Billy Joel, Pete Townshend, Chuck Berry, Ruth Brown, and Soul Asylum's Dave Pirner stuck shovels into the dirt at North Coast Harbor to officially begin construction of the long awaited shrine to rock and roll. The plans, both on-air and on-site, came off without a hitch, and it seemed as if Cleveland was finally on a fast track to the future. But it was still anyone's guess how WMMS, which had been one of the Rock Hall's most ambitious supporters, might fare in the coming months and years.

By mid-June, the powers at WNCX were feeling very confident about a morning show ratings victory, reportedly budgeting $100,000 for a Stern remote "funeral." Plans were being made to hang Kinzbach, Lanigan, Ed Ferenc, John Webster, and Jimmy Malone in effigy, and then dump their "bodies" into Lake Erie. As it turned out those plans were a bit premature, as WMJI, WGAR, and even WMMS showed strength in the rolling Arbitrends, which preceded the actual ratings book. That was quite a jolt to WNCX, but there was also speculation that although Stern got a good amount of criticism for his parody song about the deaths of two

Cleveland Indians, the subsequent publicity also generated a sizable audience tuning in to see what the fuss was about.

The Cleveland Music Group continued to present monthly forums on the state of the Northeast Ohio entertainment scene. Its July 20 gathering was significant for a number of reasons. With the topic of "The Status of Commercial Radio in Cleveland, and the Prognosis for the Future," the meeting featured WNCX operations manager Podell and music director Andrea Morris, as well as the *Plain Dealer*'s Santiago. But the real draw was the addition of WMJI's John Gorman to a panel moderated by former station head Eric Stevens.

Gorman and Stevens had been perceived as bitter rivals when their respective stations WMMS and WMJI butted heads in the 1970s and 1980s, but many got a surprise when one attendee suggested that WMJI was a failure. Gorman took immediate issue with that statement, giving serious credit to Stevens and his staff for the solid programming they put on the air every day. There were some cheap shots taken by some management toward former staff members, but Gorman made it perfectly clear that he always considered WMJI and Stevens very tough competitors, and he respected them for it.

Another rivalry took an unlikely turn as WNCX finally beat out WMMS in the spring Arbitron ratings book. But *Scene* was quick to point out that in the present scheme of things—with WMJI and WGAR beating WMMS on a fairly consistent basis—it was much like the Indians crowing about making it to fourth place. The bigger story was the delay of Stern bringing his traveling road show to town for a Cleveland funeral, as WMJI bested WNCX in morning drive by more than four points.

Across town Kinzbach, Ferenc, and company were feeling the brunt of Stern's popularity, with their number down 50 percent from a year before to just 6.1. There was talk that Shamrock Broadcasting would announce its merger with Malrite by July 28, and there could be staff changes to go along with it. The hammer finally came down for WMMS and sister station WHK on Monday, August 2, when the $300 million merger closed, ending twenty-one years of ownership by the same Cleveland company. There was a good amount of enthusiasm among the staff that positive changes were on the way. But there was also a storm cloud on the horizon, as WMJI's Carl Hirsch told his staff he would form a holding company aimed at buying another local radio station. It was generally accepted that Hirsch wanted to take over his old stomping grounds at WMMS, but it was still just a rumor in the summer of 1993. Time and lots of money would determine if that dream would be realized.

Enter the Inner Sanctum

The call for more local music on Cleveland airwaves resulted in a new show on WENZ-FM called *Inner Sanctum*, premiering on August 15, 1993. The show was originally scheduled for Sunday nights with a different guest host every week. The station said the show would be treated "like any other part of the day. Just because it's from Cleveland doesn't mean it's good. Look for good songs, and play only the good stuff."

The Agora's Linas Johansonas (better known as Johan) had been a familiar face in the Cleveland area for many years, and he was asked to be the host of the first show. Pat Johnson was brought on to produce the show.

Johansonas says the philosophy of the show was based solely on what was best for Cleveland's music scene with no influence from the sponsors, including *Scene* magazine. It sometimes led to words between the host and *Scene*'s associate publisher Keith Rathbun, which Johansonas said occasionally escalated to screaming matches, over issues such as the proposed "*Scene* Cut of the Week." Johansonas recalls he remained steadfast by saying, "I'm not going to let anyone influence what goes on the show!" But he was happy to point out articles, covers, and other features the hosts deemed relevant. They also agreed that words such as "local" and "original music" would not be used because of a prejudice some listeners had about homegrown talent.

The management was apparently impressed enough to ask Johansonas to help produce the next week's show. It wasn't long before Johansonas's weekly report on the Cleveland music scene turned into a job hosting *Inner Sanctum* on a permanent basis. The first song played on the debut show was "Everything Is Different Now" by the Exotic Birds. A song followed it by a band that was to get a huge push from *Inner Sanctum*, Dink, with "Green Mind." The tune got such a reaction from Johnson that he launched into a wild, frenetic dance that sent a chair crashing into a studio wall. The song quickly made its way up the national college charts, ranking number 30 in the respected industry magazine, the *Gavin Report*. Dink's breakout tune was still listed years later as a top request, along with bands such as Pearl Jam, Green Day, Weezer, and others. Rick Michaels started getting requests for Dink on his afternoon show, so he decided to try it out in regular rotation for about a week. The hunch paid off, because it became a huge hit on WENZ.

Johansonas, Johnson, and eventually Jim Benson developed a good on-air rapport, and all three had very discerning ears for local talent. They helped raise the bar for both production and the quality of the original songs written by the bands. "I can't tell you how many tapes I used to get from the Fifth Wheel from the beginning of the show," says Johansonas. "I just didn't hear it. But when CMG

[Cleveland Music Group's] 'Rust Belt Eruptions' came out, I decided to take that Sunday off. Benson hosted that week so I could hear how it sounded on the radio, and in the middle of the set, it was like 'Wow! What is this song?'" It was Fifth Wheel's "As Close as I Can Get." Johansonas thought, "This is a really good song. Who is this?" When Benson said it was Fifth Wheel, his reaction was, "You're kidding me!"

Johansonas lived in an apartment near the Agora, and he recalls immediately marching over to the club to pull every Fifth Wheel tape to listen to determine if he'd missed other songs. When the band's *Nothing* CD was released a short time later, Johansonas heard one potential hit after another, with most of it getting airplay on the *Inner Sanctum*. Many of the newly polished songs had already been submitted as demos to *Inner Sanctum*, but were turned down because of low production values. Johansonas later remembered the band admitting, "We used to listen to your show every week," recognizing the high standards *Inner Sanctum* had set for local musicians to get airplay.

Inner Sanctum also played a wide range of Cleveland area music, including country from Roger Martin, rap from Bone Thugs-N-Harmony, metal, power pop, and just about anything that sounded good with Northeast Ohio roots.

The show's reputation gained by leaps and bounds, most of it by word of mouth, and soon a number of national musicians made it a point to tune in when they came through town. Artists such as Trent Reznor, Dave Thomas of Pere Ubu, and others communicated to *Inner Sanctum* through Benson, with Thomas even sitting in as a guest host. Benson had established good relations with many of the artists by recognizing their early work, such as by airing the first interviews with Reznor when he was with Exotic Birds and the first version of Nine Inch Nails and with Richard Patrick when he was with The Act.

Patrick, onetime guitarist for Nine Inch Nails, was also a fan of *Inner Sanctum*, thanks in part to the push Johansonas, Johnson, and Benson had given him when he was looking for additional personnel at the time he formed his new band, Filter. Johnson debuted Filter's "Hey Man, Nice Shot" on *Inner Sanctum* weeks before it was picked up nationally. It resulted in Patrick offering *Inner Sanctum* sneak previews of some of the band's upcoming work, interviews, access to the band, and even IDs for the show.

A similar example was Marilyn Manson, who first got commercial airplay in Cleveland on Johnson's *Deep End* show. His band was rehearsing at the Agora for an upcoming tour, and Manson heard the final mixes for his *Antichrist Superstar* CD for the first time when it was aired from a tape on WENZ. When the rehearsal had finished for the night, Manson tuned into the station on the car radio. His entourage kept driving around the city of Cleveland until they heard the entire album premiered that evening, a full two months before its release. Manson once paid a visit to the station after a haircut in a salon at the Halle Building, complimenting then-program director Sean Robertson for the type of programming WENZ was airing.

The program sparked a lot of local bands to put down original tracks in the studio, hoping to get them aired. As local music vet Scott Hanson notes, "At one point there were all these local bands starting to come out with CDs, because *Inner Sanctum* started playing them. They started saying, 'We can put out our music and it'll get played on the radio!'"

By November *Inner Sanctum* was getting a lot of favorable response, so Johansonas set up a Thanksgiving show for the Odeon in the Flats, with Silicon Souls, Slackjaw, and several other bands. The response was overwhelming, with a sold-out house and a waiting line that went a good distance up the hill past the train tracks. As people left the show others were allowed in, and the crowd never let up as the night wore on.

Among the acts benefiting from *Inner Sanctum* airplay was the Sons of Elvis, as the show helped them get signed to Priority Records. Johansonas recalls beaming with pride as, "They appeared on *The Jon Stewart Show*, and [the song] 'Formaldehyde' became a huge hit." The critical acclaim, awards, and recognition the show received were vindication for Johansonas, who remembered WMMS program director John Gorman predicting in *Scene* that a local music show like *Inner Sanctum* would never fly in Cleveland. Its first anniversary was marked with a huge sell-out show with Dink at the Cleveland Agora, with live cutaways to WENZ.

The Sunday night listening audience was somewhat limited in size, but *Inner Sanctum* got respectable numbers. The show was honored by numerous publications including *Northern Ohio Live*, which named it Best Radio Show in its annual review. The program also placed second behind powerhouse Howard Stern on WNCX in the *Free Times*'s Readers Poll, placing ahead of heritage names such as WMMS's Jeff Kinzbach and Ed Ferenc.

Part of *Inner Sanctum*'s popularity came through the sense of drama that was established every week. Johansonas freely spoke his mind on any number of issues facing Cleveland's music scene, and he wasn't afraid to step on toes. Critics, writers, promoters, and even other radio stations would occasionally feel his wrath. It actually resulted in at least one lawsuit that WENZ settled out of court.

The demise of *Inner Sanctum* might well have stemmed from a disagreement between the hosts, the new program director at WNCX, and eventually the Numbers Band, which at the time had been playing gigs around Northeast Ohio for more than two decades. Some time before, Benson had traveled to the South by Southwest new music conference in Austin, with the understanding that WENZ would pick up his registration fees. He recalls, "When I got to South by Southwest, the registration had never occurred, and I made a number of calls back to the station. I was told it would be taken care of, and when I checked back with the table, found there was a fourth registration for The End for a person who had yet to show. They gave me his credentials, and asked me to check back every day, which I did the first and second days." Things started to unravel on the third day of the conference when a name unknown to Benson showed up asking for those credentials. It was the new program director at WNCX, Bob Neumann.

Looking back, Benson says, "When I went back that third day, and keep in mind that this conference is only four days long, and Neumann was upset his credentials weren't there as expected. They made up credentials for him, and when I checked in they were pretty hostile to me because they suspected I stole a badge from them, which was far from the truth. So, I just turned the badge in and walked away." That incident reportedly sparked a major confrontation between Neumann and Johansonas.

Johansonas was said to have asked Neumann, "What is the program director for a classic rock station doing at a showcase for up and coming artists? Are you going to wait twenty-five years to play one of these bands?!" The sarcastic yet truthful comment would ultimately come back to bite both Johansonas and Benson. When a new program director comes into town, it's usually best to stay in the most positive light. The South by Southwest incident was an embarrassment that didn't help anyone.

Within six months Neumann had assumed control of both WENZ and WNCX, and things didn't look good for the alternative station. Another situation lent itself to controversy and more negative press, and Johansonas and Benson were soon back in the hot seat.

Soon after the South by Southwest incident, the Numbers Band was doing a gig at the Cleveland Agora, and they had parked their truck in an alley near the venue. They reportedly left the equipment truck unattended, and their gear was stolen. The band appealed to the public for information leading to the return of their equipment. But Johansonas, Benson, and Johnson publicly questioned why a veteran band wouldn't know better than to leave their truck unattended, which is an open invitation to thieves hoping to make a quick buck. The resulting furor from those close to the band resulted in big changes on *Inner Sanctum*.

Whether the incident had any bearing on the dismissal of Johansonas and Benson could be debated, but nonetheless they soon found themselves without a radio gig. Benson recalls being told two weeks before that they were going to be let go, but a letter to *Scene* criticizing *Inner Sanctum* for its perceived insensitivity didn't help.

After Johansonas and Benson left, *Inner Sanctum* continued for a time with Johnson, but the show wasn't the same without all three sharing the mikes. As Johnson recalls, "I played music, but my base of knowledge for the local scene was nowhere near as deep as theirs." Not long after in March 1998, Johnson left the station after six productive years, with Rocco the Rock Dog taking over the local music segment. But Johnson owned the *Inner Sanctum* name, so it left with him.

Hanson was a fan of the show, but he says it takes a commitment to keep a show like that on the air. As he puts it, "*Inner Sanctum* was a great thing, but at the same time, what's the first thing that got cut? They were developing new product for people to buy, and the companies behind these radio stations don't realize the power they have. They can create a new economy, but they're so structured that they don't use their imaginations. Remember when WMMS was in its heyday? It was great because the people in charge were young and had a lot of freedom. They had a gut instinct and that's what made them great."

Johansonas says the bottom line to *Inner Sanctum*'s success was that, "We broke bands. We stimulated the music scene and brought national attention to this town. Record companies actually started coming to Undercurrents [the Cleveland music showcase] and paying attention to Cleveland. It got local music in regular rotation on The End—Slackjaw, Greenhouse 27, Fifth Wheel, Dink—on and on and on. It made people like Richard Patrick not forget where he came from. We had legitimate success, and it's well documented, and we drew people to shows. There were packed houses to see local bands."

Days of Uncertainty

WMMS countered WENZ-FM's new program *Inner Sanctum* by soliciting contributions for its *Northcoast Buzzard Tracks, Vol II* project, planned for a release on Friday, November 26. The first collection got a lukewarm reception, so it seemed a bit odd that the station would be rushing out a second collection so soon. But it still gave local artists an opportunity to get their music out to a wider audience.

A former WMMS personality was doing his part to get Cleveland music heard as well, as Matt the Cat produced *Video Jams* for WAKC-TV's Saturday night schedule. It contained a special "Home Grown Talent" segment, featuring Northeast Ohio musicians. Along with the publicity surrounding the recent Undercurrents '93 showcase and the efforts of the Cleveland Music Group, it seemed as if the local scene was finally going to get its due. As always the quality of the product available and, more importantly, listener support would determine its fate.

Alternative rock might have fared poorly in the ratings, but WENZ continued to draw sizable crowds to events such as Endfest II, which presented a series of bands, and hosting the streamlined edition of Tears for Fears at Nautica Stage. A familiar voice on the alternative scene was Sue Csendes, who was a staple of college radio and a morning producer at WHK-FM. She took a weekend job at WENZ, in addition to her fulltime duties at the Agora. Csendes had a wide knowledge of cutting-edge rock, but was best known for scooping the commercial giants by premiering U2's *Achtung Baby* album on WUJC-FM. Her defection didn't sit well with the folks in charge at the Skylight Office Tower.

By October the rumors that Carl Hirsch would take over WMMS started to surface again, supposedly for a price of $12.5 million. No one really discussed it in the halls of the station, but it was widely known that Hirsch was a serious player and if there was any station he most likely wanted, it was WMMS. The rumors became a reality on November 3.

The Long Goodbye

Shamrock Broadcasting owned WMMS for less than a year when the first rumors started up. For a staff that had seen only minor changes to its core personnel over the years, the thought of yet another transition was a bit unsettling, but still no one expected any wholesale changes. The switch from Malrite to Shamrock was fairly painless despite a good bit of belt tightening, but the staff was still intact. It was the calm before the storm.

Word soon leaked that the California-based OmniAmerica was ready to move on WMMS and WHK-AM. During his first stay in Cleveland, Carl Hirsch had been seen by some as the heir apparent to run Malrite at some future date. His departure a few years before to try his hand at station ownership was not reportedly welcomed by some higher-ups at his former employer. Many industry insiders laughed at the price he paid for his Los Angeles flagship KJOI-FM, but they stopped laughing when they saw the profit he made when he sold it. Hirsch signed the papers to buy WMMS, WHK, and WMJI-FM on a plane parked at Burke Lakefront Airport. The purchase still needed FCC approval, but it certainly appeared that the final days of the WMMS status quo were at hand.

The year of 1994 brought big changes to WMMS. In the first week of January program director Michael Luczak left the station, with consultant Jeff Pollack guiding the station until OmniAmerica's John Gorman returned in early spring. Luczak's tenure at WMMS had been a stormy one, marked by conflict with certain staff members, record companies, and even the press. *Scene* magazine's "Mouth of the Cuyahoga" column pulled no punches in announcing his departure, "It was increasingly apparent that the man was just not the right person for the job of putting 'MMS back on course. Luczak's bullshit had alienated much of the local music community, and even after three years at the helm, he still didn't have a clue as to what the station should sound like."

Luczak came to the station nearly four years before, full of enthusiasm and confidence, but faced some major challenges as early on as his first day. He would later say he never expected to come into a workplace so deeply entrenched in its own politics. He explains,

> There were so many egos, so many hidden agendas. Terrible—and I may not have been prepared for that. I came into an atmosphere where people had for many years not gone through the chain of command. If a disc jockey had a problem with a program director, he wouldn't go to that p.d. or the general manager. He'd go to the g.m.'s boss, the president of the company, who was right down the hall. That went on for years and years and years, and that's what I had to deal with almost from day one. I don't know who fostered it, but it was allowed to be that way, and I dealt with it. But I don't regret it.

Among the hurdles any program director faced at that time, and especially at WMMS, were the changing musical landscape and the changes in ownership. Luczak speculates that Malrite knew it would be selling its radio properties when they hired him, though he didn't, and he came in under the premise that he'd be working for the same owners for some time to come.

Doug Kubinski had distinguished himself as a fill-in disc jockey and a very hard worker, so he was named interim program director under Pollack's guidance. Music director Brad Hanson was relieved of that title to concentrate on his air shift. Kubinski was not prepared for the impending transition. He says, "I'm sitting in that office, and all of a sudden John Gorman calls me. 'Come on over, I want to talk to you.' So I grabbed my notebook, ran over [to WMJI], and walked into this dark conference room with Dean Thacker, John, and all these people. I was really overwhelmed by the whole thing. In the last few days, they couldn't officially take over, but they could talk to me and let me know what's going on. But having to walk back [to WMMS], put on my other hat, smile at everybody, and say, 'Hey! It's going to be cool!' was real tough."

There was much speculation on the future of the *Buzzard Morning Show*, which had been struggling badly in recent years. Jeff Kinzbach's contract had expired, Ed Ferenc's was up in a few weeks, and there were strong rumors about lingering bad blood between Kinzbach and Gorman. General manager Walt Tiburski said he would also leave when the Shamrock-OmniAmerica transition took place, but said the current morning show lineup would remain in place while he was there. There had also been talk that Hirsch would bid on Howard Stern's contract when it expired with WNCX. That would give OmniAmerica the two biggest guns in morning radio, in addition to WMJI's *Lanigan, Webster, and Malone*. It was just uncertain if WNCX would allow the Stern money machine out of its grasp.

WNCX also saw its share of changes. Jerry Shirley left the all-night shift, but he remained on the air with a one-hour, 11:00 P.M. *British Invasion* show. Shirley's public relations value, not only as a classic rocker but also for his charity work, still made him a valuable name at WNCX. Midday voice Ruby Cheeks said she knew months before that her contract wouldn't be renewed when it ran out that May, so she set her sights on a talk show if another station didn't make an offer first.

Just relieved from his music director post, WMMS's Hanson decided it was time for a change. He turned in his two-week notice to head north for a position to program sister FM stations in Madison, Wisconsin. Besides not knowing what might happen in the transition to OmniAmerica, Hanson was lured by the chance to program an alternative music format. There were a few others at WMMS who wished they had similar options.

The takeover of WMMS was the topic of conversation on the streets and on the air weeks before it happened. *Scene* recalled Kinzbach saying he'd been "whacked," and big changes were on the way. New station IDs promised "new music first," although there was speculation on a format change focusing on the so-called "Triple-A" format (Adult Alternative), which was a mix of old and new music.

The takeover date was set for Friday, April 15, 1994, and Tiburski tried to reassure the staff for a smooth transition. Everyone was told to re-apply for their positions, and interviews would follow soon after. Some people were never called, which proved very unnerving to the rest of the staff. Kinzbach didn't even apply, having no desire or expectations to ever again work with Gorman. But his long time partner Ferenc did apply, which came as a surprise to many at the station and Kinzbach himself.

Gorman says he would have preferred to keep Kinzbach and Ferenc together. He explains,

> But the fact that they weren't allowed to retaliate against Howard Stern, and especially being at a station known for taking challenges and getting in scraps, when they allowed Howard Stern to walk all over them and not even acknowledge his presence, it was wrong. It hurt Jeff and Flash, and I think if it was left up to them, they would have taken him on. They would have got some extra time. Lanigan went head on against him, and it worked.
>
> All of a sudden, we started to see a lot of Jeff's audience coming over, and it had nothing to do with Jeff Kinzbach. It had everything to do with the fact that they put the leash on Jeff. They had a muzzle on him! It didn't sound like the same show.

Interviews for WMMS jobs were held at WMJI. The new staff was set, with some current WMMS staff members retaining their positions, others being let go, and still more refusing OmniAmerica's offers. It was decided that Ferenc would be kept on for the time being. The entire staff, even those who weren't called for an interview, kept up a brave front and went about their business as usual, despite the inevitable.

In the days leading up to the final show, a petition drive to save the morning show made the rounds at Camelot Music stores, General Cinemas, and other locations. It produced about 1,700 signatures, but there was no hope that a campaign of that type would save the show or the rest of the staff.

The final day had been anticipated for some time. There was much speculation in the press and on the street about the new format and new personalities. Even the existing air staff would sometimes make mention of the change that was about to happen. April 15 came all too quickly.

The *Plain Dealer* had been running articles about the transition for some time, giving it mention on the front page, so a huge audience was expected to tune in. But the station had been such an important part of so many lives for such a long time, so it was assured that there would be audience even without the press coverage.

Early in the day former WMMS owner Milton Maltz called to congratulate Kinzbach and Ferenc for many years of successful service, especially the effort to win the Rock and Roll Hall of Fame project, as did Congresswoman Mary Rose Oakar and Governor George Voinovich.

A lot of people worked very hard to help ease the harsh reality of the situation, trying to be as upbeat as possible. At times it started to take on almost a festive atmosphere. A good friend of the morning show, Jerry's Catering, put out a huge buffet for the crowds that would pass through the radio station. A steady and seemingly endless stream of guests and phone calls flooded the studios and newsroom, with many of the staff showing up hours early to be part of what would prove to be a sad but memorable day in Cleveland broadcast history. Other friends of the station, such as Jules Belkin and Barry Gabel of Belkin Productions, stopped by in person to wish everyone their best. The mood was noticeably somber, and Belkin noted, "In the Jewish religion, after somebody dies, you sit at their home, and they call it shiva. You have all kinds of food and flowers and all that kind of stuff. I hate to say it, but it's kind of looking like shiva here right now!" That was probably the most accurate way to describe the morning. People were trying to laugh, and eat, and have a good time, but it was really a wake, and there was an underlying tone of deep sadness to the entire show.

Gabel joked about the huge cumulative audience the station had at that moment, and he plugged the upcoming Elton John–Billy Joel concert at Cleveland Municipal Stadium. There was a very heartfelt exchange about the many shows they co-sponsored, and even indulging Kinzbach's love of golf with some jokes about his game. Gabel predicted that Belkin Productions would save up to $2,500 a year in complimentary tickets for Ferenc. Kinzbach quickly recalled the day a guy in a furniture store offered him a deal like the one the salesman gave Ferenc.

"He promised [the salesman] front row tickets to Billy Joel," said Kinzbach, who then wondered aloud, "How the hell is he going to get those?! Even you guys couldn't get them! What is Flash promising everybody?" There was a lot of good natured ribbing throughout the morning, and it helped ease some of the tension, but the inevitable came closer with every tick of the clock.

The morning progressed with people milling about in the halls, taking an occasional look inside the studio, and doing some heavy reminiscing. Other local celebrities such as Benny Bonnano, Jay Westbrook, WKYC-TV's Jeff Maynor, WJW-TV's Dick Goddard, and Carl Monday, among others, stopped by to pay their respects. There was even a special introduction by NBC's Don Pardo, who announced, "It's their last morning, so let's rock!"

There was a heavy air of anticipation as the clock edged toward the final half hour. When it finally arrived, the studio filled wall to wall with fellow staff members and guests to hear the final words of what was known for a long time as the *Buzzard Morning Zoo*.

When that moment came, silence fell over the assembled group. Kinzbach gave everyone on the morning show one last introduction. He said, "We want to thank you very much for listening, and for being loyal, and being part of this family." Tiburski walked into the studio at that point and handed Kinzbach what he called, "Your final regular paycheck." At that point it started to really sink in that this was the end, and there was a lot of heavy breathing leading up to Kinzbach's final statement. Rocco the Rock Dog, Lisa Dillon, and others were all standing by to watch. After running down the list of a few people who would not be returning, Kinzbach took a final crack at the incoming management, but still saying it was a time to look ahead and not live in the past. He told the audience, "It's been a lot of fun, and I couldn't think of doing it with a better group of people."

Then it was Ferenc's turn. He read from a piece of paper,

> It's been a very emotional day for all of us. Just looking back, we were kids when we came here. Young, rebellious, restless, adventurous—with life, with love, and with our music. We had an attitude, and we were 'MMS. We were the Buzzard Nuclear Army. We stood tall, and we took no prisoners. Now we're looking back.
>
> When we came here, we came with Watergate, and we're leaving with Whitewater. We witnessed the city fall apart with default, and we saw it grow back to something new and different and exciting. We saw crime and we raised our voices on the air. We saw dirty streets and neglected neighborhoods, and on Saturday mornings we cleaned them together. We told you about the hunger, and you opened your pantry and your hearts. We saw an opportunity, and now we're building the Rock and Roll Hall of Fame.

We laughed at the jokes, hailed "the almighty weekend," we blew things up, we worked hard, and we went to "Partytown." We rocked and rolled our way through it all. We had fun, and we did it with you.

Truly the most loyal listeners, and in many cases, the most active participants any radio show could ever hope for, what existed here was a passion. It was a passion among employees, and a passion between this radio station and its listeners, and that my friends is what made it happen.

After nodding to the other staff members, Ferenc continued, "Finally, to you Mr. Kinzbach. I think it's ironic that we've been together for eighteen years, which is just about how old we were when we came here. We've known each other for twenty one years, which, by the way, is the legal age to drink!" There were a few jokes, and muted laughter, before Ferenc finished his speech. "What a long, strange trip it's been," he said, "and you're like a brother to me. A best friend, and we had our fun. We had our fights. You were my best man. Sometimes, I was your worst man. If I could point out two strong qualities that you possess, they are truly a sense of discipline, and a definite determination to win with class."

Choking back tears, Ferenc said, "Over the years, that, my man, is what we've done." He walked over to embrace Kinzbach, who replied, "Flash, I've never met anyone who could hold on to money better than you!" When the laughter died down, Kinzbach said, "Once again, I want to say thank you to everybody, and good luck. The next time around, we'll knock the hell out of 'em! Here's what I end up with. Johnny 'Guitar' Watson! 'Ain't That a Bitch?' We got fired! But you know what? We'll be all right. Ain't that a bitch, though?"

That song ended one of the longest runs in Cleveland radio, and it wasn't over yet.

Kinzbach now says, "I had my sights set on some other things, I wanted to take some time off, and to be quite frank with you, I was tired." He, along with his wife Patty, picked up a few articles in the office, said a few final goodbyes, and left the building.

All of the staff did their jobs as usual, working to make the transition as easy as possible for all involved. At 4:00 P.M., Rocco plugged in the headphones for what was about to become the final on-air newsbreak from the old staff. Rocco casually leaned over the console and joked around with his news anchor, and as the commercial ended, he turned on the mikes:

It's all over, and I do want to say "thanks a lot" to a lot of people. I can go on and on. Jeff and Flash and Laura hit a lot of people, and ditto to all of those. Lisa Dillon hit a whole list of people, and ditto to all of them, too. I can stand here and thank a lot of people for a long time, but I won't. Rock and roll is what it's about, and this is part of the business. It's part of the rock and roll attitude! I've enjoyed doing it, but it's kind of weird though . . . it's gone full circle for me.

There's something about this business that's in my blood. Seventeen years now I've been doing it. Contrary to what you may have read in certain newspapers over the years, written by certain writers, the truth is I've been doing this a long time, and I'm going to continue to do it, because I can't get a job doing anything else!

But I will tell you this. It's kind of ironic that I end up here and finish things off for this afternoon. We're just a few minutes away from the final countdown, and this station's the reason I got into this damned business in the first place! I feel good. I feel like there's a conclusion, a final chapter, all is right with the world. Karma, so to speak. I want to thank the great people we met and partied with over the years. The Mooneys, you know who you are. Brad Hanna. There's too many to mention, and I didn't even say Dana Showalter yet!

Dana and Barb Showalter had been lifelong fans of the station, entering and winning many contests and giving their prizes away to promote the station. They were at most of the station's events over the years, and were listening to Rocco's final show. It was an emotional moment, and when Rocco mentioned Dana's name, Barb broke down in tears. She probably wasn't alone.

Rocco continued, "Thanks for letting me have the airwaves. Good luck to Doug Kubinski. Dougie is the man, he's the future, and the heart that runs this radio station. I'm going to leave you with one of my favorite tunes. One of my favorite artists. Someone whom I got to watch break in this town at a *Coffeebreak Concert* at Peabody's Down Under. Melissa Etheridge. I love this town. I love you. Thanks for the memories. It's been cool. This is, and apparently always will be, Buzzard radio. WMMS."

The incoming crew was already setting up tape cartridge racks and playlists in the studio. Rocco gathered up his headphones and personal stuff, hugging his colleagues on the way out. He walked out the door with his pal Matt Mooney, and the two had a shot of tequila in their hands before the song was over. Rocco passed by some of the new staff members as he made for the elevator. Despite the talent and professional attitude of this new crew, it still seemed like pirates boarding a ship.

Kubinski recalls an emotionally draining day: "It was the end of an amazing era. But there was also this excitement. Maybe we could do it again. It could happen again! It was so exciting, but at the same time, so very sad. It was a crazy, emotional day, where you're crying one minute, and completely jubilant the next." Kubinski also says in hindsight, that if he had to do it all over again and go through those same emotions, he would likely walk away with the rest of the staff. "It ripped me in half," he says, "because you want to be excited and work with the great John

Gorman. It's like a dream come true! But at the same time, people you worked with for years are getting let go, and that was inner torture."

Melissa Etheridge faded, and the new staff cued up Van Halen's "Jump" as it readied the studio for the on-air transition. At 4:15 P.M., a taped bumper produced by incoming production director Mitch Todd was the starting gun for "The Buzzard: The Next Generation." It began with the opening strains of R.E.M.'s "It's the End of the World as We Know it (and I Feel Fine)," which seemed an odd choice considering that's how WENZ-FM brought in its alternative format.

Suddenly there was the sound of an explosion, and an electronically enhanced female voice announced, "Thank you for turning on your radio. For annoying rap music, press 92. For old boring Jurassic Rock, press 98. For a possible insomnia cure, press 102. For Paula Abdul and Madonna wannabes, press 104. For three in a row no one knows, press 107. And for today's best rock and roll, press 100.7, Buzzard Radio. WMMS!"

Then, as if to calm the fears of anyone thinking that the station might cease to exist, the sound of Len Goldberg's voice came booming out, "Serving the universe from the Rock and Roll Capital of the World. We are 100.7, Buzzard Radio, W-M-M-S, Cleveland!"

The crack of a drum and the crash of guitars introduced a parody version of Thin Lizzy's "Boys Are Back in Town," which seemed like a backhanded slap at former co-workers, enemies, and other stations on the dial. The song began with a small crowd of male voices chanting, "They're back!" and ended with the same chorus announcing, "We're ba-a-ack!" A new voice announced, "Buzzard Radio: The Next Generation. WMMS. Cleveland." The phone-option woman's voice returned, and in her best newscaster voice she stated over and over the phrase "In Cleveland tonight" as the pounding dance mix of U2's "Desire" acted as a musical bed. It was drastically different from the WMMS heard up to that time, yet strangely intriguing. Could the station keep up that kind of excitement and anticipation after the initial swell of publicity died down?

All that speculation was lost, at least for the moment, to the partygoers toasting each other at their new station. The hallways were filled with new faces, all celebrating their acquisition, with some of the WMMS veterans walking around shell-shocked from the emotional parting of their fellow employees. Just a few people remained from the old staff, finishing up their work while the new hires busily moved through the halls. Tom Rezny had established a well-deserved reputation as a consummate professional, and as easy as it would have been to just walk out at 5:00, he finished up every bit of work that he still had on his desk. As he was going over the equipment with John Filby, who was staying on to take a good part of that work load, one of the new staff members came in to ask, "Are you going to be here long?" It took a lot to get Filby mad, but that seemingly

Buzzard Radio: The Next Generation (1994). The first order of business for new owners OmniAmerica was to bring back the old Buzzard but with a whole new attitude. Brian Chalmers

insensitive attitude was the spark that set him off. He walked over to the heavy production room door, told that person in no uncertain terms he'd be told when they're finished, and angrily slammed the door in their face.

Champagne was being poured in the halls, and t-shirts were handed out to the staff with the old Buzzard wearing a *Star Trek* uniform, under a banner reading "The Next Generation." There was a great amount of enthusiasm, and the incoming management and staff held out a hand of friendship. The station was going to be okay, but it capped off a very anxious several months leading up to that moment. Some people found out as they left that day that they wouldn't be coming back the following Monday. Jo Dorado, who helped coordinate commercial traffic for WMMS, explains, "It was like kids in a divorce finding out which parent they'd be living with."

One of the holdovers from the old regime was "Spaceman Scott" Hughes, who decided not to be there during the transition festivities. He says, "I didn't want to be a part of that. I was staying on, and I felt bad enough that a lot of my friends were getting blown out." Many of those who stayed through the end of the day gathered at Shucker's Tavern later on to try to sort out what had just transpired, a far cry from the happy, crazy Friday night get-togethers at bars such as Apple Annie's. There wasn't a lot being said, but the looks on everyone's faces spoke volumes as they shared the same thought. Just a few years before, they were the "Thundering Buzzard," with a staff as tight knit as any family. But most of that family had left the station, and the staff's world came crashing down around them in the space of a few emotional hours. WMMS was being taken over by a group of solid, knowledgeable broadcasters with great track records, who wanted to take the station in an exciting new direction and make it a powerhouse once again. They also treated the people who stayed very well, but those same people had just seen most of their colleagues walk out the door to uncertain futures.

Not all the newspaper coverage centered on the emotional last show, and one front-page report in particular raised the ire of some former staff members. The *Plain Dealer* ran a story that most of the WMMS archives, many gold records, and other memorabilia had disappeared. There was even some vandalism, and some members of the outgoing staff were the prime suspects. There were reports that a one-of-a-kind, full-color Buzzard print had also vanished.

In actuality a red vinyl Van Halen award had gone missing, and it was speculated among the station personnel that one of the hosts of a brokered show on WHK was the likely new owner of that piece. Although one of the board operators did urinate in the hall near the engineering office before the final shows, the rest of the staff acted maturely and dignified during the transition. What actually happened is that much of the vinyl had been shipped out to the transmitter site because there wasn't enough room in the Tower City studios, while tapes of the *Coffeebreak Concerts* and the Agora shows were spirited away to better storage areas. When the new staff came in the gold records mysteriously reappeared with no comment to the staff or media, and the Buzzard print could be seen from the Avenue hanging prominently in the program director's office. The *Plain Dealer* reports prompted some angry former staff members to call the station, complaining that the stories put them in a bad light as they searched for new employment.

The incoming staff was starting to gel. A new morning show would soon be in place, with Ferenc staying on for the time being although he did not retain the post of news director for WMMS, WMJI, or WHK. Hughes stayed on, as did Kubinski, Goldberg, and Bill Freeman. Lou Santini and Jennifer Wylde from WRQK-FM were added to the staff, with chief engineer Barry Thomas from WENZ taking over as chief at both WMJI and WMMS. Heidi Kramer came in from Buffalo to take over the WMMS promotions job, moving Mark Bishop over to WMJI. Sally Ride, a former WMMS account executive and on-air voice, was added to the part-time staff. Cleveland native Mitch Todd joined Filby in production. For a time, Freeman did afternoon drive and middays, which exposed this longtime WMMS favorite to a new whole new audience. Art director Brian Chalmers was given even more work, and there was widespread speculation that Denny Sanders would make his return to the WMMS airwaves. However Gorman put that rumor to rest in a *Scene* interview, saying, "Denny wants to stay where he is" at WMJI. Sanders later said that Goman never asked if he wanted to be involved with his new WMMS project. Sanders says he was happy at WMJI, so he didn't really care. Gorman also said there would be no change in the call letters, though special attention would be paid to improving the air signal.

Along with the change in personnel and attitude, there was a drastic change brewing in the station's music format. The first weekend after the change was an exciting mix of the best of AOR, with some of the better alternative tunes in the mix. Gorman says a definite break with the past had to be made,

> We never used the word alternative! It was "New Rock for a New Cleveland," and when we first started there, we were playing Guns N' Roses and Aerosmith. If I had to do it all over again, I probably wouldn't be as quick to get rid of those bands. We never really were alternative.
>
> We did some research with David Tate's group, and we found we were going in the right direction with Majic, no one really knew about WHK-AM—which went all sports—and WMMS was dead. It was like, "It used to be a good station. I used to listen to it." WENZ wasn't even a player. In fact, close to 40 percent thought it was a college radio station! When they first went on the air, I thought they were a great station, and then six months into it, they fell apart. We targeted eighteen-to-twenty-five-year-olds, and as it turned out, it worked.

But *Scene*'s "Mouth of the Cuyahoga" column wondered in an article how new the "Buzzard: The Next Generation" could be if they kept on many of the old time staff members such as Freeman and Goldberg. It even questioned bringing back the old look of the Buzzard character, suggesting the station might be trying to recreate its past glory with a retro-style programming. Hughes recalls that, "John had done a great job mixing in the new shit with the AOR, kind of like what we had back in '83. But he had this thing to beat 'The End,' and all of a sudden he said, 'We're going to go totally alternative.'"

That comment threw Hughes and Kubinski off-guard. The station was doing relatively well going into its first Arbitron ratings book with a 7.7 share, and they didn't think the alternative format could sustain those numbers. Gorman also said local bands would be added to the WMMS playlist, but he doubted there would be a "Cleveland music" show, saying, "I hate that when you do a once-a-week special show, and then you never play the bands in regular rotation. If we do it, we might as well play it all the time. The only problem with local band shows is that the only people listening are the friends of the local bands, and it doesn't turn these bands on to a wider audience."

WNCX program director Bill Louis says it would have been foolish for other programmers not to see Gorman's return as a possible threat. He points out, "There was always the expectation that John had the magic potion to bring 'MMS back to the glory days. Any competitor who was in town during those glory days certainly had to take that into account."

The following Monday had an understandably strange feel to it. The interim morning show consisted of Ferenc, Hughes, Goldberg, Mr. Leonard, and Ross Brittain from WHTZ-FM. It sounded very professional, upbeat, and was a good solid team, but they had the big job of picking up the pieces of the week before. Ferenc traded quips with Brittain as easily if he were Jeff Kinzbach, which seemed a bit odd considering the emotionally charged Friday before. Brittain introduced himself to the WMMS audience, saying he'd been fired in New York and had a month and a half to kill on the non-compete clause in his contract. But he was allowed to do mornings in Cleveland, jumped right in to the mix, and sounded perfectly at home from the very first show.

There were some familiar benchmarks for the listener to grab onto. Ferenc did his news the same as always, Rezny's voice was still on many of the spots, and some of the artist IDs, such as one by Steven Tyler of Aerosmith, could still be heard throughout the morning. The music mix had changed, but not so drastically that it would cause anyone to switch stations. Familiar tunes such as Sting's "If You Love Somebody Set Them Free," Bruce Springsteen's "Streets of Philadelphia," Alice in Chains' "No Excuses," and INXS' "Suicide Blonde" could be heard, but there had been rumblings that the playlist would be going through a major transformation in the weeks and months to come.

The show promised a special *Buzzard Morning Zoo* Address to the Nation to explain who was there, who wasn't, details on events such as the upcoming Earth

Day concert, and the revamped deejay lineup, which Brittain claimed was still not set in stone. Goldberg commented that it was tough work to find new talent "looking by those dumpsters, and under those bridges and stuff," where he said he had gotten his big break. One caller, a fifty-eight-year-old man who had been a long time listener, said the station had started to sound like a jukebox. But he liked the new sound and lineup, which came as welcome news to the incoming management. Ferenc enthusiastically promised a lot of exciting things ahead from WMMS.

There was one sour note when Mr. Leonard joked that the former head zookeeper had been killed in a liquor store robbery. There was an on-air retraction and apology the next day. Even with that gaffe, the station still had its following, with or without some of its most familiar names.

Perhaps the most asked question by radio fans and the industry itself was, "Why the move to an alternative format?" Gorman says there was a lot of soul searching, but it seemed to be the most viable choice. He says,

> There was a large audience for it in Cleveland, and "The End" wasn't reaching that audience. To be perfectly fair to WENZ, they also had a signal problem, but Cleveland has a history going back to the '50s where it always embraced music beyond the commercial realm. Garage bands and one-hit wonders always did very well in Northeast Ohio. There was also a lot of crossover for R&B tunes from black stations to the Top-40s, more so than most other markets. Cleveland has that unique history of always having a different musical taste.
>
> The same thing applied to Cleveland in the '90s. You had to put your own definition of what the term "alternative" meant to the people in Cleveland. It was a decision that had to be made, because WMMS had gone through so many incarnations. It was getting into a classic rock battle, and WNCX did a better job of playing the history of Cleveland than 'MMS was at the time. Part of the problem was there was so much input coming into that station as to where it should be heading. You can't have a ten-headed leader, and that's what was going on. We also realized back then that the 'MMS that everybody knew and loved was dead, yet there was a new, younger audience coming up that wasn't being properly served. So we took a page out of *Star Trek: The Next Generation*, same show, same basic premise, but a newer cast a few years down the road. The goal was to create a new WMMS for a new audience. Why go chasing after the past? It wasn't going to work. Plus, being one of the architects of the original, I could clearly see that "building" was ready to fall down. It was time to construct a new building on that site. Part of it was building a new staff, with a link to the old.
>
> Bash and Boom were two of the oldest surviving members of the original staff, and they were the youngest in mind! Boom enjoyed listening to current music, and the same with Bash. When it was time to put the new 'MMS to-

gether, Bash was a great contributor advising us on music. They said to kill all the old stuff.

There were a lot of questions, and a lot of sleepless nights about how to do this right, because I was also conscious that the bottom line was the heavy-duty investment made in buying those stations. We had to be right the first time.

Not everyone was as confident about the change in programming. After the switch, Omni co-owner Tony Ocepek stopped by the sales department to ask how everyone liked the new format, and Errol Dengler expressed his concerns. "I said that we would be a 4.8 share [of the audience] 12+ [listeners that age and over] by the end of the year," he now recalls, "and when the fall book came out, Tony walked in, pointed down, and it was a 4.8!"

Some former WMMS staff members were already making their moves. Perhaps one of the biggest surprises in that transition came with the announcement that Walt Tiburski would be taking over the position of vice president and general manager at crosstown rival WNCX and WERE-AM. Like WMMS, WNCX parent company Metroplex Communications shared a sales staff with another station, alternative outlet WENZ. And with Howard Stern on WNCX's morning drive, WMMS clearly had a fight on its hands.

There was widespread speculation as to how many WMMS castaways might find a new roost at WNCX, joining Matt the Cat and Michael Stanley. The most suspected was Kinzbach, though WNCX was not likely to drop its top-rated *Howard Stern Show*. It soon became very clear that the "Next Generation" WMMS was bringing back the brazenly defiant attitude seen many years before in its battles with WMJI and WGCL-FM. While WNCX aired spots referring to it as "your classic rock station," and WENZ promoted itself as "a proud supporter of local music," both emphasizing programming statements that were easy to remember, WMMS's new liners proclaimed, "Thank you for listening to WMMS. Now we don't have to hunt you down and kill you." The station's promos would soon get a lot more direct, and some would even say mean-spirited.

Word spreads quickly among those in the know, like record company representatives and advertisers, and one of the worst kept secrets in Cleveland centered on who would host the new morning show on WMMS. Brian Fowler and Joe Cronauer, who had been doing their popular *Brian and Joe Radio Show* on WENZ, were a sure thing to move over to WMMS. Former WMMS engineer Frank Foti's cousin, Gina, was married to Cronauer, so they knew each other from various family functions. Foti phoned Cronauer to say he had been at WMJI that day, and he wanted to pass on some nice comments that were made about him and Fowler, suggesting it might be a good idea to send an audition tape to the management there. It came just as the pair's contract was ending at WENZ, and a meeting was set up with Gorman and Dean Thacker at Barnacle Bill's restaurant in Lakewood. Coincidentally it was set up for the same day that *Northern Ohio Live* featured the two on its cover, with a side article speculating on the direction Carl Hirsch would take with WMMS. The management said it would cover any legal hassles associated with moving the pair to the Skylight Office Tower, and all agreed that the *Brian and Joe Radio Show* could be successfully spliced onto the *Morning Zoo*, taking it into an entirely different direction.

The problem was, like with many broadcasters, Fowler and Cronauer had a ninety-day non-compete clause in their contracts, meaning they could not go to another station for three months if they were to leave WENZ. Nonetheless promotional photos with Gorman, Thacker, and Kubinski started making the rounds. They featured two unknown figures wearing WMMS bags on their heads, a "B" and a "J" on their respective shirts, and giving a thumbs-up to the camera. But the pair were sworn to secrecy, with Cronauer being put in the position of even keeping it from his in-laws because they had connections to the *Plain Dealer*. WENZ's general manager Bill Scull countered by sending out a press release stating that Fowler and Cronauer were not being re-signed to a new contract because the station wanted to find somebody who was more comfortable with the music the station played, a back-handed slap at WMMS and its direct challenge to WENZ. *Scene*'s "Mouth" column commented that Scull's argument was weak, saying if that was the case, Jerry Shirley would still be doing mornings at WNCX instead of Stern, and artists such as Pere Ubu's Dave Thomas would fit in at WENZ. WMJI's John Lanigan interviewed Fowler and Cronauer on his morning show to talk about their future, which obviously fueled even more public speculation. When the initial flurry of publicity finally died down, Fowler and Cronauer and WENZ management decided mutually to part ways when the morning show's contract expired on July 12, with the station promising an announcement about a new morning drive team in the coming days.

Fowler and Cronauer never imagined working at WMMS. It was the station that dominated the airwaves as they were growing up, and they felt the closest they could ever get to that kind of recognition was simply getting a chance to work in their old hometown. When the station started to unravel in spring of 1994, and they were approached to work there as it was being rebuilt by some of the architects of its glory days, it seemed almost too good to be true.

Howard Stern in Cleveland

The dust had barely settled from the emotional changes at WMMS a few weeks before when the station faced its worst enemy in its own backyard. Howard Stern had edged his way into the number one spot in a key demographic of eighteen-to-twenty-five-year-old men, and he was about to make good on a pledge to hold one of his infamous "funerals" in Cleveland. The new John Gorman regime at WMMS lived for those types of challenges. But no one expected the bizarre tale that would unfold on June 10, 1994.

It took Stern only eighteen months to get an overall rating of 11.4 in the winter 1994 Arbitron book, with *Lanigan, Webster, and Malone* right behind at 10.9. When the book was released, WNCX invited select media to its offices on April 22, just one week after the WMMS transition, for an interview with Stern and co-host Robin Quivers. *Sun Newspaper* writer David Sowd posed the question if Stern's rise might had been even faster if he hadn't done the parody about the tragic accident that killed two Indians players in Florida two years before. Stern agreed but also stressed, "One of the things you must understand is that the show cannot be interfered with. What you're referring to is Doug [Podell] got very nervous about things I was saying." Quivers jumped in, joking that, "One of the obstacles we've had to overleap is Doug Podell, in order to be number one!" Stern added, "Yeah, in spite of Doug we've become number one! So listen, I have three words to say about that. And these three words will sum up the answer to your question. When you ask me about the fact that we were censored over the situation with the Cleveland boating accident, I have three words to say: Doug is stupid! That's it. Thank you!" In reality, Stern had a great amount of respect for Podell, having worked with him in Detroit, and at one point even calling him a genius.

Actually it was John Lanigan, John Webster, and Jimmy Malone at WMJI-FM that kept Stern out of Cleveland's top spot until that time. The *Buzzard Morning Zoo* had gone through so many changes, and the station's musical identity was so confused, that many listeners were ready for something new. And when the *Zoo* crew failed to respond to Stern, they became sitting ducks for every shot that came over from WNCX. Stern thought WMMS would be the toughest nut to crack. Instead, it became his first victim.

Meanwhile Lanigan answered every charge from Stern with a blast of his own, and was able to hang on much longer. He came off as a real fighter, and the audience picked up on it. Lanigan even answered a letter from Jeff Kinzbach in the *Free Times* saying no one would be worried about Stern coming to Cleveland if the *Morning Zoo* had freshened up its act and had been more entertaining. At any rate, the die was cast and a date was set for a Cleveland funeral.

It wasn't going to be easy. Dave Jockers ran the board for the Stern show at WNCX, among his many other duties, and he explains,

> At first, we went to the Cleveland State Convocation Center, met with them, and we had it all mapped out. But Howard said, "I'm not putting on a show. I don't want to be on a stage surrounded by people who are expecting me to do a dog and pony act! I want it outdoors somewhere." Then we looked at the Nautica Stage, but that got nixed. At that point, Doug [Podell] was ready to cut wrists! Everybody was all bummed out, but it was also around the time that Tiffany's started doing some advertising. They had that huge parking lot that they owned, so why don't we just do it here? Walt [Tiburski] made a few phone calls, and the next thing you know, it's a go! They shut the place down and went all out ... We hired the Belkins crew to handle everything from security to audio to staging. Price was no object, and with the trade and cash value, it cost over $120,000! We didn't have a lot of lead time, but it all came together.

Gorman had worked with Lanigan at WMJI, so he was never one to shy away from a good fight. The dust had barely settled from the staff upheaval a few weeks before when Gorman and his lieutenants switched into their "war" mode. If Stern wanted to invade Cleveland, he was in for a major battle.

The first salvo was directed at *Scene* magazine. The week of Stern's appearance the paper ran a cover story about the event, which did not sit well with Gorman. In retaliation a contest was announced immediately on WMJI and WMMS offering $500 to the person who brought in the most *Scene* magazines. Since *Scene* was a free entertainment weekly, it was thought that listeners would raid the boxes and rid the streets of the offending paper. Thousands were delivered to the station's studios, and it was anyone's guess where the listeners got them. As it turned out at least one listener broke into a distribution center for his entries, about 3,000 of them. He won the contest, but it would prove to be a critical mistake.

When *Scene*'s big-money advertisers learned about the promotion, they saw their investments quickly diminishing in value and demanded the *Scene* do something. As much as ten percent of that week's run had been taken off the street, and *Scene*'s associate publisher Keith Rathbun stormed up to the Skylight Office Tower for a confrontation with Gorman. He insisted the papers be returned, but Gorman would only produce about 500. Unknown to Rathbun, the rest were in the station's prize closet. *Scene* had to go back to the presses for a second run at a cost of more than $5,000.

The following week *Scene*'s "Mouth of the Cuyahoga" column blasted Lanigan and Gorman, stating that Lanigan supported freedom of the press "as long as it serves your purposes." The column pointed out that a few weeks before, Lanigan hosted *Free Times* publisher Randy Siegel on his show to discuss the city of Cleve-

Scene in the Buzzard prize closet (1994). Hundreds of issues of *Scene* with Howard Stern on the cover were taken off the streets and stashed in the WMMS prize closet. Photo by Brian Chalmers

land's order that the paper's boxes be removed from the streets. *Scene* claimed that, "Lanigan reversed himself and advocated the same crime he so vehemently criticized Mayor White for. If that wasn't irresponsible enough, then OmniAmerica Group (the station's owner) management refused to return all the confiscated *Scenes*. (At least the Mayor returned the *Free Whines* to Siegel for redistribution;

thus it was the boxes and not the papers the Mayor wanted off the streets.)" It went on to claim,

> When *Scene* asked where the remaining papers were, a smug John Gorman, operations director for both WMMS and WMJI, denied that any more papers were collected. The stations' denials persisted despite a Tower City Center security officer's claim that he helped the winner "load at least 50 bundles of papers onto a dolly and take them to the WMMS offices." *Scene* ordered another 6,000 papers to be printed while making one final plea to WMJI promotions director Mark "Munch" Bishop, who probably thought he was done defending such third-grade shenanigans when Michael Luczak left as WMMS program director a few months ago.
>
> Gorman, Munch, and OmniAmerica Group VP Dean Thacker laughed about their promotion and offered up that "it was just a funny little thing." But the intent of the promotion was the equivalent of knocking their stations off the air for a week, because *Scene* has one chance a week to speak to its readers. Think about that, Mr. Lanigan, next time you blindly go along with a promotion so obviously against your principles.

In an ominous side note *Scene* warned, "This story isn't over, so stay tuned... but not to either WMMS or WMJI if you want to know what's in the *Scene*. Our weekly spots, (you know, where Murray Saul bellows 'In S-C-E-N-E') will continue on WNCX (98.5 FM), WENZ (107.9 FM) and WRQK (106.9 FM) and other stations forthcoming." Lanigan and Gorman didn't seem concerned at all about the article or the threatened pull of advertising money. But Gorman still had to deal with his biggest problem, and that was Stern.

Gorman speculated that Stern was a control freak. The challenge would be to wrest control from WNCX, or at least make it appear that way, and use the media coverage to embarrass Stern and promote WMMS. He also knew that Walt Tiburski would be prepared for just about anything. Doug Podell and Jockers traveled over to the Ritz-Carlton for a meeting with Stern and his crew, and Podell stressed the potential for some kind of retaliation from WMMS. Stern is said to have calmly asked Podell, "Tell me this much. Let's say something major did happen, that 'MMS did do something. What's to stop us from just packing up and going home?" A dead silence fell over the room. Jockers recalls, "All of a sudden, Howard just gets this smile on his face. I looked over at Doug, and you could see these little beads of sweat on his forehead. He was dumbfounded at what he just heard! I just kind of reached up at his balding head and said, 'Losing a few more hairs there on your shoulders, Doug?!'" The room burst out laughing. The station spent big money to cover the event, and the thought of Stern canceling was not something Podell wanted to hear. Stern himself later said on the air that the night before the event, a very nervous Podell paced the floor in his hotel room expecting WMMS to somehow try to derail the Cleveland funeral. Clearly WNCX management was bracing for the unexpected. But WMMS management never expected the eventual outcome either.

WMMS had a new influx of money from the OmniAmerica takeover, and the station combined it with promotional cash from WMJI. Now it was time to develop strategy. Gorman called a disc jockey meeting to bounce around some ideas. One thought was to have a helicopter dump water on Stern while he was on stage, but chief engineer Barry Thomas warned that with all the high-voltage equipment on that platform, there could be serious or possibly fatal results. At this point Gorman stressed to the deejays that no one could do anything illegal, a critical point that would act in his defense. It was decided that instead the helicopter would buzz Stern like the one that followed Ray Liotta in *Goodfellas*, and a plane would drag a WMMS banner overhead during his appearance in the Flats. The promotions team would paper cars with handbills and stickers showing a devil Buzzard stabbing Stern in the flames of hell. People left the conference room thinking it wasn't much, but at least they would do what they could. There were other ideas, too, but not everyone would be privy to those plans.

Both WNCX and WMMS claimed to have spies in each other's camps, so most information was on a need-to-know basis. WMMS learned that Milano Monuments had been commissioned by WNCX to make a tombstone announcing the deaths of Lanigan and the *Buzzard Morning Zoo*. OmniAmerica contacted Milano's Willy Ramos, who ended up delivering a stone showing Stern's demise instead, but that was intercepted before it could be unveiled to the crowd. Gorman also had Stern's entire Cleveland schedule well ahead of his arrival, so security was critical. Neither side wanted to tip its hand. As the funeral date neared, dozens of Stern fans would stand outside the studios overlooking the Avenue at Tower City Center with signs and giving the finger to anyone who chanced to look out the window. The WMMS studios became a bunker for the remainder of Stern's time in Cleveland.

One of the stops on Stern's Cleveland itinerary was a morning press conference at Tiffany's Cabaret, the city's premier strip club where the funeral was set to take place in the parking lot. WNCX planned to screen all media outlets by requesting credentials. WMMS had plans to attend, submitting a request on phony letterhead for a magazine called *New Age Rock Review* out of Cuyahoga Falls. Jockers was in charge of the screening process, and later told a reporter for the *Small Business News* that it was faxed just minutes before Stern's plane was about to touch down, so he didn't have time to check the legitimacy of that request. Gorman had his backstage access. As it turned out, he wouldn't need it.

This was not a pleasant time to be at WMMS. The bunker mentality might have been invigorating to some at the station, but for the rest it was a major headache.

The stress of losing so many trusted friends a few weeks before, the demands of a new format and management, and the arrival of Stern made it very difficult for the staff at the Skylight Office Tower.

There had been some unsettling rumors about the upcoming event. The most prevalent story concerned a possible appearance at the funeral by former *Zoo* co-host Jeff Kinzbach, whom Stern was expected to bring on stage to embarrass WMMS and Gorman. There actually were conversations between Tiburski and Kinzbach about such an appearance, but Kinzbach explained he was leaving for a vacation in Green Bay. He says he also felt, "It would not have been a good thing for me to get involved in that. I'm only going to get in the ring if I have a good chance to win the fight, and at that particular point, I didn't have that chance." The morning of the broadcast, Kinzbach and his wife were on the road and listened to the show in his car.

Friday, June 10 arrived, and Stern and his crew triumphantly ascended the stage of Tiffany's to a thunderous roar from fans who crowded the club for a look at their hero. The WMMS spies were there as well, wisely staying in the background while Stern fielded questions from the press. This was not a place for the weak of heart. Stern was at the top of his game, invigorated by the applause of his Cleveland fans, and was ready to slice apart anyone who tried to top him. WJW-TV's Mike DiPasquale made a comment that earned him a verbal broadside from Stern. But the one person who would help define the day had also arrived, and he never had to open his mouth.

Earlier that day, engineer Bill Alford stopped by the WMMS studios and was greeted by "Spaceman Scott" Hughes. As Ed Ferenc was doing his 6:30 newscast in the adjoining studio, the two chatted about Stern's appearance. Hughes recalls Alford casually mentioning, "You never know what could happen." Then he turned and walked out the door.

Rocco the Rock Dog had found work at WNCX after his departure from WMMS. Like a lot of people at WMMS, he was a closet fan of Stern. The running joke among the engineering staff was to call each other "Gus," because almost every time a crew was working on a road there was a guy with that name directing traffic. Rocco put aside a red all-access laminate pass for his longtime friend Mike Knisely, who had also left WMMS during the April purge and planned to join the festivities. The third member of the old WMMS engineering crew was Alford. When he found out that Knisely couldn't make it to the funeral, he showed up at Tiffany's dressed in a green Army fatigue jacket, much like Robert DeNiro's in *Taxi Driver*. He told the person in charge of checking credentials to tell Rocco that "Gus" had arrived. When he returned he handed Alford the red laminate. Alford put it around his neck and walked into the restricted area.

Rocco was enjoying Stern's press conference when he got a tap on his shoulder from a smiling Alford. Rocco's eyes widened, his face dropped, and he said, "What the hell are you doing here?!" Recalling that day, Rocco admits, "I lost it! He had a smile a mile wide! What I'm thinking is I just got fired, I got a new job, and I need the money! I have a wife and kids, and a mortgage!" He ripped the pass from Alford's neck and ordered him out of the area for his own good. Alford no longer had a pass, but he had a plan and everything he needed to put it into action.

After the press conference, Stern let the crowd in Tiffany's parking lot continue to grow before his public appearance. They were entertained by acts such as Roger Clinton, David Lee Roth, and America, the last of which was promptly dumped from WMJI's playlist. Other show regulars such as Joey Buttafuoco, "Fred the Elephant Boy" Schreiber, and plenty of strippers kept the thousands on hand occupied while Stern and his crew joined the media for a bus tour of Cleveland that was broadcast live to the rest of the country. There were some unsettling moments such as a disruption in the live signal, but the WNCX engineers were able to deal with them. Meanwhile, Alford says he casually walked up to the broadcast center, reviewed the situation, and thought, "This is like a dream come true! I can make all the points in the world here!" He took out a pair of wire clippers, snipped the cable, and strolled toward the door. About halfway through the broadcast, Podell's prediction came true. The signal finally went dead.

Oddly enough, Alford might have gotten away with it if he had kept walking, but he still heard the live broadcast coming over some nearby speakers. He didn't realize it was the feed to the public address system and not the air sound, so he decided to go back and finish the job. He laughingly recalls, "You know, I'm a bad criminal. I stuck around to see what happened!"

Luckily a massage and pool party had been held in the Stern crew's honor the day before at a home in Seven Hills, so they put that twenty minute taped piece on the air until they could find out the source of the problem. A back-up frequency was quickly secured, but it was clear that someone was trying to sabotage the show. Stern was scheduled for an appearance on WEWS-TV's *Morning Exchange*. Sensing the drama that was unfolding, he switched it to a remote live shot at Tiffany's to announce that he was "under attack by WMMS." Jockers was standing backstage at Tiffany's with two Cleveland police officers, when chief engineer Sam Cappas came running up announcing, "We're off the air! Somebody cut the wires!"

They all ran toward the scene, with Jockers jumping a wooden fence. They found a familiar face from a competing station, just leaning against a wooden barrier and casually talking to someone else. Jockers stopped dead in his tracks and told the cops, "That guy right there works for WMMS, and he's not supposed to be here! This is a secured area!" The cops walked up to him.

"What do you want from me? I didn't do anything! I didn't do it!" Alford told them.

"Do what?" the officers calmly asked.

"I didn't cut any wires."

Stern takes the stage (1994). The "King of All Media" addresses thousands of his loyal subjects. Video capture by Dave Little

"Well, we never said you cut any wires."

At the site amateur videographers taped the arrest of the suspect accused of cutting the wires of the satellite link. Witnesses say he became agitated, swinging his arms around, and the police pushed him up against the fence and warned him to calm down. That's when they found wire cutters in his pocket. Stern's producer Gary "Baba Booey" Dell'Abate took to the airwaves to announce that an engineer from a competing station had been nabbed, and his name was William Alford.

Doug Kubinski had seen the tension leading up to Stern's arrival, but had planned for some time to attend a wedding in Connecticut. He wasn't in Cleveland when Stern came in, but he was listening to it during airport layovers. It was at Washington's Dulles International Airport that he heard about the wire cutting, spitting out a spray of coffee when Stern's show mentioned Alford's name.

Witnesses claimed Alford told the arresting officers that he acted alone. The security police very likely saved Alford from the crowd, because if they had grabbed hold of him it would have gotten ugly very quickly. Tiburski was said to be livid, and he angrily told Alford to take a message back to his bosses: They were in for the court fight of their lives.

Stern played the incident to its fullest. He arrived to take the stage, and the crowd was driven to a new height of frenzy. After poking out the eye of a Cyclops, meant to represent Lanigan, and chopping the head off a Buzzard, he left with a huge grin on his face. It was his best funeral ever.

Back at WMMS, Gorman was unlocking the door of his office. He was also wearing a huge smile.

Aftermath

There was never really a cooling off period following the Howard Stern fiasco. Television stations led their newscasts with the wire cutting, it commanded front-page headlines, and became a topic of conversation on Stern's show for weeks. Before heading back to the airport, Stern's producer, Gary "Baba Booey" Dell'Abate, commented that while it was still not confirmed that a WMMS employee disrupted the broadcast, the incident was still "incredibly low, stupid, and makes for great radio!" That one sentence spoke volumes about all of the stations involved.

Bill Alford's brother bailed him out of jail, and WMMS management proclaimed him "innocent until proven otherwise." Alford didn't need any more legal problems. He was already fighting for custody of his son, and this wasn't going to work in his favor. Alford was allowed to return to work, and that Monday he walked into the newsroom to declare he "didn't do it!" Tom Bush had returned to radio, and he was taking a break from his show on WHK-AM. He looked Alford in the eye and said, "You're an asshole, and because there's a kid involved, you're a bigger asshole! But I'm praying for you." Alford paused for a moment, let out a sigh, and retreated to the engineering office.

Alford was a rabid WMMS fan. He worked his way up from the mailroom to man the Buzzard Rock and Roll Machine, the station's remote disc jockey service, and he was a mainstay at station promotions. At times he was known to be somewhat reckless, but he was very devoted to his son, and even those who didn't care for him admitted that he was a loving and attentive father. Alford and his wife had split some time before, and there was a very real possibility she would move his son to the West Coast.

In its June 16 issue, *Scene* magazine's "Mouth of the Cuyahoga" column wondered if WMMS management might have been behind the plot, saying,

> If the court finds Alford guilty as charged, these are some of the questions that will need to be answered. WMMS, naturally, denies any knowledge of or involvement with the crime. And Mouth would find that easier to believe if it hadn't been for the events that transpired last Thursday and WMMS's lack of cooperation with *Scene* to resolve that conflict. Even so, WMMS had so much

to lose (i.e., its license) and so little to gain that Mouth begrudgingly gives them the benefit of the doubt, at least for now.

Or, does Alford have such a passion for the Buzzard that he acted on his own volition with a kamikaze-like mentality. One former WMMS DJ told Mouth that Alford was a hardline Buzzardist who even sports a Buzzard tattoo (probably wears it on his ass, too). It's hard to believe he worked alone, but it's possible.

Doug Kubinski is among those doubting that WMMS management had any involvement in that incident. He says, "Gorman planned radio combat maneuvers that weren't illegal. He would walk right up to the edge, but not step over. John was a motivator, and everybody was into it, but no one was told to go in and cut wires or anything like that. No one."

Scene also considered whether WNCX or Stern might have had a hand in the incident, saying the publicity would be a windfall for both parties. But, with so much at stake with the FCC, it was an unlikely scenario. At least one reader wrote to *Scene* suggesting that it put Stern on the cover of the next few issues just to annoy John Gorman, John Lanigan, and their respective stations.

But that war wasn't over yet. The highly critical report by the "Mouth of the Cuyahoga" prompted WMMS and WMJI-FM to again call for listeners to gather up *Scenes*, but now the paper monitored its distribution stations. The stations returned the papers to *Scene* management, prompting the paper to suggest in an article that both the stations and *Scene* "Give Peace a Chance."

A lengthy investigation ensued with a number of present and former WMMS employees giving depositions. The Cuyahoga County grand jury eventually handed down indictments on March 13, 1995, against Heidi Kramer and assistant Greg Smith, who worked with Kramer in Buffalo and came to Cleveland in time for the Stern visit. The indictments included a series of charges, including receiving stolen property, forgery, and uttering against Kramer, and breaking and entering against Smith, with both facing charges of disrupting a public service. Kramer's charges were connected to receiving *Scene* magazines stolen from a distribution center, as well as reportedly forging the document to get a press pass for the Stern event. Alford had already pleaded guilty, and was cooperating with authorities in the continuing investigation, which delayed his sentencing.

Reviewing those events years later, Alford now sums it saying it was a reaction to the pressure of the last sale, Stern's arrival, and the departure of so many staff members he considered family. "It was a really, really rough time," according to Alford, "and for those of us who weren't fired in the transition, we were in a worse position than those who left. We felt alienated from the people coming in and from our friends who lost their jobs. I knew why they wanted me there at the station. I was cheaper than Mondo, who was obviously much more qualified than I was. But where else was I going to go? I was in heaven. That was my gig, my perfect job right there, and I couldn't ask for anything more. Then everything changed."

While he doesn't directly blame anyone for his actions, other than himself, Alford says he knew what could be done in that atmosphere. He explains, "I wanted these people to be impressed with what I could do for them. I knew what Gorman was all about, because his reputation preceded him. Heidi knew what Gorman was all about, and I knew they wouldn't condone anything that I did, but at the same time, there was an environment in which they were very careful in their wording so that they might have wanted to see something happen, but didn't want to be a part of it. I also knew there would be a benefit if I could do something and get away with it." Alford also expressed disgust at the show's treatment toward onetime radio rival John Debella of Philadelphia when his ex-wife appeared on the Stern show.

Alford was also inspired by the aggressive reputation WMMS had built up under Gorman's previous reign. The night before Stern's rally, he took a walk along a breakwall on the Lake Erie shore near Wildwood Park, trying to sort out the confusion of the past few weeks and decide what his future at the station might be. As the waves crashed angrily on the rocks, Alford was listening to the station on a headset radio. The voice of Len Goldberg burst through the airwaves with a top of the hour ID, "Serving forty counties, three states, two countries, and you! We are W-M-M-S, Cleveland!" At that point, as if on cue, a bolt of lightning flashed across the sky, followed by the roar of thunder. Alford said it seemed obvious what he should do.

In the end, despite widespread speculation that top management at WMMS would fall, only Kramer would face any serious retribution. On the day of the wire cutting, Kramer watched afternoon television coverage of the funeral and Alford's arrest. She kept saying, "Oh, my god! How could he have been so stupid?" Several months later, Kramer was not only implicated, but also found guilty of accepting stolen property, the *Scene* magazines, and she apologized to WNCX at her sentencing. WNCX tapes rolled in the courtroom, and her voice soon became part of a humiliating on-air promo. She soon left her job at WMMS. Alford pleaded guilty and spent two weeks in jail for disrupting a public service, but there was a silver lining. After a heartfelt discussion with his wife, Alford was able to get custody of his son.

Despite the negative publicity, WNCX, WMMS, and even Alford came out winners. Stern talked about the incident for months, often mentioning WMMS, and the news reports did the same. Alford went to work as an engineer for WOIO-TV and WUAB-TV, and he later bragged that even Stern wanted to hire him. But Kubinski also points out the publicity had a negative effect, especially coming after the staff upheaval just two months before. He says, "It absolutely killed the immediate momentum of WMMS. It cut the legs out. For the next year, it distracted the entire staff, and I still say it was the biggest waste of taxpayer money."

In less than two years after the Stern visit, Gorman would leave WMMS when it was taken over by Nationwide Broadcasting, and he soon ended up in Detroit to program an alternative music station. The station's morning show was none other than that of Howard Stern, who lauded Gorman on-air for his confrontational style.

The New Zoo

It's not easy to follow an act like the *Buzzard Morning Zoo*. Despite its ever-changing lineup, Jeff Kinzbach and company had many loyal fans. But with Kinzbach's departure, John Gorman had to unveil a new morning lineup, and it had to be good. "Spaceman Scott" Hughes had been part of the interim team with Ross Brittain, Ed Ferenc, and Mr. Leonard, and he saw Brian Fowler and Joe Cronauer as the logical next step.

Hughes explains, "I think Jeff and Flash were much better in their heyday, and that's how you would compare Brian and Joe. Everybody was pretty pumped up about them." Joined by Ferenc on news, Fowler and Cronauer brought a different type of entertainment to the morning drive. Each had a quick sense of humor, which was developed from years of working together at various stations, and each had a distinct and likable personality. Fowler came off as the slightly more "serious" of the two, while the musically talented Cronauer could act silly without sounding foolish. Perhaps their most impressive qualities were how they communicated ideas to each other and how closely they worked together.

At the same time, because of his long association with Kinzbach, Ferenc seemed out of place, and even a bit out of step with the pace set by Fowler and Cronauer. From day one it was *The Brian and Joe Radio Show on the Buzzard Morning Zoo*, with Ferenc limited to a few minutes of news because he stood as a reminder of the old regime. Within a couple of months Ferenc left to work with Kinzbach on the AM dial.

It wasn't long before the station's aggressive nature was turned up a notch with on-air salvos targeting its enemies, or anyone even thought to be an enemy. Suspecting the competition was keeping a close ear on its programming, WMMS produced scathing liners that soon made their way into regular rotation. They usually started out, "Buzzard radio asks the question: Who has the highest IQ? Forrest Gump or . . . ?" Among those singled out were Walt Tiburski, his staff and sales crew, Doug Podell, "those clowns that listen to Stern," *Scene* magazine's "Mouth of the Cuyahoga" column, and even the cleaning crew at the Agora. It was followed by the production voice asking something to the effect of, "Or does anyone give a flying [cuckoo]?"

Brian and Joe meet the *Morning Zoo* (1995). Despite the pressure of having to replace the *Morning Zoo* Brian and Joe were able to draw a loyal following. Art by Brian Chalmers

Perhaps the biggest surprise came with the jab at the station's former cornerstone that said, "We're the radio station that guarantees absolutely and positively no Jerk and Flush, now and forever! Sanitized for your protection. Buzzard radio: The Next Generation." It was obviously a slam aimed squarely at Kinzbach and Ferenc.

The pair in turn took occasional shots at WMMS starting with their very first afternoon show on WWWE-AM, on which they interviewed wire cutter Bill Alford about his latest legal battles. It was surprising because Ferenc seemed to be getting along famously with Gorman, Fowler, and Cronauer before he left to rejoin Kinzbach.

Fax machines at stations and businesses all over town started humming with anonymous transmissions, again targeting former WMMS employees. One pictured an aging lounge act clad in tuxedoes, with a hand pointing to a familiar face in the back, and asking, "What ever happened to Jeff Kinzbach?" Yet another featured a sketch that looked suspiciously like Tiburski promising clients the world by stating "They promise . . . we deliver!" with a separate Buzzard drawing and the "Next Generation" logo. The on-air and fax assault meant nothing to the listeners, as they were meant to unnerve their targets and undermine the rival sales departments.

Buzzard Paloozas

There was still a great deal of enthusiasm among the "Next Generation" WMMS staff entering the fall of 1994 ratings wars. In a confidential memo to the staff, John Gorman congratulated the station for an excellent summer Arbitron, saying:

> Those doomsday critics like the *Plain Dealer*'s Roberto Santiago can run and hide in their caves. They can have their lowest common denominator WNCX and no-ratings WENZ. They can have all those living-on-borrowed-time deejays who have spent their time and energy predicting doom and gloom for our Next Generation Buzzard Radio. They'll never understand. To them, yesterday isn't history, and today and tomorrow is too much of a mystery. Let them choke on their own venom. Let them drown in a vat of classic rock and yesterday's schlock. The majority of our audience is new, younger, and better educated than the (many) formats the previous owners of this station had. True, we did take a great risk by putting a new, untried current-music based rock-based format on the air. True, we were taking a major risk by swapping WMMS's old, fragmented audience for new, younger alternative-rock leaning listeners. It's also true that most of the so-called rock and radio critics in this market felt we would not succeed. True, most of our competition did anything and everything to thwart every concert and promotion we were involved with over the summer months ... especially Buzzard-Palooza (which actually was our turning point—when we began to attract our new audience). So, while Santiago and the staff of WENZ and WNCX drink and drown in their drinks at Tiffany's, we are light years ahead and truly taking radio to the next generation.

Later in that same memo, under "The Bottom Line," Gorman continued,

> Over the last few months, the competition has become obsessed with our music, our promotions, and our plans. The best they can do is imitate. Our programming, promotion, and marketing is original. They can imitate, but they cannot duplicate. As you know, Stern, his personal press agent *Plain Dealer* Stern groupie Roberto Santiago, the management of WNCX along with Jerk and Flush at WeeWee are doing their best to zap us via Bill Alford's wire cutting incident. Since they cannot fight us on a level playing field, these losers are manipulating the legal system in an attempt to discredit our successes. This is unfortunate because, ultimately, only the lawyers will make out on this one. What these people fail to understand in their attempt to gain free publicity from this is that no one (who counts) cares! Our listeners don't care ... the

Plain Dealer readers don't care . . . WNCX's listeners don't even know who the President is let alone this story. Like every other obstacle put before us, we will get through it, get it behind us, and move forward.

Finally, in a nod to the very important fall ratings period, Gorman wrote, "Now we begin the real war. Cleveland has always been a competitive market—but, these days the dirty tricks seem to be hitting an all time low. Everyone from WENZ (unless they get smart and change format to country or polka), WNCX . . . even Jammin' 92 will be aiming their weapons directly at us. The good news is that we are sitting at the top of the mountain . . . and it's much easier to shoot them down than for them to shoot up at us. On our end, we must continue to do what we do best. Music, news, information, a tightly run board, and supporting our listeners."

The memo ended with a promise to celebrate this and all upcoming victories in the very near future. One of those victories was Buzzard Palooza, the all-day alternative rock festival that drew a sell-out crowd to Blossom Music Center based mostly on word-of-mouth advertising.

Prior to this event WENZ-FM had presented Endfest, which also drew enthusiastic crowds. Then-WENZ deejays Brian Fowler and Joe Cronauer had performed there as well. Gorman saw the response they got there, and he knew the WMMS event, like Lollapalooza, would be another major hit.

Buzzard Palooza drew more than twenty thousand people. It sounded bigger than life on air, with one act after another praising WMMS and its personalities for the incredible response. But with that kind of mob mentality, it eventually got out of hand, and soon large patches of turf were being set afire and launched into the pavilion, and assorted bonfires damaged the grounds as well. It was the first of a series of spectacular shows featuring some of the biggest names in alternative music, and it became a feather in the "Next Generation's" collective hats.

Thirty Days in the Hole

Promotions are a critical part of the radio game, and over the years WMMS had set the bar pretty high with its marketing machine. It had mastered this over the years, and by the early 1990s had closely aligned itself with St. Augustine's annual holiday food drive, which greatly helped the church in its efforts to feed the hungry.

WNCX had a similar campaign. On the surface, Jerry Shirley's Thirty Days in the Hole was a phenomenal success. It collected tons of food over the years, won honors for Shirley and the station, and got extensive media coverage. But after a time, that press coverage became a curse.

Every December Shirley would stay in a small trailer in the parking lot of a shopping center, broadcasting updates from the site, and vowing to stay there until a trailer load of food was collected for the Salvation Army Hunger Center. Donations ranging from hundreds of pounds to a simple can of food were equally welcomed. Cash donations were also encouraged, and a good amount of money was collected for the Salvation Army. It was a tremendous success, and Shirley was honored time and time again for his efforts. Mayor Michael White declared December 11, 1995, to be Jerry Shirley Day in Cleveland. Shirley may not have had the ratings power of Howard Stern, but his public relations value was considerable.

But all good things come to an eventual, and in this case tragic, end. "Party Marty" Young was doing fill-in air work and promotions, had helped the food drive at the site for the first couple of years, and had become pretty good friends with Shirley. He said, "The last year, Jerry didn't seem to be Jerry! He just wasn't his usual self, at least the person I had known." There was talk that Shirley hadn't spent all of his nights at the trailer as expected, opting instead for a nearby motel. It was actually understandable, since the house trailer parked at Great Northern Mall was said to have had occasional heating problems.

Walt Garrett, best known as "Mr. Classic" on the *Saturday Night House Party*, remembers the chain of events that led to Shirley's perceived fall from grace:

On Saturday night, for the last call back to the radio station, Jerry sounded [pause] pretty crappy. He sounded down, like he was in pain. I asked him what was wrong, and he said, "Ah, I'm just tired. My back is killing me," and on and on. I said, "I'll tell you what. How about if I come out tomorrow morning when you open? I'll stack the stuff in the truck, and all the canned goods, and shake people's hands." It was getting kind of cold, so Jerry could stay in and heal his back a little bit. People could still come into the trailer and see him, but he doesn't have to sit out there. Jerry said that would be a great idea.

The next day rolled around, and I showed up in the morning at about 9:20, 9:30. I walk in and two people are still in there, a brunette with a leather coat and a tall guy with a Cleveland police hat. He said, "Good luck getting him up! You can baby-sit him now." The place was trashed, there were two broken chairs and beer bottles, pop cans, pizza cartons all over the place. I packed everything up, straightened the place up, and went to get Jerry up. He wanted ten more minutes, and I gave it to him and cleaned up a little more. Jerry finally gets up, after I nagged him out of bed! Probably about 9:50, because we were just about to miss the first break, and by the time 10 o'clock came around, he basically decided he wasn't going to do anything. He bounces off the walls to the point where he drops down on his hands and knees, crawls to the equipment, turns it all on, and says, "Okay! You do it! I'm going back to sleep!" Then he used a few expletives, "F— you! You're an idiot! Who the hell are you? Get the hell out of here!" And I came out there to help him!

Basically, what happened after that was I called Bill Louis. He comes down, and goes into the trailer to have a conversation with him, and you hear all this yelling and screaming! The next thing you know, Bill leaves the mobile home and Jerry's right behind him in nothing but a pair of pants. He's yelling, "Come back, you! Come back like a man!" Bill just put his hands in the air, gets in his car, and drives away. I realized this was probably not the place for me anymore!

It was not the kind of thing Louis expected to deal with in his first week as program director.

Shirley was reportedly told to move his gear and to show up at the office the next day, a Monday. Garrett recalls, "What happens later in the day is that I get a phone call about five, six o'clock, and there's concern that there's a missing six, seven, eight hundred dollars. The next day, we find out the North Olmsted Police Department had been out there two hours before I was, because they got an anonymous phone call there was money missing from the charity drive. The money was missing before I even got to the trailer."

By the time Young had arrived, Shirley was said to be very distraught. He told Young that after he took painkillers for his back and went to bed, he had been robbed and someone drew strange markings all over his leg with a Magic Marker and poured alcohol over him. After Shirley left, Young kept an eye on the trailer for the rest of the day.

Mike Onesky, who worked promotions for the station, got a call from a fellow staff member. They told him that Shirley had been found disoriented and covered with writing, and that the caller would be collecting food and cash with some of the disc jockeys who were filling in at the site. Onesky remembers the trailer as something of an open house, with friends showing up at all hours, and he says the news came as little surprise to him. Onesky recalls, "The night he got busted, I know he had a party that night, because he invited me, but we had another promotion going on."

This put WNCX in a very difficult position from a public relations standpoint. Shirley was put on an indefinite leave by the station pending the results of an investigation into the alleged missing funds. Walt Tiburski called it a "timeout," telling the newspapers that Shirley's suspension was not an accusation but a "neutral position which allows for fair analysis." Tiburski promised that the station would reimburse any money verified to be missing from the donations, and he vowed that the food drive would continue as planned, but without Shirley's participation at this time. On December 13, Shirley stated he had been fired, though his lawyer added that the dismissal was a business decision on the part of WNCX and not connected to the investigation. Shirley also went on record to say that he was innocent and he wanted his name cleared, even offering to take a polygraph test. The case caught the interest of WNCX's own morning host Stern, who voiced fears that the homeless might eat Shirley if he was left to fend for himself on the streets.

The *Plain Dealer*'s *Sunday Magazine* ran a feature story on Shirley in its April 13 issue. The very next day, he filed a $10 million wrongful-dismissal suit against WNCX, claiming that the firing caused his financial ruin and that he was existing on unemployment benefits. Shirley also revealed he had been going through an alcohol rehabilitation program at the Salvation Army's Harbor Light center, and he felt he was perfectly capable of resuming work. He was eventually able to find a job at a local auto dealership while his lawsuit was considered.

A year after Shirley's suit was filed it was quietly dropped with no further action. As Louis puts it, "My greatest wish was that this could have ended differently, because of the contributions [Shirley] made during his time, and his efforts toward charities and some very worthwhile causes, were obscured by the way things turned out. It was very unfortunate."

Change Again

WMMS had the same core staff for fifteen years, but that was with Malrite. Shamrock Broadcasting kept the station for about a year, and OmniAmerica gave way to Nationwide Broadcasting. Ratings had started to flag some time before, John Gorman quietly left the station, and the popular team of Brian Fowler and Joe Cronauer were moved to afternoons to make way for an ill-fated attempt to beat Howard Stern at his own game.

Fowler and Cronauer were understandably very concerned about their futures, but they kept plugging away. Cronauer lost an on-air bet about an expansion-team Super Bowl, which resulted in him running down Ontario Street naked. They had also heard that Drew Carey was planning a special episode of his show with Cleveland Mayor Michael White, Joe Walsh, Bernie Kosar, and other local luminaries. They called the comedian at his Los Angeles home to see if they could be part of it as well. Carey said that if they could be on the set the very next day, they were in. So the two boarded planes as Cronauer's naked run was plastered across the evening news. When they returned, Fowler and Cronauer were moved to afternoons to make way for the next version of the *Buzzard Morning Zoo*. That experiment ended after six months, and fired talk show hostess Liz Wilde came back to sue the station.

In the meantime, Nationwide was actually considering inviting Jeff Kinzbach back to help restore some of the station's credibility. Needless to say, the publicity would have been like nothing Cleveland radio had ever seen. Kinzbach recalls, "I was talking to Spaceman, and we were going to play some golf. We started talking about the morning show, and he said, 'Would you be interested?' I thought it would be cool, and I told Space, 'I would want you to do it with me,' because I always liked him, and thought he was funny. If you know how to handle Spaceman, he's hilarious! It would have been a damned good morning show, but you get out of the business, you remember all the great times. You don't remember the bad stuff." For whatever reason, those plans were eventually scrapped.

In late 1997, Strongsville native Danny Czekalinski was called in to take over the morning slot at WMMS. He had made impressive gains in Kansas City and other markets, and was a radio fan since his very early days, even helping weekend air personality "Uncle Vic" during his time at WGCL-FM. Czekalinski was very comfortable in Kansas City, and he said he would only move if he got an offer in his old hometown. He was joined in Cleveland by Darla Jaye, a likable sort who had already co-hosted with Czekalinski in Flint, Michigan, and Cory Lingus, a holdover from the show's previous incarnation. The three had chemistry and the public seemed to like them, especially at station promotions. They hosted an early morning Saint Patrick's Day broadcast to a packed house at Dick's Last Resort in the Flats, calling back to the glory days of the *Morning Zoo*, and drew hundreds of partiers to a '70s-style Last Chance Prom at Gray's Armory. They gave away thousands of items at "Drive By Shirtings" that were broadcasted live from locations around Northeast Ohio. In one of the most bizarre promotions ever in Cleveland radio, they played to a standing-room-only house at Dick's Last Resort for a Third Trimester Bikini Contest, which as the title describes, invited women in the final stages of pregnancy to pose in skimpy swim wear. It received a good amount of television coverage, although no station would send any of its personnel as celebrity judges.

Program director Bob Neumann scored on-air coups by bringing in members of Aerosmith and Queensryche to co-host with Fowler and Cronauer. With a simple phone call he also brought in Led Zeppelin's elusive Jimmy Page for an hour-and-a-half visit with WMMS midday host "Crankin' Craig" Dori, even taking phone calls from listeners.

The station brought back a Cleveland tradition, if only for one day, by hosting Metallica for a live broadcast from the Odeon after the band sold out two shows at Gund Arena. Nationwide Broadcasting reportedly cut a huge check to secure that concert. The station handed out WMMS t-shirts to anyone who would wear them for the over-twenty hours they stood in line for a WNCX ticket promotion, and then also wear them into the Odeon. Very few people said no. A WMMS t-shirt was like cigarettes in prison. No one turned them down and everyone was willing to do just about anything to get one, no matter what the station stood for at the time.

Fowler and Cronauer continued to do well. With producer Cheryl Zivich as their lovable on-air foil, the duo was promised that they could basically do a morning show in the afternoon drive. They brought in a variety of guests such as guitar virtuoso Joe Satriani, Bon Jovi's Richie Sambora, and Megadeth's Dave Mustaine, who explained his role as a soccer dad as well as a heavy metal guitarist. Other musical guests such as Satriani and Sambora would join the hosts in performing "The Beer Song," the show's kick off to the weekend. Based on the song "Do-Re-Mi" from *The Sound of Music* ("Dough, the stuff that buys me beer . . ."), the parody lyrics weren't exactly an original concept. In fact a number of the show's comedic bits were inspired by jokes that would make the rounds in such ways as forwarded e-mails. But often the pair would give even the most ordinary of jokes their own flair, such as punctuating "The Beer Song" with a medley of Homer Simpson yelling "D'oh!" and then his line, "To alcohol, the cause of, and solution to, all of life's problems!" The new tradition was not unlike Kid Leo's weekly playing of Bruce Springsteen's "Born to Run" to start the weekend. Fowler and Cronauer's afternoon block quickly became the most fun show on WMMS in quite some time.

But in 1997 the station was waiting for yet another ownership change when Nationwide Broadcasting decided to sell its radio stations. Promotion money was scarce, and some people in power didn't understand the station the way Clevelanders did. Word-of-mouth and civic pride can only go so far. It doesn't matter how innovative hosts and programming are. Unless advertising can get out to people who took the station off their dial, it can be a long and fruitless struggle.

After Jacor Broadcasting assumed control of several Cleveland stations from Nationwide, the inevitable changes quickly began. Lingus, Jaye, Dori, and deejay Brett Hart left the station. *The Brian and Joe Radio Show* was switching to mornings at the revamped WMVX-FM, formerly WLTF.

WMMS told listeners that there would be a special announcement on October 1. At noon on the appointed day, the announcement was a bombshell. The longtime voice of the station, Len "The Boom" Goldberg, took to the airwaves. He didn't sound happy.

The announcement was quick and to the point, and totally unexpected: "Hi, folks. This 'The Boom,' Len Goldberg. As you know, the Buzzard has been part of Cleveland for thirty years. However, the time has come to say goodbye. So, on November 1, a new station will be on 100.7. I ask you to join us as we bid the Buzzard a fond farewell. So, listen through the month of October as we re-live some great WMMS memories. A tip of the Buzzard's wing to you, our thanks for supporting 100.7, WMMS, for the past thirty years."

An ominous bell toll and the end tag "Thirty days 'til we bury the Buzzard" ended the statement. Queen's hit "Another One Bites the Dust" followed the announcement. Listeners new and old, casual or diehard, were no doubt stunned by the news.

The familiar voices of Kinzbach, Leo, Denny Sanders, Betty Korvan, Matt the Cat, and others were played in quick succession to share memories and anecdotes with the listeners. Dia Stein related a bizarre story via telephone about Joan Jett befouling the station restroom, which she later asked to have taken off the air. No WMMS flashback could have been complete without the mention of Springsteen and a vintage clip of Leo announcing Springsteen's summer Municipal Stadium show from the mid-80s boomed over the airwaves.

Off the air, the news wasn't any less shocking. The WMMS staff had heard about the change at a meeting just minutes before the broadcasted announcement. Goldberg knew about it the day before. Still the question remained as to why a major company such as Jacor would buy a history-making station such as WMMS, and then turn around and scrap the format. It was very similar to the fate that had befallen WLTF, when Jacor took over that property and changed the staff, format, and call letters to WMVX. "Mix 106.5" took off with music from the "Eighties, Nineties, and Seventies" and a lot of promotion.

If Jacor was planning to kill a station, then it was curious why it wanted WMMS to constantly reflect on its glory days. It was suspected in the WMMS offices that it was probably the way most people would have wanted it at the end.

WNCX reacted swiftly, promising that it would always be there for its listeners. Personalities at other stations openly spoke on-air about the announcement, forgetting the old rule about not mentioning competition. The news made front-page headlines in the *Akron Beacon Journal* and the *Cleveland Plain Dealer*. The big rumor was that the station would be changing its call letters and be referred to as "Kiss 101," with a Top-40 format.

True to its word, WMMS aired "best of" segments every hour. It re-ran key programs such as Bruce Springsteen's Agora concert for the station's tenth anniversary. Sanders and Leo came back for farewell shows, though even they had some reservations about reports of a format change. Sanders agreed to do the show, never mentioning the end of the Buzzard, but saying he was back "one last time" as part of the station's thirtieth anniversary. It also appeared Kid Leo had his "eyes wide open," going so far as to state, "I don't know if the Buzzard is actually going away!" Even so, two different t-shirts suddenly appeared to commemorate the end of the "Buzzard," although they didn't feature any specific dates finalizing the station's demise.

Despite the genuine attitude of the WMMS veterans, other Cleveland stations were skeptical that the management was serious about going through with this change. It takes time for a station to change call letters, and applications to the FCC are public record. It's not just a matter of "Bam! We're different!"

The Buzzard Unplugged (1998). Two special t-shirt designs were commissioned to be the last of the series when WMMS announced its plans to change format and call letters.
Brian Chalmers

WNCX went on the air to say that no FCC filing had taken place. The station challenged that WMMS was putting the public on, offering a $5,000 prize to back up the claim. WNCX turned WMMS's announcement into a promotion of its own.

As the final day approached, a new announcement from Goldberg was aired. This time he said that the powers at Jacor never expected the overwhelming public response, and the Buzzard and call letters would stay. He went on to say that every effort would be made to make the station one Cleveland could be proud of again.

Some longtime WMMS listeners said they felt used and betrayed, but they did tune in. The winter Arbitron ratings book showed WMMS with 30,000 more listeners per week than just a few months before, catapulting it back into the top ten stations at number seven. The *Plain Dealer* summarized it this way: The scam worked! The paper pointed out that the station now sounded a lot more focused, which may have kept more listeners tuned in. Scam or not, the announced change caused so much controversy over the course of a month that if station management decides on a definite change in the future, it will likely happen overnight with very little initial fanfare.

The WNCX staff was elated by the backlash WMMS received after it announced there would not be a change. WNCX stressed its credibility and vowed to never fool the audience with that type of promotion. But the celebration was short lived.

Just a few weeks after the WMMS announcement, a mandatory staff meeting was called at WNCX, and Walt Tiburski told the disc jockeys assembled in the conference room that parent company Clear Channel would be merging with Jacor. WNCX, WERE-AM, and WENZ-FM would very likely have to be sold.

The mood of the room suddenly plummeted. There was a period of silence. It was now very evident that no station, no matter what its history was, was immune from corporate America. WNCX was picked up by Infinity, and the Radio One group bought WENZ and WERE, with plans to change the FM's format to urban.

As WMMS entered 1999, the station prepared an on-air imaging campaign aimed at stressing attitude. The early and ongoing success of WMMS going into the mid-90s was about showing aggressive pride in being from Northeast Ohio. Some of the new taped IDs had already been aired prior to the announced change in format. However the new campaign was a bit unsettling to many die-hard fans, some of whom remained loyal to the station through the decades and changing formats. The music was pretty safe—some new songs, but mostly classics that withstood the test of time and were familiar to the audience—but that didn't draw a lot of young listeners.

But the campaign likened the station itself to a "bad case of scabies," boasting that it was the station of choice for the majority of "deadbeat dads" and "liquor store gunmen." The "no bullshit" attitude had always been implied, but now WMMS was just coming right out and saying it. The station offered a "knee to the groin of Cleveland radio," but some members of the audience were starting to think after all the previous changes, the knee might be directed at them. One of the liners suggested the call letters could have been translated to "We Might Make it Someday." In reality the station was very likely hoping to make it back to where it had once been. Considering the long rough-and-tumble history of the station, and the state of the radio industry, anything was possible.

That same October Sanders traveled to Seattle for the National Association of Broadcasters convention. He had been program director at WMJI-FM for a little more than two years, and he could look back on a long history of helping to shape Cleveland's radio landscape. While he was in Seattle the summer ratings book

Veteran broadcaster Denny Sanders looks back on a long and very successful run in Cleveland radio (2000). Photo by Ken Blaze

had been released and the results were phenomenal. WMJI had rocketed to number one in age categories twelve-and-older, twenty-five-to-fifty-four, and thirty-five-to-sixty-four in both the Cleveland and Akron markets. It was already a time for victory, and later that night WMJI won the coveted Marconi Award for Best Large Market Radio Station in America. That award is usually reserved for older news or talk stations, but that year WMJI had triumphed.

Sanders had seen FM rock radio in its infancy, helped it grow, saw dozens of people come and go, and now he was reaping the rewards of a long, sometimes difficult, but always fascinating trip through radio history. WMJI was honored for its staff and sound, playing a lot of music that came into popularity in the first years of the FM revolution. It seemed like a fitting coda to the first thirty years of FM rock radio in Northeast Ohio.

Meanwhile WMMS—the station that Sanders, Kid Leo, the *Morning Zoo*, Betty Korvan, David Helton, Matt the Cat, Len Goldberg, and Bill Freeman built and helped define into a ratings monster—continued, for some as a fond memory. It

Art by Brian Chalmers

had never again scored the ratings seen in its heyday, and since 1994 averaged a new morning show per year. But that doesn't mean it couldn't happen again. Radio is a volatile medium, subject to change, and WMMS could re-emerge to its former glory and beyond. It wouldn't be easy, but it's certainly not impossible.

After all these years, the Buzzard is still circling.

Epilogue

What a difference three decades make!

Things sure have changed since I left WMMS in 1994. In the years to follow, pressured by a Republican-controlled Congress, President Bill Clinton signed the Telecommunications Act of 1996 into law. It pretty much lifted the restrictions on how many stations a company could own in one city, and now only a few broadcasting conglomerates own the majority of stations nationwide.

Clear Channel Communications took over the rock radio market by buying up hundreds of stations and then taking controlling interest of SFX Entertainment, the most prominent concert-promotion and venue-management company in the United States. The firm further cemented its position as corporate owner of all rock outlets by also purchasing television stations, a radio research company, program syndication firms, the trade magazine *Album Network,* consulting firms, and hundreds of thousands of outdoor billboards.

The original rules on ownership were meant to assure diverse programming, but now every city has a "classic rock" station, a "mix" station, "extreme" radio, and other similarly named formats. They often feature out-of-town voice-tracked personalities and music programmed at one central location.

This homogenization of radio no doubt saved some stations that had been struggling and bouncing among owners for years, and of course it turned tremendous profits for Clear Channel. But as expected, it also drew its fair share of critics, many of who felt that radio was losing its unique voice. I never thought I would see the day when *Rolling Stone* would run an article examining, "Why Radio Sucks."

As for me, on September 29, 2002, I sat on a dais with Jack Riley, Jeff Baxter, and the "Real Bob" James, along with many other distinguished names, as we were inducted along with Gary Dee, Ron "Captain Penny" Penfound, and Gene Carroll into the Radio/Television Broadcasters Hall of Fame. My dear friend Linn Sheldon introduced me, and it's times like these when you get a real sense of history about where you've been, and wonder about the future. I would have never met Linn, the folks at WMMS, or so many other radio and television people who have become a part of my life without my career in the media. For that, I'm eternally grateful.

I sometimes think back to that spring day in 1971, when I was still at Bedford High, and I walked more than fifty blocks down Euclid to the old WMMS studios. Those first steps off the elevator are still fresh in my memory, and I wonder what course my life would have taken if I had started my radio career earlier than I did. They say everything happens for a reason, and I believe that.

I look at some folks who were with the station when it was barely a blip to anyone outside the counterculture, but grew with it and reaped incredible rewards as a result. Some of them also saw divorce and other problems linked to the "excesses of the rock and roll lifestyle," and came to the crashing realization that after many years at a proven long time winner, they were suddenly thrust into the real world of fighting for a job like everyone else—and that was a bitter pill for them to swallow.

I, however, went to school and sometimes worked in factories and warehouses, where I met some of my oldest friends who we've embraced as family. I would have never met my wife, Janice, who I've been with since 1971, and in May of 2003 we celebrated twenty-five years of marriage. We would have missed out watching nephews and godchildren, and as we see now, a new generation of godchildren, grow up before our eyes. Yeah, everything happens for a good reason. I chose the wrong day—a reception-less Saturday—to try for an internship that would have put me in the job I wanted a lot earlier than I realized. But looking back, I would never have had it any other way.

I have had the absolute pleasure of working with people I respected, and I was able to form friendships that I treasure and continue to this day. I recall the indescribable heartache of seeing my radio family leave the Buzzard's nest and go on to other jobs. It's the same when any group of co-workers leaves together for one reason or another. You know no matter where they land or how you see them in the future, it will never be the same.

Rocco walked into the newsroom the afternoon of April 15, 1994, and said, "Come on. You're part of this." We walked into the studio to begin the transition that would become "The Next Generation." Just a few years before, we agreed that there had to be a better place than the station where we both worked, and we found it at WMMS. After we brought the curtain down on that era with the final moments of his show, Rocco grabbed me in a bear hug like the one he greeted me with on my very first day. It was now a different era of radio, and we had changed as well.

The new management simply had a different vision. To be quite honest, they treated me very, very well. I stayed on for a few months to work with some extremely talented people, including Brian Fowler and Joe Cronauer, who are still my friends to this day. But sometimes you know when it's time to move on. The Rock and Roll Hall of Fame was set to open soon, and with all the attention I had paid to that story over the years I thought it would be best if I could bring that excitement to television. The opportunity arose to move on to television, and I met with Dean Thacker and John Gorman to explain my decision. As I was leaving his office, I paused at the door and looked back saying, "You know, the only place I ever wanted to really work was WMMS!" We all smiled, and I took the first steps toward a new phase of my working life.

WMMS was a high point in my own radio career, as well as in the careers of many of my colleagues. It was the reason I decided to pursue that profession. When I started at the Buzzard, I became part of an indefinable energy, a powerful force in radio and the community, one that was more than just information and entertainment. It was literally a way of life for so many, as well as a stable routine that people embraced and depended on as an important part of themselves and the way they went about their day.

I just can't help think about all of the wonderful memories I've had there. The crazy nights at Apple Annie's and other spots with a staff that never seemed to tire of each other's company—the Coffeebreak Concerts—rock stars visiting the station on a regular basis—interviewing Paul McCartney in Washington, New York, and especially here at Municipal Stadium—conversations with Mick Jagger and Pete Townshend—broadcasting from Hollywood and meeting people like Bob Hope and other legends—the Rock and Roll Hall of Fame ceremonies—and memories of my wife, Janice, nervously chatting with one of her musical heroes, David Bowie.

I'm not the only one who reflects longingly on this station. Every year in April since 1995, the old staff of WMMS gets together to share thoughts and memories about a special time in their lives. Matt the Cat, T. R., Rocco, Mondo, Tom Bush, John Filby, and myself, not to mention the office staff and many others, join to celebrate the past and look to the future. Inevitably, at some point in the evening, they raise a glass to toast each other and those who moved on.

But they also remember the people who embraced the old Buzzard, who listened in their cars, came out to the clubs and shows, and applauded its success. The audience was the real heart and soul of the station, and the reason that Cleveland radio, and all of our FM memories, will be an important part of our lives for many years to come.

We may have moved on, but the FM attitude is still with us and stronger than ever.

Bibliography

Articles

"50,000 Tickets Left to Jackson Concert; Mail Sales Pondered." *Cleveland Plain Dealer,* October 5, 1984.
"A Progressive Veteran Reflects on WMMS' Growth." *Record World,* February 3, 1979.
"About Face." *Cleveland Plain Dealer,* May 20, 1981.
Albrecht, Brian E. "DJ Put on Leave as Inquiry Begins on Food Drive Funds." *Cleveland Plain Dealer,* December 13, 1996.
"Bacha Goes Bananas." *Great Swamp Erie da da Boom* 1, no. 21, November 2–16, 1971.
Ball, Chris. "Questionable Tactics Make Radio Waves." *Small Business News,* March, 1995.
Barrett, Bill. "Stokes Tries to Halt Sale of Radio Stations." *Cleveland Press,* October 8, 1981.
———. "Time Is Running Out." *Cleveland Press,* February 10, 1977.
———. "WNCR Aims Its Rock Music at 'Sophisticated Listeners.'" *Cleveland Press,* April 28, 1971.
———. "WNCR May Make You Gasp as It Zings In on the Young." *Cleveland Press,* April 27, 1971.
Biddle, Daniel, R. "Kid Leo: The Voice of Cleveland." *Rolling Stone,* February 8, 1979.
"Billy Bass Quits WNCR." *Great Swamp Erie da da Boom* 1, no. 18, September 21–October 5, 1971.
Black, Judy. "A New Wave and More Classic Rock." *Scene,* September 20–26, 1990.
———. "*Scene* Breaks away from 'Breakout.'" *Scene,* November 1–7, 1990
Bornino, Bruno. *Cleveland Press,* April 21, 1967.
———. "WMMS Is Tuned in as City's Concert Connection." *Cleveland Press,* March 23, 1978.
Brown, Roger. "Buzzard Scam Gets WMMS What It Wanted: More Listeners." *Cleveland Plain Dealer,* January 18, 1999.
Budin, David. "David Spero—Rock around the Block." *Cleveland Plain Dealer,* February 13, 1994.
Burkhardt, Karl R. "Sudden Changes Due in Radio." *Cleveland Plain Dealer,* July 20, 1969.
———. "WMMS Flips Out." *Cleveland Plain Dealer,* November 15, 1968.
"C. Miller Refiles $74 Million Suit." *Cleveland Press,* January 21, 1982.
Cali, Joe. "Radio Activity: WMMS Always Wants to Improve." *Scene,* April 30–May 6, 1987.
"Cleveland Update." *Variety,* February 10, 1971.
"Cleveland Update." *Billboard,* March 6, 1971.
"Concert Bummer." *Great Swamp Erie da da Boom* 1, no. 6, March 23–April 5, 1971.

Crea, Joe. "Rockin' in the Buzzard Nest." *Cleveland Plain Dealer Sunday Magazine*, 1978.

Denver, Joel. "WGCL Beats the Buzzard." *Radio & Records,* November 25, 1983.

———. "WMMS Regroups for the 90s." *Radio & Records,* November 10, 1989.

———. "WPHR's Power Play." *Radio & Records,* November 10, 1989.

———. "WRQC Hot to Trot." *Radio & Records,* November 10, 1989.

"Dig the Underground?" *Cleveland Plain Dealer,* April 5, 1968.

"Disc Jockey Spero Leaves WWWM-FM." *Cleveland Plain Dealer,* May 30, 1979.

Dolgan, Robert. "Ethnic-power Radio Rally Attracts 2,000 Downtown." *Cleveland Plain Dealer,* February 7, 1977.

Dyer, Bob. "A Visit to the Buzzard's Nest: Cleveland Giant WMMS Rocking along at Top of World." *Wooster Daily Record,* February 19, 1982.

———. "Kid Leo: Godfather of Rock." *Wooster Daily Record,* February 19, 1982.

———. "Stern Shakes Up Status Quo in Cleveland with Blunt Apathy." *Akron Beacon Journal,* September 6, 1992.

Eszterhas, Joseph. "Hippies Given It Straight: Language Is Going to Pot." *Cleveland Plain Dealer,* December 22, 1968.

Evans, Christopher. "Rock On—Jerry Shirley Eats Humble Pie as He Struggles to Recover from WNCX Firing." *Cleveland Plain Dealer Sunday Magazine,* April 13, 1997.

Ewinger, James. "Gary Dee Slips to No. 2 in Radio." *Cleveland Plain Dealer,* October 10, 1981.

———. "M105 to Switch to Soft-Rock Beat." *Cleveland Plain Dealer,* June 12, 1982.

———. "WLYT-FM Declares Album-Rock War on Leader WMMS-FM." *Cleveland Plain Dealer,* August 28, 1982.

Eyman, Scott. "The Disco Nite Moves." *Cleveland Plain Dealer Magazine,* May 27, 1979.

Feran, Tom. "CBS Weatherman Shines on City." *Cleveland Plain Dealer,* December 2, 1988.

"FM Station Denies Sponsoring Party Marred by Arrests." *Cleveland Plain Dealer,* November 14, 1985.

Frolik, Joe. "Buzzards Clipped in Ratings." *Cleveland Plain Dealer,* May 1, 1987.

———. "Malrite Seeks Channel 19 Control; Detroiters to Buy WERE, WGCL." *Cleveland Plain Dealer,* June 6, 1986.

———. "Top Buzzards Leaving Nest at WMMS-FM." *Cleveland Plain Dealer,* August 19, 1986.

———. "WRQC Takes Off White Gloves, Dons Leather." *Cleveland Plain Dealer,* March 31, 1986.

Gerdel, Thomas W. "Daffy Dan's Firm Resists Shrinking." *Cleveland Press,* July 24, 1980.

Girard, Jim. "Billy Bass and Joyce Sell-Out." *Scene,* March 22–28, 1973.

Glasier, David. "Kinzbach Admits He's Unhappy with Format." *News Herald,* October 7, 1988.

Goldstein, Patrick. "Cleveland Is on a (Rock 'n') Roll." *Los Angeles Times,* June 1, 1986.

Gomez, Brian. "The WMMS Buzzard: Born to Be Active." *Scene* (supplement), April 28–May 4, 1983.

———. "WMMS Marketing & Promotion: Behind the Scene Performances." *Scene* (supplement), April 28–May 4, 1983.

Guyette, Jim. "Spero Back in Radio after Four Years." *News Herald TGIF*, February 3, 1978.

Hagan, John F. "WNCX Fires DJ After Charity Money Theft." *Cleveland Plain Dealer*, December 14, 1996.

Halonen, Doug. "FCC Planning Stern Reprimand on 'Indecencies.'" *Electronic Media*, December 7, 1992.

Hart, Raymond P. "Dialing Around." *Cleveland Plain Dealer*, September 26, 1970

———. "Hallaman, McLean, Reynolds Fired by WHK." *Cleveland Plain Dealer*, February 8, 1973.

———. "Morrow Takes Straight Path." *Cleveland Plain Dealer*, February 9, 1975.

———. "Musical Pyrotechnics on July 4 to Launch WNCR Rock Format." *Cleveland Plain Dealer*, June 29, 1970.

———. "Sale of WHK to Malrite OK'd; Chain to Move Headquarters Here." *Cleveland Plain Dealer*, November 1, 1972.

———. "Three More Quit Staff at WMMS." *Cleveland Plain Dealer*, December 29, 1972.

———. "WCLV, WGAR Set FM 'Step Beyond.'" *Cleveland Plain Dealer*, February 22, 1970.

———. "WMMS-FM Sale Won't Squelch Hard Rock." *Cleveland Plain Dealer*, September 14, 1972.

———. "WMMS-FM Will Drop Band Sound." *Cleveland Plain Dealer*, October 27, 1969.

———. "WNCR Has No Plans to Sit Still." *Cleveland Plain Dealer*, December 23, 1970.

———. "WNOB-FM Will Join 'Rock" Battle Tomorrow." *Cleveland Plain Dealer*, June 29, 1969.

Heaton, Michael. "Back When the Buzzard Flew Free." *Cleveland Plain Dealer*, February 26, 1988.

———. "Howard, King of Sternrovia." *Cleveland Plain Dealer*, September 11, 1992.

Heaton, Michael, and David Sowd. "Analysts Circling as Buzzards Look for Answers." *Cleveland Plain Dealer*, February 20, 1989.

———. "Discord at WMMS Leads to Massive Staff Shakeups, Departures." *Cleveland Plain Dealer*, February 20, 1989.

———. "'We Were A Family' Kid Leo Says." *Cleveland Plain Dealer*, February 20, 1989.

Hickey, William. "Uggams Hour Is in Variety Show Rut." *Cleveland Plain Dealer*, September 30, 1969.

———. "WWWE Terminations Explained." *Cleveland Plain Dealer*, November 22, 1972.

Holan, Mark. "From Underground to the Penthouse." *Scene* (supplement), April 28–May 4, 1983.

———. "John Gorman/Kid Leo: The Buzzard High Command." *Scene* (supplement), April 28–May 4, 1983.

———. "The On-Air Staff—WMMS Sounds Like." *Scene* (supplement), April 28–May 4, 1983.

———. "WMMS: Overhauling the Buzzard for the '90s." *Scene,* April 21–27, 1994.
"How WNCR-FM Became a Showcase 'By Accident.'" *Billboard,* December 19, 1970.
Joyce, Tim. "Getting Down with Murray Saul." *Exit,* September 10, 1975.
Kojan, Harvey. "Still No. 1 . . . WMMS in the '90s." *Radio & Records,* December 6, 1991.
Lawrence, Joe. "WMMS 'Buzzard' Flying High." *Cleveland Press,* February 15, 1982.
Ludlow, Liz. "WNCX's 'Those Guys in the Morning.'" *Scene,* July 20–26, 1989.
Madison, Shari. "WMMS' Long, Rocky Journey" *News-Herald,* November 6, 1987.
———. What's Changed: Rock 'n Roll, WMMS, or You?" *News-Herald,* November 6, 1987.
Mahoney, Mike. "Exec Here Tunes in Record Radio Deal in LA." *Cleveland Plain Dealer,* October 26, 1985.
McCarty, James. "Fired Disc Jockey Sues WNCX." *Cleveland Plain Dealer,* April 15, 1997.
McGee, David. "Kid Leo: Growth of a Cleveland Legend." *Record World,* February 7, 1979.
———. "Peter Schliewen Remembers When." *Record World,* February 3, 1979.
Mouth of the Cuyahoga. "Circulation Wars, Lame Ducks and the Slack Jawed." *Scene,* June 9–15, 1994.
———. "Jeff & Flash Finale, Bleacher Naming and Traffic Jams." *Scene,* April 14–20, 1994.
———. "Money for Music, Getting There, and the Next Generation." *Scene,* April 21–27, 1994.
———. "Sexism, Shopping Mauls and Silver Buzzards." *Scene,* December 24, 1992.
———. "Yanked Off the Streets, Howardgate, and an Official Opinion" *Scene,* June 16–22, 1994.
Mr. Nobody. "Underground Radio in Our Swamp." *Great Swamp Erie da da Boom* 1, No. 19, October 5–19, 1971.
Neus, Elizabeth. "Cleveland Station Praised as Leader." *Daily Record,* March 5, 1985.
Norman, Michael. "Stanley Relives Brush with Eternity in Las Vegas." *Cleveland Plain Dealer,* January 7, 1992.
———. "WMMS' Historic Rock Archives Lost in Owner Shuffle." *Cleveland Plain Dealer,* December 22, 1994.
Peters, Harriet. "Bass Leaves WMMS Despite Getting Raise." *Cleveland Press,* December 30, 1972.
———. "Carolyn Has Proper Spirit for Radio." *Cleveland Press,* October 6, 1972.
———. "Debbie Ullman: Free in Form and Spirit."
Cleveland Press, January 18, 1974.
———. "Disc Jockey Shauna May Take a Turn for the Verse." *Cleveland Press,* September 15, 1972.
———. "Our Man Friday: Bloody Murray." *Cleveland Press,* May 14, 1976.
———. "Philosopher Korvan Has a DJ Degree." *Cleveland Press,* April 2, 1976.
———. "Radio Scene Causes Static." *Cleveland Press,* January 19, 1973.
———. "Tiny Jane Disko is Busy with Radio and Theater." *Cleveland Press,* August 8, 1980.

———. "WMMS Is Becoming Stile Conscious." *Cleveland Press,* August 22, 1975.
"Powerhouse." *Cleveland Plain Dealer Action Tab,* October 3, 1969.
Powers, Tanja. "Buzzard Flying into Higbee to Market New Clothing Line." *Crain's Cleveland Business,* July 23, 1989.
Rathbun, Keith. "Malrite Communications Group, Inc: Solidly Behind WMMS Success." *Scene* (supplement), April 28–May 4, 1983.
———. "Much Ado about Nothing." *Scene,* March 3–9, 1988.
———. "The Demise and Projected Rise of 105." *Scene,* June 24–30, 1982.
———. "To Those about to Rock . . . WMMS Salutes You." *Scene* (supplement), April 28–May 4, 1983.
———. "WMMS Sales: Surrounding Itself with Strength . . . and Utilizing (Strength)." *Scene* (supplement), April 28–May 4, 1983.
Rogers, Tim. "Them's Fightin' Words, Podnar." *Cleveland Press,* May 14, 1981.
Santiago, Roberto. "Malrite to Merge with Emmis or Infinity?" *Cleveland Plain Dealer,* December 3, 1992.
———. "Mouth-to-Mouth Smut Is A.M. CPR." *Cleveland Plain Dealer,* September 10, 1992.
———. "Rights Groups Attack Radio Stations Deal." *Cleveland Plain Dealer,* December 2, 1992.
———. "Stanley Adds to WNCX Role." *Cleveland Plain Dealer,* June 1, 1992.
———. "WENZ Program Director Fiddling Every Hour to Find Right Mix." *Cleveland Plain Dealer,* May 15, 1992.
———. "WNCX Signs Up for Howard Stern's 'Shock' Radio Show." *Cleveland Plain Dealer,* August 12, 1992.
Sawicki, Stephen. "The Buzzard Goes to War." *Cleveland Magazine,* December, 1986.
Scott, Jane. "Brandy Krellogg—'Ohio' in Japan." *Cleveland Plain Dealer,* January 20, 1984.
———. "Breakfast on the Air." *Cleveland Plain Dealer,* April 15, 1978.
———. "Chicago, San Francisco Are Rock Hall Die-Hards." *Cleveland Plain Dealer,* May 10, 1986.
———. "Dancin' Danny Adds Right Touch to 'Rockspot.'" *Cleveland Plain Dealer,* January 13, 1984.
———. "Disc Jockeys Being Taken." *Cleveland Plain Dealer,* January 2, 1976.
———. "Hall of Fame: Pros and Cons." *Cleveland Plain Dealer,* November 1, 1985.
———. "Lee Andrews Takes to the Air in His Spare Time, Too." *Cleveland Plain Dealer,* July 30, 1971.
———. "Let's Have Another Cup of Coffee." *Cleveland Plain Dealer,* April 7, 1972.
———. "Rock Poll on a Roll; Voters' Calls Put City Way Ahead on Hall of Fame." *Cleveland Plain Dealer,* January 21, 1986.
———. "Rock Reverberations." *Cleveland Plain Dealer,* November 28, 1975.
———. "Ron and Kaye: Two by Two." *Cleveland Plain Dealer,* January 15, 1971.
———. "The Buzzards Have Their Day." *Cleveland Plain Dealer,* January 21, 1992.
———. "The Happening." *Cleveland Plain Dealer,* March 29, 1974.

———. "The Happening/Jelly Tight." *Cleveland Plain Dealer,* January 16, 1970.
———. "The Third Time is the Charm for M-105." *Cleveland Plain Dealer,* March 3, 1978.
———. "Tonight's the Night to Howl." *Cleveland Plain Dealer,* March 21, 1986.
———. "TV Is Giving Viewers Bigger Piece of the Rock." *Cleveland Plain Dealer,* September 22, 1978.
———. "What's Your Stand on the Walkout?" *Cleveland Plain Dealer,* September 25, 1970.
———. "WMMS-FM Fights Air and Earth Pollution." *Cleveland Plain Dealer,* March 19, 1971.
———. "WNCR." *Cleveland Plain Dealer Action Tab,* July 24, 1970.
"Secret?" *Cleveland Plain Dealer,* February 26, 1988.
Serbin, Ken. "Larry Robinson Plans to Acquire Radio Stations Here." *Cleveland Press,* July 21, 1981.
Strassmeyer, Mary. "Mary, Mary." *Cleveland Plain Dealer,* April 21, 1981.
———. "Mary, Mary." *Cleveland Plain Dealer,* April 29, 1982.
"Street Talk." *Radio & Records,* December 11, 1992.
Sowd, David. "All-Jovi Day Makes Ears Listen." *Cleveland Plain Dealer,* July 7, 1989.
———. "Barefoot DJ Gets Attention." *Cleveland Plain Dealer,* January 20, 1989.
———. "'Best Station' Returns to Poll." *Cleveland Plain Dealer,* November 16, 1988.
———. "MMS Airs Little-Heard Rock." *Cleveland Plain Dealer,* March 31, 1989.
———. "MMS Plays Fair, Drops to 4th." *Cleveland Plain Dealer,* February 24, 1989.
———. "Much Ado about Nothing?" *Cleveland Plain Dealer,* November 4, 1988.
———. "NPR Staff Due in Cleveland." *Cleveland Plain Dealer,* Summer, 1988.
———. "No. 1 Station Keeps Slipping." *Cleveland Plain Dealer,* January 13, 1989.
———. "Pool Helps Stations Report." *Cleveland Plain Dealer,* September 7, 1990.
———. "Ratings Soar for Country Station." *Cleveland Plain Dealer,* January 25, 1991.
———. "Station Takes on Big City Voice." *Cleveland Plain Dealer,* December 2, 1988.
———. "Talk Radio' Says It In Print." *Cleveland Plain Dealer,* January 27, 1989.
Stringfellow, Eric. "Critics Just Too Serious, WZAK Says." *Cleveland Plain Dealer,* March 29, 1987.
Sutherland, Sam. "Billy Bass on the Early Years." *Record World,* February 3, 1979.
Swindell, Mary. "Rock Rears Revolution." *Cleveland Press,* July 25, 1969.
Trakin, Roy. "Dialogue: Cleveland Rocks as WMMS' Kid Leo Rolls On." *Hits,* April 13, 1987.
"WMMS Sold . . . But Not Lost Yet." *Great Swamp Erie da da Boom* 1, no. 23, December 14–28, 1971.
"WNCR Birthday Bash Draws Cleveland's Largest Mellowest Crowd." *Great Swamp Erie da da Boom* 1, no. 13, July 12–26, 1971.
"WNCR Gets It On." *Great Swamp Erie da da Boom* 1, no. 2, January, 1971.
"WNCR Moves to New Studios." *Great Swamp Erie da da Boom* 1, no. 23, December 14–28, 1971.

"WRUW Announces News Focus." *Great Swamp Erie da da Boom* 1, no. 19, October 5–19, 1971.

"Woman's Slavery Never Ends . . . and Cleveland Radio Says." *Cleveland Plain Dealer*, April 28, 1972.

Radio Broadcasts

WMMS—Buzzard Morning Zoo Final Broadcast, April 15, 1994
WMMS—"Kid Leo" Final Broadcast, December 16, 1988
WMMS—Lou "King" Kirby / Station Returns to Live Format, September 11, 1970
WMMS—Rocco, the Rock Dog Final Broadcast, April 15, 1994
WNCX—Howard Stern "Cleveland Debut," August 31, 1992
WNCX—Howard Stern "Cleveland Funeral," June 10, 1994
WNCX—Howard Stern "LA Funeral," November 1993

Index

Abramson, Roger, 49
Aerosmith, 112, 145–46, 243
Abell, Lyn, 381
Agora, The, 99, 110, 124–26, 132–34, 143, 150, 172, 181, 207, 242, 269, 388, 394, 401, 415, 432, 441; fire at, 253; *Live at the Agora* TV series from, 157–58; toga parties at, 158
Akron Beacon Journal, 441
Albertine, Charles, 13
Alden, Nancy, 288
Alexander, Ted, 191
Alford, Bill, 371, 382, 426–31, 433–34
Allen, Don, 19
Allen, Jim, 37
Allen, Jerry, 37
Amato, Frank, 260
Amato, Jim, 260
Ambrozic, Jack, 25–26, 31, 38, 46, 53, 55, 57, 60, 64–65. 77
America, 427
American Federation of Radio and Television Artists (AFTRA), 343
Anderson, Ernie "Ghoulardi," 5, 136
Anderson, Jennifer. *See* Cheeks, Jennie
Anderson, Jon, 175, 241
Andrews, Lee, 20, 21, 37, 53, 57, 77, 103
Armstrong, Jack, 22
Armstrong, Tom, 25, 136
Avner, Chuck, 7, 21, 44
Avsec, Mark, 260
Awarski, John, 246, 259

BP Riverfest, 374
Baechle, Cliff, 219
Bacha, 218
Bailey, Bill, 107
"Ballad of John and Yoko," 72–73
Barker, Len, 181
Barnett, Rich, 256
Baron, Dave, 259–60
Barrett, Bill, 45–46
Bass, Billy, 21–22, 23, 27–28, 32, 35, 39, 46, 48–50, 60–61, 63, 65–66, 71, 73, 77–78, 84–85, 90, 98–99, 103, 107–8, 388; background of, 14; debut on WHK, 12; departure from WIXY, 42; departure from WNCR, 52–57; Milton Maltz and, 86–88; *Scene* interview with, 92–93; WMMS employment and, 58, 88; WMMS–WVIZ-TV Glass Harp simulcast with, 67
Bassette, John, 71, 113
Batiuk, Tom 16, 185, 250, 309
Batts, Sharon, 310
Bauer, Gary, 177, 195
Baxter, Jeff, 446
Becker and Heggs, 352
Behemoth, Crocus. *See* Thomas, David
Belew, Adrian, 356
Belkin, Jules, 23, 105, 271, 275, 396, 409
Belkin, Michael, 378
Belkin, Mike, 23, 155, 182, 275
Belkin Productions, 20, 25, 49, 53, 86, 115, 125, 143, 149, 155, 181, 201, 324, 346, 395, 409, 422; Rock and Roll Hall of Fame and, 262; World Series of Rock and, 106, 136, 149, 154, 174
Bennett, Ric "Rocco, the Rock Dog," 330, 338, 340, 346–47, 362–63, 382, 392, 395, 404, 410, 446, 448; *Buzzard B-Movies* and, 373; departure from WMMS, 411–12; *Gonzo Radio* and, 359–60, 369–70, 373, 376; Howard Stern and, 426–27; Powercord and, 324
Benson, Jim, 401–5
Benz, Mike, 270–71, 273
Beres, Tom, 218
Bevan, Alex, 58, 71, 85, 113, 137, 156, 182, 260, 340, 371
Bishop, Mark "Munch," 362, 416, 424
Blackwood, Nina, 158, 257
Blossom Music Center, 184, 198, 200, 435
Boc, Victor, 12, 13, 18–19
Bole, Larry, 215, 225
"Be-Bop" Kirby, 221–22, 224
Bonda, Ted, 118
Bondage, Boobie, 293, 307
Bone Thugs-N-Harmony, 402
Bonfanti, Jim, 308
Bortnick, Chuck, 347, 352, 358, 363, 391–92, 394
Bowie, David, 86, 99–100, 104, 108, 110, 113, 116, 123–24, 153, 177, 188, 201, 220, 248, 265, 301–302, 351, 354, 398, 448; with Iggy Pop band, 134–35; Richfield Coliseum concerts of, 355–57; U.S. concert premiere at Cleveland Music Hall, 79–83, 188

Brady, Pat, 301, 305, 309
Brandt, Todd, 344, 348, 355, 360
Brittain, Ross, 417, 432
Brown, Bob, 259
Bryson, Wally, 260, 308, 354
Buchwald, Don, 385, 391
Buddhist 13th Street Junk Mail Oracle, 10
Burnstein, Cliff, 105
Bush, Jerry, 182
Bush, Tom, 346, 362, 429, 448
Butler, Jim, 218, 328
Buttafuoco, Joey, 427
Buzzard. *See* WMMS-FM
Buzzard B-Movies. *See* Bennett, Ric "Rocco, the Rock Dog"
Buzzard Morning Zoo. *See* Kinzbach, Jeff
Byrd, Tim "Birdman," 166, 168, 296,

CKLW-AM, 24
Caen, Herb, 265
Campbell, George W., 26, 37
Canfora, Alan, 354
Capen, Steve, 60
Cappas, Sam, 427
Carey, Drew, 438
Carmen, Eric, 273, 307
Carr, Carolyn, 167
Carroll, Gene, 446
Carroll, Joe, 248,
Caroline, 167–68
Cartellone, Michael, 379
Chaffee, John, 90, 114, 126, 128, 156, 284, 323, 394; decision to play Michael Jackson by, 204; early days of Malrite's WMMS and, 96; John Gorman and, 282–83, 114; programming influence on, 111
Chalmers, Brian, 237–39, 347, 361–62, 370, 380, 416
Charboneau, Joe, 181
Charles, Jeffrey, 266, 300, 312
Cheeks, Jennie (Jennifer Anderson), 167, 168, 191, 192
Cheeks, Ruby, 259, 265, 305, 312, 326, 345–46, 348, 372, 382, 385, 393, 408; afternoon shift at WMMS of 338, 342; departure from WMMS, 347; joins WNCX, 373–74; morning-show ratings dominance of, 268–69, 298
Chenot, Jim, 256, 313
Christopher, Allen "Frank," 326
Church, Steve, 253–55, 281–82, 286, 288, 290, 292, 301–2, 337
Ciulla, John, 355
Ciulla, Mark, 355
Ciulla, Tony, 355

Clark, Dick, 270
Clean, Kenny, 202, 231, 253, 259, 305
Clear Channel, 443, 446
Clemons, Clarence, 197
Cleveland after Dark, 24
Cleveland Artists Recording for Ethiopia (CARE), 259, 264
Cleveland Edition, 339
Cleveland International Records, 197
Clinton, Roger, 427
Coburn, Bob, 261
Coffeebreak Concerts. See WMMS-FM
Cohen, Ted, 154
Colebrook, Kim, 240, 271, 384–85, 350–51
Coliseum. *See* Richfield Coliseum
Comfort, Jeff, 247
Conrad, Bob, 9, 29, 272
Cooper, Alice, 306–307, 346–47
Coughlin, Dan, 207
Crews, Tim, 396
Crocker, Jim, 300
Cronauer, Joe, 362, 438–40, 44; and the *Drew Carey Show*, 438; paired with Brian Fowler, 324; WENZ morning slot and, 383, 419–20, 435; WMMS morning slot and, 419, 432; WMVX morning slot and, 440
Crumb, Robert, 234
Csendes, Sue, 405–6
Czekalinski, Danny, 439

D'Amico, Rick, 12, 17, 20–21, 347
Daffy Dan's, 200
Damnation of Adam Blessing, 24, 32, 50, 71
Daniels, Ray, 104
Darden, Thom, 159, 247
Dardis, Ken, 300
Daugherty, Tim, 256
Davis, Clive, 274
Day, Charles, 29, 217
Dazz Band, 260
Dean, Jerry, 28, 31, 36
DeBella, John, 390, 431
Dee, Gary, 123, 136, 139, 147, 170, 177, 182, 239, 244, 446
Deeley, Dan, 264
DeFrasia, Diane, 174
Dell'Abate, Gary, 428, 429
DeMarne, Phil, 191
Dengler, Errol, 107, 419
Denver, Joel, 350
DePompeii, Joanne "Froggy," 149–50
Deutsch, Harvey, 286, 287, 296, 299
DiLeo, Frank, 247

458 Index

Dillon, Lisa, 331, 349, 359, 372, 388, 393, 410; and afternoon drive, 347; joins WMMS 323–24; and midday slot at WMMS, 391
Dink, 401, 403, 405
DiPasquale, Mike, 426
Dixon, Hanford, 310
Dobeck, Tommy, 182
Donley, Mark, 24
Donohue, Tom, 3, 171
Dorado, Jo, 415
Dori, Craig "Crankin' Craig," 439, 440
Douglas, Jack, 243
Doyle, Lynn, 50, 77, 106
Duffy, Maureen, 282, 314, 321
Dunaway, Chuck, 19, 22
Dyer, Bob, 187, 386–87

Eddinger, Mark, 37, 39
Edgewater Park, 49, 63, 74–75, 123–24
Eduardos, Charles. *See* Edwards, Gary
Edwards, Gary (Charles Eduardos), 33, 43, 66, 117
Eliot, Chris, 194
Elliot, "Trapper Jack," 301, 348
Elmore, Dave, 29, 36
Embrescia Broadcasting Corporation, 180
Embrescia, James, 180, 186
Embrescia, Thomas, 180, 186, 193–94
Emery, Brian "Sparky," 353
Empire, The, 355, 362–63, 370, 382
Endfest, 435
Ertegun, Ahmet, 270–74

Farina, Maria, 299, 341–42, 381, 397
Farrell, Laura, 393
Fassinger, Charles, 354
Fastway, 207
Ferenc, Ed "Flash," 131–32, 150, 163, 199–200, 218, 241, 268, 298, 302, 305, 318, 320, 322–23, 369, 376, 379–80, 390, 392, 403, 407–8, 416–18, 426, 432–33; the Agora and, 158; Cadillac Beach and, 307, final *Morning Zoo* show of, 409–11; and Jeff 'n' Flash's Monopolies, 177, 195; joins WHK, 104; Noisemakers and, 290; with Steve Lushbaugh, 126; WWWM talks and, 160; and unofficial Rock and Roll Hall of Fame groundbreaking, 276
Ferguson, Ted, 33, 44
Ferry, Bryan, 104
15-60-75 (The Numbers Band). *See* Numbers Band, The
Filby, John, 349–50, 371–72, 413–14, 416, 448
Fisher, Hal, 88, 101–2, 110
Fisher, Lee, 274

Fleetwood Mac, 137, 153–54, 166
Forward, Dewey, 268
Foster, Ron, 194, 196,
Foti, Frank, 172, 184, 225, 243, 314, 375
Fowler, Brian, 313, 362, 447; and afternoon drive at WMMS, 438–40; background of, 256; the *Drew Carey Show* and 438; morning drive at WENZ and, 383–84, 419, 435; morning drive at WMMS and, 419, 432; morning drive at WMVX and, 440; paired with Joe Cronauer, 324
Fox, Michael J., 269, 295–96
Frampton, Peter, 360
Franke (and the Knockouts), 182
Franklin, Pete, 355
Free Clinic, The, 49, 58, 73, 75, 98
Free Times, 403, 422–23
Freed, Alan, 9, 14, 269, 272–73
Freeman, Bill, 116, 340, 359, 383, 392, 416, 419–20, 444; arrival in Cleveland, 128–31; B. L. F. Bash and, 128–29, 268, 306

G. C. C. Communications of Cleveland, 278
Gabel, Barry, 324, 346, 409
Gabriel, Peter, 303
Gaines, Ken, 24
Gale, Roberta, 305, 310, 312, 322–23, 372, 393
Garcia, Bill, 60, 174
Garfinkel, Dan, 123, 131–32, 156, 187; "exclusive cume" and, 178; marketing the Buzzard by, 237; Morning Mind Exercises and, 136; WMMS employment of, 115–16, 182–83
Garland, Wayne, 182
Garrett, Walt "Mr. Classic," 436–37
Gatz, Tom, 85
Gee, Stan, 102
Gelb, Jeff, 47–48, 51, 72, 77, 88, 89, 171, 178, 211–12
George, Carl, 26–28, 42
George, Geoff, 177, 195
Gersh, Gary, 201
Giraldo, Neil, 335
Glass Harp, 67–70, 74, 77, 85, 115
Glazier, David, 329
Goddard, Dick, 410
Godfrey, Pam, 300
Golden, Lorraine, 286, 287, 296
Goldberg, Judy, 314–15, 320–21
Goldberg, Len "Boom," 86, 93, 100, 147, 181, 265, 268, 298, 305, 322, 324, 347, 354, 373, 392, 413, 416–18, 431, 442. 444; hosts Hilarities Friday night, 290; *Solid Gold Sunday* and, 159; WMMS employment of, 75–77, 108; WMMS format and, 440–41

Golic, Bob, 335
Gonzo Radio. See Bennett, Ric "Rocco, the Rock Dog"
Goodman, Mark, 258
Gorman, John, 101, 103, 114, 116, 120, 124, 140–41, 159, 170, 178, 184, 189, 213–15, 225–27, 233, 236, 238, 245, 259–60, 265–66, 278, 286, 288–90, 292, 299, 304, 311, 341, 347, 348, 351, 372, 373, 383, 403, 406–8, 413, 416, 418, 420, 432–33, 438, 447; advertising for WMMS by, 110; decision to play Michael Jackson of, 202–4; departure from WMMS, 279–85, 337; Eric Stevens and, 192; exclusives and, 221–25; format for "The Buzzard: The Next Generation" and, 416–19; Gorman Media and, 137–38, 282, 400; hiring philosophy of, 100; Howard Stern and, 424–25, 429–30; improvements to WMMS of, 96–97; joins WMMS, 94–95; *Morning Zoo* concept of, 305; as program director at WMJI, 369; "radio as war" mentality of, 143–49, 154–55, 161–70; and *Rolling Stone* ballot controversy, 314–15; and *Scene* magazine, 422–24; sabotaging Slade show by, 248–50; telephone conference call with John Lennon of, 111–12
Gott, Bob, 241
Graham, Bill, 49, 270, 275
Grande Ballroom, 16, 19, 20
Grant, Mudcat, 181
Grateful Dead, 395–96
Gray, Chris, 29, 31, 36
Gray, John, 107
Great Swamp Erie da da Boom, 11, 23, 42, 49, 52–57, 85, 218; on the sale of WMMS to Malrite Communications, 62–63
Griffin, Mike, 32, 44
Griffith, Fred, 85
Gronek, Lonnie, 202–3, 245, 282, 305, 315–16, 321, 375
Gullett, Mark, 355
Gunderman, Joe, 303

Halasa, Joyce, 63, 86, 88, 92, 117
Hall, Bill, 313
Hallaman, Ted, 198
Halper, Donna, 99–100, 104–5, 109–10, 216–17, 220–21
Hanson, Brad, 349, 357, 364, 395, 407; *Modern Music Show* and, 381, 397; Powercord and, 325; WMMS employment of, 311–12, 408
Hanson, Kurt, 390,
Hanson, Scott, 379, 403, 405

Hargrove, Mike, 181
Harken, Patty ("Janet from Another Planet"), 380, 399
Harper, Jim, 268, 282, 286–87, 292, 296, 302
Hart, Brett, 440
Hart, JoAnne "Mother Love," 259, 299
Heaton, Michael, 340–41, 360–61; and Howard Stern coverage, 387; joins WNCX, 352; and *Rolling Stone* ballot controversy, 315, 317–20
Helton, David, 145, 156, 174, 178, 199, 202, 233–38, 271, 310, 322, 343, 361, 370, 373, 380, 444; departs Malrite Communications, 347; returns to Cleveland, 305
Henke, Jim, 315, 319, 330–31
Hendrix, Jimi, 388, 397; Cleveland Music Hall concert of, 4
Herman, Skip, 363, 373, 384
Hinckley Buzzard Day, 112–13
Hirsch, Carl, 123, 252, 284–85, 339, 341, 359, 363, 369, 400, 406, 420; advertising criteria for WMMS of, 110; as general manager at WMMS and WHK, 105; departure from Malrite Communications, 264
Holan, Mark, 349, 359, 364
Holmes, Ward, 256, 312
Holston, Jim, 26–28
Hoolihan and Big Chuck, 7, 139
Hope, Bob, 188, 305, 374, 448
Hroblak, Jerry, 195
Hughes, Scott "Spaceman Scott," 248–49, 259, 268, 289, 297–98, 340, 416–17, 426, 432, 439; joins WMMS, 202; and *Rolling Stone* ballot controversy, 313, 315; WMMS transition to OmniAmerica comments on, 415
Hudson, Jay, 300
Hunter, Alan, 257, 279
Hunter, Ian, 164, 168, 373

Imus, Don, 26, 31, 44–45, 63
Inner Sanctum, 401–5
Infinity Broadcasting, 390, 443
Ingles, Paul, 343, 363
Iorillo, Gina, 245, 288, 301
Iris, Donnie, 259
Ivers, Mike, 194

Jackson, Linda, 200
Jackson, J. J., 258
Jackson, Michael (singer), 188, 244, 246–47, 259, 298, 316, 320, 338; "Bad" premiere on WMMS and, 303–4; "Beat It" played on WMMS and, 202–6

Jackson, Michael (WERE talk-show host), 274
Jacksons, 251–53
Jacobs, Jeff, 383
Jacobs, Maria, 399
Jacor Communications, 264, 359, 363–64, 440, 442–43
Jagger, Mick, 186, 204, 261, 348, 398, 448
James, "Real Bob," 446
Jansen, Dick, 18
Jaye, Darla, 439–40
Jeffries, Tom, 247, 266–67
Jett, Joan, 268, 294–95
Jockers, Dave, 396, 422, 424, 427, 425
Joel, Billy, 335
Johan. *See* Linus Johansonas
Johansonas, Linas, 401–405
Johnson, Doug, 143, 149, 156–57, 160–61, 174, 176
Johnson, Jeff, 252
Johnson, Pat, 401–2, 404–5
Joos, Steve, 286, 297

KAZY-FM, 362
KBEQ-FM, 258
KFOG-FM, 265, 267, 318
KIQQ-FM, 340
KISS-FM, 316
KJOI-FM, 264, 406
KLAC-FM, 264
KLOL-FM, 197
KLOS-FM, 261
KMET-FM, 3, 60, 75, 106, 298
KMPX-FM, 3
KNAC-FM, 293
KRLD-AM, 63
KSAN-FM, 3, 61, 67–68, 86, 145, 171
KSHE-FM, 261, 267
KTSP-TV, 347
KYW-AM, 94, 388
KZEW-FM, 126
KZLA-FM, 264
Kahler, Dean, 354
Kasem, Casey, 279
Kaye, Johnny, 24
Keaggy, Phil, 68–70, 85
Kellogg, Brandy. *See* Severson, Catrina
Kelly, Jim, 21, 24
Kelly, "Shotgun Tom," 116
Kelly, Tom "Tree," 66, 75, 93, 171–72
Kemp, Dick "the Wilde Childe," 22, 33, 43, 44, 49, 93
Kendall, Charlie, 114, 126, 221
Kennedy, Royal, 153

Kent State University, 30, 38–39, 83–84, 219, 352–54
Kent, Tom, 194
Kerwin, Michael, 69
Kerwin, Phil, 21
Kiefer, Rhonda, 288
Kilpatrick, Craig "Killer," 343, 347, 359, 365
Kimble, Bernie, 288, 300, 301
King, Jim, 241
Kinzbach, Jeff, 98, 109, 114, 116, 128, 131, 154–58, 171, 174, 176, 181–82, 196, 198–200, 241, 268, 279, 302, 318, 340, 369, 376, 379–80, 385, 390, 392, 397, 403, 407, 417, 419, 421, 433, 441; appearances at the Agora, 158; *Buzzard Morning Zoo* and, 264, 238, 276, 278, 300–301, 305, 309–10, 312–13, 322–23, 326, 329, 335, 338–39, 344, 346–47, 352–54, 356–57, 362–63, 365–68, 370, 371–73, 375, 379, 382–84, 389, 393, 397, 408, 425, 432, 439, 444; Cadillac Beach and, 307; decision to play Michael Jackson of, 205–6; defection to WNCX of, 284; drug-free stand of, 214–15; Easy Money Contest and, 331; feud with Randy Youngman, 179; final Morning Zoo show on WMMS and, 409–11; Howard Stern and, 386, 426; and Jeff 'n' Flash's Monopolies, 177, 195; and Jeff Kinzbach and Friends, 247; and morning show 132, 135–36; Noisemakers and, 290, 328; ratings dominance by, 266, 298; and *Rolling Stone* ballot controversy, 316–17, 320; reaction to *Cleveland Edition* article of, 339–40; reaction to *Lake County News Herald* article of, 329–30; reaction to *Scene* comments of, 397; and unofficial Rock and Roll Hall of Fame groundbreaking, 276
Kinzbach, Patty, 411
Kirby, Lou "King," 33, 35, 44
Klawon, Danny, 308
Knapp, Chuck, 22
Knight, Bob, 20
Knisely, Mike "Mondo," 353, 371, 382, 426, 430, 448
Korvan, Betty "Crash," 116, 143, 158, 168, 213–14, 243, 253, 348, 441, 444; joins WMMS, 103; leaves WMMS, 267–68; Pete Townshend and, 120
Kosar, Bernie, 438
Koski, Al "The Bear" (Jeff Koski), 137, 174, 187, 191
Koslen, Jonah, 377
Kramer, Heidi, 416, 430–31
Krane, Brigid, 361
Kubinski, Doug, 391, 407, 412–13, 416–17, 420, 428, 430–31
Kubiszewski, Andy, 379
Kucinich, Dennis, 75, 143–45, 150, 168, 170

Kuiper, Duane, 181
Kunes, J. D., 256–57, 313

La Cave, 10, 15, 86
Lake County News Herald, 329
Lanigan, John, 63, 123, 177, 197–98, 245, 251, 267, 269, 300, 302, 305–6, 310, 346, 363, 372–73, 376, 382, 384–85, 389, 393, 397, 399, 407–8, 420–21, 425, 429–30; departure from WGAR, 246; Howard Stern and, 385; Jerry Seinfeld and 374–75; rumors of return to Cleveland, 262; *Scene* magazine and, 422–24; WMJI mornings and, 264; and WMMS-*Rolling Stone* ballot controversy, 315–17; WQHS-TV simulcast and, 342–43; WZAK simulcast and, 343
Lansing, Chuck, 29, 36
Lapczynski, Matthew "Matt the Cat," 113, 137, 150, 156–57, 174, 176, 187, 214, 220, 244, 298, 348, 359, 388, 405, 419, 441, 444, 448; Bruce Springsteen and, 125; joins WMMS, 103; joins WNCX, 395; departure from WMMS, 391–92; premieres "Bad," 303
La Rose, Tim, 271
Laughner, Peter, 85, 106, 111, 117, 124, 139
Lauper, Cyndi, 244, 258, 346
Legacy Broadcasting Partners, 339
Legacy Communications, 359
Legerski, Steve, 321
Leo's Casino, 15, 153
Lennon, John, 72–73, 101–2, 112, 123, 134, 159, 175, 243
Levine, Phil, 375
Levy, D. A., 10
Lewis, Bob "Bob-a-Loo," 24
Lewis, Huey, 200
Liberatore, Dick, 19
Lingus, Cory, 439, 440
Linhart, Buzzy, 48
Light of Day, 268, 295–96
Live at the Agora. See Agora, The
Locke, Susan, 303
LoConti, Hank, 143, 175, 253; Bruce Springsteen show and, 152; David Bowie and, 135; and Rock and Roll Hall of Fame, 270–72
Lombardi, Vince, 247
Louis, Bill, 346, 385, 437–38; and John Gorman, 417; joins WNCX, 304
Lucas, Dave, 154
Luczak, Michael, 358–59, 363, 364, 370, 376, 380–81, 389, 391, 424; *Buzzard Morning Zoo* and, 393; joins WMMS, 357; departure from WMMS, 406

Lushbaugh, Steve, 93, 100, 116, 126–28, 131–32; Dennis Wilson and, 228–29; John Gorman and, 114–15
Lux, Ted, 300
Lyall, Lori, 315
Lyles, Harry, 298
Lyon, John "Southside Johnny," 157, 197, 289, 335, 388

M-105. *See* WWWM-FM
MTV, 189, 199, 200, 202, 244, 253, 257, 259, 261, 279, 290, 303, 390
McCartney, Jeff, 313, 323, 328–29, 339
McCartney, Paul, 125, 159, 214, 223, 296, 352, 354, 357, 398, 448
McCormack, Tom, 21
McCoy, Pat, 11, 12–20,
McCrone, Mary, 377
McEwen, Mark, 331–32, 347
McGovern, George, 354
McKay, Deeya, 256, 313
McKay, Jan, 273
McKean, Michael, 268, 295
McKee, Jeff, 93, 304
McVay, Mike, 192–93, 195–96, 261, 195, 296–97
Macoska, Janet, 246, 259
Mad Dogs and Englishmen. *See* Shirley, Jerry
Maddock, Buddy, 66
"Maggot Brain" (Funkadelic), 138
Malmsteen, Yngwie, 360
Malone, Jimmy, 305–6, 310, 343, 389, 399, 407, 421
Malrite Communications, 84, 90, 126, 136, 202, 252, 264, 268, 278, 280, 288, 301, 305, 310, 331, 335, 339–41, 347, 358–59, 361, 373, 386, 390–91, 438; buys WMMS from Metromedia, 61–62; corporate headquarters in Cleveland of, 66–67; and the FCC, 86; and Henry Speeth Jr., 85; merger with Shamrock Broadcasting of, 393–95, 397, 400; Operation: Payroll and, 385, 394; and *Rolling Stone* ballot controversy, 314–16; success of WHTZ of, 246–47
Maltz, Milton, 61–62, 66, 84, 89–90, 116, 264, 284, 338–39, 390–91, 394, 409; Billy Bass and, 86–88; Denny Sanders and, 283; Howard Stern and, 391; and *Rolling Stone* ballot controversy, 316, 321
Manson, Marilyn, 402–3
Marchysyn, Jim, 125, 187, 199, 259, 261
Martin, Roger, 401
Masky, Walt, 174
Mason, Seth, 60, 73, 77
May, Brian, 394–95

Maynor, Jeff, 410
Mercer, William "Rosko," 35
Medlocke, Rickie, 260, 365
Melendez, "Stuttering John," 389
Mellencamp, John, 335
Metallica, 439
Metroplex Communications, 245, 330, 350, 384–85, 391, 419
Metropolis Broadcasting, 278–79, 281, 283, 287, 292, 296, 302, 304, 330, 350
Michael Stanley Band, 117, 142, 148–49, 183–85, 201–2, 253, 257, 259–60, 269, 274, 278; final shows at Front Row Theater of, 290–92; *PM Magazine* segments with, 299
Michaels, Mitch, 29, 31, 36
Michelli, Mike, 256, 313
Mileti, Nick, 78, 186
Miller, Dan, 199
Miller, Ed, 177
Minard, Jim, 89
Mintz, Leo (Record Rendezvous), 269
Monday, Carl, 353, 410
Money, Eddie, 303, 335, 389
Moon, Keith, 120, 157, 195
Mooney, Matt, 412
Moore, Norman, 89, 217
Moorehead, David, 33, 35, 58, 214
Morgan, Ken, 196–97
Morgan, Matt, 259
Morris, Andrea, 400
Morrow, Larry, 9, 266
Moss, Ralph, 69
"Mother Love." *See* Hart, JoAnne
Mottola, Tommy, 335
Moyes, Bill, 194
"Mr. Classic." *See* Garrett, Walt
"Mr. Leonard." *See* Rio, John
"Mr. R," 139
Muni, Scott, 326
Myers, Pete "Mad Daddy," 5

Nationwide Communications, 25, 37, 56, 77, 432, 438–40
Nelson, J. R., 116, 369
Nemo, Doc (Steve Nemeth), 93, 110, 147, 388; Barry Weingart and, 4–9; debut on WHK, 12; departure from WMMS, 18; and "Doc Nemo" character, 6; "Doc Nemo's Mind Blowing Concert #1" and, 7–8; returns to WXEN, 20; WIXY contest fixing and, 7–8
Neuhoff, Paul, 78
Neumann, Bob, 404, 439

New Age Rock Review, 217–19, 425
Newton, Wayne, 377
Nine Inch Nails. *See* Reznor, Trent
Nite, Norm N., 270–71, 273, 349
Norman, Michael, 398
Novello, Don, 281
Numbers Band, The (15–60–75), 112, 137, 403, 404

O'Brien, Skip, 116
O'Brien, Tom, 347
Ocepek, Tony, 358, 419
Oden, Murray, 352, 356, 366, 370
Oktavec, Jim, 338
Olbrys, Kate, 42–43
Olin, Steve, 396
OmniAmerica, 238, 423–24, 438
Onesky, Mike, 437
Orr, Benjamin, 260
Owens, Tom, 376

Pardo, Don, 410
Patrick, Richard, 402, 405
Payola, 216–17
Peabody's Down Under, 259, 268, 290, 295, 342, 344, 355, 364, 395, 412
Peanuts, 324
Pecchio, Daniel, 68–70, 77
Pecchio, Ted "Toto," 70
Pei, I. M., 275, 294
Pelander, Bob, 260
Penfound, Ron "Captain Penny," 446
Perdue, Jimmy, 106, 113
Perk, Ralph, 75
Perlich, Martin, 9, 21, 46, 56, 60, 66, 69, 84, 132, 217; Billy Bass and, 58–59; departure from WMMS, 75, 171; early life of, 9–10; Henny Youngman and, 212; joins WMMS, 58; joins WNCR, 28, 39–40; WCLV and, 9, 12; WHK-FM and, 12
Perry, Joe, 243
Peters, Suzi, 167–68
Phantasy, 174, 331, 355, 364
Phillips, Brian, 293, 310–11, 325
Pinkney, Arnold, 118
Piombino, Rich, 321, 350, 352, 359, 357
Plain Dealer, 141, 195, 252, 258, 261–62, 270, 273, 278, 282, 290, 292, 300, 328, 335, 351–52, 355, 363, 381–82, 393–94, 397, 399, 409, 415–16, 420, 434–35, 438, 441, 443; Howard Stern and, 385, 387; and *Rolling Stone* ballot controversy, 313–16, 330–31, 340, 341, 398
Plasmatics. *See* Williams, Wendy O.

Pollock, Larry, 193
Podell, Doug, 357, 360, 363, 373, 382, 384, 386, 396, 400, 421–22, 432; Howard Stern and, 424–25
Poole, Ralph, 343
Popovich, Dave, 245, 300–301, 304,
Popovich, Pam, 197
Popovich, Steve, 152, 197
Powers, Danny, 260
Preston, Jeanne, 20
Price, Mark, 85
Price, Roger, 381–82
Psychic Rock Phenomenon, A, 24

Quadrophonic sound, 25, 71, 78, 118
Quinn, Martha, 202, 257
Quivers, Robin, 386, 396, 421

RAP Productions. *See* Price, Roger
Radio One, 443
Radio-Television Broadcasters Hall of Fame, 446
Raleigh, Kevin, 184, 260
Ramos, Willy, 425
Randle, Bill, 9, 99, 191, 219–20
Rathbun, Keith, 193, 318, 324, 401, 422
Reid, Tim, 187
Reineri, Mike, 22
Reese, Carl, 191
Regency Communications, 264
Reznor, Trent, 199, 349, 379, 402
Rezny, Tom "T. R.," 191, 267, 300, 339, 391, 413, 417, 448; and the American Cancer Society, 118–19; joins WWWM, 110–11
Richards, Ed, 104
Richfield Coliseum, 120, 125, 137, 143, 172, 182, 184–85, 195, 214, 250, 326, 355–56, 395
Ride, Sally, 373, 416
Riley, Jack, 446
Rio, John "Mr. Leonard," 229–32, 278, 298, 310, 320, 322, 324, 328, 343, 346, 358, 369–70, 417, 418, 432; and billboard incident, 365–68; Ritz Niteclub and, 355; Scott Shannon and, 340
Roberts, John, 66
Robinson Communications, 180, 186
Robinson, Larry "J. B.," 180, 186, 191, 193–94, 196, 251, 264, 330
Rock and Roll Hall of Fame, 262, 269–77, 294, 303, 317–18, 349–51, 358, 390, 409–10, 447–48; groundbreaking ceremony for, 397–99
Rockline, 199, 214, 261
Rolling Stone, 116, 188, 325, 446; and ballot stuffing controversy, 313–21, 330–31, 340, 388; and Kid Leo, 159; and Radio Station of the Year, 173, 176, 187, 197, 245, 247, 251, 258, 260, 267, 312, 351; and Readers and Critics Awards Show, 267
Rolling Stones, 115, 185, 187, 222, 345, 348, 350, 397
Rosas, Rick "The Bass Player," 307–9
Rosenberg, Arnie, 172
Rosenwald, Gil, 126, 268, 280–82, 285, 323, 331, 338, 352
Rosko. *See* William "Rosko" Mercer
Roth, David Lee, 427
Rothman, Archie, 99
Roxy Music, 104, 168, 177, 336
Rufus, Dean, 259
Rundgren, Todd, 172, 184
Runyon, Jim, 94
Rush: and Donna Halper, 104–5
Ryan, Bruce, 300
Rydell, Rick, 344, 348, 355, 360
Rydgren, Rev. John, 23

SFX Entertainment, 446
Sanders, Denny, 75, 77–78, 86, 89, 90–91, 93, 98, 100, 103, 105, 116, 139, 155–56, 158, 163, 181–82, 184, 198–99, 221, 225, 233–35, 246, 248, 259, 261, 278, 283, 285–88, 292, 341, 348, 369, 383, 416, 441, 443–44; brings John Gorman to WMMS, 95; Bruce Springsteen and, 153; departure from WMMS, 279–83, 299, 37; John Lennon interview with, 101–2, 175; joins WMJI, 329; joins WMMS, 59; and *USA Today* Rock Hall poll, 273; and WGCL, 242
Santiago, Roberto, 387, 397, 400, 434
Santini, Lou, 416
Saul, Murray, 117, 119, 124, 135, 137, 165, 211, 328, 349, 424; "Get Down" salute and, 132, 225–29; Hinckley Buzzard Day and, 113
Scene (magazine), 92–93, 193, 305, 307, 324, 328–29, 363–65, 374, 392, 394, 397, 399–401, 403–4, 408, 416–17, 420, 431–32; Breakout and, 349, 355, 359–60; Howard Stern and, 422–24, 429–30, Kid Leo and, 339, 394; and *Rolling Stone* ballot controversy, 318–20
Schliewen, Peter, 42, 105
Schreiber, "Fred, the Elephant Boy," 389, 427
Scorsese, Martin, 197
Scott, "Wild Bill," 293, 307
Scott, Jane, 261, 292, 303
Scott, Kim, 264
Scott, Randy, 21
Scott, Steve, 29
Scull, Bill, 420
Seinfeld, Jerry, 374–75
Sells, Alan, 191
Severson, Catrina (Brandy Kellogg), 301

Sferra, John, 68–70
Shamrock Broadcasting, 393–95, 397, 400, 406, 438
Shannon, Scott: and Mr. Leonard, 229, 231, 340, 343; and the *Morning Zoo* concept, 202, 341
Sharp, Dave, 191–93
Sheldon, Linn, 251, 446
Sherard, Bill, 42,
Sheridan, Phil, 89
Sherill, Judy, 77, 407–8
Shirley, Jerry, 207, 352, 358, 385, 420; joins WNCX, 341; *Mad Dogs and Englishmen* and, 360, 363, 373; Thirty Days in the Hole promotion with, 363, 365, 393, 435–38
Showalter, Barb, 412
Showalter, Dana, 412
Siegel, Randy, 422
Simmons, Gene, 360
Sinton, Steve. *See* Sutton, Ginger
Skrovan, Steve, 378
Slade, 247–50
Smiling Dog Saloon, 77, 98–99, 108
Smith, Bill, 271
Smith, Bingo, 182
Smith, Greg, 430
Snodgrass, J. P., 53–54
Sobol, Marty, 191, 325
Sogg, Wilton, 193
Sohio Riverfest, 262, 279, 326, 346, 374, 383
Solly, Dale, 259, 261, 358
South by Southwest (music conference), 403–4
Sowd, David, 315, 322–23, 355, 340–41, 421
Speeth, Henry, Jr., 63, 84–85
Speizel, Eddie, 53, 270
Spencer, Tim, 195
Spero, David, 50–52, 59, 60, 65–66, 71–72, 78, 85–86, 93, 100, 102, 104, 124, 149, 157, 170, 182, 192, 220, 388; Fleetwood Mac and, 153–54; Henny Youngman and, 212; Joe Walsh and, 370–71; joins WMMS, 58; joins WNCR, 44; joins WWWM, 142, 159–60; and Keith Richards drug trial, 218; Michael Stanley and, 103
Spero, Herman, 44, 155, 212
Springsteen, Bruce, 104, 116, 118, 124, 145, 157, 168, 174, 177, 197, 199, 250, 259, 289, 294–95, 324, 355, 417; "Born to Run" and 112, 173, 223–24, 389, 440; "Brilliant Disguise" debut on WMMS, 304–5; Cleveland Municipal Stadium shows of, 262, 441; "Darkness on the Edge of Town" and 148; Kid Leo and, 335–36; Stockholm worldwide broadcast of, 326; visits WMMS, 125; WMMS 10th anniversary show and, 150–53, 388, 441
St. John, Tracy, 300

Stagg, Jim, 94
Stallings, Bill, 117, 158, 176, 185, 191–92
Stanley, Michael, 102, 155, 177, 182, 201–2, 220, 250, 253, 260, 267, 290, 293–94, 325, 340, 360, 370, 373, 379, 384–85, 390; heart problems of, 377–78; WNCX afternoon drive and, 382, 419
Stanley, Paul, 360
Stanton, Mike, 181
Stein, Dia, 225, 277, 299–300, 305, 346, 352, 360, 372, 441; "Cleveland Breakout" segments on WEWS-TV with, 290; departure from WMMS, 347–348; joins WMMS, 175–76
Stevens, Eric, 108, 110, 143, 159–60, 170, 192–93, 227, 400; early radio background of, 94; departure from WWWM, 176; WMMS tricks and, 165
Stern, Howard, 383, 391, 393, 408, 434, 436, 438; Cleveland funeral held by, 399–400, 421–22, 424–32; Cleveland Indians controversy with, 396–97; comments on Jerry Shirley by, 438; Lakewood rally against, 389; rumored arrival of, 382; WNCX and, 384–90, 407, 419
Stile, Shelly, 101, 114, 116, 131, 138, 214, 218
Stokes, Carl B., 180, 186
Summers, Bret, 256
Sutherland, Doug, 300–301
Sutton, Ginger (Steve Sinton), 29, 32–33, 36, 44, 50, 66, 91, 211
Swingo's, 85, 120, 186, 222
Swoboda, Robin, 330

T. R. *See* Rezny, Tom
Talbert, Steve, 290
Tapie, Paul, 245, 288, 343, 373, 384
Taylor, Brian, 312
Taylor, Sharon, 261
Taylor, Tim, 12
Tedesco, Carmen, 219
Tegreene, Joe, 142–43, 145, 270
"The Wave." *See* WNWV-FM
Thacker, Dean, 359, 369, 407, 420, 424, 447
Thal, Michael, 85
Thayer, Jack, 26, 36, 64, 214
Thomas, Barry, 416, 425
Thomas, Carolyn, 66, 77
Thomas, David (Crocus Behemoth), 117, 402, 420
Thompson, Ron "Ugly," 41–43, 47, 344
Tiburski, Walt, 27, 40, 46, 64, 77, 150, 175, 197, 226, 341, 358–59, 395–97, 407–8, 422, 424–25, 432–33, 437, 443; as general manager of WNCX, 419; WMMS employment and, 247, 394
Todd, Mitch, 413, 416
Tokai, Robert "Drac," 326

Tolliver, Lynn, 79, 189, 265–66, 268, 343
Townshend, Pete, 265, 371, 399, 448; and Betty Korvan, 120
Trapp, Victor, 18
Travagliante, Jackie, 175
Travagliante, Lawrence "Kid Leo," 98, 103–4, 114, 116, 124, 141, 143, 150–51, 163, 178, 187, 189, 197–98, 214, 220–22, 248, 261, 267, 274, 278–79, 284, 292, 294, 298, 302, 304–6, 309, 312–13, 323, 324–325, 330, 346–47, 350, 360, 383, 440–41, 444; the Bookie Joint and, 299; Bryan Ferry and, 104; the "Cleveland sound" and 177; departure from WMMS, 328, 332, 335–39; hired by WMMS, 91–92; Keith Moon and, 120; "One Trick Pony" and, 174; *Plain Dealer Sunday Magazine* interview with, 351; *PM Magazine* segments with, 155; return visit to Cleveland, 352, 365, 388–89; role in *Grease* by, 251; and *Rolling Stone* ballot controversy, 315, 320–21; *Rolling Stone* profile of, 159; Spring-steen Stockholm broadcast with, 326; WMMS 10th anniversary show and, 152
Travis, Bob, 168, 247
"Tree." *See* Kelly, Tom
Trivisonno, Mike, 343, 355, 360, 373, 384
Tuna, Charlie, 116

Ullman, Debbie 98, 108–9
United Broadcasting Company, 195
Urbach, Dave, 344, 350

VH-1, 296, 325
Vai, Steve, 360
Van Zandt, Steve, 125, 157, 197, 246
Vaughn, James, 85
Viking Saloon, 77, 117
Vincent, Debbie, 313
Vitale, Joe, 260
Voinovich, George, 173, 176, 187, 197, 262, 273, 409

WAAF-FM, 347–48
WABC-AM, 22
WABC-FM, 174, 176
WABQ-AM, 73, 106
WABU-FM, 23
WAKC-TV, 405
WAKR-AM, 326
WBAI-FM, 3
WBBG-AM, 180, 186, 196, 251, 264
WBCN-FM, 80, 298
WBEA-FM: change to WCZR-FM (Z-Rock), 293
WCLQ-TV, 186, 261

WCLV-FM, 9, 25–26, 39, 132, 250, 272; Perlich Project on, 9–10, 52
WCOL-FM, 28, 29
WCSB-FM, 103
WCUE-FM, 114
WCUY-FM, 26
WCZR-FM, 293, 307, 324
WDBN-FM, 285
WDGO-FM, 9
WDMT-FM, 200, 256, 259
WDOK-FM, 25, 62, 92, 198, 262
WDTX-FM, 302
WEBN-FM, 357, 376
WELW-AM, 220
WENZ-FM, 381, 393, 394, 399, 413, 416, 418, 424, 434, 443; bid for Howard Stern, 38; Brian Fowler and Joe Cronauer on, 383, 419–20
WERE-AM, 77–78, 93, 105, 123, 170, 219, 280, 285, 302, 328, 419, 443; announces Cleveland Rock and Roll Hall of Fame, 274; sold, 278
WEWS-TV, 12, 44, 85, 153–55, 174, 246, 261, 290, 427
WGAR-AM, 31, 44, 63, 92, 123, 133, 177, 194, 197–98, 245, 264
WGAR-FM, 25, 30, 351, 381, 399
WGCL-FM, 79, 92, 107, 157, 163, 167–68, 170, 175, 178, 184, 194, 198, 200, 202, 250–51, 258–59, 261, 264, 266, 269, 271, 279, 280, 285, 296, 320, 350, 439; as CHR station, 189; Danny "Dancin' Danny" Wright and, 197–98, 246; final hours of, 289; ratings win over WMMS, 239–42, 245; sold, 278
WGTR-FM, 363
WHK-AM, 12, 35, 73, 92, 97, 102, 104–5, 123, 135–36, 139, 166–67, 170, 174, 177, 218, 239, 244–45, 267, 328, 415–16, 429; format change of, 4; Powercord and, 324; and *Rolling Stone* ballot controversy, 314
WHK Auditorium, 50, 65, 69
WHK-FM, 12–13, 21, 88; 388
WHTZ-FM (Z100), 202, 224, 229, 310, 314, 316, 325, 340, 341, 417
WIXY-AM, 7, 12, 19, 20, 30–32, 41, 46, 62, 66, 72, 93–94, 103, 192, 194, 220, 245, 330, 369, 388; early programming of, 3, Doc Nemo and Barry Weingart on, 6
WJLA-TV, 358
WJMO-AM, 20, 35,
WJW-AM, 20, 72, 93, 269
WJW-TV, 12, 14, 273, 330, 353, 410, 426
WJKW-TV, 155, 173, 174
WKDD-FM, 202
WKNT-FM, 105

WKYC-AM, 4, 19, 72, 78
WKYC-FM, 78
WKYC-TV, 78, 153, 177, 218, 247, 251, 261, 410
WLTF-FM, 256, 298, 301, 303–4, 315, 346, 348, 363, 440–41; and Coats for Kids promotion, 300
WLYT-FM, 103, 166, 195, 200
WMGG-FM, 246, 264
WMJI-FM, 194, 196, 245–46, 251, 261, 264, 267, 269, 270, 300–302, 312, 315, 329–30, 339, 346, 348, 359, 363, 369, 383, 385, 443–44; *Knuckleheads in the News: The Book* and, 309–10; Day in Cleveland, 373
WMMR-FM, 61, 315, 341
WMMS-FM, 20, 30, 43, 66, 73, 77, 105, 108, 111, 118, 123–25, 139, 141, 143, 158–59, 168–69, 174–75, 177, 184, 186–87, 194, 197–202, 213–16, 218–19, 229–31, 245, 262, 266, 278, 290, 293, 302–306, 310–13, 324, 326, 329, 332, 341, 346–50, 355, 388, 443, 446; *Billboard* magazine and, 21; the Buzzard and, 106, 187, 211, 233–39, 241, 300, 320, 322, 326, 329–30, 343, 347, 361, 363, 380, 416, 429, 440; Buzzard Beatle Blitz and, 159; and Buzzard Nuclear Army and, 139, 144, 155, 200, 410; Buzzard-Palooza and, 434–35; Buzzard Record and Filmworks and, 156; Buzzard t-shirts and apparel and, 98, 139, 158, 199, 237–38, 290, 320, 353, 380, 415, 439; change from WHK-FM, 14; change to adult contemporary format, 21; change to CHR station, 189; Classic Rock Saturday Night and, 253, 257, 307, 376; Cleveland Police drug-testing promotion and, 211; *Coffeebreak Concerts,* 70–71, 72, 77–78, 85, 90, 174, 176, 182, 187, 215, 244, 253, 258, 327, 342, 344, 346, 360, 412, 415, 439, 448; Edgewater Park concert by, 74–75; exclusives of, 220–25; Glass Harp simulcast with, 68–70; Live Wire and, 253–55; Michael Jackson and, 202–6, 246; mushroom logo of, 65; Northcoast Buzzard Tracks and, 370, 378, 405; as official Rolling Stones tour station, 185; and the Rock and Roll Hall of Fame, 269–73, 275–77; and *Rolling Stone* ballot controversy, 313–22, 330–31; and Smokestock, 123–24; tongue logo of, 65; TV commercials for, 132, 156, 158; *Victory* Tour sponsorship of, 247, 250–53, 258, 303, 338; Who show tickets and, 195–96
WMVX-FM, 440–41
WNCI-FM, 27, 89
WNCR-FM, 26, 30–58, 62, 66, 70–71, 73, 77–78, 94, 147, 211–12, 214, 220; country format and, 101; change from WGAR, 26–29; Edgewater Park concert with, 49, 63; news and, 217–18; quadrophonic broadcasting by, 71; Renaissance Fair and, 77; tri-sonic simulcasting by, 32, 71
WNCX-FM, 283, 288, 290, 292–93, 296, 299, 302–4, 310–11, 328–30, 337, 339, 341, 346–48, 350, 355, 359, 363, 370, 383, 396, 418, 424, 441, 443; challenges to WMMS by, 442; "Classic Cat" and, 325; Howard Stern and, 384–88, 399–400, 403, 419, 421–22, 424; Live Wire and, 255, 301; and *Rolling Stone* ballot controversy, 314–16
WNOB-FM, 21, 23
WNEW-FM, 3, 35, 61, 298, 315, 326, 341
WNTN-FM, 59
WNWV-FM ("The Wave"), 307
WOIO-TV, 267, 273, 301, 326, 330, 361, 373, 431
WONE-FM, 256, 312, 315, 324, 362, 383 256
WPHR-FM, 310, 348–49, 352, 381
WQAL-FM, 92, 170, 177, 198, 266, 300, 312
WQCD-FM, 359
WQFM-FM, 343
WQHS-TV, 342–43
WREO-FM, 20
WRJW-FM, 311
WRQC-FM, 200, 273, 299, 303
WRQK-FM, 416, 424
WRQX-FM, 246
WRUW-FM, 12, 77
WSKS-FM, 325
WTBS-FM, 59
WTOD-AM, 21
WTTO-FM, 20
WUJC-FM, 37
WVBF-FM, 114
WVIZ-TV, 69, 85, 142
WWWE-AM, 78, 192, 316, 346, 352, 355, 433
WWWM-FM (M-105), 78, 108, 111–12, 115–18, 140, 143, 155, 157–58, 160–63, 165–68, 174, 178, 184–85, 187, 194–95; exclusives on, 221, 223–25; format change of, 110; "$5000 Rock Guarantee" of, 189–91; Larry Robinson and, 180; as Paul McCartney tour stop, 125
WXEN-FM, 13, 20, 176
WZAK-FM, 6, 13, 19, 21, 189, 256–57, 259, 315, 348, 363; early programming of, 4; *For Lovers Only* and, 300; Lynn Tolliver and, 265–66; Tim Reid commercials with, 187; urban programming of, 176–77
WZZP-FM, 157, 161, 168, 194, 196
WIN Communications, 247
Wagner, Stirling, 85

Wain, Norman, 245, 384, 396; Billy Bass and, 22–23; Doc Nemo and, 6; "Doc Nemo's Mind Blowing Concert #1" and 7; at WIXY, 3–7; WNCX and, 330
Walsh, Joe, 41, 102, 158, 307–9, 370–71, 390, 438
Waple, Ben, 26
Warfield, Paul, 181
Warren, Verdell, 102, 325
Waybill, Fee, 182
Webster, John, 245, 261, 301–2, 305, 310, 342–43, 363, 372–73, 382, 389, 399, 407, 421
Weinberg, Max, 197
Weingart, Barry "Buttons," 4, 21, 44, 93; Doc Nemo and, 6–9; "Doc Nemo's Mind Blowing Concert #1" and, 7–8; WIXY "High-Low" contest and, 8–9
Weiss, Bob, 3–7, 4, 245, 330, 384, 396
Weisz, Sharon, 154
Welch, Digby, 58, 77
Wendell, Bud, 220
West, Leslie, 384
Who, The 120, 195, 225, 338, 344–46, 348, 397
White, Michael, 383, 423–24, 436, 438
Wild Horses, 182
Wilde, Liz, 438

Williams, Wendy O., 180–83
Wilson, Brian, 237
Wilson, Dennis, 228–29, 237
Wilson, Dick, 72–73, 105
Wood, Jim, 280
Wright, Bob, 328–29
Wright, "Dancin' Danny," 197–98, 240–42, 246, 251, 255–60, 279, 287–87, 299, 303
Wright, Don, 369
Wright, Steven, 360
Wylde, Jennifer, 416
Wyman, Bill, 185–86

Young, "Party Marty," 436–37
Youngman, Henny, 212
Youngman, Randy, 179

Z100. *See* WHTZ-FM
Z-Rock. *See* WCZR-FM
Zapis, Lee, 176, 189, 315, 347–48
Zeppelin, 235
Zingale, Joe, 3, 4, 186
Zivich, Cheryl, 440
Zurbrugg, Shauna, 47–48, 50–61, 66, 77, 86, 89